Lecture Notes in Mathematics

Edited by A. Dold and B. Eckmann

882

Integral Representations and Applications

Proceedings of a Conference held at Oberwolfach, Germany, June 22–28, 1980

Edited by Klaus W. Roggenkamp

Springer-Verlag
Berlin Heidelberg New York 1981

Editor

Klaus W. Roggenkamp
Mathematisches Institut B, Universität Stuttgart
Pfaffenwaldring 57, 7000 Stuttgart 80, Federal Republic of Germany

AMS Subject Classifications (1980): 12A57, 12A62, 15A36, 16A18, 20C10

ISBN 3-540-10880-7 Springer-Verlag Berlin Heidelberg New York
ISBN 0-387-10880-7 Springer-Verlag New York Heidelberg Berlin

Printing and binding: Beltz Offsetdruck, Hemsbach/Bergstr.
2141/3140-543210

INTRODUCTION

Integral representation theory and applications thereof have over the past years made considerable progress.

The intention of the Oberwolfach meeting in 1980 was to demonstrate the various applications of orders to number theory, topology, geometry and crystallography and also to show the influences these various subjects have had on integral representation theory. Apart from these, new trends in integral representation theory were presented; Almost split sequences and Auslander-Reiten graphs in connection with classification of indecomposable lattices, and ζ-functions on lattices.

Thus we had several survey talks:

"Orders in geometry, topology and number theory";

"Algebraic aspects of crystallography";

"Graham Higman's thesis" (it was surprising to hear of the wealth of results in Higman's 1940 thesis which escaped the attention of people working on the subject - Higman's thesis never being published - and thus were reproved between 1960 and 1980);

"Zeta-functions of orders";

"The class group à la Fröhlich"

"Integral representations in the study of finite Poincaré complexes";

"Poset representations" (unfortunately Ludmilla Nazarova was the only one of the soviet mathematicians working in representation theory who was able to attend the meeting);

"Preprojective lattices over orders over Dedekind domains";

"Projective modules and resolutions for finite groups".

After some hesitation - being aware of the complications - the contributions in these notes appear not alphabetically but according subjects.

Part I contains the historical aspects leading to crystallography and via units to finite conjugate subrings of orders and via personal connections to constructive methods in ideal theory.

Part II deals with ζ-functions of orders - influenced by number theory - and Galois module structure - as applications of integral representations to number theory - in its various aspects.

Part III gives applications of integral representation theory to topology, differential geometry and algebraic geometry.

Part IV deals with representation theory eo ipso i.e. classification of indecomposable integral representations, partly in a general setup, partly in concret computations for special group rings and blocks of cyclic defect two.

Part V turns to projective resolutions including relation modules and units in group rings.

Finally I have included in these notes an extended version of two talks, that I gave at "Workshop on Representation Theory of Finite-Dimensional Algebras" in Bielefeld, which presents in a uniform fashion some of the main applications of almost split sequences and Auslander-Reiten quivers.

Since the meeting took place in Oberwolfach there was apart from a tight schedule enough time for a "leisurely" conversation of both personal nature and serious mathematical nature. Hence it is my prime pleasure to express my gratitude to the Forschungsinstitut Oberwolfach. Next I would like to thank the participants for making it a successful meeting. Finally my thanks go to Frau Annemarie Selbor for typing part of the manuscripts and for her secretarial assistance.

CONTENTS

LIST OF LECTURES

J.L.ALPERIN *	Projective modules and resolutions for finite groups
M.AUSLANDER *	Preprojective lattices over orders over Dedekind domains
H.BASS	Lenstra's calculation of $G(R[\pi])$, and Morse-Smale diffeomorphisms
E.A.BEHRENS	A non-commutative arithmetic for semigroups and its application to Asano orders
CH.BESSENRODT	On blocks of finite lattice type
J.BRZEZINSKI	Algebraic geometry of ternary quadratic forms and orders in quaternion algebras
C.J.BUSHNELL - I.REINER *	Zeta-functions of orders
M.C.R.BUTLER	Grothendieck groups and almost split sequences
J.F.CARLSON	The varieties of a module over an elementary abelian group
G.CLIFF	On units of integral group rings
E.DIETERICH	Representation types of group rings over complete discrete valuation rings
A.FRÖHLICH	Hermitian class groups
W.H.GUSTAFSON *	Orders in geometry, topology and number theory
W.KIMMERLE	Relative relation modules as generators
B.A.MAGURN	Uses of units in the Whitehead-groups
L.R.McCULLOH	Stickelberger relations in class groups of orders in abelian group rings
L.A.NAZAROVA *	Poset representations
W.PLESKEN *	Algebraic aspects of crystallography
H.-G.QUEBBEMANN	Integral orthogonal and symplectic representations of the cyclic group of prime order

J.QUEYRUT	S-Grothendieck groups and Galois module structure of rings of integers
I.REINER *	(s.BUSHNELL)
J.RITTER *	The class group à la Fröhlich
R.SANDLING *	Graham Higman's thesis
L.SCOTT	Hecke actions on Picard groups
H.R.STEIN	K_2 of integral group rings
O.TAUSSKI-TODD	Some facts concerning integral representations of ideals in an algebraic number field
ST.V.TAYLOR	Fröhlich's conjecture and logarithmic methods
TH.THEOHARI-APOSTOLIDI	On integral representations of twisted group rings
CH.THOMAS *	Integral representations in the study of finite Poincaré complexes
ST.V.ULLOM	Galois module structure for intermediate extensions
A.WIEDEMANN	Auslander-Reiten graphs of orders and blocks of cyclic defect two
J.WILLIAMS	Prime graph components of finite groups
S.M.J.WILSON	Miyata's theorem on the transfer map from the class group of a dihedral group to that of a cyclic group
H.ZASSENHAUS	On F.C.subrings of rings

Some of the lectures do not appear in these proceedings, partly since the results will be published elsewhere.

---

*invited lecturers

LIST OF PARTICIPANTS

ALPERIN, J., University of Chicago, Chicago, Illinois U.S.A.

AUSLANDER, Maurice, Brandeis University, Waltham, Mass. U.S.A.

BÄCKSTRÖM, Karl-Johan, University of Göteborg, Göteborg, Sweden

BASS, Hyman, Columbia University, New York, U.S.A

BEHRENS, E.A., McMaster University, Hamilton, Ontario, Canada

BESSENRODT, Christine, Univerität Essen, Essen, West Germany

BRZEZINSKI, J., University of Göteborg, Göteborg, Sweden

BUSHNELL, C., King's College, London, United Kingdom

BUTLER, M., University of Liverpool, United Kingdom

CARLSON, J.F., University of Georgia, Athens, Georgia, U.S.A.

CLIFF, G., University of Alberta, Edmonton, Alberta, Canada

DENNIS, K., Cornell University, Ithaka, New York, U.S.A.

DIETERICH. Ernst, Universität Bielefeld, Bielefeld, West Germany

FRÖHLICH, A., King's College, London, United Kingdom

GEEL, J. van, Universiteit Antwerpen, Wilrijk, Belgie

GRUENBERG, K.W., Queen Mary College, London, United Kingdom

GUSTAFSON, W.H., Texas Tech University, Lubbock, Texas, U.S.A.

HALTER-KOCH, F., Universität Essen, Essen, West Germany

JONES, Alfredo, Universidade de Sao Paulo, Sao Paulo, Brasil

KEATING, M.E., Imperial College, London, United Kingdom

KIMMERLE, W., Universität Stuttgart, Stuttgart, West Germany

KUKU, A.O., University of Ibadan, Ibadan, Nigeria

LORENZ, F., Universität Münster, Münster i.W., West Germany

MAGURN, B., University of Oklahoma, Norman, Oklahoma, U.S.A.

McCULLOH, L.R., University of Illinois, Urbana, Illinois, U.S.A.

MEYER, J., Universität Stuttgart, Stuttgart, West Germany

MICHLER, G., Universität Essen, Essen, West Germany

NAZAROVA, Ljudmilla, Akad.Nauk, Kiev, U.S.S.R.

OLIVER, R., Aarhus Universitet, Aarhus, Denmark

PLESKEN, W., RWTH Aachen, Aachen, West Germany

QUEBBEMANN, H.G., Universität Münster, Münster i.W., West Germany

QUEYRUT, J., Université de Bordeaux I, Talence, France

REINER, I., University of Illinois, Urbana, Illinois, U.S.A.

REITEN, Idun, Universitetet i Trondheim, Trondheim, Norway

RITTER, J., Universität Heidelberg, Heidelberg, West Germany

ROGGENKAMP, K.W., Universität Stuttgart, Stuttgart, West Germany

SANDLING, R., University of Manchester, Manchester, United Kingdom

SCHARLAU, R., Universität Bielefeld, Bielefeld, West Germany

SCHMID, P., Universität Tübingen, Tübingen, West Germany

SCOTT, L., University of Virginia, Charlottesville, Virginia, U.S.A.

STEIN, M.R., Northwestern University, Evanston, Illinois, U.S.A.

STROOKER, J.R., Rijksuniversiteit, Utrecht, Netherlands

TAUSSKI-TODD, Olga, California Institute of Techn.,Pasadena, U.S.A.

TAYLOR, M., Université de Besançon, Besançon, France

THEOHARI-APOSTOLIDI, Theodora, Univ.of Thessaloniki, Greece

THOMAS, Ch., SFB 40, Universität Bonn, Bonn, West Germany

ULLOM, S.V., University of Illinois, Urbana, Illinois, U.S.A.

WEBB, P., Girton Coll., University of Cambridge, Cambridge, U.K.

WIEDEMANN, Alfred, Universität Stuttgart, Stuttgart, West Germany

WILLIAMS, J., Adolphi University, Long Island, New York, U.S.A.

WILSON, S., University of Durham, Durham, United Kingdom

ZASSENHAUS, H., Ohio State University, Columbus, Ohio, U.S.A.

REMARKS ON THE HISTORY AND APPLICATIONS OF INTEGRAL REPRESENTATIONS

William H. Gustafson

Introduction.

This report gives a brief outline of the history of orders and integral representations, together with brief sketches of the connections between these notions and some other parts of mathematics. In particular, we look at composition of quadratic forms, algebraic number theory (especially integral normal bases), elliptic curves and complex multiplication, and crystallography. This is in no way a complete survey of the subject, nor is it an introductory text. For surveys, the reader is refered to Reiner [175,179], while good texts are Curtis and Reiner [46], Roggenkamp and Huber-Dyson [186,187] and Swan and Evans [221].

I am pleased to acknowledge the help of several people who supplied useful remarks on an earlier version of this paper. They were Karl-Johan Bäckström, Hyman Bass, Keith Dennis, Andreas Dress, Albrecht Fröhlich, Wilhelm Plesken, Irving Reiner, Klaus Roggenkamp, Jan Strooker, Olga Taussky, John Todd and Stephen Ullom. I mention separately Martin Taylor, who very kindly supplied an extensive write-up on the integral normal basis problem; I have followed it closely in the section on algebraic number theory, and I thank him for it.

Finally, I would like to take the opportunity to acknowledge the indirect assistance of several people. Like any other mathematician, I have learned both the substance and the style of mathematics from many people. Among them, a few have been most responsible for my interest in integral representations, and I hereby thank them for all they have taught me about this fascinating subject: Maurice Auslander, Sam Conlon, Everett Dade, Andreas Dress, Heinz Jacobinski, Klaus Roggenkamp, Dick Swan and above all, Irv Reiner. I also want to thank three nonmathematical friends, whose encouragment and receptivity were most sustaining during the preparation of this paper: Jeff Scott, Brook Hall and Gary Horby.

History

To the best of my knowledge, orders were first discussed by Dedekind [50] in 1871, in the famous work where the rigorous foundations of algebraic number theory and of the theory of ideals were first set down.

Among the several needs that motivated him, one can clearly point to two classical problems that originate with Fermat (or, perhaps, with Diophantus). We are all familiar with Fermat's Last Theorem, and with the attempts to prove it over the years. Interest in this problem was particularly intense in 1847, when Lamé, with the enthusiastic support of Cauchy, proposed to give a proof using factorizations in the ring of cyclotomic integers. However, Liouville communicated the news that Kummer, while studying higher reciprocity laws in 1844, had already discovered that such factorizations are not unique. (There is a widespread belief that Kummer had prepared a proof of Fermat's Last Theorem based on a false assumption of unique factorization of cyclotomic integers. However, Edwards [73] casts serious doubt on the veracity of this story. Edwards' book is an excellent source of information on the development of algebraic number theory in the nineteenth century). In any event, Kummer was aware of the failure of unique factorization, and he introduced his theory of <u>ideal numbers</u>, by which he could restore enough factorization theory to prove the Fermat theorem for exponents that are regular primes. Dedekind's main intention in 1871 was to reformulate Kummer's approach in a more workable form and in a more general context.

As a by-product of this investigation, Dedekind was also able to give a nicer formulation of Gauss' theory of composition of quadratic forms. As we will explain later in the section on that subject, the idea again goes back to Fermat, this time to his discussion of integers that are sums of two squares.

Our interest will focus mainly on the consequences of this second aspect, as a discussion of algebraic number theory <u>per se</u> would take us too far afield. So, we will adopt the conventional attitude of those who study orders that algebraic number theory is essentially known. Hence, when looking at orders in fields (or in commutative algebras in general), we focus on the nonmaximal ones. This allows us to skip far ahead, for nonmaximal orders and orders in noncommutative algebras were not studied much at first. We should, however, note a later work of Dedekind [51], wherein class numbers of nonmaximal orders in number fields are discussed, and the class group of projective ideals of these orders is introduced. From a different point of view, the early crystallographers worked with integral representations of finite groups (in some sense), while Jordan made similar studies in order to examine groups of motions and automorphisms of integral quadratic forms.

A development of considerable interest in the late nineteenth century was the discussion of integral quaternions. In 1886, Lipschitz [142] attempted to give an arithmetic discussion of quaternions, but met with considerable difficulty. Hurwitz [119,120], ten years later, had much more success. Here is where the difference lay: Lipschitz worked with the obvious Z-order spanned by 1, i, j, k (in analogy with Gaussian integers), while Hurwitz used the maximal order obtained by adjoining $(1+i+j+k)/2$ to Lipschitz' order. As later experience has shown, the use of a maximal order makes all the difference in discussing multiplicative ideal theory and factorization of elements.

After a pause of more than a decade in the development, Hurwitz' student du Pasquier [71] made a series of proposals for the definition of arithmetic in an algebra. For an algebra over the rational field, he proposed that a suitable arithmetic structure should be a unital subring that has a finite Z-basis and is maximal. Dickson was critical of this definition (and of another due to Hurwitz) in his influential book [57] of 1923, as it cannot be used in the case of nonsemisimple algebras. He proposed replacing the finite basis condition with the requirement that all elements have integral minimal polynomial. He then went on to show that the resulting arithmetic of an algebra (as far as factorization theory goes) reduced to that of the algebra modulo its radical (or, more properly, to that of a Wedderburn complement). Despite this, experience has shown that orders in nonsemisimple algebras are of limited interest, so du Pasquier's definition is vindicated, and is in fact the contemporary definition of a maximal Z-order.

Under the influence of Dickson's work, the ideal theory of maximal orders quickly developed at the hands of Artin [4,5], Brandt [24,25,26] and Speiser [213,214]. Some of this work dealt with attempts to generalize the ideal theory of algebraic number fields by looking at a single maximal order, while other parts looked at several orders at once, as is useful in discussing the composition of quaternary quadratic forms. Expositions of this theory can be found in Deuring [53] and Reiner [177]. There followed an explosion of interest in the arithmetic of algebras, as one can see by examining the bibliographies of Deuring [53] and Albert [2].

One of the most interesting papers in this era was the elegant work of Hasse [110], wherein the properties of maximal orders in local division algebras were exhaustively discussed. This work led to further results that Hasse obtained in collaboration with Brauer, Noether and Albert, giving a very precise structure theory for central simple algebras over algebraic number fields [3,111]. These results have, in turn, strong consequences in classfield theory [240]. One should also note the deep work of Eichler [74,75] on class numbers and reduced norms.

Up to this point, the emphasis has been on the <u>internal structure</u> of orders. In the late thirties began the emergence of interest in <u>representations</u> or orders. Of course, this had its precedents. Perhaps the earliest recognizable work along these lines was that of Frobenius and Stickelberger on finitely generated abelian groups [85]. Steinitz [216] had extended the theory to the case of modules over rings of integers of number fields (with some antecedents in Dedekind [52]), and in 1936, Chevalley [41] gave the results for arbitrary Dedekind domains. In the more general context, Zassenhaus [249] gave in 1938 his proof of the Jordan-Zassenhaus theorem, and later that year, his student, Diederichsen [60] discussed the basic properties of integral representations of finite groups, and gave the classification of ZG-lattices, where G is of prime order (a more modern discussion of this classification was given by Reiner [172]). Latimer and MacDuffee [138] investigated the connections between ideal classes and similarity classes of integral matrices. Brauer [28] began the investigation of the connections between integral and modular representations, Taussky and Todd [225,226] remarked on integral representations of cyclic groups, and G. Higman [117] studied the units of integral group rings (see Sandling [202] for a review of the unpublished parts of Higman's thesis).

During the tumult of World War II and the remaining years of the forties, the subject was rather dormant, save for some work oriented toward modular representations, and the development of Zassenhaus' algorithm [250] for the determination of space groups (see the section on geometric and topological applications below). Intensive work resumed in the fifties, when D. Higman, Maranda, Reiner and Takahashi wrote several papers dealing with the behavior of integral representations under localization and reduction modulo powers of primes, with extensions and cohomology, and with genera. These works gave the foundations for the prolific efforts of the sixties. Around 1960, three major events occured that gave tremendous impetus to the development of integral representations and orders. Auslander and Goldman [9] reworked and extended the theory of maximal orders, using modern ideas from homological algebra. Their work introduced new techniques to the subject and stimulated much interest in the global dimension of orders. Swan [218,219] adapted the theory of induced representations and Bass' algebraic K-theory to the setting of integral group rings, and thereby obtained very deep results on projective modules over integral group rings. The connections between K-theory methods and the theory of maximal orders were investigated by Strooker [217]. The third important development was the discovery by Troy [236] and Roiter [198] that Diedrichsen [60] had been incorrect when he asserted that the cyclic group of order four has infinitely many indecomposable integral representations. This led to the problem of determining the group rings, and, more generally, the orders, of finite representation type. The case of integral group rings was solved by Jones [130], using local results of Heller and Reiner [113] (see also Borevič and Faddeev [23], Berman and Gudivok [20] and Dade [47]). A characterization by Jacobinski [123] and Drozd and Roiter [69] of commutative orders of finite representation type resolves the problem for group rings with coefficients in the ring of integers of a number field. The general problem for orders is still open, and has consistently been a source of great interest.

Through the rest of the sixties and into the seventies, there came many developments that arose primarily from these seminal works. Hereditary orders (of which maximal orders form a subclass) were investigated by Brumer [30], Harada [106] and Williamson [243] (later, Jacobinski [127] studied hereditary orders from a different point of view; alternate proofs of some of his results were given by Reiner [176]). Maximal orders over more general base rings were studied by R. Fossum [82,83] and Ramras [171]. The works of Bass [14,15,16,17] on modules and rings were utilized heavily in the theory of orders, especially his work on Gorenstein rings (see Drozd, Kiričenko and Roiter [70] and Roggenkamp [190], for instance).

The use of methods from K-theory developed extensively as more and more variations were introduced. Heller and Reiner [114] gave an explicit formula for the Grothendieck group of an order (it is rather difficult to use, however. For instance, it is only recently that it has been possible to determine $K^{\circ}(ZG)$, G a finite abelian group, in easily understood terms. See Reiner [179] and Lenstra [139]). Class groups were studied early on by Rim [182,183], and their properties have since been extensively investigated; see below. Reiner [174] introduced the integral representation ring, following the lead of Green's modular studies. Relative Grothendieck rings were examined by Conlon [43], Dress [63,65,67] and Gustafson [103,104,105]. The use of induced representations in K-theory of finite groups was codifed by Dress [66], Lam [136] and Green [102]. The functor K^1 and others related to it were used in topology (see Milnor [154]), and Bass [18] used it to formulate a generalization to orders of Dirichlet Unit Theorem.

Along other lines, great advances were made in the theory of genera by Faddeev [79,81], Jacobinski [124,125,126] and Roiter [199,200,201]. The integral representations of the Klein four-group were calculated by Nazarova [157]; this was one of the first examples of a full discussion of an instance of tame representation type. Maximal orders over function fields were discussed by Mattson [152]. Multiplicative ideal theory of commutative orders was pursued; see Dade, Taussky and Zassenhaus [49] and Singer [210] (see also Fröhlich [86] and Jacobinski [123] for more on modules over commutative orders). Vertices and sources for integral representations were considered by Dress [64] and Thompson [233]. The use of adèles in the classification of lattices was noted by Drozd [68], Faddeev [80] and Takahashi [222]. Stancl [215] and Obayashi [163] studied the multiplicative properties of Grothendieck rings of finite groups. Taussky continued the Latimer-MacDuffee program extensively; see [224] and many other papers cited by Reiner [175]. The annihilator of $\mathrm{Ext}^1_\Lambda(M,N)$ was studied by D. Higman [116], T. Fossum [84], Jacobinski [122], Reiner [173], Roggenkamp [189] and Sjörstrand [210].

In more recent times, many of these investigations have continued, and new trends have started as well. A renewal of interest in class groups of orders was led by Fröhlich, Reiner and Ullom. Great progress was made in this area, because of new formulae and computational methods. Reiner and Ullom [181] introduced the effective use of the Mayer-Vietoris sequence of algebraic K-theory, while Fröhlich [92,94] gave various explicit formulae for the class group (see Joly [129], Matchett [151] and Wilson [244] for extensions and alternatives; Ritter [186] for an exposition of Fröhlich's approach). Also, the difficult theory of the Picard group of invertible bimodules was attacked; see Fröhlich [90] and Fröhlich, Reiner and Ullom [97]. These matters are surveyed more extensively in Reiner [177,178]. The applications of integral representations to presentations of finite groups were developed by Roggenkamp and Gruenberg; see Roggenkamp [193] for a survey.

Methods from the representation theory of algebras have recently been adapted to the case of integral representations. These include diagrammatic approaches to the classification of representations (see Bäckström [13], Butler [37,38], Green and Reiner [101], Ringel and Roggenkamp [185]), the method of almost split sequences (see Auslander [10], Reiten [182], Roggenkamp [191,192], Roggenkamp and Schmidt [197], Schmidt [203] and Wiedemann [242]), and the ideas of preprojective and preinjective modules; see Auslander [11].

The analytic number theory of orders, originally studied by Hey [115] and Zorn [251] (see also Deuring [53]) has recently been revived by Solomon [211,212] and by Bushnell and Reiner; see Bushnell and Reiner [34,35,36], Galkin [99], Reiner [180] and Weil [240].

Deep results on blocks of p-adic group rings have been derived by Plesken [167,168], Jacobinski (unpublished), Roggenkamp [194,196] and Wiedemann [242]. New results on the representation type of orders are emerging; see Bessenrodt [21], Dieterich [61] and Theohari-Apostolidi [232].

Composition of quadratic forms.

Formulae such as

(1) $$(a^2+nb^2)(c^2+nd^2) = (ac-bdn)^2+n(ad+cb)^2$$

have presumably been known since ancient times (at least in special cases such as $n = \pm 1$). In this section, we show how the theory of orders yields formulae like (1), as well as more exotic ones like

(2) $$(3a^2+14b^2)(2c^2+21d^2) = 6(ac-7bd)^2+7(2bc+3ad)^2.$$

In 1640, Fermat used the first formula above, in the case $n = 1$, to prove his famous result on integers representable as a sum of two squares. The formula shows that a representation of a composite number $k = mn$ as a sum of two squares can be found if such representatives are known for the factors m and n. By careful analysis of this process, Fermat showed that one need only solve the genus problem; i.e. one need only determine which primes p are sums of two squares, in order to solve the problem for all integers. Of course, he discovered the necessary and sufficient condition $p \not\equiv 3 \pmod 4$ for $p = a^2+b^2$.

Similar isolated results were discovered over the years, but it remained for Gauss [100] to attack the general problem in 1801. In order to discuss the representability of integers by binary integral quadratic forms, Gauss divided the work into two parts, as Fermat had done. If the form we are interested in is $Q(x,y)$, one must first find all representations of Q as a composition of other forms. We should make this precise: if Q_1 and Q_2 are forms, we say that Q is obtained from Q_1 and Q_2 by composition if there is an equation.

$Q_1(x,y)Q_2(w,z) = Q(F_1(x,y,z,w),F_2(x,y,z,w))$, where F_1 and F_2 are bilinear forms with integral coefficients. Formulae (1) and (2) above exhibit examples. The composition of several forms is defined by iteration. Once these representations are found, one must solve the genus problem, i.e. one must discuss solutions of the equations $Q_i(x,y) = p$, with p prime. We will not consider the genus problem any further.

When Gauss considered the composition problem, he worked on it by direct computation, in what Kaplansky [131] describes as a "tour de force that makes remarkable reading to this day". Here, we will examine Dedekind's formulation in terms of ideals of quadratic fields. To get started, let us see where (2) comes from. Consider the maximal order $R = Z[\alpha]$ in $Q(\alpha)$, where $\alpha^2 = -42$. We have the norm mapping $N\colon R \to Z$, given by $N(a+b\alpha) = a^2+42b^2$. We observe that this gives a nice integral quadratic form, but not one of the ones involved in (2). Now, the class group of R is a Klein group, with the nonprincipal classes represented by the ideals $J_1 = (3,\alpha)$, $J_2 = (2,\alpha)$ and $J_1J_2 = (6,\alpha)$. A typical element of J_1 has the form $3a + \alpha b$, and its norm is $9a^2 + 42b^2$. If we now divide by $N(J_1) = 3$, we obtain the form $3a^2 + 14b^2$, which does appear in (2). In a similar way, J_2 and J_1J_2 give rise to the forms $2a^2 + 21b^2$ and $6a^2 + 7b^2$, where in each case, we use the element norm divided by the ideal norm. Formula (2) is now revealed as

$$\frac{N(3a+\alpha b)}{N(J_1)} \cdot \frac{N(2c+\alpha d)}{N(J_2)} = \frac{N((3a+\alpha b)(2c+\alpha d))}{N(J_1J_2)},$$

which is obviously valid by the multiplicative properties of the norm. In this way, we see that the composition of forms is closely related to the multiplication of ideals of orders in quadratic fields. There are some difficulties associated with Dedekind's approach, as one must use ordered bases of ideals; and a stronger equivalence of ideals is also required (for I and J to be equivalent, one requires $I = xJ$, where $N(x) > 0$). So, we will sketch the exposition by Kaplansky [131], wherein these problems are circumvented.

Kaplansky gives his results for any base ring that is a Bézout domain whose characteristic is not two. We will confine our attention to the case of a principal ideal domain R, with field of fractions K. Let L be a separable quadratic extension of K, and let $x \to x^*$ be the nontrivial automorphism of L/K; note that $x^{**} = x$. Further, the norm and trace maps are given by $N(x) = xx^*$ and $T(x) = x + x^*$. If we represent L as $K(\alpha)$, $\alpha^2 \in K$, then a typical element of L has the form $a + \alpha b$, with $a,b \in K$, and its norm is $a - \alpha^2 b$. So, the norm gives a quadratic form over K. We want to restrict it to full R-lattices in L, which are evidently R-free of rank two. If A is such a lattice, its <u>discriminant</u> is $DA = (xy^* - x^*y)^2$, where $A = Rx \oplus Ry$. Note that DA belongs to K, and is defined only up to the square of a unit of R. If B is another full R-lattice in L, we call A and B <u>equivalent</u> if $B = xA$, for a nonzero $x \in L$. To avoid the problems of Dedekind's approach, Kaplansky refines things a bit at this point. The basic objects are now pairs [A,a], where A is a full R-lattice in L and $a \in K$ is nonzero. We then define $[A,a]^* = [A^*,a]$, $N[A,a] = NA/a$, $[A,a][B,b] = [AB,ab]$ and $D[A,a] = DA/a^2$. (Here, $A^* = \{x^* | x \in A\}$, NA is the fractional R-ideal generated by all Nx, $x \in A$ and AB is the usual product of lattices in an algebra). [A,a] and [B,b] are <u>equivalent</u> if $B = xA$, where $x \in L$ satisfies $N(x)a = b$.

We want to connect these things with "concrete" binary quadratic forms over R. Such a form is an expression $Q(x,y) = ax^2 + bxy + cy^2$, with $a,b,c \in R$. This can be viewed as a quadratic form on a free R-module of rank two with a distinguished basis; a change of basis leads to an equivalent form. The discriminant of such a form is the familiar

$b^2 - 4ac$. Now, let Δ be the discriminant of some pair. As it has the form $(xy^* - x^*y)^2/a^2$, it has a square root $\delta \in L$ (fix one square root arbitrarily). We want to look at various pairs of discriminant Δ; let $[A,a]$ be one such. It has an R-basis $\{x,y\}$ that is <u>admissible</u>, in that $(xy^* - x^*y)/a = \delta$. Relative to this basis, we have the concrete form $Q(r,s) = N(rx + sy)/a$ (whose coefficients lie in R if $[A,a]$ is primitive; see below). If we change to another admissible basis, a form properly equivalent (= $SL(2,R)$-equivalent) to Q will be obtained. Also, if $[A,a]$ and $[B,b]$ are equivalent, then both yield the same concrete form. Conversely, given $ax^2 + bxy + cy^2$, we construct $[A,a]$, where A is spanned over R by a and $(b - \delta)/2$. These constructions give a bijective correspondence between the equivalence classes of pairs of discriminant Δ and the proper equivalence classes of binary quadratic forms of discriminant Δ.

To define the composition of forms, we thus want to pass to the classes of pairs and perform ideal multiplication as in the example treated earlier. Unfortunately, if $[A,a]$ and $[B,b]$ have discriminant Δ, it needn't follow that $[AB,ab]$ does also. However, it will if $[A,a]$ and $[B,b]$ are <u>primitive</u>, i.e. if their norms are R. For the corresponding concrete forms, this means that the coefficients lie in and generate R. One then obtains the central result on composition: the primitive pairs of given discriminant Δ form a group under multiplication. The proof involves the use of invertible ideals for an appropriate order.

Choose for A a basis $\{a,z\}$, with $a \in K$. Let $T(z) = b$, $N(z) = c$. Then $N(A) = (a^2, ab, c)$ is a principal ideal, say $N(A) = (e)$. For $t = az/e$, the R-module $S = R \cdot 1 \oplus R \cdot t$ turns out to be an order whose discriminant equals that of $[A,e]$. Further, A is an invertible ideal for this order and for no other. This order S is the unique one of its discriminant (even up to squares of units of R), and one sees that the primitive pairs of this discriminant form a group with $[S,1]$ as identity element. It is called the extended class group $H(S)$ of S, and is related to the usual class group $C(s)$ by an exact sequence

$$0 \to U(R)/N(U(S)) \to H(S) \to C(S) \to 0,$$

where $U(-)$ is the units functor. Observe that $U(R)/N(U(S)) = H^2(\langle * \rangle, U(S))$. See Towber [235] for an extensive discussion of the method of "united" forms.

This is a sketch of the story for binary forms; one could ask about forms in a larger number of variables. As it turns out, composition can only occur when the number n of variables is 1, 2, 4 or 8 (see [128]), and each of these cases has been examined. Now, $n = 1$ is trivial, so let us go to $n = 4$, i.e. the case of quaternary quadratic forms.

Composition of such forms was studied extensively by Brandt, starting in 1913. At first, he did much explicit calculation in the style of Gauss. In 1928, he put his techniques in the framework of ideals of orders in quaternion algebras, and obtained results similar in spirit to those above. Of course, he had to work with the reduced norm in order to get quadratic forms, and the answers are more complicated. Since the left and right orders of a lattice may differ, and there isn't a unique order of given discriminant, more care must be exercised. In particular, products AB are useful only in the concordant case, i.e. when the right order of A is the left order of B. So, one no longer gets a group from the primitive forms, but rather, the famous Brandt groupoid. This is a group-like structure in which not all products are defined, and there are several "local" identity elements. A modern exposition has been given by Kaplansky [132]. A summary of Brandt's work was given by Aeberli [1], but Kaplansky has pointed out some errors. However, it provides a useful bibliography of Brandt's works. In the case of eight variables, things get even harder, as one must now work in the nonassociative Cayley algebras. A discussion in Kaplansky's style was given by his student Epp [77].

For more about orders in quaternion algebras and Cayley algebras, see Brandt [24,25,26,27], Brzezinski [31], Coxeter [44], Deuring [53], Dickson [57,58,59], Eichler [76], Estes and Pall [78], Hashimoto [107, 108], Hurwitz [119,120], Kirmse [133,134], Lamont [137], Mahler [145], Martinet [147,148], Mipp [160], Pall [164], Peters [165] and van der Blij and Springer [237] (note: Kirmse's work contains serious errors; see Coxeter's paper).

Algebraic number theory.

Of course, the theory of orders uses and generalizes algebraic number theory. However, some aspects of algebraic number theory benefit from the theory of orders. Recall that if K/k is a finite Galois extension of fields, with Galois group G, then K has a normal basis over k. That is, there is an element $a \in K$ such that $\{g(a) \mid g \in G\}$ is a k-basis for K. To put this in invariant form, note that the actions of k and G on K make K a kG-module; the normal basis theorem just says that $K \cong kG$ as kG-modules (see Berger and Reiner [19] for an interesting proof). It is natural to ask whether a corresponding result holds for integral elements, when k is the field of fractions of a Dedekind domain R, and we view the integral closure S of R in K as an RG-module. If it does turn out that $S \cong RG$ as RG-modules, we say that K has an integral normal basis. It turns out to be rather difficult to determine when this occurs. However, we shall see below that the deviation of S from being a projective RG-module is measured by familiar arithmetic invariants.

The first result discovered about this situation seems to be Hilbert's Theorem 132 [118]: let K be an abelian Galois extension of the rationals, with [K:Q] relatively prime to the discriminant of K. Then K has an integral normal basis. The proof is rather deep, in that it uses the Kronecker-Weber Theorem of classfield theory [238]. Fortunately, a different point of view allows (weaker) generalizations, without such heavy machinery. Let S denote the ring of all algebraic integers in K. If P is a prime ideal of K lying over the rational prime p, we know that S/P is a separable extension of Z/pZ, and the hypotheses guarantee that p does not divide the ramification index of P. This is the situation we abstract: let R be a Dedekind domain with field of fractions k, and let S be the integral closure of R in a finite extension field K of k. Let P be a prime ideal of S, and let $p = R \cap P$. We say that the extension K/k is <u>tamely</u> <u>ramified</u> <u>at</u> <u>P</u> if S/P is a separable extension of R/p, and the characteristic of R/p doesn't divide the ramification index e defined by $P^e \| pS$. The extension K/k is <u>tamely</u> <u>ramified</u> (or, for brevity, <u>tame</u>) if it is tamely ramified at each prime.

Noether [162] showed that in the above context, if K is Galois over k with group G, then K/k is tamely ramified if and only if S is a projective RG-module. Let us sketch a proof of this in the local case, following Fröhlich [88]. So, we assume k is a complete, discretely valued field, K/k is Galois with group G and K is equipped with an extension of the valuation on k. We show that if the extension is tamely ramified, then there is an integral normal basis (the converse is an exercise for the reader). Let R and S be the valuation rings of k and K respectively. In order to show S = RG, it suffices by a theorem of Swan [46, Theorem 77.14] to show that S is RG-projective, since we already know that $K \cong kG$. By [206; p. 132] it suffices to show that S has an R-endomorphism λ with $\sum_{g \in G} g\lambda g^{-1} = \text{identity}$. Let P be the maximal ideal of S, $p = R \cap P$. We write $pS = P^e$, so e is the ramification index. Let $\bar{K} = S/P$, $\bar{k} = R/p$. Then $\bar{K}/\bar{k}$ is separable, and the characteristic of $\bar{k}$ doesn't divide e. The trace map $t_{K/k}$ maps S into R, and so induces a $\bar{k}$-linear map $\bar{t}: S/pS \to \bar{k}$. Now, we identify S/pS with S/P^e, and see that it has a $\bar{k}$-filtration

$$S/P^e \supseteq P/P^e \supseteq \cdots \supseteq P^{e-1}/P^e \supseteq \{0\},$$

with factors $(P^i/P^e)/(P^{i+1}/P^e)$ each isomorphic to $S/P = \bar{K}$. So, for $s \in S$, multiplication by $s + pS$ on S/pS can be represented by a matrix of the exe block form

$$\begin{pmatrix} \tilde{s} & * & * & \dots & * \\ & \tilde{s} & * & \dots & * \\ & & & \dots & \\ & & & & \tilde{s} \end{pmatrix},$$

where $\tilde{s}$ is a matrix over R representing multiplication by $s + P$ on $\bar{K}$. Hence, $\bar{t}$ is given by $\bar{t}(s + pS) = e \cdot t_{\bar{K}/\bar{k}}(s + P)$. As $\bar{K}/\bar{k}$ is separable and char $\bar{k} \nmid e$, it follows that $\bar{t}$ is surjective. This easily implies that $t_{K/k}$ maps S onto R. So, there is $t \in S$ with $t_{K/k}(t) = 1$, i.e. $\sum_{g \in G} g(t) = 1$. Let λ be the R-endomorphism of S given by $\lambda(s) = ts$. Then we see that $\Sigma g \lambda g^{-1}(s) = \Sigma g(t \cdot g^{-1}(s)) = \Sigma g(t) \cdot s = s$, for all $s \in S$, and the proof is complete.

Note that the above easily implies the global result of Noether. Here is a generalization due to Fröhlich [87] (an independent proof by Miyata [155] can be found in most libraries). Suppose we have a Galois extension of local fields as above, but not necessarily tame. S can be written as a direct sum $\oplus M_i$ of indecomposable RG-modules, and we let $V_i \subseteq G$ be a vertex of M_i. Define the vertex V of S to be the normal subgroup of G generated by all the V_i. Then V is the first ramification group of K/k, i.e. the subgroup of G corresponding under Galois theory to the maximal tame extension of k in K.

There have recently been great advances in the global integral normal basis problem. Now let K/k be a Galois extension of algebraic number fields, with Galois group G. Let S,R be the rings of algebraic integers in K,k respectively. One would like to be able to decide when $S \cong RG$ as RG-modules. Unfortunately, S may even fail to be R-free (see MacKenzie and Scheuneman [143] for a simple example). Hence, one asks the weaker question of whether S is ZG-free. One finds that locally, RG-freeness and ZG-freeness are equivalent for S, so tame ramification is required even to start on the problem, by Noether's result above. As we mentioned earlier, Hilbert showed that tame abelian extensions of the rationals have integral normal bases. A striking generalization has recently been found by Taylor [228]; to wit, if G is abelian and K/k is tame, then S is ZG-free. By use of induction methods, one can show from this that in the tame nonabelian case $[S]^{A(G)}$ is trivial in the locally free class group of ZG, where A(G) is the Artin exponent of G (see Lam [136] for the definition). However, the nonabelian global theory really got started in the work of Martinet [146], wherein he considered the case where k is the rationals and G is dihedral of order 2p, p an odd prime. He showed that S is always ZG-free in this case. He later used his methods [147,148] to show that freeness could fail, when G is the quaternion group H_8 of order eight.

The general discussion of existence of integral normal bases was clarified by the study of Artin root numbers, which arise in analytic number theory. Associated to any character χ of G, there is an extended Artin L-function $\Lambda(s,\chi)$, which satisfies a functional equation

$$\Lambda(s,\chi) = W(\chi)\Lambda(1-s,\bar{\chi})$$

where $\bar{\chi}$ is the conjugate of χ and $W(\chi)$ is a constant called the <u>Artin root number</u>. It is easily seen to be ± 1. Fröhlich and Queyrut [96] showed that $W(\chi) = 1$ when χ is orthogonal. On the other hand, examples were found where G is H_8, χ is a symplectic character, and $W(\chi) = -1$. The basic connection began to emerge when Fröhlich [89] showed that for a tame H_8-extension of the rationals, S is ZG-free if and only if $W(\chi) = 1$, for χ the irreducible character of degree two.

There ensued a complete revolution in the area of Galois module structure of rings of integers. This commenced with Fröhlich's description of the locally free class group of ZG as a subquotient of the group of homomorphisms from the virtual characters of G to the ideles of a suitably large number field. In order to describe the class of S in these terms, Fröhlich had to invent a far-reaching generalization of Lagrange resolvents, and connect it with Gauss sums and Artin root numbers; see [94]. As a striking illustration of the power of his methods, Fröhlich verified Martinet's conjecture for tame extensions (this is the conjecture that ΓS is stably Γ-free, where Γ is a maximal order containing ZG. It is not generally true for nontame extensions; see Cougnard [45]).

Returning to the tame situation, we find the integral normal basis problem virtually solved by Taylor's logarithmic description [230] of determinants and reduced norms. This approach aids greatly in checking whether a given homomorphism on virtual characters represents the trivial class. Using this, Taylor [231] verified a conjecture of Fröhlich to the effect that vanishing of the class of S in the locally free class group of ZG is obstructed only by the Artin root numbers of the irreducible symplectic characters. In particular, that class has order at most two. The class of S can be described precisely in terms of invariants introduced by Cassou-Noguès [40]. Taylor's method can be used in other class group computations as well; see [229].

Some other papers on interest in this area are Auslander and Rim [12], Fröhlich [91,93], Jacobinski [121], Leopoldt [141], Miyata [156], Nelson [158], Queyrut [170], Taylor [227] and Yokoi [248] (most of these predate the recent revolution). The Durham Symposium Proceedings [95] contain interesting survey articles by Martinet, Tate, Fröhlich and Ullom on Artin root numbers, Galois module structure and locally free class groups.

For the structure of the units of S as a Galois module (local case), see Pieper [166] and Wingberg [245]. For the action of G on the class group of S, see Kuroda [135], Leopoldt [140] and Yamamoto [247].

Let us make a few remarks about a different sort of application. It has long been an open question whether every finite group appears as the Galois group of some Galois extension of the rational field. Noether [161] observed that an affirmative answer would follow, if one could verify the following conjecture: let a subgroup G of the symmetric group S_n act on a function field $K = k(x_1,\dots,x_n)$ by permuting the variables. Then the fixed field K^G is a purely transcendental extension of k. Swan [220] showed that if n is 47 or 113 or 233, and G is the subgroup of S_n generated by an n-cycle, then K^G is not purely transcendental over k. He obtained this result by studying a certain ideal of an inegral group ring. Note that the above special values of n are primes. In general, Swan considers a prime p, and the group $\pi = \mathrm{Aut}(Z/pZ)$, which is cyclic of order $p-1$. The natural action makes Z/pZ a $Z\pi$-module, and there is a unique $Z\pi$-surjection $f\colon Z\pi \to Z/pZ$, with $f(1) = 1$. Swan shows that the ideal $I = \ker(f)$ is not principal for $p = 47$ or 113 or 233. Now, Masuda [150] had considered Noether's conjecture in the following way: let ζ be a primitive pth root of unity over a field F, and assume $[F(\zeta): F] = p-1$. Let $x_1,\dots,x_p$ be indeterminates cyclically permuted by a group G of order p. Let $K = F(x_1,\dots,x_p)$ and let $L = K^G$. We want to decide whether L is purely transcendental over F. Let $\bar{F} = F(\zeta)$. Put $\bar{K} = \bar{F}K = \bar{F}(x_1,\dots,x_p)$ and $\bar{L} = \bar{F}L = \bar{K}^G$. Let $y_i = \sum_{j=1}^{p} \zeta^{ij}x_j$. Then $\bar{K} = \bar{F}(y_0,\dots,y_{p-1})$. Let π be the Galois group of $\bar{F}/F$. π is cyclic of order $p-1$, and it is also the Galois group of $\bar{K}/K$ and of $\bar{L}/L$. π fixes y_0 and permutes $y_1,\dots,y_{p-1}$ freely. Now, the subgroup $\bar{M}$ of $\bar{K}^{\times}$ generated by $y_1,\dots,y_{p-1}$ is $Z\pi$-free on y_1, and $M = \bar{M} \cap \bar{L}$ is a $Z\pi$-submodule. Masuda shows that if M is $Z\pi$-free, then L is purely transcendental over F, and that M is isomorphic to the ideal $I = \ker(f)$ described earlier. Swan shows that conversely, if L is purely transcendental, then M must be $Z\pi$-free. Since he had shown that I is not principal for the specific values of p, it follows that counterexamples exist.

Incidentally, Swan was able to use this same calculation to solve a problem in algebraic topology that had been posed by Steenrod: if a finite group π, a Z-finitely generated $Z\pi$-module A and an integer $n > 0$ are given, is there a finite complex K such that $\tilde{H}_i(K)$ (the reduced simplicial homology group) is 0 for $i \neq n$, $H_n(K) = A$, and π acts simplically on K so as to induce the given $Z\pi$-module structure on $A = H_n(K)$? In the case where $A = Z/pZ$ and $\pi = Aut(Z/pZ)$, D. Kahn (unpublished) had shown that it would suffice that the ideal I be principal. Swan showed for $p = 47$ that the failure of I to be principal also caused the failure of Steenrod's conjecture.

Elliptic curves and complex multiplication.

Elliptic integrals $\int_a^b P(t)^{-1/2}dt$, where P is a polynomial of degree three or four arose in geometric investigations. For instance, Fagnano showed in 1718 that

$$\int_0^a (1-t^4)^{-1/2}dt = 2\int_0^b (1-t^4)^{-1/2}dt,$$

where b can be expressed algebraically in terms of a. Hence, given a lemniscate, he could construct another with twice the arc length. Later, it was shown that Fanano's transformation could be obtained in several steps, one of which gives a formula

$$\int_0^a (1-t^4)^{-1/2}dt = (1+i)\int_0^c (1-t^4)^{-1/2}dt,$$

where again, c is an algebraic function of a (of course, c is complex, and the integral must be taken over a suitable path; see [208, p. 6]). This is one of the earliest instances of complex multiplication. In order to see how this is related to the theory of orders and to class-field theory, we must develop some connections between these classical integrals and more modern geometric and algebraic objects.

Let L be a Z-lattice in the complex field C, so $L = Zw_1 \oplus Zw_2$, where $\{w_1, w_2\}$ is an basis for C over the real number field. Then C/L is a complex Lie group of dimension one, and is topologically a 2-torus. Replacing L by zL, where $0 \neq z \in C$ yields an isomorphic quotient C/zL, so we may assume that $w_1 = 1$ and w_2 is an element τ of the upper half plane. As C/L is a complex manifold, it is interesting to determine the field of meromorphic functions on it. These can clearly be identified with those meromorphic functions f on C with the property $f(z+w) = f(z)$, for $c \in C$, $w \in L$. Such functions are called elliptic functions for L. One such function is the Weierstrass Pe-function

$$P(z) = z^{-2} + \Sigma'\{(z-w)^{-2} - w^{-2}\},$$

where Σ' means that the sum is taken over all nonzero $w \in L$. The derivative $P'(z) = -2\,\Sigma(z-w)^{-3}$ is another example. These two examples yield all others, for it can be shown that the field elliptic functions for L is $C(P(z), P'(z))$. One can further show that

$$(P')^2 = 4P^3 - g_2P - g_3,$$

where $$g_2 = 60\Sigma' w^{-4}$$

and $$g_3 = 140\Sigma' w^{-6}.$$

Hence, the elliptic function field for L is isomorphic to the quotient field of $C[X,Y]/(Y^2 - 4X^3 + g_2X + g_3)$, i.e. to $C(X, \sqrt{4X^3 - g_2X - g_3})$. Further, the mapping φ: $C\backslash L \to C \times C$ given by $\varphi(z) = (P(z), P'(z))$ obviously has image contained in the algebraic curve X_0 with equation $y^2 = 4x^3 - g_2x - g_3$. It can be shown that the image fills up the entire curve. By sending the elements of L to a point at infinity, one obtains (thanks to the periodicity of φ with respect to L) an identification of C/L with the nonsingular projective cubic curve X with homogeneous equation $y^2z = x^3 - g_2xz^2 - g_3z^3$. This gives C/L the structure of an algebraic curve of genus one, or an <u>elliptic curve</u>. We note that X can be made into an algebraic group, with the point at infinity as identity element. The group law is given by $P_1 + P_2 + P_3 = 0$ if P_1, P_2 and P_3 are collinear. With this structure, φ becomes a group isomorphism.

On the other hand, if an affine complex curve is given by an equation $y^2 = P(x)$, where P is a polynomial of degree three or four, without repeated roots, then its projective completion has genus one, and the Riemann-Roch Theorem implies that one can in fact take P cubic (up to birational equivalence). Further, one can assume that P is in Weierstrass form:

$$P(x) = 4x^3 - ax - b, \text{ with } a^3 - 27b^2 \neq 0.$$

Hence, our projective curve can be <u>identified</u> with the Riemann surface R of the (2-valued) function $y = \sqrt{4x^3 - ax - b}$. This is a 2-sheeted covering space of the Riemann sphere S^2. (In terms of the curve representation, the covering map is $(x,y) \to x$). Note that this covering is ramified of order 2 at four points: ∞ and the three zeroes of $4x^3 - ax - b$. Now we can head back toward the beginning. The differential form $dx/y = (4x^3 - ax - b)^{-1/2}dx$ turns out to be a holomorphic form on R, so path integrals of it can be considered. Given a point (x_0, y_0) on R, form

$$\int_{\infty}^{(x_0,y_0)} \frac{dx}{y}$$, where the integral is taken over any convenient path. The value is not independent of the path, but is determined up to an element of a lattice $L = Zw_1 \oplus Zw_2$, where w_1 and w_2 are path integrals around closed curves in R that represent the two generators of the singular homology group $H_1(R)$ (recall that R is a torus, because its genus is one). Indeed, L itself can obviously be identified with $H_1(R,Z)$. Further, it can be shown that the above integration process yields an

identification of R with C/L. Hence, the old elliptic integrals yield the connection between curves of genus one and lattices in C.

So, we regard the following types of objects as essentially the same:

i) Elliptic integrals $\int P(x)^{-1/2}dx$, P(x) of degree 3, with three distinct roots.

ii) Complex Lie groups of the form C/L, with L a lattice as above.

iii) Projective algebraic curves of genus one. Note that function fields of the form $C(X,\sqrt{P(x)})$ with P(x) as in i) could be included in this list, as well as one-dimensional abelian varieties.

Now let us look again at complex multiplication. Given X = C/L and X' = C/L', we want to consider Hom(X,X'). It can be shown that any holomorphic function $f: X \to X'$ with $f(0) = 0$ is a group homomorphism, and it lifts to a holomorphic homomorphism $\tilde{f}: C \to C$ of the universal covering spaces. Hence, $\tilde{f}$ is multiplication by an element $\lambda \in C$, with the property that $\lambda L \subseteq L'$. It follows that if f is nonzero, it is an <u>isogeny</u>, i.e. it is surjective with finite kernel isomorphic to $L'/\lambda L$. So, we identify $\text{Hom}(X,X') = \{\lambda \in C | \lambda L \subseteq L'\}$. In particular, $\text{End}(X) = \{\lambda \in C | \lambda L \subseteq L\}$. If we assume that L has Z-basis $\{1,\tau\}$, with τ in the upper half plane, we see that $\text{End}(X) \subseteq L$, so End(X) has Z-rank one or two (rank one occurs when L is not itself a ring). Hence, it is isomorphic to Z or it is an order in a quadratic number field. Since τ is nonreal, this quadratic number field must be an imaginary one $Q(\sqrt{-d})$. This second possibility occurs if and only if $Q(\tau) = Q(\sqrt{-d})$, and we then say that X has complex multiplication.

How does this turn up in the other descriptions of the context? Complex multiplication on X is equivalent to a formula of the form

$$\int_{\infty}^{(x_0,y_0)} dx/y = \lambda \int_{\infty}^{(x_1,y_1)} dx/y,$$

where λ is nonreal, (x_1,y_1) is algebraic in (x_0,y_0), and dx/y is the holomorphic form described earlier. Alternatively (see Masser [149]), it amounts to the algebraic dependence of the functions $\mathcal{P}(z)$ and $\mathcal{P}(\lambda z)$.

Henceforth, let X be an elliptic curve admitting complex multiplication. Then, as we have noted, End(X) is an order $\mathcal{O}$ in an imaginary quadratic field $K = Q(\sqrt{-d})$. Then, one can obtain X as $C/\mathcal{A}$, where $\mathcal{A}$ is a projective fractional ideal of $\mathcal{O}$. Clearly, if $\mathcal{A}$ is varied in its ideal class, the isomorphism class of X is not changed. Conversely, any such $C/\mathcal{A}$, $\mathcal{A}$ projective has $\mathcal{O}$ as its endomorphism ring, and one can show that $C/\mathcal{A}$ and $C/\mathcal{B}$ are isomorphic if and only if $\mathcal{A}$ and $\mathcal{B}$ lie in the same ideal class. So, the isomorphism classes of elliptic curves with endomorphism ring $\mathcal{O}$ are in one-to-one correspondence with the elements of the class group $C(\mathcal{O})$. Now, the maximal order in $Q(\sqrt{-d})$ is $Z[w]$, where $w = \sqrt{-d}$ or $(1+\sqrt{-d})/2$, depending on the congruence class of $-d$ modulo 4, and it is easy to show that any order in $Q(\sqrt{-d})$ is of the form $Z[nw]$, with n an integer (see Cohn [42] or Shimura [207]). Hence, the elliptic curves admitting complex multiplication would be classified, if we could solve the class group problem for all $Z[nw]$. This has not been done, but there is at least a class number formula for $Z[nw]$, due to Dedekind [51] (it unfortunately requires knowledge of the class number of the maximal order).

Now, we turn to constructive classfield theory. The famous Kronecker-Weber Theorem (see Weber [238]) states that the maximal abelian extension of Q is obtained by adjoining all roots of unity. Unfortunately, this result is restricted to Q. Kronecker hoped (the "Liebster Jugendtraum"; see Hasse [112]) that maximal abelian extensions of other number fields could be obtained by adjoining algebraic values of transcendental functions, as done with the exponential function in the case of Q. Not much progress has been made on this project. The only notable exception is the case of imaginary quadratic fields, where complex multiplication yields the results.

For an elliptic curve X with affine equation $y^2 = x^3 - g_2x - g_3$, we define the <u>j-invariant</u> $j_X = g_2^3/(g_2^3 - 27g_3^2)$. It is a complete isomorphism invariant of X. For a projective ideal A of an order $\mathcal{O}$ in an imaginary quadratic field K, we define $j(A) = j_{C/A}$. Then: i) $j(A)$ is an algebraic number, ii) the Galois group of $K(j(A))$ over K is isomorphic to the class group $C(\mathcal{O})$, iii) $[K(j(A)): K] = [Q(j(A)): Q]$ and iv) if $\mathcal{O}$ is the maximal order in K, then $K(j(A))$ is the Hilbert classfield of K.

By adjoining a bit more, one can obtain the maximal abelian extension of K. First, it is necessary to classify the elliptic curves by their automorphism groups. The automorphisms of an elliptic curve X are just the units in End(X). Since this is an order in an imaginary quadratic field, its only units are ±1, unless End(X) = Z[i], which has four units or $\mathrm{End}(X) = Z[(1+\sqrt{-3})/2]$, which has six units. Hence, we put the curve X in the class E_i if $|\mathrm{Aut}(X)| = 2i$, for i = 1,2 or 3. It can be shown that $X \in E_2$ if and only if $j_X = 1$, and $X \in E_3$ if and only if $j_X = 0$. Now, letting X have affine equation $y^2 = x^3 - g_2x - g_3$, we put $\Delta = g_2^3 - 27g_3^2$, and define functions on X by

$$h_X^1((x,y)) = (g_2g_3/\Delta)x$$

$$h_X^2((x,y)) = (g_2^2/\Delta)x^2$$

$$h_X^3((x,y)) = (g_3/\Delta)x^3.$$

(Note that x determines y up to sign on X, so the dependence on x alone is reasonable).

Let $\mathcal{O}$ be the maximal order in an imaginary quadratic field, A an ideal of $\mathcal{O}$. Put $X = C/A$, and suppose $X \in E_i$. Then the maximal abelian extension of K can be obtained by adjoining j_X (as before), and all values $h_X^i(t)$, where $t \in X$ is of finite order in the group structure of X.

Observe that these adjoined values are values of transcendental functions. For, the powers of x in the definition of the function h_X^i can be viewed as powers of values of the Pe-function, in light of the parameterization $z \to (\mathcal{P}(z), \mathcal{P}'(z))$ of X. As for j_X, if we represent X (up to isomorphism) as $C/Z \oplus Zz$, we can view j as a function of z rather than X, and as such, it is a modular function of level one on the upper half-plane.

In this exposition, I have followed Siegel [208], Shimura [207] and Serre's part of [22]. For another exposition, see Serre [205]. The results on complex multiplication are due to Weber [239], Fueter [98] and Hasse [109]. See also Deuring [54,55,56]. Historical remarks on elliptic functions and the works of Fagnano can be found in Todd [234].

Geometric and topological applications.

Let us consider briefly the problem of tiling the plane with decorative tiles that are identical to one another. We assume that each tile has on it a picture that has no symmetry, and that the tiles are of a transparent material, so that the picture can be seen from either side. For instance, we might use a square tile, and obtain the following picture:

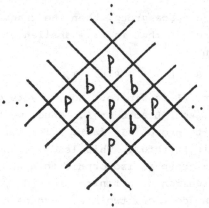

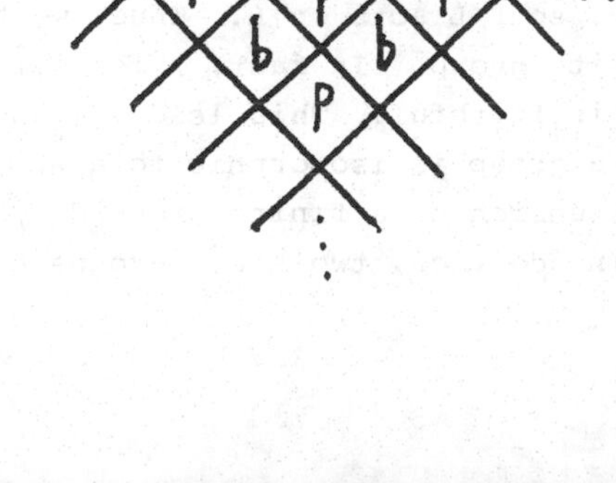

We observe that certain rigid motions of the plane carry this pattern onto itself. For instance, there are clearly horizontal and vertical translations preserving the pattern, as well as glide reflections: translate halfway horizontally, then reflect vertically, for instance. Of course, the set of such motions preserving the pattern is a group, and it is the object of mathematical crystallography to classify these sorts of patterns and groups, in the plane and in spaces of other dimensions as well. The case of most basic importance is that of three dimensions, where applications abound in the physical sciences, especially mineralogy, chemistry and solid-state physics. For, the classification yields information on macroscopic crystals (quartz, diamonds, etc.), as well as microscopic structures; notably, the arrangement of atoms in molecules.

The classification of three-dimensional structures was carried out by primarily geometric means by several people in the nineteenth century. The history of these early efforts has been presented at Oberwolfach in 1966 by Burckhardt [33].

Let us now be more technical, in order to see the relationship with integral representations. Following Brown et al. [29], we define a crystal structure as a subset C of n-dimensional real Euclidean space E^n with the following properties:

i) among the mappings of E^n to itself that map C into itself, there are n linearly independent translations.

ii) there is a positive real number d such that any translation of E^n mapping C onto itself has magnitude at least d.

The set of all rigid motions of E^n mapping a crystal structure C onto itself is the space group of C. Alternatively, if R(n) is the group of rigid motions of E^n with its usual topology, space groups are characterized as discrete subgroups Γ of R(n) such that the homogeneous space $R(n)/\Gamma$ is compact.

Now, if $\Gamma \subseteq R(n)$ is a space group, then the translations contained in Γ form a normal subgroup L that is free abelian of rank n. Hence, we see Γ as an extension

$$0 \to L \to \Gamma \to \Gamma/L \to 1,$$

and it can be shown that Γ/L is finite. Now, Γ acts on L by conjugation, with L acting trivially, so Γ/L acts on L. Thus, we have an integral representation of a finite group. In fact, L is self-centralizing in Γ, so this representation is faithful. This leads to another characterization of space groups: a group is isomorphic to a space group on E^n if and only if it is an extension of a finite group G by a G-faithful ZG-lattice of Z-rank n. In addition, two space groups are conjugate in the affine group if and only if they are isomorphic as abstract groups. These considerations lead to the following algorithm of Zassenhaus [250] for the determination of all isomorphism classes of space groups on E^n:

1) choose one element G from each conjugacy class of finite subgroups of $GL(n,Z)$, and calculate $H^2(G,Z^n)$.
2) let the normalizer of G in $GL(n,Z)$ act on $H^2(G,Z^n)$. The orbits of this action correspond to the isomorphism classes of space groups Γ on E^n with Γ/translations $\cong G$.

In order to implement this program, one must know the finite subgroups of $GL(n,Z)$ up to conjugacy. That is, one must determine all faithful representations of all finite groups on Z^n. This can be done by hand for $n = 1,2,3$, but becomes more suitable for computer calculation in large dimensions. Indeed, the Zassenhaus algorithm was carried out by computer for $n = 2,3,4$; see Brown et al. [29]. For a geometric discussion of the cases $n = 2$ and 3, see Burckhardt [32]. To understand the rate at which the difficulties grow, note that there are 17 space groups in dimension 2, 230 in dimension 3 and 4895 in dimension 4. It should be pointed out that there are only finitely many space groups in any given dimension. This assertion was part of the eighteenth problem in Hilbert's famous list of 1900, and was established soon thereafter by Bieberbach. (This is now trivial, of course. By a theorem of Minkowski (see Newman [159; p. 175]), any finite subgroup of $GL(n,Z)$ is isomorphic to a subgroup of $GL(n,Z/3Z)$. Hence there are finitely many possibilities for Γ/translations. Each of these has finitely many representations on Z^n by Jordan-Zassenhaus. The groups $H^2(G,Z^n)$ are finite, so the proof is complete).

To give some of the flavor of the subject, let us take an incomplete look at the case $n = 2$. By the above, a finite subgroup G of $GL(2,Z)$ has order dividing $|GL(2,Z/3Z)| = 48$. We can restrict the possibilities for G even more by using the following theorem attributed by Weyl [241] to Leonardo da Vinci: any finite group of 2×2 real matrices is either cyclic or dihedral. Finally, if $g \in GL(2,Z)$ has order n, then the cyclic group of order n is faithfully represented on $Z^{(2)}$. As the faithful integral representation of least degree of this group is the natural one on $Z[\sqrt[n]{1}]$, we must have $\varphi(n) \leq 2$ (φ is the Euler totient), so $n = 1,2,$ 3,4 or 6. This last conclusion is called the <u>crystallographic restriction</u>. Hence, G must be one of $C_1, C_2, C_3, C_4, C_6, D_1, D_2, D_3, D_4$ or D_6, where C_i is cyclic or order i and D_i is dihedral of order $2i$. (note: D_1 and C_2 are isomorphic. However, crystallographers distinguish between them. C_2 is viewed as being generated by a half-turn, while D_2 is generated by a reflection). All these groups do have faithful representations on $Z^{(2)}$

(some have more than one).

Let us have a look at $G = D_2 = \langle g \rangle$, where $g = \begin{pmatrix} 1 & 0 \\ 0 & -1 \end{pmatrix}$. Let us denote $Z^{(2)}$ with the natural action of G by L. Then we have $H^2(G,L) \cong L^G/NL$, where L^G denotes G-fixed points, and NL is the image of the norm map $N: L \to L$ given by $N(x) = (1+g)x$. So, $H^2(G,L) \cong \left\{ \begin{pmatrix} a \\ 0 \end{pmatrix} \middle| a \in Z \right\} \Big/ \left\{ \begin{pmatrix} 2a \\ 0 \end{pmatrix} \middle| a \in Z \right\} \cong Z/2Z$. Therefore, there are two extensions of G by L. The split extension is represented faithfully in R(n) as $\Gamma_1 = \{(\gamma;t) \mid \gamma \in G,\ t \in Z^{(2)}\}$, where $(\gamma;t)(v) = \gamma v + t$. To get the other extension, we follow MacLane [144; Chapter IV, §3]: pick an element, say $\begin{pmatrix} 1 \\ 0 \end{pmatrix}$, of L^G that represents the nontrivial element of $H^2(G,L)$. Then the extension group is abstractly presented as

$$\Gamma_2 = \langle \gamma, \begin{pmatrix} a \\ b \end{pmatrix} \mid a,b \in Z, \gamma^2 = \begin{pmatrix} 1 \\ 0 \end{pmatrix}, \gamma \begin{pmatrix} a \\ b \end{pmatrix} \gamma^{-1} = \begin{pmatrix} a \\ -b \end{pmatrix},$$
$$\begin{pmatrix} a \\ b \end{pmatrix} \begin{pmatrix} c \\ d \end{pmatrix} = \begin{pmatrix} a+c \\ b+d \end{pmatrix} \rangle.$$

We see that

$$\Gamma_2 \cong \langle \left(\begin{pmatrix} 1 & 0 \\ 0 & -1 \end{pmatrix}; \begin{pmatrix} 1/2 \\ 0 \end{pmatrix} \right),\quad (I;t) \mid t \in Z^{(2)} \rangle,$$

a subgroup of R(2). Note that Γ_1 and Γ_2 are nonisomorphic, as Γ_2 is torsionfree, while Γ_1 contains the reflection g. What are the geometric patterns that correspond to these? In order to find them, we produce for each of Γ_1 and Γ_2 a <u>fundamental</u> <u>domain</u>, i.e. a subset S of the plane whose images under the elements of Γ_i cover the plane and have disjoint interiors.

Now, Γ_1 is generated by

$$\alpha = \left(\begin{pmatrix} 1 & 0 \\ 0 & -1 \end{pmatrix}; 0 \right), \quad \beta = \left(I; \begin{pmatrix} 1 \\ 0 \end{pmatrix} \right) \quad \text{and} \quad \gamma = \left(I; \begin{pmatrix} 0 \\ 1 \end{pmatrix} \right).$$

reflection in x-axis — x-translation — y translation

Equally well, $\Gamma_1 = \langle \alpha, \gamma\alpha, \beta \rangle$, and $\gamma\alpha$ is the reflection over the line $y = 1/2$. Hence, we can take $S = \{(x,y) \mid 0 \le x \le 1,\ 0 \le y \le 1/2\}$ as fundamental domain.

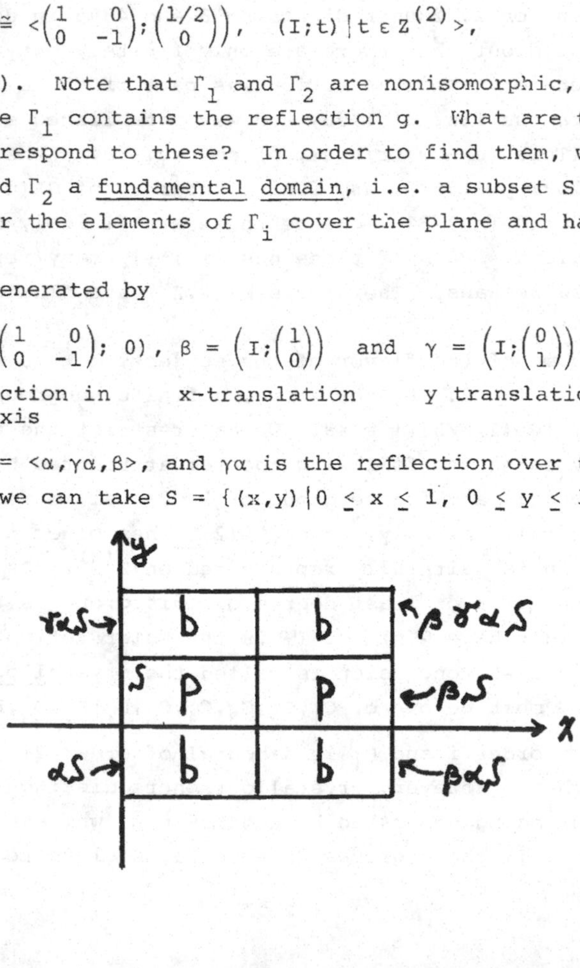

Similarly, Γ_2 is generated by the glide-reflections $\alpha = \left(\begin{pmatrix} 1 & 0 \\ 0 & -1 \end{pmatrix}; \begin{pmatrix} 1/2 \\ 0 \end{pmatrix}\right)$ and $\beta = \left(\begin{pmatrix} 1 & 0 \\ 0 & -1 \end{pmatrix}; \begin{pmatrix} 1/2 \\ 1 \end{pmatrix}\right)$. As a fundamental domain, we can take a square with corners at (0,1/4), (1/2,3/4), (1/1/4) and (1/2,-1/4). The reader is invited to convince himself that the complete pattern is the one indicated at the beginning of this section.

We now take a quick look at the Clifford-Klein space form problem. A <u>space form</u> is a complete, connected, riemannian manifold of constant curvature K. A space form is <u>spherical</u> if $K > 0$, <u>euclidean</u> if $K = 0$ or <u>hyperbolic</u> if $K < 0$. According to a theorem of Killing and Hopf (see Wolf [246; p. 69]), any Euclidean space form of dimension $n \geq 2$ is an orbit space E^n/Γ, where Γ is a subgroup of R(n) that acts freely and properly discontinuously on E^n. (Γ acts <u>freely</u> if the isotropy subgroup of each point is trivial; it acts properly discontinuously if each point $x \in E^n$ has a neighborhood U with $|\{\gamma \in \Gamma | \gamma(U) \cap U \neq \emptyset\}| < \infty$). It turns out that the acceptable groups Γ are precisely the torsionfree space groups, together with some analogous groups that don't contain enough translations to be space groups. Again, we look at the case of two dimensions. From Brown et al. [29], there are two torsionfree space groups in R(2). One is the group Γ consisting only of translations, in which case E^2/Γ is the flat torus. The other torsionfree space group on E^2 is the group Γ_2 described above; E^2/Γ_2 is the Klein bottle. The remaining two-dimensional space forms are the plane itself ($\Gamma = \langle e \rangle$), the cylinder ($\Gamma = Z$, viewed as translations in a single direction) and the Moebius band ($\Gamma = \langle \left(B; \begin{pmatrix} a \\ b \end{pmatrix}\right) \rangle$, where $B^2 = I \neq B$, $B\begin{pmatrix} a \\ b \end{pmatrix} = \begin{pmatrix} a \\ b \end{pmatrix}$ and $\left(B; \begin{pmatrix} a \\ b \end{pmatrix}\right)^2 = \left(I; \begin{pmatrix} 2a \\ 2b \end{pmatrix}\right)$). Note that this gives a classification up to affine diffeomorphism; the isometric classification is much more complicated. Note also that the space groups are just the ones that yield compact space forms.

On E^3, there are ten classes of torsionfree space groups; the corresponding compact Euclidean space forms are described in Wolf [246]. There one also finds a description of the (infinitely many) noncompact ones. I do not know if the space forms corresponding to the 74 torsionfree space groups on E^4 have been constructed. Finally, we note that methods of this sort can be used to discuss fundamental groups of other manifolds, as well, see Auslander and Kuranishi [8]. For further discussion see Ascher and Janner [67], Maxwell [153], Dade [48], Plesken [169] and Schwarzenberger [204]. Plesken and Neübuser are preparing an algebraically-oriented exposition. Pictures for the two-dimensional case are shown in Durbin [72].

REFERENCES

1. G. Aeberli: Der Zusammenhang zwischen quaternären quadratischen Formen und Idealen Quaternionenringen, Comment. Math. Helv. 33 (1959), 212-239.

2. A. Albert: Structure of Algebras, American Mathematical Society, New York, 1939.

3. A. Albert and H. Hasse: A determination of all normal division algebras over an algebraic number field, Trans. Amer. Math. Soc. 34(1932), 722-726.

4. E. Artin: Zur Theorie der hyperkomplexen Zahlen, Abh. Math. Sem. Univ. Hamburg 5(1928), 251-260.

5. E. Artin: Arithmetik hyperkomplexen Zahlen, Abh. Math. Sem. Univ. Hamburg 5(1928), 261-289.

6. E. Ascher and A. Janner: Algebraic aspects of crystallography - Space groups as extensions, Helv. Phys. Acta 38(1965), 551-572.

7. E. Ascher and A. Janner: Algebraic aspects of crystallography II. Non-primitive translations in space groups, Comm. Math. Phys. 11(1968), 138-167.

8. L. Auslander and M. Kuranishi: On the holonomy group of locally euclidean spaces, Ann. of Math. (2)65(1957), 411-415.

9. M. Auslander and O. Goldman: Maximal orders, Trans. Amer. Math. Soc. 97(1960), 1-24.

10. M. Auslander: Existence theorems for almost split sequences, Proc. Conf. on Ring Theory II (Norman, Okla.), Marcel Dekker, New York, 1977, 1-44.

11. M. Auslander: Preprojective lattices over orders over Dedekind domains, Oberwolfach Tagungsbericht 29/1980, 4.

12. M. Auslander and D. Rim: Ramification index and multiplicity, Ill. J. Math. 7(1963), 566-581.

13. K. Bäckström: Orders with finitely many indecomposable lattices, dissertation, Chalmers University of Technology, Göteborg, 1972.

14. H. Bass: Finitistic dimension and a homological characterization of semi-primary rings, Trans. Amer. Math. Soc. 95(1960), 466-488.

15. H. Bass: Projective modules over algebras, Ann. of Math. (2)73 (1961), 532-542.

16. H. Bass: Torsion free and projective modules, Trans. Amer. Math. Soc. 102(1962), 319-327.

17. H. Bass: On the ubiquity of Gorenstein rings, Math. Z. 82(1963), 8-28.

18. H. Bass: The Dirichlet unit theorem, induced characters, and Whitehead groups of finite groups, Topology 4(1966), 391-410.

19. T. Berger and I. Reiner: The normal basis theorem, Amer. Math. Monthly 82(1975), 915-918.

20. S. Berman and P. Gudivok: Indecomposable representations of finite groups over the ring of p-adic integers, Izv. Akad. Nauk SSSR, Ser. Mat. 28(1964), 875-910 = Amer. Math. Soc. Transl. (2) 50(1966), 77-113.

21. C. Bessenrodt: On blocks of finite lattice type, Arch. Math. (Basel) 33(1979/80), 334-337.

22. A. Borel, S. Chowla, C. Herz, K. Iwasawa and J. P.-Serre: Seminar on complex multiplication, Springer Lect. Notes in Math. 21(1966).

23. Z. Borevič and D. Faddeev: Theory of homology in groups II, Vestnik Leningrad Univ. 14(1959), no. 7, 72-87.

24. H. Brandt: Idealtheorie in Quaternionenalgebren, Math. Ann. 99 (1928), 1-29.

25. H. Brandt: Idealtheorie in einer Dedekindschen Algebra, Jber. Deutsch. Math.-Verein. 37(1928), 5-7.

26. H. Brandt: Zur Idealtheorie Dedekindschen Algebren, Comment. Math. Helv. 2(1930), 13-17.

27. H. Brandt: Zur Zahlentheorie der Quaternionen, Jber. Deutsch. Math.-Verein. 53(1943), 23-57.

28. R. Brauer: On modular and p-adic representations of algebras, Proc. Nat. Acad. Sci. USA 25(1939), 252-258.

29. H. Brown, R. Bülow, J. Neubüser, H. Wondratschek and H. Zassenhaus: Crystallographic groups of four-dimensional space, John Wiley and Sons, New York, 1978.

30. A. Brumer: Structure of hereditary orders, Bull. Amer. Math. Soc. 69(1963), 721-724; addendum, ibid. 70(1964), 185.

31. J. Brzezinski: Algebraic geometry of ternary quadratic forms and orders in quaternion algebras, preprint.

32. J. Burckhardt: Die Bewegungsgruppen der Kristallographie, second edition, Birkhäuser, Basel, 1966.

33. J. Burckhardt: Zur Geschichte der Entdeckung der 230 Raumgruppen, Arch. Hist. Exact Sci. 4(1967/68), 235-246.

34. C. Bushnell and I. Reiner: Solomon's conjectures and the local functional equation for zeta functions of orders, Bull. Amer. Math. Soc. (New Series) 2(1980), 306-310.

35. C. Bushnell and I. Reiner: Zeta functions of hereditary orders and integral group rings, Texas Tech Univ. Math. Series, to appear.

36. C. Bushnell and I. Reiner: Zeta functions of arithmetic orders and Solomon's conjectures, to appear.

37. M. Butler: The 2-adic representations of Klein's four group, Proc. Second Internat. Conf. Theory of Groups, Canberra, 1973, 197-203.

38. M. Butler: On the classification of local integral representations of finite abelian p-groups, Springer Lect. Notes Math. 488(1975), 54-71.

39. J. Cassels and A. Fröhlich (editors): Algebraic Number Theory, Thompson Book Company, Washington, 1967.

40. P. Cassou-Noguès: Quelques théorèmes de base normale d'entiers, Ann. Inst. Fourier (Grenoble) (3)28(1978), 1-35.

41. C. Chevalley: L'arithmétique dans les algèbres de matrices, Act. Sci. et Ind. 323, Hermann, Paris, 1936.

42. H. Cohn: A second course in number theory, John Wiley and Sons, New York, 1962. New edition, retitled Advanced number theory, Dover, New York, 1980.

43. S. Conlon: Decompositions induced from the Burnside algebra, J. Algebra 10(1968), 102-122; correcticns,ibid. 18(1971), 608.

44. H. Coxeter: Integral Cayley numbers, Duke J. Math. 13(1946), 561-578.

45. J. Cougnard: Un contre-exemple a une conjecture de J. Martinet, in [95].

46. C. Curtis and I. Reiner: Representation theory of finite groups and associative algebras, Interscience, New York, 1962.

47. E. Dade: Some indecomposable group representations, Ann. of Math. (2)77(1963), 406-412.

48. E. Dade: The maximal finite groups of 4 × 4 integral matrices, Ill. J. Math. 9(1965), 99-122.

49. E. Dade, O. Taussky and H. Zassenhaus: On the theory of orders, in particular on the semigroups of ideal classes and genera in an algebraic number field, Math. Ann. 148(1962), 31-64.

50. R. Dedekind: Über die composition der quadratischen formen, Supplement X in [62].

51. R. Dedekind: Über die Anzahl der Ideal-Klassen in den verschiedenen Ordnungen eines endlichen Körpers, Festschrift der Technischen Hochschule in Braunschweig zur Säkularfeier des Geburtstages von C. F. Gauss, Braunschweig, 1877, 1-55.

52. R. Dedekind: Über eine Erweiterung der Symbols (A,B) in der Theorie der Moduln, Nachr. Königl. Gesell. Wiss. Göttingen Mathem.-phys. Klasse, 1895, 183-188.

53. M. Deuring: Algebren, second edition, Springer-Verlag, Berlin, 1968. (First edition appeared in 1935).

54. M. Deuring: Algebraische Begründung der komplexen Multiplikation, Abh. Math. Sem. Univ. Hamburg 16(1947), 32-47.

55. M. Deuring: Die Struktur der elliptischen Funktionenkörper und die Klassenkörper der imaginär-quadratischen Körper, Math. Ann. 124(1952), 393-426.

56. M. Deuring: Die Klassenkörper der komplexen Multiplikation, Enz. Math. Wiss., Band I-2, Heft 10, Teil II, Stuttgart, 1958.

57. L. Dickson: Algebras and their arithmetics, Chicago, 1923.

58. L. Dickson: A new simple theory of hypercomplex integers, J. Math. Pures Appl. (9)2(1923), 281-326.

59. L. Dickson: Algebren und ihre Zahlentheorie, Orell-Füssli, Zürich, 1927.

60. F. Diederichsen: Über de Ausreduktion ganzzahliger Gruppendarstellungen bei arithmetischer Äquivalenz, Abh. Math. Sem. Univ. Hamburg 14(1938), 357-412.

61. E. Deiterich: Representation types of group rings over complete valuation rings, Oberwolfach Tagungsbericht 29/1980, 10-11.

62. P. Lejeune Dirichlet: Vorlesungen über Zahlentheorie, second edition, Vieweg und Sohn, Braunschweig, 1871.

63. A. Dress: On relative Grothendieck rings, Bull. Amer. Math. Soc. 75(1969), 955-958.

64. A. Dress: Vertices of integral representations, Math. Z. 114 (1970), 159-169.

65. A. Dress: On integral and modular relative Grothendieck rings, multicopied notes of the Summer Open House for Algebraists, Aarhus University (1970), 85-108.

66. A. Dress: Contributions to the theory of induced representations, Springer Lect. Notes Math. 342(1973), 183-240.

67. A. Dress: Relative Grothendieckringe über semilokalen Dedekindringen, Surjektivität des Reductionshomomorphismus und ein Theorem von Swan, preprint.

68. Ju. Drozd: Adèles and integral representations, Izv. Akad Nauk SSSR, Ser. Mat. 33(1969), 1080-1088 = Math. of USSR Izv. 3(1969), 1019-1026.

69. Ju. Drozd and A. Roiter: Commutative rings with a finite number of indecomposable integral representations, Izv. Akad Nauk USSR, Ser. Mat. 31(1967), 783-798 = Math. of USSR Izv. 1(1967), 757-772.

70. Ju. Drozd, V. Kiričenko and A. Roiter: On hereditary and Bass orders, Izv. Akad. Nauk SSSR, Ser. Mat. 31(1967), 1415-1436 = Math. of USSR Izv. 1(1967), 1357-1376.

71. L. DuPasquier, Vjschr. Naturforsch. Ges. Zürich 54(1909), 116-148.

72. J. Durbin: Modern Algebra, John Wiley and Sons, New York, 1979.

73. H. Edwards: Fermat's Last Theorem, a genetic introduction to algebraic number theory, Springer-Verlag, New York, 1977.

74. M. Eichler: Über die Idealklassenzahl total definiter Quaternionenalgebren, Math. Z. 43(1938), 102-109.

75. M. Eichler: Über die Idealklassenzahl hyperkomplexen Zahlen, Math. Z. 43(1938), 481-494.

76. M. Eichler: Quadratische Formen und orthogonale Gruppen, Springer-Verlag, Berlin, 1952.

77. S. Epp: Submodules of Cayley algebras, J. Algebra 24(1973), 104-126.

78. D. Estes and G. Pall: Modules and rings in the Cayley algebra, J. Number Theory 1(1969), 163-178.

79. D. Faddeev: On the semigroup of genera in the theory of integer representations, Izv. Akad. Nauk SSSR, Ser. Mat. 28(1964), 475 - 478 = Amer. Math. Soc. Transl. (2)64(1967), 97-101.

80. D. Faddeev: An introduction to the multiplicative theory of modules of integral representations, Trudy Mat. Inst. Steklov 80 (1965), 145 - 182 = Proc. Steklov Inst. Math. 80(1965), 164-210.

81. D. Faddeev: Equivalence of systems of integer matrices, Izv. Akad. Nauk SSSR, Ser. Mat. 30(1966), 449 - 454 = Amer. Math. Soc. Transl. (2)71(1968), 43-48.

82. R. Fossum: Maximal orders over Krull domains, J. Algebra 10 (1968), 321-332.

83. R. Fossum: Injective modules over Krull orders, Math. Scand. 28 (1971), 233-246.

84. T. Fossum: On symmetric orders and separable algebras, Trans. Amer. Math. Soc. 180(1973), 301-314.

85. G. Frobenius and L. Stickelberger: Über Gruppen von vetauschbaren Elementen, Crelle 86(1879), 217-262.

86. A. Fröhlich: Invariants for modules over commutative separable orders, Quart. J. Math. (2)16(1965), 193-232.

87. A. Fröhlich: Some topics in the theory of module conductors, Oberwolfach Berichte 2(1966), 59-83.

88. A. Fröhlich: Local fields, in [39].

89. A. Fröhlich: Artin root numbers and normal integral bases for quaternion fields, Invent. Math. 17(1962), 143-166.

90. A. Fröhlich: The Picard group of non-commutative rings, in particular of orders, Trans. Amer. Math. Soc. 180(1973), 1-45.

91. A. Fröhlich: Module invariants and root numbers for quaternion fields of degree $4\ell^r$, Proc. Cambridge Phil. Soc. 76(1974), 393-399.

92. A. Fröhlich: Locally free modules over arithmetic orders, Crelle 274/275(1975), 112-138.

93. A. Fröhlich: A normal integral basis theorem, J. Algebra 39 (1976), 131-137.

94. A. Fröhlich: Arithmetic and Galois module structure for tame extensions, Crelle 286/287(1976), 380-440.

95. A. Fröhlich (editor): Algebraic Number Fields, Academic Press, New York, 1977.

96. A. Fröhlich and J. Queyrut: On the functional equation of the Artin L-function for characters of real representations, Invent. Math. 14(1971), 173-183.

97. A. Fröhlich, I. Reiner and S. Ullom: Class groups and Picard groups of orders, Proc. London Math. Soc. (3) 29(1974), 405-434.

98. R. Fueter: Vorlesungen über die singulären Modules und die komplexe multiplication der elliptischen Funktionen I, II, Teubner, Leipzig, 1924 and 1927.

99. V. Galkin: ζ- functions of some one-dimensional rings, Izv. Akad. Nauk SSSR, Ser. Math. 37(1973), 3 - 19 = Math. of USSR Izv. 7(1973), 1-17.

100. C. F. Gauss: Disquisitiones arithmeticae, Leipzig, 1801.

101. E. Green and I. Reiner: Integral representations and diagrams, Michigan Math. J. 25(1978), 53-84.

102. J. Green: Axiomatic representation theory for finite groups, J. Pure and Appl. Algebra 1(1971), 41-77.

103. W. Gustafson: Integral relative Grothendieck rings, J. Algebra 22(1972), 461-479.

104. W. Gustafson: On an induction theorem for relative Grothendieck groups, Proc. Amer. Math. Soc. 35(1972), 26-30.

105. W. Gustafson: The theory of relative Grothendieck rings, Springer Lect. Notes Math. 353(1973), 95-112.

106. M. Harada: Hereditary orders, Trans. Amer. Math. Soc. 107(1963), 273-290.

107. K. Hashimoto: On Brandt matrices associated with the positive definite quaternion hermitian forms, J. Fac. Sci. Univ. Tokyo (Sec. IA) 27(1980), 227-245.

108. K. Hashimoto: Some examples of integral definite quaternary quadratic forms with prime discriminant, Nagoya Math. J. 77 (1980), 167-175.

109. H. Hasse: Neue Begründung der komplexen Multiplikation I, Crelle 157(1927), 115-139; II, ibid. 165(1931), 64-88.

110. H. Hasse: Über p-adische Shiefkörper und ihre Bedeutung für die Arithmetik hyperkomplexen Zahlsysteme, Math. Ann. 104(1931), 495-534.

111. H. Hasse, R. Brauer and E. Noether: Beweis eines Hauptsatzes in der Theorie der Algebren, Crelle 167(1932), 399-304.

112. H. Hasse: History of class field theory, in [39].

113. A. Heller and I. Reiner: Representations of cyclic groups in rings of integers I, Ann. of Math. (2) 76(1962), 73-92; II, ibid. (2) 77(1963), 318-328.

114. A. Heller and I. Reiner: Grothendieck groups of integral group rings, Ill. J. Math. 9(1965), 349-359.

115. K. Hey: Analytische Zahlentheorie in Systemen hyperkomplexer Zahlen, dissertation, Hamburg, 1929.

116. D. Higman: On orders in separable algebras, Canad. J. Math. 7 (1955), 509-515.

117. G. Higman: The units of group-rings, Proc. London Math. Soc. (2) 46 (1940), 231-248.

118. D. Hilbert: Die Theorie der algebraischen Zahlkörper, Jber. Deutsch. Math.-Verein. 4 (1897), 175-546.

119. A. Hurwitz: Über die Zahlentheorie der Quaternionen, Nachr. Akad. Wiss. Göttingen 1896, 313-334.

120. A. Hurwitz: Vorlesungen über die Zahlentheorie der Quaternionen, Berlin, 1919.

121. H. Jacobinski: Über die Hauptordnung eines Körpers als Gruppenmodul, Crelle 213 (1963/64), 151-164.

122. H. Jacobinski: On extensions of lattices, Michigan Math. J. 13 (1966), 471-475.

123. H. Jacobinski: Sur les ordres commutatifs avec un nombre fini de réseaux indecomposables, Acta Math. 118 (1967), 1-31.

124. H. Jacobinski: Über die Geschlecter von Gittern über Ordnungen, Crelle 230 (1968), 29-39.

125. H. Jacobinski: Genera and decompositions of lattices over orders, Acta Math. 121 (1968), 1-29.

126. H. Jacobinski: On embedding of lattices belonging to the same genus, Proc. Amer. Math. Soc. 24 (1970), 134-136.

127. H. Jacobinski: Two remarks about hereditary orders, Proc. Amer. Math. Soc. 28 (1971), 1-8.

128. N. Jacobson: Composition algebras and their automorphisms, Rend. Circ. Mat. Palermo 7 (1958), 55-80.

129. J.-R. Joly: Groupe de classes de certains ordres d'entiers algébrique, C. R. Acad. Sci. Paris (Ser. A) 266 (1968), 553-555.

130. A. Jones: Groups with a finite number of indecomposable integral representations, Michigan Math. J. 10 (1963), 257-261.

131. I. Kaplansky: Composition of binary quadratic forms, Studia Math. 31 (1968), 523-530.

132. I. Kaplansky: Submodules of quaternion algebras, Proc. London Math. Soc. (3) 19 (1969), 219-232.

133. J. Kirmse: Über die Darstellbarkeit natürlicher ganzen Zahlen als Summen von acht Quadraten und über ein mit diesem Problem zusammenhängendes nichtkommutatives und nichtassociatives Zahlensystem, Ber. Verh. Sächs. Akad. Wiss. Leipzig 76 (1924), 63-82.

134. J. Kirmse: Zur Darstellbarkeit natürlicher ganzer Zahlen als Summen von acht Quadreten, Ber. Vehr. Sächs. Akad. Wiss. Leipzig 80 (1928), 33-34.

135. S. Kuroda: Über den allgemeinen Spiegelungssatz für Galoissche Zahlkörper, J. Number Theory 2(1970), 282-297.

136. T. Lam: Induction theorems for Grothendieck groups and Whitehead groups of finite groups, Ann. Sci. École Norm Sup. (4)1(1968), 91-148.

137. P. Lamont: Arithmetics in Cayley's algebra, Proc. Glasgow Math. Assoc. 6(1963), 99-106.

138. C. Latimer and C. MacDuffee: A correspondence between classes of ideals and classes of matrices, Ann. of Math. (2)34(1933), 313-316.

139. H. Lenstra: Grothendieck groups of abelian group rings, preprint.

140. H. Leopoldt: Zur Struktur der ℓ-Klassengruppe galoisscher Zahlkörper, Crelle 199(1958), 165-174.

141. H. Leopoldt: Über die Hauptordnung der ganzen Elemente eines abelschen Zahlkörpers, Crelle 201(1959), 119-149.

142. R. Lipschitz: Untersuchungen über die summen von Quadraten, Bonn, 1886. French translation in J. Math. Pures et Appl. (4)2 (1886), 393-439.

143. R. MacKenzie and J. Scheuneman: A number field without a relative integral basis, Amer. Math. Monthly 78(1971), 882.

144. S. MacLane: Homology, Springer-Verlag, Berlin, 1963.

145. K. Mahler: On ideals in the Cayley-Dickson algebra, Proc. Royal Irish Acad. (Ser. A) 48(1943), 123-133.

146. J. Martinet: Sur l'arithmétique des extensions galoisiennes à groupe de Galois diédral d'ordre 2p, Ann. Inst. Fourier (Grenoble) 19(1969), 1-80.

147. J. Martinet: Modules sur l'algèbre du group quaternionien, Ann. Sci. École Norm. Sup. (4)4(1971), 399-408.

148. J. Martinet: Sur les extensions à groupe de Galois quaternionien, C. R. Acad. Sci. Paris (Sér. A) 274(1972), 933-935.

149. D. Masser: Elliptic functions and transcendence, Springer Lect. Notes Math. 437(1975).

150. K. Masuda: Application of the theory of the group of classes of projective modules to the existence problem of independent parameters of invariant, J. Math. Soc. Japan 20(1968), 223-232.

151. A. Matchett: Exact sequences for locally free class groups, preprint.

152. H. Mattson: A generalization of the Riemann-Roch theorem, Ill. J. Math. 5(1961), 355-375.

153. G. Maxwell: The crystallography of Coxeter groups, J. Algebra 35(1975), 159-177.

154. J. Milnor: Whitehead torsion, Bull. Amer. Math. Soc. 72(1966), 358-426.

155. Y. Miyata: On a characterization of the first ramification group as the vertex of the ring of integers, Nagoya Math. J. 43(1971), 151-156.

156. Y. Miyata: On the module structure of the ring of all integers of a p-adic number field, Nagoya Math. J. 54(1974), 53-59.

157. L. Nazarova: Unimodular representations of the four group, Dokl. Akad. Nauk SSSR 140(1961), 1011-1014 = Sov. Math. Dokl. 2(1961), 1304-1307.

158. A. Nelson: Module resolvents, preprint.

159. M. Newman: Integral matrices, Academic Press, New York, 1972.

160. G. Nipp: Quaternion orders associated with ternary lattices, Pac. J. Math. 53(1974), 525-537.

161. E. Noether: Gleichungen mit vorgeschriebener Gruppe, Math. Ann. 78(1916), 221-229.

162. E. Noether: Normalbasis bei Körpern ohne höhere Verzweigung, Crelle 167(1932), 147-152.

163. T. Obayashi: On the Grothendieck ring of an abelian p-group, Nagoya Math. J. 26(1966), 101-113.

164. G. Pall: On generalized quaternions, Trans. Amer. Math. Soc. 59 (1946), 280-332.

165. M. Peters: Ternäre und quaternäre quadratische Formen und Quaternionenalgebren, Acta Arith. 15(1969), 329-365.

166. H. Pieper: Die Einheitengruppe eines zahm-verzweigten galoisschen lokalen Körpers als Galois-Modul, Math. Nachr. 54(1972), 173-210.

167. W. Plesken: Projective lattices over group orders as amalgamations of irreducible lattices, Carleton Math. Lect. Notes 25(1980), 1901-1911.

168. W. Plesken: Gruppenringe über lokalen Dedekindbereichen, Habilitationsschrift, Rheinisch-Westfälischen Technischen Hochschule, Aachen, 1980.

169. W. Plesken: Applications of the theory of orders to crystallographic groups, preprint.

170. J. Queyrut: K-théorie algébrique et structure galoisienne des anneaux d'entiers, Thesis, Bordeaux.

171. M. Ramras: Maximal orders over regular local rings, Trans. Amer. Math. Soc. 155(1971), 345-352.

172. I. Reiner: Integral representations of cyclic groups of prime order, Proc. Amer. Math. Soc. 8(1957), 142-146.

See end of list for [173].

174. I. Reiner: The integral representation ring of a finite group, Michigan Math. J. 12(1965), 11-22.

175. I. Reiner: A survey of integral representation theory, Bull. Amer. Math. Soc. 76(1970), 159-227.

176. I. Reiner: Hereditary orders, Rend. Sem. Mat. Univ. Padova 52 (1974), 219-225.

177. I. Reiner: Maximal orders, Academic Press, London, 1975.

178. I. Reiner: Class groups and Picard groups of group rings and orders, CBMS Regional Conference Series in Mathematics 26, American Mathematical Society, 1976.

179. I. Reiner: Topics in integral representation theory, Springer Lect. Notes Math. 744(1979), 1-143.

180. I. Reiner: Zeta functions and integral representations, Comm. Alg. 8(1980), 911-925.

181. I. Reiner and S. Ullom: A Mayer-Vietoris sequence for class groups, J. Algebra 31(1974), 305-342.

182. I. Reiten: Almost split sequences, Workshop on permutation groups and indecomposable modules, Giessen, 1975.

183. D. Rim: Modules over finite groups, Ann. of Math. (2)69(1959), 700-712.

184. D. Rim: On projective class groups, Trans. Amer. Math. Soc. 98 (1961), 459-467.

185. C. Ringel and K. Roggenkamp: Diagrammatic methods in the representation theory of orders, J. Algebra 60(1979), 11-42.

186. J. Ritter: The class group à la Fröhlich, preprint.

187. K. Roggenkamp and V. Huber-Dyson: Lattices over orders I, Springer Lect. Notes Math. 115(1970).

188. K. Roggenkamp: Lattices over orders II, Springer Lect. Notes Math. 142(1970).

189. K. Roggenkamp: Projective homomorphisms and extension of lattices, Crelle 246(1971), 41-45.

190. K. Roggenkamp: Bass-orders and the number of nonisomorphic indecomposable lattices over orders, Proc. Symp. Pure Math. XXI, 1971.

191. K. Roggenkamp: Almost split sequences for group rings, Mitt. Math. Sem. Giessen 121(1976), 1-25.

192. K. Roggenkamp: The construction of almost split sequences for integral group rings and orders, Comm. Alg. 5(1977), 1363-1373.

193. K. Roggenkamp: Integral representations and presentations of finite groups, Springer Lect. Notes Math. 744(1979), 149-275.

194. K. Roggenkamp: Integral representations and structure of finite group rings, Les Presses de l'Université de Montréal, 1980.

195. K. Roggenkamp: Indecomposable representations of orders of global dimension two, J. Algebra 64(1980), 230-248.

196. K. Roggenkamp: Representation theory of blocks of defect 1, Carleton Math. Lect. Notes 25(1980), 2001-2024.

197. K. Roggenkamp and J. Schmidt: Almost split sequences for integral group rings and orders, Comm. Alg. 4(1976), 893-917.

198. A. Roiter: On the representations of the cyclic group of fourth order by integral matrices, Vestnik Leningrad Univ. 15(1960), no. 19, 65-74 (in Russian).

199. A. Roiter: Categories with division and integral representations, Dokl. Akad. Nauk SSSR 153(1963), 46 - 48 = Sov. Math. Dokl. 4(1963), 1621-1623.

200. A. Roiter: On integral representations belonging to a genus, Izv. Akad. Nauk SSSR, Ser. Mat. 30(1966), 1315 - 1324 = Amer. Math. Soc. Transl. (2) 71(1968), 49-59.

201. A. Roiter: On the theory of integral representations of rings, Mat. Zametki 3(1968), 361 - 366 = Math. Notes USSR 3(1968), 228-230.

202. R. Sandling: Graham Higman's thesis "Units in Group Rings", preprint.

203. J. Schmidt: A construction of large lattices by almost split sequences, Arch. Math. (Basel) 29(1977), 481-484.

204. R. Schwarzenberger: N-dimensional crystallography, Pitman, San Francisco, 1980.

205. J.-P. Serre: Complex multiplication, in [39].

206. J.-P. Serre: Représentations linéaires des groupes finis, second edition, Hermann, Paris, 1967. English translation, Springer-Verlag, New York, 1977.

207. G. Shimura: Introduction to the arithmetic theory of automorphic functions, Iwanami Shoten, Tokyo and Princeton University Press, Princeton, 1971.

208. C. Siegel: Topics in complex function theory, volume 1: Elliptic functions and uniformization theory, Interscience, New York, 1969.

209. M. Singer: Invertible powers of ideals over orders in commutative separable algebras, Proc. Cambridge Phil. Soc. 67(1970), 237-242.

210. D. Sjöstrand: Conductors with respect to hereditary orders, Michigan Math. J. 20(1973), 181-185.

211. L. Solomon: Zeta functions and integral representation theory, Advances in Math. 26(1977), 306-326.

212. L. Solomon: Partially ordered sets with colors, Proc. Symp. Pure Math. 34(1979), 309-329.

213. A. Speiser: Allgemeine Zahlentheorie, Vjschr. Naturforsch. Ges. Zürich 71(1926), 8-48.

214. A. Speiser: Idealtheorie in rationalen Algebren, in [59].

215. D. Stancl: Multiplication in Grothendieck rings of integral group rings, J. Algebra 7(1967), 77-90.

216. E. Steinitz: Rechteckige Systeme und Moduln in algebraischen Zahlenkörpern I, Math. Ann. 71(1911), 328-354; II, ibid. 72 (1912), 297-345.

217. J. Strooker: Faithfully projective modules and clean algebras, dissertation, Utrecht, 1965.

218. R. Swan: Induced representations and projective modules, Ann. of Math. (2)71(1960), 552-578.

219. R. Swan: The Grothendieck ring of a finite group, Topology 2 (1963), 85-110.

220. R. Swan: Invariant rational functions and a problem of Steenrod, Invent. Math. 7(1969), 148-158.

221. R. Swan and E. Evans: K-theory of finite groups and orders, Springer Lect. Notes Math. 149, 1970.

222. S. Takahashi: Arithmetic of group representations, Tôhoku Math. J. (2)11(1959), 216-246.

223. O. Taussky: Matrices of rational integers, Bull. Amer. Math. Soc. 66(1960), 327-345.

224. O. Taussky: Ideal matrices I, Arch. Math. (Basel) 13(1962), 275-282; II, Math. Ann. 150(1963), 218-225.

225. O. Taussky and J. Todd: Matrices with finite period, Proc. Edinburgh Math. Soc. (2)6(1940), 128-134.

226. O. Taussky and J. Todd: Matrices of finite period, Proc. Roy. Irish Acad. Sect. A 46(1941), 113-121.

227. M. Taylor: On the self-duality of a ring of integers as a Galois module, Invent. Math. 46(1978), 173-177.

228. M. Taylor: Galois module structure of integers of relative abelian extensions, Crelle 303/304(1978), 97-101.

229. M. Taylor: Locally free classgroups of groups of prime power order, J. Algebra 50(1978), 462-487.

230. M. Taylor: A logarithmic approach to class groups of integral group rings, J. Algebra 60(1980), 321-353.

231. M. Taylor: On Fröhlich's conjecture for rings of integers of tame extensions, Invent. Math., to appear.

232. T. Theohari-Apostolidi: On integral representation of twisted group rings, Oberwolfach Tagungsbericht 29/1980, 21-22.

233. J. Thompson: Vertices and sources, J. Algebra 6(1967), 1-6.

234. J. Todd: The lemniscate constants, Comm. ACM 18(1975), 14-19.

235. J. Towber: Composition of oriented binary quadratic form-classes over commutative rings, Advances in Math. 36(1980), 1-107.

236. A. Troy: Integral representations of cyclic groups of order p^2, thesis, University of Illinois, Urbana, 1961.

237. F. van der Blij and T. Springer: The arithmetics of octaves and of the group G_2, Nederl. Akad. Wentensch. Proc. 62A(1959), 406-418.

238. H. Weber: Theorie der Abel'schen Zahlkörper I, Acta Math. 8 (1886), 193-263; II, ibid. 9(1887), 105-130.

239. H. Weber: Lehrbuch der Algebra III, Braunschweig, 1908.

240. A. Weil: Basic Number Theory, Springer-Verlag, New York, 1967.

241. H. Weyl: Symmetry, Princeton University Press, Princeton, 1952.

242. A. Wiedemann: Auslander-Reiten graphs of orders and blocks of cyclic defect two, Oberwolfach Tagungsbericht 29/1980, 24-25.

243. S. Williamson: Crossed Products and hereditary orders, Nagoya Math. J. 23(1963), 103-120.

244. S. Wilson: Reduced norms in the K-theory of orders, J. Algebra 46(1977), 1-11.

245. K. Wingberg: Die Einseinheitengruppe von p-Erweiterungen regulären *p*-adischer Zahlkorper als Galoismodul, Crelle 305(1979), 206-214.

246. J. Wolf: Spaces of constant curvature, McGraw-Hill, New York, 1967.

247. S. Yamamoto: On the rank of the p-divisor class group of Galois extensions of algebraic number fields, Kumamoto J. Sci. (Math.) 9(1972), 33-40.

248. H. Yokoi: On the ring of integers in an algebraic number field as a representation module of the Galois group, Nagoya Math. J. 16(1960), 83-90.

249. H. Zassenhaus: Neuer Beweis der Endlichkeit der Klassenzahl bei unimodularer Äquivalenz endlicher ganzzahliger Substitutionsgruppen, Abh. Math. Sem. Univ. Hamburg 12(1938), 276-288.

250. H. Zassenhaus: Über einen Algorithmus zur Bestimmung der Raumgruppen, Comm. Math. Helv. 21(1948), 117-141.

251. M. Zorn: Note zur analytischen hyperkomplexen Zahlentheorie, Abh. Math. Sem. Univ. Hamburg 9(1933), 197-201.

173. I. Reiner: Extensions of irreducible modules, Michigan Math. J. 10(1963), 273-276.

APPLICATIONS OF THE THEORY OF ORDERS TO CRYSTALLOGRAPHIC GROUPS

Wilhelm Plesken

I. INTRODUCTION

A. Basic definitions and results. Let $\langle , \rangle$ denote the standard scalar product of $\mathbb{R}^n$ and let $E_n = T_n \cdot O(\mathbb{R}^n)$ denote the Euclidean group of all motions of $\mathbb{R}^n$, where T_n is the normal subgroup of E_n consisting of all translations $\tau_t : \mathbb{R}^n \to \mathbb{R}^n : x \mapsto x + t \quad (t \in \mathbb{R}^n)$ and where $O(\mathbb{R}^n)$ is the orthogonal group of $\mathbb{R}^n$ (with respect to the positiv definite scalar product $\langle , \rangle$ of $\mathbb{R}^n$). The faithful matrix representation

$$E_n \to GL(n+1, \mathbb{R}) : \tau_t \varphi \mapsto \left(\begin{array}{c|c} {}_B\varphi_B & {}_Bt \\ \hline 0 \ldots 0 & 1 \end{array}\right),$$

where ${}_Bt$ is the coordinate column of $t \in \mathbb{R}^n$ with respect to a basis B of $\mathbb{R}^n$ and ${}_B\varphi_B$ the matrix of the orthogonal transformation φ with respect to B, equips E_n with a topology induced from $\mathbb{R}^{(n+1)\times(n+1)}$ and defines a Lie group structure on E_n. A (crystallographic) space group R is a discrete subgroup of E_n such that the topological factor space E_n/R is compact.

By the fundamental result by Bieberbach (cf. [Bie 10a], [Bie 10b], [Bie 12] or [BBNWZ 78] Appendix) a space group R contains n ($\mathbb{R}$-) linearly independent translations and can therefore be defined as follows.

(I.1) DEFINITION. (i) *A subgroup L of $\mathbb{R}^n$ is a (vector) lattice, if there exist n $\mathbb{R}$-linearly independent elements $x_1,\ldots,x_n \in \mathbb{R}^n$ such that*

$$L = \{ \sum_{i=1}^{n} \alpha_i x_i \mid \alpha_i \in \mathbb{Z},\ 1 \le i \le n\} =: \langle x_1,\ldots,x_n\rangle_{\mathbb{Z}} \ .$$

(ii) *A subgroup T of the translation group T_n is called (translation) lattice, if there exists a vector lattice $L \le \mathbb{R}^n$ with $T = \{\tau_t \mid t \in L\}$.*

(iii) A subgroup R of the Euclidean group E_n is called (crystallographic) space group, if the translation subgroup $T(R) := R \cap T_n$ is a translation lattice, i.e. $L(R) := \{t \in \mathbb{R}^n \mid \tau_t \in R\}$ is a vector lattice in $\mathbb{R}^n$.

(iv) A subgroup X of E_n is called crystallographic, if there exists a space group $R \le E_n$ containing X.

Certainly the space groups are the most important crystallographic groups. Their algebraic structure was first described by Bieberbach [Bie 10a], [Bie 12].

(I.2) THEOREM. *Let $R \le E_n$ be a space group. Then the following holds:*

(i) $T(R) = T_n \cap R$ is the unique maximal normal abelian

subgroup of R *, i.e., every normal abelian subgroup of* R *is contained in* $T(R)$ *, and* $R/T(R)$ *operates faithfully on* $T(R)$ *by conjugation, i.e. the centralizer* $C_R(T(R))$ *of* $T(R)$ *in* R *is* $T(R)$ *.*

(ii) $R/T(R)$ *is a finite group.*

Proofs of the theorems stated explicitely in this introduction are provided in the appendix. Because of (I.2) one defines abstract space groups.

(I.3) DEFINITION. *Let* R *be a group.* R *is called an abstract space group (of degree n), if there exists an exact sequence of groups*

$$0 \to T \to R \to P \to 1$$

where (i) T *is free abelian of rank* n *,*

(ii) P *is finite and acts faithfully on* T *;*

i.e. R *has a free abelian, normal subgroup of rank* n *, which is its own centralizer and has finite index in* R *.*

Abstract space groups and their isomorphisms are not very far removed from crystallographic space groups and their isomorphisms induced by conjugation with affine transformations. Let $\mathrm{Aff}_n = T_n \cdot GL(\mathbb{R}^n)$ denote the group of all affine transformations of $\mathbb{R}^n$.

(I.4) THEOREM. *Let* R *be an abstract space group of degree* n *. Then the following holds:*

(i) [*Zas 48*] *R is isomorphic to a crystallographic space group in E_n.*

(ii) [*Bie 12*] *Any two crystallographic space groups in E_n isomorphic to R are conjugate under Aff_n. More precisely let $R_1, R_2 \leq Aff_n$ be two abstract space groups of degree n and $\alpha : R_1 \to R_2$ an isomorphism. Then there exists an affine transformation $\bar{\alpha} \in Aff_n$ with $\alpha(r) = \bar{\alpha}^{-1} r \bar{\alpha}$ for all $r \in R$.*

By Theorem (I.4) the geometric-group theoretical task of classifying crystallographic space groups up to affine equivalence (i.e. conjugation under the affine group) is reduced to a purely group theoretical problem, namely classifying the extensions of finitely generated free abelian groups T by finite subgroups P of the automorphism group $Aut_{\mathbb{Z}}(T) \cong GL(n,\mathbb{Z})$ where n is the $\mathbb{Z}$-rank of T.

(I.5) DEFINITION. *In the situation of Definition (I.3) $T(R) := T$ is called the translation subgroup of R and P, more precisely the image of P in $Aut_{\mathbb{Z}}(T)$ defined by the action of P on T, is called the point group of R and denoted by $P(R)$.*

The term "point group" here is a slight (but convenient) abuse of language, since in the crystallographic situation $P(R)$ (after obvious identifications) only turns up as a stabilizer of a point in the Euclidean space, if R

splits over T (i.e. if R is symmorphic, as crystallographers say). Nevertheless the symmorphic case shows that point groups are crystallographic groups (again after the identification of $T(R)$ with $L(R)$ and $Aut_{\mathbb{Z}}(T(R))$ with a subgroup of $GL(\mathbb{R}^n)$).

<u>B. The relevance of the theory of orders for crystallographic groups.</u> The simple but nevertheless basic observation that $T(R)$ is a $\mathbb{Z}P(R)$-lattice for every space group R opens a wide range of applications of the theory of lattices over orders to crystallographic groups. So the basic step in Bieberbach's fundamental theorem, that there are only finitely many isomorphism (or affine equivalence) classes of space groups in a given dimension is the application of the Jordan-Zassenhaus-Theorem (cf. Chapter II.A). For the purposes of construction of crystallographic groups and distinguishing them according to the various equivalence relations, which are usually defined in crystallography, as well as understanding these equivalence relations from an arithmetic (and not only geometric) point of view the following purely order-theoretic problems are relevant:

For a given finite group G

(i) find all $\mathbb{Z}G$-lattices L up to isomorphism, usually with restrictions of the kind that L is contained in a given $\mathbb{Q}G$-module, on which G acts faithfully;

(ii) decide isomorphism of two given $\mathbb{Z}G$-lattices and construct isomorphisms.

In Chapters II, III some problems of the kind indicated here will be discussed.

There is, however, a second type of orders apart from group rings of finite groups that play a rôle in the theory of crystallographic groups. These are the centralizer rings of finite subgroups of $GL(n,\mathbb{Z})$ in $\mathbb{Z}^{n\times n}$, which are orders over $\mathbb{Z}$. Actually one wants to construct finite generating sets of the normalizer of a finite subgroup of $GL(n,\mathbb{Z})$ in $GL(n,\mathbb{Z})$. But the unit group of the order above is of finite index in this normalizer. Hence the third problem can be formulated:

(iii) find a finite set of generators of the unit group of a given $\mathbb{Z}$-order, namely the order of all matrices in $\mathbb{Z}^{n\times n}$ commuting with a given finite subgroup of $GL(n,\mathbb{Z})$.

That unit groups of $\mathbb{Z}$-orders are finitely generated was proved by Siegel in [Sie 43]. Solutions of problem (ii) and (iii) together yield finite sets of generators for the normalizers mentioned above. This and its relevance to the construction of all space groups with given point group is discussed in Chapter II.B.

Generally speaking, there are two possible approaches to problems (i) - (iii). On the one side one can prove that these things can in principle be done without restrictions. However such general methods often turn out to be impracticable already for low dimensions. On the other hand one can design algorithms in order to actually construct the groups in question (and implement them), which solve these problems efficiently for large classes of groups - for instance with restrictions on the degrees involved or with structural restrictions. Both points of view have to be taken into account (sometimes the borderlines between the two aspects are not very sharp), the first being more theoretic, the second more to the needs of users of crystallographic groups.

In this paper some relevant problems in the theory of crystallographic groups will be sketched and the rôle of the theory of orders for solving them will be discussed. But first it seems to be adequate to give a brief list of applications of crystallographic groups.

C. Applications of crystallographic groups. The classical application of crystallographic groups is their use for classifying crystals: The symmetry group of an idealized crystal (i.e. a crystal without defects expanding all over the Euclidian 3-space) is a three dimensional space group (cf. [Hen 69], [Bue 56], [Bue 70]).

In solid state physics the finite dimensional unitary representations of three dimensional space groups are used to interprete experimental data from spectroscopy. (For the theory and construction of these representations cf. [Bac 80]; for tables used in laboratories cf. [Cra 80] for the physical background cf. [Bir 80]). In recent years up to six dimensional (abstract) space groups are used for a finer analysis of the experimental data, e.g. for explaining "satellite reflections" cf. [Jan 80].

Within mathematics space groups (of arbitrary dimension) turn up in discrete geometry, for instance as automorphism groups of coverings and packings of the Euclidean space, but also as an important building block for general discrete subgroups of the Euclidean group E_n, cf. [Bie 10b] and [BBNWZ 78] Appendix. In differential topology fixed point free space groups (i.e. space groups, in which the stabilizer of any point in $\mathbb{R}^n$ consists only of the identity) are used to construct locally Euclidean manifolds , i.e. they are isomorphic to the fundamental groups of these topological spaces cf. [Wol 67] and [Max 77]. Finally certain types of space groups (so called p-uniserial space groups) have recently been used to construct series of finite p-groups of constant coclass as factor groups of space groups, cf. [LeN 80a], [FiNP 80], [LeN 80b], [FNP 80]. This led for instance to counterexamples to the long standing class breadth conjecture in [FNP 80]. The hope is that these ideas lead to a reasonable classification principle for p-groups.

This list of applications is by no means complete; it might however give an idea of the wide range of applications.

For the historical background the reader is referred to [BBNWZ 78].

II. SPACE GROUPS

A. Bieberbach's finiteness theorem. In reponse to Hilbert's 18th problem Bieberbach proved the following theorem cf. [Bie 10b], [Fro 11], [Bie 12]. The proof given here is a reduction of Bieberbach's Theorem to the Jordan-Zassenhaus Theorem. Its understanding will be crucial for the rest of this paper.

(II.1) THEOREM. *(i) There are only finitely many isomorphism classes of space groups of given dimension* n *.*
(ii) There are only finitely many (crystallographic) space groups of given dimension n *up to conjugacy in* Aff_n.

PROOF. By (I.4) the two statements are equivalent. Therefore it suffices to prove (i). By (I.3) the following statement has to be proved:

There are only finitely many isomorphism classes of groups, which are extensions of $\mathbb{Z}^n$ by some finite subgroup G of $GL(n,\mathbb{Z})$ (= $Aut(\mathbb{Z}^n)$).

For each such finite unimodular group G there are only finitely many isomorphism classes of extensions, since the second cohomology group $H^2(G,\mathbb{Z}^n)$ (= group of extensions) is finite with G being finite and $\mathbb{Z}^n$ being finitely generated.

The proof could now proceed as follows: First one checks that conjugate subgroups of $GL(n,\mathbb{Z})$ yield isomorphic groups as extension. Then one quotes (or proves) Jordan's theorem asserting that there are only finitely many conjugacy classes of finite subgroups of $GL(n,\mathbb{Z})$ cf. [Jor 80]. In view of what follows it is worthwhile to point out a natural way to view such an extension of $\mathbb{Z}^n$ by G as an affine group or an $(n+1)\times(n+1)$-matrix group. (Note that the matrix representation of the Euclidean group E_n at the beginning of the introduction can easily be extended to the full affine group Aff_n.) Let G act on $\mathbb{R}^n$ in the natural way by left matrix multiplication. Then the short exact sequence of G-modules

$$0 \to \mathbb{Z}^n \to \mathbb{R}^n \to \mathbb{R}^n/\mathbb{Z}^n \to 0$$

yields a long exact sequence proving that $H^2(G,\mathbb{Z}^n)$ and $H^1(G,\mathbb{R}^n/\mathbb{Z}^n)$ are isomorphic. Let $v: G \to \mathbb{R}^n/\mathbb{Z}^n$ be a 1-cocycle representing $\bar{v} \in H^1(G,\mathbb{R}^n/\mathbb{Z}^n)$, and let $c \in H^2(G,\mathbb{Z}^n)$ correspond to $\bar{v}$ by the connecting isomorphism. Then the extension class of G by $\mathbb{Z}^n$ may be represented as the matrix group

$$R(G,v) = \left\{ \left(\begin{array}{c|c} g & v_r(g)+t \\ \hline 0\ldots0 & 1 \end{array} \right) \middle| \; g \in G,\ t \in \mathbb{Z}^n \right\},$$

where $v_r(g) \in \mathbb{R}^n$ is some representative of the coset $v(g) \in \mathbb{R}^n/\mathbb{Z}^n$. ($v$ resp. v_r is called vectorsystem in crystallography. It should be noted that Schur's summation argument in cohomology makes the transition from some extension described by c to the isomorphic extension $R(G,v)$ quite explicit , cf. [Zas 48].)

The faithful matrix representation

$$\varphi_B : \mathrm{Aff}_n \to GL(n+1,\mathbb{R}) : \tau_t\varphi \mapsto \left(\begin{array}{c|c} B^{\varphi_B} & B^t \\ \hline 0\ldots 0 & 1 \end{array}\right)$$

defined as the corresponding representation of E_n at the beginning of the introduction, may now be used to turn $R(G,v)$ into an affine group.

The translation subgroup of $R(G,v)$ is given by

$$\left\{ \left(\begin{array}{c|c} I_n & t \\ \hline 0\ldots 0 & 1 \end{array}\right) \middle| \; t \in \mathbb{Z}^n \right\} \quad (I_n = n{\times}n\text{-unit matrix})$$

and the defining equation for 1-cocycles

$$v(gh) = v(g) + g\,v(h) \quad \text{resp.}$$

$$v_r(gh) \equiv v_r(g) + g\,v_r(h) \pmod{\mathbb{Z}^n}$$

for $g, h \in G$, is just the condition which makes $R(G,v)$ a group.

Conjugation by $\left(\begin{array}{c|c} I_n & t \\ \hline 0 & 1 \end{array}\right)$ with $t \in \mathbb{R}^n$ amounts to replacing v by a different (translationally equivalent) cocycle (vectorsystem) in $\bar{v} \in H^1(G,\mathbb{R}^n/\mathbb{Z}^n)$.

Finally, if $H \leq GL(n,\mathbb{Z})$ is conjugate to G under $GL(n,\mathbb{Z})$, say $G = x^{-1}Hx$ for some $x \in GL(n,\mathbb{Z})$, then $xv : H \to \mathbb{R}^n/\mathbb{Z}^n : h \to xv(x^{-1}hx)$ is a 1-cocycle for H, and $R(G,v)$ and $R(H,xv)$ are isomorphic, the isomorphism being induced by conjugation with $\begin{pmatrix} x & 0 \\ 0 & 1 \end{pmatrix} \in GL(n+1,\mathbb{Z})$.

Because of the last remark the proof is complete, if one proves that there are only finitely many conjugacy classes of finite subgroups of $GL(n,\mathbb{Z})$. Lemma (II.2) below shows immediately that there are only finitely many isomorphism types of finite subgroups of $GL(n,\mathbb{Z})$. Hence by standard arguments of classical representation theory there are finitely many finite subgroups of $GL(n,\mathbb{Z})$ up to conjugacy in $GL(n,\mathbb{Q})$. (In fact, for each isomorphism class in question this number is the number of $Aut(G)$-orbits on the characters of G, which belong to a faithful rational representation of degree n, where G is some group of the given isomorphsm type.)

Now one is left with the most difficult problem of the proof: Given a finite subgroup G of $GL(n,\mathbb{Z})$. Show that there are - up to conjugacy in $GL(n,\mathbb{Z})$ - only finitely many subgroups of $GL(n,\mathbb{Z})$ of the form $X^{-1}GX$ for some $X \in GL(n,\mathbb{Q})$.

To this aim the $\mathbb{Z}G$-lattice $X\mathbb{Z}^n = \{Xt \mid t \in \mathbb{Z}^n\}$ is associated with the group $X^{-1}GX$, where G acts on $X\mathbb{Z}^n$ in the obvious way by multiplication from the right. Clearly, if $Y \in GL(n,\mathbb{Q})$ with $Y^{-1}GY \leq GL(n,\mathbb{Z})$, such that $X\mathbb{Z}^{n\times 1}$ and $Y\mathbb{Z}^{n\times 1}$ are isomorphic $\mathbb{Z}G$-lattices, then $X^{-1}GX$ and $Y^{-1}GY$ are conjugate under $GL(n,\mathbb{Z})$. (The converse does not hold!) All $\mathbb{Z}G$-lattices turning up in this context generate the same $\mathbb{Q}G$-module, namely $\mathbb{Q}^n$. By the Jordan-Zassenhaus Theorem [Zas 38], [Rog 70] only finitely many $\mathbb{Z}G$-sublattices of $\mathbb{Q}^n$ of $\mathbb{Z}$-rank n exist up to isomorphism; hence the above statement follows and there are only finitely many conjugacy classes of finite subgroups of $GL(n,\mathbb{Z})$.

q.e.d.

<u>(II.2) LEMMA [Min 87]:</u> *For finite subgroups* G *of* $GL(n,\mathbb{Z})$ *, the homomorphism*

$$\varphi_p : G \to GL(n,\mathbb{Z}/p\mathbb{Z}) : (g_{ij}) \mapsto (g_{ij} + p\mathbb{Z})$$

is injective, in case p *is a prime number* $\neq 2$ *.*

<u>PROOF.</u> It suffices to prove the lemma for $G \cong Z_q$, q a prime number. Obvious reductions (such as replacing the natural $\mathbb{Z}G$-lattice $\mathbb{Z}^n$ (with G acting by left matrix multiplication) by factor lattices and these by other $\mathbb{Z}G$-lattices in the same genus) show that one only has to check the following:
The accompanying matrix of the q-th cyclotomic polynomial is not congruent to the unit matrix I_{q-1} modulo $p\mathbb{Z}^{(q-1)\times(q-1)}$. But this is clear.

q.e.d.

A historical comment on the proof of (II.1) might be appropriate. The most difficult part is the finiteness of the number of conjugacy classes of finite subgroups of $GL(n,\mathbb{Z})$. This is exactly the contents of Jordan's theorem [Jor 80]. Jordan proves this by first showing that each finite subgroup of $GL(n,\mathbb{Z})$ is conjugate to a subgroup of $GL(n,\mathbb{Z})$, which fixes a reduced quadratic form cf. [Min 05]. The second and most complicated step in that proof is, to show that there are altogether only finitely many matrices in $GL(n,\mathbb{Z})$, which transform a reduced quadratic form into a (possibly different) reduced quadratic form, cf. also [Min 05]. Though explicite bounds for the entries of these transforming matrices are known cf. [BiS 28], [Sie 40], it does not seem to be easy to obtain the groups in this way, cf. however [Rys 72a,b] for geometric approaches. Therefore the above proof was given in the seemingly inefficient way. But this - as can be seen later - certainly pays, when one tries to actual construct the groups, since one then can apply the heavy machinery of representations of finite groups.

The number of isomorphism classes of space groups are known up to dimension 4. They are 2, 17, 219 resp. 4783 isomorphism classes of 1,2,3 resp. 4-dimensional space group. Complete lists of the 2-4-dimensional groups in the form introduced in the proof of (II.1) can be found in [BBNWZ 78]. A geometric derivation of the 2- and 3-dimensional space groups can be found in [Bur 66] and [Sch 80a]. Crystallographers, physicists, and chemists usually use [Hen 69]. For dimensions bigger than 4 Schwarzenberger proves that there are already 9608 resp. 1540944 isomorphism classes of space groups of degree 5 resp. 6 with the group of all diagonal matrices in $GL(n,\mathbb{Z})$ for $n = 5,6$ as point group by using enumerations of directed graphs with n vertices, cf. [Sch 76].

B. The Zassenhaus algorithm. Further analysis of the cohomological part of the proof of Bieberbach's finiteness theorem (II.1) leads to the Zassenhaus algorithm, cf. [Zas 48], which constructs a set of representatives of all isomorphism classes of space groups with a given point group G, which can be thought of as a finite subgroup of $GL(n,\mathbb{Z})$. The notation of the proof of (II.1) will be used.

(II.3) ZASSENHAUS ALGORITHM.

(i) Determination of $H^1(G,\mathbb{R}^n/\mathbb{Z}^n)$.

Required data: (α) *a set of generators of* G :

$\{g_1,\dots,g_k\}$

(β) *defining relations for* G :

$w_1(g_1,\dots,g_k),\dots,w_l(g_1,\dots,g_k)$.

Algorithm: A 1-cocycle $v : G \to \mathbb{R}^n/\mathbb{Z}^n$ *is determined by* $v(g_1),\dots,v(g_k)$ *because of the defining rule* $v(gh) = v(g) + gv(h)$ $(g,h \in G)$. *By repeated application of the same rule, one derives a system of equation for* $v(g_1),\dots,v(g_k)$ *from the defining relations* $w_1,\dots,w_l$, *which are necessary and sufficient for* $v(g_1),\dots,v(g_k)$ *to define a cocycle. The equations are solved by the standard algorithm for solving linear homogeneous congruences (well known from the matrix-theoretic proof of the main theorem of finitely generated abelian groups). The group* $V(G)$ *of solutions consists of k-tuples of elements of* $\mathbb{R}^n/\mathbb{Z}^n$. *To get* $H^1(G,\mathbb{R}^n/\mathbb{Z}^n)$ *one has to factor* $V(G)$ *by the subgroup* $B(G) = \{(g_1x-x,\dots,g_kx-x) \mid x \in \mathbb{R}^{n\times 1}/\mathbb{Z}^{n\times 1}\}$, *which comes from the 1-coboundaries and is the maximal divisible subgroup of* $V(G)$.

(ii) Distinguishing the isomorphism types.

Required data: Finite set of generators of the normalizer $N = N_{GL(n,\mathbb{Z})}(G)$ *of* G *in* $GL(n,\mathbb{Z})$.

Algorithm: N *operates on* $H^1(G,\mathbb{R}^n/\mathbb{Z}^n)$.
Namely if $v : G \to \mathbb{R}^n/\mathbb{Z}^n$ *is a 1-cocycle and* $n \in N_{GL(n,\mathbb{Z})}(G)$, *then* $nv : G \to \mathbb{R}^n/\mathbb{Z}^n : g \to nv(n^{-1}gn)$ *is again a 1-cocycle. This defines an action of* N *on the group of 1-cocycles, which induces an action on the finite group* $H^1(G,\mathbb{R}^n/\mathbb{Z}^n)$. *The standard algorithm for computing orbits yields a set of representatives* $\{\bar{v}_1,\dots,\bar{v}_l\}$ *of the orbits of* N *on* $H^1(G,\mathbb{R}^n/\mathbb{Z}^n)$.
Let v_i *be a 1-cocycle representing* $\bar{v}_i \in H^1(G,\mathbb{R}^n/\mathbb{Z}^n)$ *for* $i = 1,\dots,l$. *Then* $R(G,v_1),\dots,R(G,v_l)$ *form a set of representatives of the isomorphism (or affine equivalence) classes of space groups with point group* G .

The last claim of the algorithm is easily verified by means of Theorem (I.4)(ii) and the cohomological part of the proof of (II.1). The first implementation of the Zassenhaus algorithm goes back to Brown in 63, cf. [Bro 69]. Later Köhler generalized the Zassenhaus algorithm to crystallographic groups of degree n , which have a translation subgroup of rank m with $1 \leq m \leq n$, cf. [Köh 73], [Köh 80]. The following example might demonstrate the Zassenhaus algorithm:

<u>(II.4) EXAMPLE.</u> Let $G = \langle g_1 = \begin{pmatrix} -1 & 0 \\ 0 & 1 \end{pmatrix}, g_2 = \begin{pmatrix} 1 & 0 \\ 0 & -1 \end{pmatrix} \rangle$

(i) $g_1^2 = 1$, $g_2^2 = 1$, $(g_1g_2)^2 = 1$ are defining relations for G. They yield the following equations for $v(g_1)$ $v(g_2) \in \mathbb{R}^2/\mathbb{Z}^2$

$(I_2+g_1)\ v(g_1) = 0,\ (I_2+g_2)\ v(g_2) = 0,$

$(I_2+g_1g_2)\ v(g_1) + (g_1+g_1g_2g_1)\ v(g_2) = 0$

i.e.

$$\begin{pmatrix} 0 & 0 \\ 0 & 2 \end{pmatrix} v(g_1) = 0,\ \begin{pmatrix} 2 & 0 \\ 0 & 0 \end{pmatrix} v(g_2) = 0\ ,\quad 0 = 0\ .$$

Hence

$$v(g_1) = \begin{pmatrix} \alpha \\ i \end{pmatrix} + \mathbb{Z}^2 \quad \text{and} \quad v(g_2) = \begin{pmatrix} j \\ \beta \end{pmatrix} + \mathbb{Z}^2$$

with $\alpha, \beta \in \mathbb{R}$ and $i,j \in \{0,\frac{1}{2}\}$. Hence $H^1(G,\mathbb{R}^2/\mathbb{Z}^2)$ is of order 4 and exponent 2; the elements might be represented by the cocycles $v_{i,j}$ with $v_{i,j}(g_1) = \begin{pmatrix} 0 \\ i \end{pmatrix} + \mathbb{Z}^2$ and $v_{i,j}(g_2) = \begin{pmatrix} j \\ 0 \end{pmatrix} + \mathbb{Z}^2$, where $i,j \in \{0,\frac{1}{2}\}$.

(ii) The normalizer N of G in $GL(2,\mathbb{Z})$ is generated by $\langle \begin{pmatrix} -1 & 0 \\ 0 & 1 \end{pmatrix}, \begin{pmatrix} 0 & 1 \\ 1 & 0 \end{pmatrix} \rangle$. There are three orbits of N on $H^1(G,\mathbb{R}^2/\mathbb{Z}^2)$, because $\overline{v}_{\frac{1}{2},0}$ and $\overline{v}_{0,\frac{1}{2}}$ are interchanged. Hence one ends up with $R(G,v_{0,0})$ (split extension), $R(G,v_{0,\frac{1}{2}})$ and $R(G,v_{\frac{1}{2},\frac{1}{2}})$ as representatives for the isomorphism classes of space groups with point group G.

The critical point about algorithm (II.3) is that it requires a finite set of generators for the normalizer $N = N_{GL(n,\mathbb{Z})}(G)$, which might be infinite. A necessary and sufficient criterion for the finiteness of N is given in [BNZ 73]. Since the automorphism group of G is finite, the index of the centralizer $C = C_{GL(n,\mathbb{Z})}(G)$ in N is finite. Therefore one can proceed in two steps to obtain a finite set of generators of N (cf. [BNZ 73]):

(i) Find a finite set of generators for C;

(ii) test for each outer automorphism of G whether it is induced by some element of N (and add this element to the list of generators of N in case it exists).

Step (i) amounts to a solution of problem (iii) of chapter IB, since C can be viewed as the unit group of a $\mathbb{Z}$-order, namely of the ring of all matrices in $\mathbb{Z}^{n\times n}$ commuting with the elements of G . There are constructive proofs known for the generalization of Dirichlet's unit theorem that the unit group of a $\mathbb{Z}$-order is finitely generated cf. e.g. [Sie 43], [Zas 72] (or even finitely presented cf. [Beh 62], [Zas 72]; compare also [GrS 80] for generalizations). However, it seems that these methods which involve the construction of fundamental domains have hardly been used algorithmically. The methods employed for finding generators of the infite centralizers (and normalizers) of the finite subgroups of $GL(4,\mathbb{Z})$ are described in [BNZ 73]. They essentially come down to the use of known unit groups from number theory and using techniques similar to the Euclidean algorithm in non commutative sitations. Before develloping these techniques Brown, Neubüser, and Zassenhaus prove some results on the possible "reduction schemes" of centralizers and normalizers in the same paper. Refinements of some of these results can be found in [Ple 77].

Solution of problem (ii) in chapter IB can be used for step (ii) above. Namely let α be a non-inner automorphism of G. The natural $\mathbb{Z}G$-lattice $L = \mathbb{Z}^n$ can be turned into a new $\mathbb{Z}G$-lattice L' on which the elements of G act in the original way after α has been applied to them. L and L' are isomorphic $\mathbb{Z}G$-lattices if and only if α is induced by some element of $N = N_{GL(n,\mathbb{Z})}(G)$. The matrix of each such isomorphism (if it exists) with respect to the standard basis of $\mathbb{Z}^n$ lies in N and induces α. The actual approaches towards a solution of problem (ii) will be discussed in the next chapter.

At the end of the discussion of space groups two recent results might be of interest. Meyer [Mey 79] considers all space groups with an isomorphic point group and proves the following statement:

For any finite group G there exist a finite number of "prototypes" of extensions of $\mathbb{Z}G$-lattices by G (not necessarily space groups) such that every space group R whose point group is isomorphic to G can be obtained in two steps from one of the prototypes X: First form the split extension S of $L := T(R)$ (viewed as $\mathbb{Z}G$-lattice) with G and let X_L be the subdirect product of S with X amalgamated (in the obvious way) over the common factor group G. Then X_L is factored by a certain normal subgroup contained in the translation lattice which - as a $\mathbb{Z}G$-lattice - is isomorphic to the $\mathbb{Z}G$-lattice yielding the extension X. The quotient group is isomorphic to R. Note that generally one need not know all $\mathbb{Z}G$-lattices up to isomorphism in order to find all the "prototypes" of extensions which have encoded all the possible extension types of all space groups with point group isomorphic to G . However the procedure is not unique for every space group, i.e. several prototypes might yield the same isomorphism type of space groups.

Finally in [FiNP 80] the question is investigated in how far the isomorphism type of a space group is determined by that of all resp. one sufficiently large factor group. It turnes out that Maranda's results for lattices [CuR 62] can be carried over to the space group situation with somewhat worse constants since the extensiontype has to be taken care of and the image of the translation subgroup has to be recognizable in the factor group. (Note, under certain aspects there is not a big difference between $\mathbb{Z}G$-lattices L , on which G acts faithfully, and space groups which are split extensions of L by G .)

III. FINITE UNIMODULAR GROUPS

A. Geometric and arithmetic equivalence. Finite subgroups of $GL(n,\mathbb{Z})$, in the sequel abbreviated as f.u. groups (finite unimodular groups), play a central rôle in mathematical crystallography since they can be viewed as point groups of space groups (after suitable choice of a $\mathbb{Z}$-basis of the translation subgroup). The most important equivalence relations of crystallographic groups besides that of isomorphism and/or affine equivalence for space groups are therefore the following:

<u>(III.1) DEFINITION.</u> *Let G,H be f.u. groups of degree n.*
(i) G and H are called arithmetically equivalent or $\mathbb{Z}$*-equivalent, if they are conjugate under* $GL(n,\mathbb{Z})$.
(ii) G and H are called geometrically equivalent or $\mathbb{Q}$*-equivalent, if they are conjugate under* $GL(n,\mathbb{Q})$.

The two equivalence relations made their appearance already in the second part of the proof of Bieberbach's finiteness theorem (II.1). $\mathbb{Z}$-equivalent f.u. groups lead to isomorphic sets of space groups and the point group of a space group - if viewed as integral matrix group - is only determined up to arithmetic equivalence by the space group, since the translation subgroup has more than one lattice basis. Roughly speaking two f.u. groups are $\mathbb{Q}$-equivalent if they can be viewed as the same groups acting on the vectorspace $\mathbb{R}^n$ or $\mathbb{Q}^n$. (Note, by the Deuring-Noether theorem conjugacy in $GL(n,\mathbb{Q})$ is the same as in $GL(n,\mathbb{R})$ [CuR 62]). Hence both of them contain the same number of reflections, elements of determinant 1 etc., i.e. they are identical from the geometric point of view and differ only by a base change of $\mathbb{Q}^n(\mathbb{R}^n)$, i.e. by conjugation by an element of $GL(n,\mathbb{Q})$ $(GL(n,\mathbb{R}))$. In the case of $\mathbb{Z}$-equivalence the lattice $\mathbb{Z}^n$ is taken into consideration in addition to the vectorspace $\mathbb{Q}^n$, which contains many lattices on which the group acts. For example the two groups $\left\langle \begin{pmatrix} -1 & 0 \\ 0 & 1 \end{pmatrix} \right\rangle$ and $\left\langle \begin{pmatrix} 0 & 1 \\ 1 & 0 \end{pmatrix} \right\rangle$ are geometrically equivalent - both of them consist of the identity and a reflection. But they are not arithmetically equivalent, as can be seen by taking the matrix entries modulo 2. (If one introduces an Euclidean metric on the lattices, which is respected by the group action, one gets a rectangular type of lattice in the first case, i.e. the two standard basis vectors are perpendicular, and a rhomb-shaped lattice in the second case, i.e. the two standard basis vectors have the same length.)

A geometric equivalence class of f.u. groups splits up into finitely many arithmetic equivalence classes by the Jordan-Zassenhaus theorem. The theory of orders provides means to gain some insight in how far the $\mathbb{Z}$-classes within a $\mathbb{Q}$-class are interrelated.

<u>(III.2) DEFINITION.</u> *Let G be a finite group and L a $\mathbb{Z}G$-lattice.*

(i) The $\mathbb{Z}G$-lattices in $\mathfrak{Z}(L) = \{M \underset{\mathbb{Z}G}{\leq} L \mid [L:M] \text{ finite}\}$ are called the centerings of L.

(ii) The $\mathbb{Z}G$-lattices in $\mathfrak{Z}(\mathbb{Q}L) = \{M \underset{\mathbb{Z}G}{\leq} \mathbb{Q}L \mid M \text{ finitely generated}, \mathbb{Q}M = \mathbb{Q}L\}$ are called the centerings of $\mathbb{Q}L$ $(:= \mathbb{Q} \otimes_{\mathbb{Z}} L)$.

$\mathfrak{Z}(L)$ and $\mathfrak{Z}(\mathbb{Q}L)$ form modular lattices with respect to sum and intersection. Clearly each isomorphism type of lattice contained in $\mathfrak{Z}(\mathbb{Q}L)$ occurs already in $\mathfrak{Z}(L)$ and $\mathfrak{Z}(\mathbb{Q}L)$ can easily be obtained from $\mathfrak{Z}(L)$. Assume now that G is an f.u. group and $L = \mathbb{Z}^n$ the natural $\mathbb{Z}G$-lattice. Every $M \in \mathfrak{Z}(\mathbb{Q}L)$ gives rise to a $\mathbb{Z}$-class of f.u. groups, which are $\mathbb{Q}$-equivalent to G, namely the $\mathbb{Z}$-class of all $X^{-1}GX$, where $X \in GL(n,\mathbb{Q})$ describes the basis transformation of the standard basis of $L = \mathbb{Z}^n$ to some basis of M. As mentioned at the end of the proof of (II.1) isomorphic lattices give rise to the same $\mathbb{Z}$-classes of f.u. groups. Here is the exact relation between $\mathbb{Z}$-equivalence of groups and isomorphism of lattices.

<u>(III.3) PROPOSITION.</u> *Let G be an f.u. group with natural lattice $L = \mathbb{Z}^n$.*

(i) The isomorphism classes of $\mathbb{Z}G$-lattices in $\mathfrak{Z}(\mathbb{Q}L)$ are the orbits of the centralizer $C_{GL(n,\mathbb{Q})}(G)$ of G in $GL(n,\mathbb{Q})$ on $\mathfrak{Z}(\mathbb{Q}L)$.

(ii) The $\mathbb{Z}$-classes of f.u. groups which are $\mathbb{Q}$-equivalent to G are in bijection with the orbits of the normalizer $N_{GL(n,\mathbb{Q})}(G)$ of G in $GL(n,\mathbb{Q})$ on $\mathfrak{Z}(\mathbb{Q}L)$.

($C_{GL(n,\mathbb{Q})}(G)$ and $N_{GL(n,\mathbb{Q})}(G)$ act in the obvious way on $\mathfrak{Z}(\mathbb{Q}L)$: for $n \in N_{GL(n,\mathbb{Q})}(G)$ and $M \in \mathfrak{Z}(\mathbb{Q}L)$ one defines $nM := \{nx \mid x \in M\}$.)

<u>PROOF.</u> (i) Obvious.

(ii) If two centerings lie in the same orbit under the action of $N_{GL(n,\mathbb{Q})}(G)$, they certainly give rise to the same $\mathbb{Z}$-class of f.u. groups. Conversely, if M and N are centerings of $\mathbb{Q}L$ which give rise to the same

$\mathbb{Z}$-class of f.u. groups, then there are matrices $X,Y \in GL(n,\mathbb{Q})$ with $XL = M$, $YL = N$, and $X^{-1}GX = Y^{-1}GY$. But then $Z = XY^{-1} \in N_{GL(n,\mathbb{Q})}(G)$ and $ZN = M$.

q.e.d.

The index of $C_{GL(n,\mathbb{Q})}(G)$ in $N_{GL(n,\mathbb{Q})}(G)$ is finite, since G has a finite outer automorphism group. Nevertheless, once in a while it happens that the normalizer has fewer orbits on $\mathfrak{Z}(\mathbb{Q}L)$ than the centralizer. For instance, the $\mathbb{Q}$-class of $G = \langle \begin{pmatrix} -1 & 0 \\ 0 & 1 \end{pmatrix}, \begin{pmatrix} 0 & 1 \\ 1 & 0 \end{pmatrix} \rangle \leq GL(2,\mathbb{Z})$ (dihedral group of order eight) consist of exactly one $\mathbb{Z}$-class, whereas there are two $\mathbb{Z}G$-isomorphism types of centerings of the natural lattice. For $G \otimes G = \{g \otimes h \mid g,h \in G\} \leq GL(4,\mathbb{Z})$ there are two $\mathbb{Z}$-classes of groups and eight isomorphism classes of lattices, and for $G \otimes G \otimes G = \{g \otimes h \otimes k \mid g,h,k \in G\} \leq GL(8,\mathbb{Z})$ there are six $\mathbb{Z}$-classes of groups and 192 isomorphism classes of lattices ($g \otimes h$ denotes the Kronecker product of the two matrices g and h). The preceeding discussion shows that a solution of problem (i) and (ii) in the introduction yields a set of representatives of the $\mathbb{Z}$-classes within the $\mathbb{Q}$-class of a given f.u. group G. The problem splits into two parts:

(α) Find sufficiently many centerings of the natural lattice $L = \mathbb{Z}^n$ of G.

(β) Decide, which are isomorphic resp. lie in the same orbit under the action of $N_{GL(n\mathbb{Q})}(G)$.

For solving problem (α) the following general idea going back to [Ple 74] has been applied successfully in [PlP 77,80] and [Ple 80a]: Assume that one is only interested in those centerings M of $\mathfrak{Z}(L)$ where the index of M in L is only divisible by certain primes (usually prime divisors of $|G|$). By a well known result of Brauer (cf. [CuR 62] Thm. 82.1) the (finite) quotient module L/M has only composition factors, which are isomorphic to the $\mathbb{Z}/p\mathbb{Z}G$-composition factors of L/pL , where p is one of the relevant prime numbers. Let $A_1,\ldots,A_k$ be representatives of the isomorphism classes of the $\mathbb{Z}G$-composition factors of L/pL with p ranging over the prime numbers in question. Any of the centerings M in $\mathfrak{Z}(L)$ for which the index of M in L has only the above prime divisors, can be obtained layer by layer first by taking the kernels of epimorphisms of L onto $A_1,\ldots,A_k$, then by taking the kernels of the epimorphisms of these kernels onto $A_1,\ldots,A_k$ etc. Since all the composition factors of L/M are among $A_1,\ldots,A_k$, M will be obtained in the l-th layer, where l is the $\mathbb{Z}G$-composition length of L/M . Computing the epimorphisms of a lattice N onto A_i amounts to solving a system of linear equations over $\mathbb{Z}/p\mathbb{Z}$ $(pA_i = 0)$, and computing the kernel of such an epimorphism again means solving linear equations over $\mathbb{Z}/p\mathbb{Z}$.

Two examples might demonstrate how part (β) of the problem can be approached efficiently. Assume first that the natural $\mathbb{Z}G$-lattice L is absolutely irreducible, i.e. $\mathrm{End}_{\mathbb{Z}G}(L) \cong \mathbb{Z}$.

<u>(III.4) PROPOSITION [Ple 74].</u> *Let L be an absolutely irreducible $\mathbb{Z}G$-lattice. Then*

$$\mathfrak{R}(L) = \{M \underset{\mathbb{Z}G}{\leq} L \mid M \not\subseteq pL \text{ for any prime number } p\}$$

is a set of representatives of the isomorphism classes of the centerings of L .

<u>PROOF.</u> Clearly $\mathfrak{R}(L)$ contains a set of representatives. Let $M_1, M_2 \in \mathfrak{R}(L)$ and assume that M_1 and M_2 are isomorphic. By (III.3)(i) there exists a nonzero rational number α with $\alpha M_1 = M_2$. Since $M_1 \in \mathfrak{R}(L)$, there exists an $l \in M_1$, such that $\mathbb{Z}l$ is $\mathbb{Z}$-direct summand of M_1 as well as of L . Hence α is integral. But also $M_2 \in \mathfrak{R}(L)$. Therefore $\alpha = \pm 1$ and hence $M_1 = M_2$.

q.e.d.

From the computational point of view (III.4) gives an ideal description of representatives of isomorphism classes of lattices. Namely, whenever a new centering is computed (by the above procedure), one only has to check whether it is contained in $\mathfrak{R}(L)$ and need not compare it with all other lattices obtained before. Once $\mathfrak{R}(L)$ is computed, usually is relatively easy to obtain the $\mathbb{Z}$-classes of f.u. groups in which the $\mathbb{Q}$-class of G is devided, i.e. the orbits of $N_{GL(n,\mathbb{Q})}(G)$ on $\mathfrak{Z}(\mathbb{Q}L)$. Note that the automorphisms of G induced by this normalizer are those which transform the character belonging to L into itself. That this action, however, must not be neglected was already demonstrated after (III.3) by the three extraspecial 2-groups of degree 2,4, and 8 of order 8,32, and 128 respectively. For instance, the second of these groups has the following lattice of centerings:

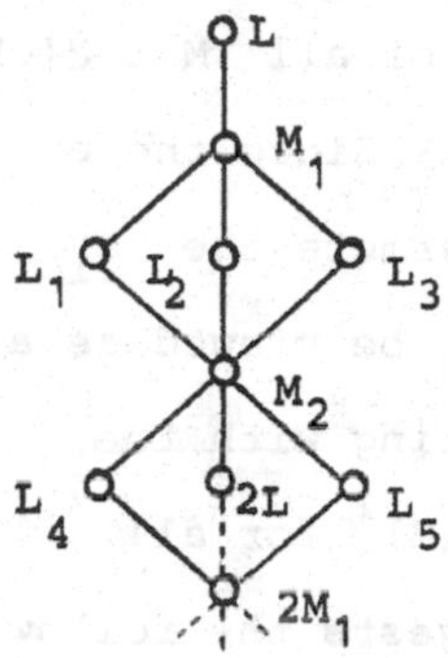

Here $\mathfrak{R}(L) = \{L, L_1, \ldots, L_5, M_1, M_2\}$. The L's and their multiples as well as the M's and their multiples form orbits of $\mathfrak{Z}(\mathbb{Q}L)$ under the normalizer in $GL(4,\mathbb{Q})$. The L's yield a monomial matrix group and the M's a nonmonomial one, cf. [PlP 77,80] part IV.

A generalization of (III.4) to arbitrary irreducible lattices in the complete local case, i.e. with $\mathbb{Z}G$ replaced by an order over a complete valuation ring, is given in [Ple 80b], Satz (IV.3).

As a second example of an approach to problem (β) reducible f.u. groups G having more than one ($\mathbb{Q}$-) irreducible constituent in its natural character will be discussed. Then the enveloping algebra of G in $\mathbb{Q}^{n\times n}$, consisting of all rational linear combinations of the matrices of G , has $k > 1$ central primitive idempotents $e_1,\dots,e_k$. Note, for each i, $1 \le i \le k$, there exists a $\mathbb{Q}$-irreducible character χ_i with $e_i = \frac{\chi_i^o(1)}{|G|} \sum_{g \in G} \chi_i(g^{-1})g$, where χ_i^o is a $\mathbb{C}$-irreducible constituent of χ_i , cf. [CuR 62].

Following [Ple 80a], each centering M of $\mathbb{Q}L$ can be associated with a decomposable lattice $D(M) := \bigoplus_{i=1}^{k} e_i M$ in $\mathfrak{Z}(\mathbb{Q}L)$. Let $\mathfrak{D}(\mathbb{Q}L)$ denote the set of all $M \in \mathfrak{Z}(\mathbb{Q}L)$ with $D(M) = M$. (Note $D(D(M)) = D(M)$). Since the elements of the normalizer $N_{GL(n,\mathbb{Q})}(G)$ permute the e_i among themselves by conjugation, D can be viewed as a mapping from $\mathfrak{Z}(\mathbb{Q}L)$ onto $\mathfrak{D}(\mathbb{Q}L)$ commuting with the operation of the normalizer: $D(nM) = nD(M)$ for all $n \in N_{GL(n,\mathbb{Q})}(G)$, $M \in \mathfrak{Z}(\mathbb{Q}L)$. This suggests the following strategy for computing the $\mathbb{Z}$-classes within the $\mathbb{Q}$-class of G , which are in bijection with the $N_{GL(n,\mathbb{Q})}(G)$-orbits on $\mathfrak{Z}(\mathbb{Q}L)$:

(i) Find the $N_{GL(n,\mathbb{Q})}(G)$-orbits on $\mathfrak{D}(\mathbb{Q}L)$;

(ii) For each orbit in $\mathfrak{D}(\mathbb{Q}L)$ choose a representative M and decide which centerings in $D^{-1}(M) = \{M' \in \mathfrak{Z}(M) \mid D(M') = M\}$ lie in the same $N_{GL(n,\mathbb{Q})}(G)$-orbit, i.e. yield $\mathbb{Z}$-equivalent f.u. groups.

Step (i) can be a very hard problem, in case $e_i L$ is not irreducible for some $i, 1 \le i \le k$. However, as pointed out in [Ple 80a], cf. also [Ple 78], if the $\mathbb{Z}$-rank of e_iL is less than eight, there are only very few cases, when e_iL is not fully decomposable, i.e. not direct sum of irreducible lattices. This can only occur, when the epimophic image of G acting faithfully on e_iL is isomorphic to D_8 (dihedral group of order 8), A_4 (alternating group of 4 elements), or $C_2 \times A_4$. Therefore step (i) usually is very easy in small dimensions.

As for step (ii), assume without loss of generality M = L (natural lattice of G). Then the following holds, cf. also [Ple 78], [Ple 80a], [Ple 80b]:

__(III.5) PROPOSITION.__ *(i) $D^{-1}(L)$ is a finite set of centerings of L.*

(ii) Two lattices in $D^{-1}(L)$ lie in the same orbit under the action of $N_{GL(n,\mathbb{Q})}(G)$ (i.e. yield $\mathbb{Z}$-equivalent f.u. groups) if and only if they lie in the same orbit under $N_{GL(n,\mathbb{Z})}(G)$.

__PROOF.__ (i) This follows immediately from the above formula for the central primitive idempotents e_i: each $M \in D^{-1}(L)$ contains $|G|L$; but $L/|G|L$ is finite.

(ii) $N_{GL(n,\mathbb{Q})}(G)$ acts on $\mathfrak{Z}(\mathbb{Q}L)$; the stabilizer of L is $N_{GL(n,\mathbb{Z})}(G)$. Because D commutes with the action of the normalizer, the result follows.

q.e.d.

If one is interested in isomorphism classes of lattices rather than $\mathbb{Z}$-classes of f.u. groups, one only has to replace the normalizers in the above discussion by the corresponding centralizers. Both normalizers and centralizers of f.u. groups in $GL(n,\mathbb{Z})$ are discussed in [BNZ 73] and [Ple 77]. Note that it is very easy to check whether a centering is in $D^{-1}(L)$ and that $D^{-1}(L)$ can be computed by the algorithm solving problem (α) described above. Proposition (III.5) certainly does not give quite as convenient conditions as (III.4) for computational purposes, since it requires (at worst) the knowledge of a finite set of generators of $N_{GL(n,\mathbb{Z})}(G)$ resp. $C_{GL(n,\mathbb{Z})}(G)$. But it seems to be difficult to avoid this. Two examples might demonstrate the procedure.

(III.6) EXAMPLE [Sch 74],[Sch 80b]. *Let o_n be the number of $\mathbb{Z}$-classes of f.u. groups of degree n which are $\mathbb{Q}$-equivalent to the group G_n of all diagonal matrices in $GL(n,\mathbb{Z})$ (of order 2^n). Then o_n is equal to the number of orbits of the symmetric group S_n on those subspaces of $\mathbb{F}_2{}^n$ which do not contain a standard basis vector. (S_n operates on $\mathbb{F}_2{}^n$ by permuting the standard basis vectors in the natural way.) Hence $o_2 = 2, o_3 = 4,$..*

PROOF. Let $L = \mathbb{Z}^n$ be the natural $\mathbb{Z}G_n$-lattice. Then $N_{GL(n,\mathbb{Q})}(G_n)$ acts transitively on $\mathfrak{D}(\mathbb{Q}L)$, because L splits into a direct sum of 1-dimensional lattices e_iL , where e_i is the i-th diagonal idempotent $\mathrm{diag}(0,..,\underset{i}{1},0,..)$.

$D^{-1}(L)$ is the set of those subgroups M of L with $e_iM = e_iL$ for $i = 1,2,\ldots,n$ and $M \supset 2L$, because $2e_i$ is in the $\mathbb{Z}$-span of the matrices of G_n $(i=1,2,\ldots,n)$. Hence $D^{-1}(L)$ is in 1-1-correspondence with the subspaces of $\mathbb{F}_2^{\,n}$ ($\cong$ dual space of $L/2L$) described above. The normalizer of G_n in $GL(n,\mathbb{Z})$ consists of all monomial matrices (having ± 1 in exactly one position of each row and column and zero elsewhere) and is isomorphic to the wreath product $C_2 \sim S_n$. The action induced on $L/2L$ resp. $\mathbb{F}_2^{\,n}$ is that of the natural permutation module of S_n. Hence the result follows from (III.5).

q.e.d.

<u>**(III.7) EXAMPLE [Ple 78].**</u> *Let k (= number of central primitive idempotents of the enveloping algebra of G in $\mathbb{Q}^{n\times n}$) be equal to two. Then the centerings in $D^{-1}(e_1L \oplus e_2L)$ can be obtained as pullbacks of the diagrams $e_1L \overset{\varphi_1}{\rightarrow} A \overset{\varphi_2}{\leftarrow} e_2L$, where $\varphi_i : e_iL \rightarrow A$ $(i=1,2)$ are $\mathbb{Z}G$-epimorphisms onto finite $\mathbb{Z}G$-modules A. In particular $D^{-1}(e_1L \oplus e_2L) = \{e_1L \oplus e_2L\}$ if and only if e_1L and e_2L have no isomorphic factor modules $\neq 0$ in common.*

Note that by the last example a sufficient and necessary condition for $\mathfrak{D}(\mathbb{Q}L) = \mathfrak{Z}(\mathbb{Q}L)$ can easily be obtained from the (Brauer)-decomposition numbers of G, namely the condition that e_1L/pe_1L and e_2L/pe_2L have no common constituents for all prime numbers p . This follows immediately from Brauer's well known theorem (cf.[CuR 62] Thm 82.1).

B. Bravais classification. The second important classification principle for f.u. groups, besides that of geometric and arithmetic equivalence, is due to the fact that a (crystallographic) space group is defined as a subgroup of the Euclidean group E_n . Therefore quadratic forms come into the game. Namely, if R is a (crystallographic) space group, then the vector lattice $L(R)$ associated with R (cf. Definition (I.1)(iii)) carries two kinds of structure: The point group $P(R)$ of R acts on $L(R)$, i.e. $L(R)$ is a $\mathbb{Z}P(R)$-lattice (which was the starting point of the discussion in III.1); and secondly $L(R)$ is embedded into the Euclidean space $\mathbb{R}^n$, i.e. $\langle\, ,\, \rangle$ induces an $\mathbb{R}$-valued bilinear form on $L(R)$. Moreover the action of $P(R)$ respects this bilinear form. This is the background for the following discussion, which essentially follows [BNW 71] and [BNZ 73].

Let G be an f.u. group with natural lattice $L = \mathbb{Z}^n$. Which are the $\mathbb{R}$-valued positive definite quadratic forms defined on L which are left invariant by G ? Dually, let S be a set of (positive definite) quadratic forms on L . What is the biggest f.u. group (i.e. subgroup of $Aut_{\mathbb{Z}}(L)$) which respects all these quadratic forms? These two questions give rise to the Bravais classification of f.u. groups defined below. Instead of quadratic forms the matrices representing them with respect to the standard basis of $\mathbb{Z}^n$ are used.

<u>(III.8) DEFINITION.</u> *Let* $G \leq GL(n,\mathbb{Z})$ *be finite and* $S \subseteq \mathbb{R}^{n\times n}$ *be a set of symmetric matrices.*

(i) $S(G) := \{X \in \mathbb{R}^{n\times n} \mid X^{tr} = X,\ g^{tr}Xg = X \text{ for all } g \in G\}$ *is called the space of forms of* G *. (*X^{tr} *denotes the transposed matrix of* X *.)*

(ii) $B(S) := \{g \in GL(n,\mathbb{Z}) \mid g^{tr}Xg = X \text{ for all } X \in S\}$ *is called the Bravais group of* S *, in case* $B(S)$ *is finite.* G *is called Bravais group, if there exists a set* S *of symmetric matrices as above, with* $G = B(S)$ *.*

(iii) $B(G) := B(S(G))$ *is called the Bravais group of* G *. All f.u. groups with* $\mathbb{Z}$*-equivalent Bravais groups belong to the same Bravais flock.*

As G is finite, $S(G)$ contains a positive definite matrix, for instance $\sum_{g\in G} g^{tr}g$. On the other hand, if S contains a positive definite matrix, then $B(S)$ is finite, since it is a discrete subgroup of a compact (orthogonal) group. Therefore the Bravais group $B(G)$ of an f.u. group G is finite. Clearly $G \leq B(G)$ and $B(B(G)) = B(G)$. One has a Galois correspondence between the (full) form spaces on the one side and the Bravais groups on the other side: The bigger the form space the smaller the Bravais group and vica versa. The positive definite matrices in $S(G)$ form a convex open cone $S^+(G)$ in the $\mathbb{R}$-vector space S(G) (with the usual topology). The dimension of S(G) can easily be computed from the character of the natural representation of G , cf. [Ple 80a] Lemma (IV.1). That Bravais groups can be viewed as the full automorphism group of a vector lattice in the Euclidean space $\mathbb{R}^n$, follows from the following proposition, cf. [NPW 80] and also [Sch 74].

<u>(III.9) PROPOSITION.</u> *Let G be a Bravais group. Then there exists an* $X \in S(G)$ *with* $G = B(\{X\})$.

<u>PROOF.</u> Since $GL(n,\mathbb{Z})$ is a countable group, it has only countably many finite subgroups. Therefore G is contained in at most countably many finite f.u. groups $H_i, i = 1,2...$. Each of the H_i has a form space properly contained in that of G , i.e. $\dim S(H_i) < \dim S(G)$. Hence $\bigcup_{i=1}^{\infty} S(H_i)$ is a subset of measure zero in S(G) . Hence $S^+(G) \setminus \bigcup_{i=1}^{\infty} S(H_i)$ is not empty and for each X in this set $B(\{X\}) = G$ holds.

q.e.d.

To get a feeling for Bravais groups, it might be helpful to look at the 2-dimensional ones, and their lattices in $\mathbb{R}^2$, which they classify.

Table 1:

Bravais Groups of Degree 2

Representative	Order of Bravais group	Number of $\mathbb{Z}$-classes in Bravais flock	Form space	Sample of lattice
$\langle \begin{pmatrix} -1 & 0 \\ 0 & -1 \end{pmatrix} \rangle$	2	2	$\{ \begin{pmatrix} a & c \\ c & b \end{pmatrix} \mid a,b,c \in \mathbb{R} \}$	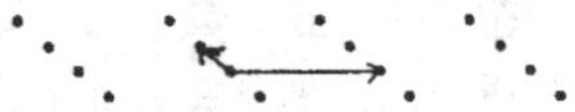
$\langle \begin{pmatrix} -1 & 0 \\ 0 & -1 \end{pmatrix}, \begin{pmatrix} 1 & 0 \\ 0 & -1 \end{pmatrix} \rangle$	4	2	$\{ \begin{pmatrix} a & 0 \\ 0 & b \end{pmatrix} \mid a,b \in \mathbb{R} \}$	
$\langle \begin{pmatrix} -1 & 0 \\ 0 & -1 \end{pmatrix}, \begin{pmatrix} 0 & 1 \\ 1 & 0 \end{pmatrix} \rangle$	4	2	$\{ \begin{pmatrix} a & b \\ b & a \end{pmatrix} \mid a,b \in \mathbb{R} \}$	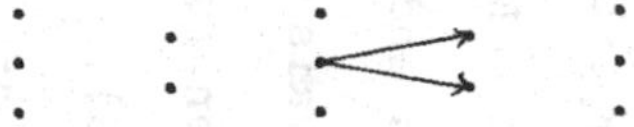
$\langle \begin{pmatrix} -1 & 0 \\ 0 & 1 \end{pmatrix}, \begin{pmatrix} 0 & 1 \\ 1 & 0 \end{pmatrix} \rangle$	8	2	$\{ \begin{pmatrix} a & 0 \\ 0 & a \end{pmatrix} \mid a \in \mathbb{R} \}$	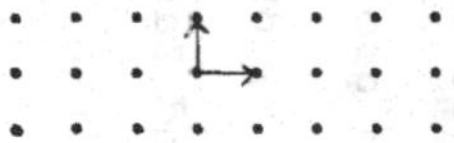
$\langle \begin{pmatrix} 0 & 1 \\ -1 & 1 \end{pmatrix}, \begin{pmatrix} 0 & 1 \\ 1 & 0 \end{pmatrix} \rangle$	12	4	$\{ \begin{pmatrix} 2a & -a \\ -a & 2a \end{pmatrix} \mid a \in \mathbb{R} \}$	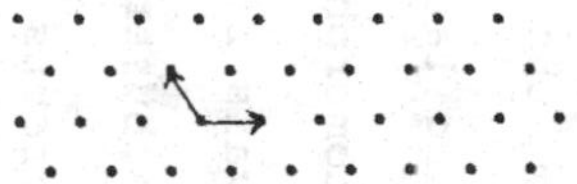

Apart from the obvious importance of Bravais groups for classifying lattices in the Euclidean space according to their automorphism groups and for subdividing f.u. groups into Bravais flocks, Bravais groups have been used by Brown, Neubüser and Zassenhaus for computing normalizers of f.u. groups in [BNZ 73]. Namely, for every f.u. group G the normalizer of G in $GL(n,\mathbb{Z})$ is contained in the normalizer of $B(G)$ in $GL(n,\mathbb{Z})$ with finite index. This is due to the easily verified fact that $N_{GL(n,\mathbb{Z})}(B(G)) = \{g \in GL(n,\mathbb{Z}) \mid g^{tr} S(G) g = S(G)\}$.

Integral representation theory of finite groups opens efficient possibilities for classifying and computing Bravais groups, cf. [PlP 77,80],[Ple 77],[Ple 80a]. The following result, which was proved in [Ple 77] for the field of rational numbers, is the algebraic basis for the application of the theory of orders to this problem. The enveloping algebra $E_K(G) = \{\sum_{g \in G} a_g g \mid a_g \in K\} \subseteq K^{n \times n}$ of an f.u. group $G \leq GL(n,\mathbb{Z})$ over a field $K \subseteq \mathbb{R}$ is compared with the enveloping algebra of the Bravais group $B = B(G)$ of G .

<u>(III.10) PROPOSITION.</u> *Let G be an f.u. group with Bravais group $B = B(G)$ and natural lattice $L = \mathbb{Z}^n$, and let K be a subfield of $\mathbb{R}$.*

(i) If $KL = \bigoplus_{i=1}^{s} V_i$ is a decomposition of KL $(= K \otimes_{\mathbb{Z}} L)$ into irreducible KG-modules, then V_i is also an (irre-

ducible) KB-submodule of KL *for* $i = 1,\ldots,s$. *Moreover* V_i *and* V_j $(1 \leq i,j \leq s)$ *are KG-isomorphic if and only if they are KB-isomorphic.*

(ii) Let $e_1,\ldots,e_r$ *be the central primitive idempotents of* $E_K(G)$. *Then* $e_1,\ldots,e_r$ *are also the central primitive idempotents of* $E_K(B)$. *For each* $l, 1 \leq l \leq r$ *the simple component* $e_l E_K(B)$ *of* $E_K(B)$ *contains* $e_l E_K(G)$ *in such a way that* $e_l E_K(G)$ *acts irreducibly on each irreducible* $e_l E_K(B)$*-module. In case* $e_l KL$ *is reducible as KG-module, one has* $e_l E_K(G) = e_l E_K(B)$.

<u>PROOF.</u> In case $K = \mathbb{R}$, the proof is considerably simpler than for general K . Therefore the proof is first given in this case, and afterwards the modifications for general K are described. Hence, assume $K = \mathbb{R}$.

(i) After a suitable transformation of G , i.e. after replacing the standard basis of KL by the union of orthonormal basis of the V_i $(i=1,\ldots,s)$ with respect to some G-invariant positive definite scalar product, one may assume that the diagonal idempotents $f_1,\ldots,f_s \in K^{n\times n}$ representing the projections of $KL = \bigoplus_{i=1}^{s} V_i$ onto $V_1,\ldots,V_s$ lie in $S(G)$. Since $I_n = f_1+\ldots+f_s$, the unit matrix I_n lies in $S(G) = S(B)$, hence $g^{tr}g = I_n$ or $g^{-1} = g^{tr}$ for all g in the (transformed) Bravais group B . Therefore $S(B) = S(G)$ is contained in the commuting algebra $C_K(B) = \{X \in K^{n\times n} \mid gX = Xg \text{ for all } g \in B\}$

of B in $K^{n\times n}$. But $C_K(B)$ represents the KB-endomorphism ring of KL , which therefore contains the projections corresponding to the idempotents $f_1,\ldots,f_s$. Hence $V_i = f_iKL$ is B-invariant. Assume $V_i \underset{KG}{\cong} V_j$ for some $i,j \in \{1,\ldots,s\}$. To prove $V_i \underset{KB}{\cong} V_j$ assume without loss of generality that the i-th diagonal block g_i of $g = \mathrm{diag}(g_1,\ldots,g_s) \in G$ (with $g_i \in GL(n_i,K), n_i = \dim_K V_i$) describing the action of G on V_i is equal to the j-th diagonal block g_j . Let $G_i (\leq GL(n_i,K))$ be the group of all $g_i \in GL(n_i,K)$ which turn up as i-th block in some block diagonal matrix $g = \mathrm{diag}(g_1,\ldots,g_s) \in G$. Since $g^{tr} = g^{-1}$ for all $g \in G_i$, one has $C_K(G_i)^{tr} = C_K(G_i)$. Therefore for any $X \in C_K(G_i)$ the block matrix $\tilde{X}$ with $(\dim_K V_k \times \dim_K V_l)$-blocks X_{kl} is in $S(G) = S(B)$, where $X_{kl} = 0$ for $1 \leq k,l \leq s$, $(k,l) \neq (i,j)$ and $(k,l) \neq (j,i)$, $X_{ij} = X$, $X_{ji} = X^{tr}$. But $S(B) \subseteq C_K(B)$ and for any nonsingular $X \in C_K(G_i)$ the matrix $\tilde{X}$ maps V_i onto V_j and induces a KB-isomorphism. The converse (namely that $V_i \underset{KB}{\cong} V_j$ implies $V_i \underset{KG}{\cong} V_j$) is obvious.

(ii) The first statements are immediate consequences of (i). To prove the last, assume e_1KL is a reducible KG-module. Then there exist indices $i,j \in \{1,\ldots,s\}$ with $V_i, V_j \subseteq e_1KL$. If $g = \mathrm{diag}(g_1\ldots,g_s) \in B$, then for any $X \in C_K(G_i)$, one has $\tilde{X}g = g\tilde{X}$, i.e. $Xg_i = g_jX$. Choosing X to be the unit matrix, one has $g_i = g_j$ (compare also (i)). Therefore g_i lies in the centralizer of $C_K(G_i)$ in $K^{n_i \times n_i}$ which is $E_K(G_i)$. This implies $e_1E_k(G) = e_1E_K(B)$ $(\cong E_K(G_i))$.

If $K \neq \mathbb{R}$, one can no longer choose orthonormal bases of V_i , but has to choose arbitrary bases. Let $S_K(G) = S(G) \cap K^{n \times n}$. Then $D^{-1}S_K(G)$ (instead of $S(G)$) contains idempotents $f_1,\ldots,f_s$ with $f_iKL = V_i$ and lies in $C_K(B)$ for any nonsingular $D \in S_K(G)$. (It is convenient to choose D as a block diagonal matrix with blocks of degree $\dim V_i$.) The rest is straight forward, cf. [Ple 77].

q.e.d.

Clearly, Proposition (III.10) also yields restrictions for those components $e_iE_K(B)$ of the enveloping algebra of the Bravais group $B = B(G)$ in terms of the component $e_iE_K(G)$ in case e_iKL is irreucible $(1 \leq i \leq r)$. For instance, if $K = \mathbb{R}$ one has the following possibilities: if $e_iE_{\mathbb{R}}(G) \cong \mathbb{R}^{m \times m}$ then $e_iE_{\mathbb{R}}(B) \cong \mathbb{R}^{m \times m}$; if $e_iE_{\mathbb{R}}(G) \cong \mathbb{C}^{m \times m}$ then $e_iE_{\mathbb{R}}(B) \cong \mathbb{C}^{m \times m}$ or $e_iE_{\mathbb{R}}(B) \cong \mathbb{R}^{2m \times 2m}$; finally if $e_iE_{\mathbb{R}}(G) \cong Q^{m \times m}$, where Q is the algebra of real quaternions, then $e_iE_{\mathbb{R}}(B)$ is isomorphic to $Q^{m \times m}$, $\mathbb{C}^{2m \times 2m}$ or $\mathbb{R}^{4m \times 4m}$. In low dimensions there seems to be a tendency for the last ones of these possibilities in each case.
Another way to look at Proposition (III.10) is this:
If $f_1+\ldots+f_s = I_n$ is a decomposition of the I_n into primitive orthogonal idempotents in the commuting algebra

$C_K(G)$, then $I_1 = f_1+\ldots+f_s$ is also such a decomposition of I_n in $C_K(B)$. This proves the following important arithmetic corallary, which also has a predecessor in [Ple 77].

(III.11) CORALLARY. *Under the assumptions of (III.10) let $f_1,\ldots,f_s$ be orthogonal primitive idempotents of the commuting algebra $C_{\mathbb{Q}}(G)$ of G in $\mathbb{Q}^{n\times n}$ with $I_n = f_1+\ldots+f_s$. Then the following holds:*

(i) The Bravais group $B = B(G)$ acts on f_iL for $i = 1,2,\ldots,s$ and hence $\bigoplus_{i=1}^{s} f_iL$ is a completely decomposable $\mathbb{Z}B$-lattice.

(ii) Let $X \in \mathbb{Q}^{n\times n}$ such that the columns of X form a $\mathbb{Z}$-basis of $\bigoplus_{i=1}^{s} f_iL$, i.e. $XL = \bigoplus_{i=1}^{s} f_iL$. Then $B(G) = XB(X^{-1}GX)X^{-1} \cap GL(n,\mathbb{Z})$; i.e. $B(G)$ is rationally equivalent to the biggest subgroup of the Bravais group of the fully decomposable group $X^{-1}GX$ which leaves L invariant.

PROOF. (i) follows immediately from (III.10) and (ii) is an easy consequence of (i).

q.e.d.

This corollary reduces the determination of the Bravais group of an f.u. group G to that of a completely decomposable f.u. group $\tilde{G}$ which is $\mathbb{Q}$-equivalent to G . The next result reduces the problem to irreducible f.u. group.

<u>(III.12) COROLLARY [Ple 77].</u> *Let* $r > 1, n_1, \ldots, n_r$ *be natural numbers and* $G_1, \ldots, G_r$ *f.u. groups of degree* $n_1, \ldots, n_r$. *Furthermore let* $n = n_1 + \ldots + n_r$ *and* $G \leq GL(n, \mathbb{Z})$ *be a subgroup of* $G_1 \otimes \ldots \otimes G_r := \{diag(g_1 \ldots, g_r) \mid g_i \in G_i, i=1,\ldots,r\}$ *such that the projections* $\pi_i : G \to G_i : diag(g_1 \ldots, g_r) \mapsto g_i$ *are surjective for* $i = 1, \ldots, r$.

(i) Assume that no two of the π_i, *viewed as representations of* G *into* $GL(n_i, \mathbb{Q})$, *have irreducible constituents in common. Then* $B(G) = B(G_1) \otimes \ldots \otimes B(G_r)$.

(ii) Assume all of the π_i *are* $\mathbb{Q}$*-irreducible and equivalent, i.e. there exist* $A_2, \ldots, A_r \in GL(n_1, \mathbb{Q})$ *with* $\pi_i(g) = A_i \pi_1(g) A_i^{-1}$ *for* $i = 2, 3, \ldots r$. *Then*

$B(G) =$

$\{diag(h, A_2 h A_2^{-1}, \ldots, A_r h A_r^{-1}) \mid h \in B(G_1) \cap E_{\mathbb{Q}}(G_1)\} \cap GL(n, \mathbb{Z})$.

<u>PROOF.</u> (i) Obviously $\{diag(X_1, \ldots, X_r) \mid X_i \in S(G_i)\} \subseteq S(G)$. By the proof of (III.10) equality holds. Since also by (III.10) $B(G) \leq GL(n_1, \mathbb{Z}) \otimes \ldots \otimes GL(n_r, \mathbb{Z})$, the result follows.

(ii) One has

$E_{\mathbb{Q}}(G) = \{diag(Y, A_2 Y A_2^{-1}, \ldots, A_r Y A_r^{-1}) \mid Y \in E_{\mathbb{Q}}(G_1)\}$.

By (III.10) $E_{\mathbb{Q}}(G) = E_{\mathbb{Q}}(B(G))$. Hence $B(G)$ is contained in the group described above. Conversely it is clear by the proof of (III.10) that each element g of this group fixes all forms in $S(G)$, i.e. $Xg = (g^{-1})^{tr} X$ for all

$X \in S(G)$, because both groups, G and the group in the statement of (ii), have the same commuting algebra in $\mathbb{Q}^{n \times n}$.

q.e.d.

Proposition (III.10) and its corollaries suggest and allow the following procedure for finding the $\mathbb{Z}$-classes of Bravais groups in a given degree n : Find all $\mathbb{Q}$-classes of f.u. groups G of degree n with the property

(*) $\quad H \lneqq G$ implies $B(H) \lneqq B(G)$.

Since (*) is equivalent to $H \lneqq G$ implies $\dim_{\mathbb{R}} S(H) > \dim_{\mathbb{R}} S(G)$, the set of all subgroups of $GL(n,\mathbb{Z})$ with property (*) consists of full $\mathbb{Q}$-classes. Starting out from certain decomposable groups one computes the $\mathbb{Z}$-classes in the spirit of part A of this chapter and then computes the Bravais groups of (certain of) the groups obtained. This program has been developed and carried out for the 5-dimensional and irreducible 6-dimensional Bravais groups in [Ple 80a]. In a modified form it was used already in [PlP 77,80] to find the maximal finite absolutely irreducible subgroups of $GL(n,\mathbb{Z})$ for $5 \leq n < 10$. Note that maximal f.u. groups, i.e. f.u. groups which are not contained in other f.u. groups of the same degree, are examples of Bravais groups. The reason why one has to go through such a complicated procedure to find all $\mathbb{Z}$-classes of Bravais groups is the following typical difficulty of the subject. For two $\mathbb{Q}$-equivalent f.u. groups G and $\tilde{G}$ it need not be true that their Bravais groups $B(G)$ and $B(\tilde{G})$ are $\mathbb{Q}$-equivalent.

For this one has already 3-dimensional counterexamples. A particularly drastical counterexample is given in [Ple 78] , namely an example of a reducible maximal f.u. group of degree 6, the 3-dimensional (irreducible) constituent groups are no longer maximal finite in $GL(3,\mathbb{Z})$. It also need not be true in general, that one of the groups $B(G)$ and $B(\tilde{G})$ is $\mathbb{Q}$-equivalent to a subgroup of the other. Examples for this and the crystallographic consequences are discussed in [NPW 80].

IV. STATISTICS

Table 2 below represents the state of knowledge of classification of crystallographic groups with respect to those equivalence relations which have been discussed in this paper, as far as full classifications for a given degree is concerned. Apart from the dimensions the table is incomplete in at least two respects. Firstly it does not contain all conventional equivalence relations for the groups discussed above. For instance various other equivalence relations may be composed out of $\mathbb{Q}$-equivalence and Bravais flocks, cf. [NPW 80]. The corresponding class numbers in some higher dimensions as well as asymptotic behaviour is discussed in [Jar 80].

Also orientations of the Euclidean space have not been taken into account. They lead to the concept of enantiomorphism, cf. [BBNWZ 78]. Secondly not all of the crystallographic groups turn up. In particular so called subperiodic groups, i.e. subgroups of n-dimensional space groups with a translation subgroup of rank k, $o < k < n$, do not turn up, cf. [Köh 80]. Also colour groups are missing, cf. e.g. [Sch 80a], [Sen 79].

NOTATION:

r_n : number of affine equivalence (or isomorphism) classes of space groups in n dimensions

a_n : number of $\mathbb{Z}$-classes of f.u. groups of degree n (= finite subgroups of $GL(n,\mathbb{Z})$)

g_n : number of $\mathbb{Q}$-classes of f.u. groups of degree n

b_n : number of $\mathbb{Z}$-classes of Bravais groups (or Bravais flocks of f.u. groups) of degree n

ib_n : number of $\mathbb{Z}$-classes of irreducible Bravais groups of degree n

m_n : number of $\mathbb{Z}$-classes of maximal f.u. groups of degree n

aim_n : number of absolutely irreducible maximal f.u. groups of degree n

Table 2

n	1	2	3	4	5	6	7	8	9
r_n	2	17	219	4783	?				
a_n	2	13	73	710	?				
g_n	2	10	32	227	?				
b_n	1	5	14	64	189	?			
ib_n	1	2	3	9	7	22	7	?	20
m_n	1	2	4	9	17	?			
aim_n	1	2	3	6	7	17	7	26	20

Only in dimension ≤ 4 one has a complete classification. All these groups can be found in [BBNWZ 78]. The four-dimensional groups were obtained as follows. First E.C. Dade determined the maximal f.u. groups of degree 4 up to $\mathbb{Z}$-equivalence [Dad 65]. R. Bülow used these results to determine the 710 $\mathbb{Z}$-classes of all finite subgroups of $GL(4,\mathbb{Z})$ by using group-theoretical programs implemented by V. Felsch [Bül 67,70, Fel 63]. He checked $\mathbb{Z}$-equivalence of two f.u. groups G, H, which "looked $\mathbb{Z}$-equivalent", by producing matrices $X \in \mathbb{Z}^{4\times 4}$ such that $\alpha(g)X = Xg$ for some isomorphism $\alpha: G \to H$, and all $g \in G$. The groups were proved to be $\mathbb{Z}$-equivalent as soon as an X with determinant ± 1 turned up. This procedure worked surprisingly well, probably due to the fact that two four dimensional $\mathbb{Z}G$-lattices are isomorphic iff they lie in the same genus. 1966 H. Brown implemented Zassenhaus' algorithm [Bro 69] and in 73 he determined [BNZ 73] the normalizers of the four dimensional f.u. groups so that also in 73 the 4783 affine classes of space groups could be computed by a new version of Brown's program implemented by K.-J. Köhler [Köh 73], cf. [BBNWZ 78]. m_5 was determined by S.S. Ryskov by geometric methods [Rys 72a,b]. Independently R. Bülow computed aim_5 (= im_5), cf. [Bül 73]. aim_n for $5 \leq n \leq 9$ and ib_7, ib_9 were computed by the methods indicated in Chapter III, cf. [PlP 77,80], as well as b_5 and ib_6, cf. [Ple 80a]. Note that irreducible representations of prime degree are absolutely irreducible, cf. [BNZ 72] or [PlP 77,80], and therefore $ib_n = aim_n$ for prime numbers n.

Because of Schwarzenberger's results, quoted at the end of Chapter IIA, it is probably unfeasable to compute all 5-dimensional space groups. Determining a_5 and g_5 lies probably in the range of computational possibilities. b_6, b_7 and ib_8 certainly can be obtained by the methods from [Ple 80a], discussed in Chapter III. m_6 and m_7 could be obtained from the 6- and 7-dimensional Bravais groups. Unfortunately there does not exist an algebraic method for finding the maximal f.u. groups in dimension ≥ 5 without computing the Bravais groups first.
Finally it is certainly possible to compute aim_{10} by the methods of [PlP 77,80], once the primitive finite complex matrix groups are classified. aim_{11} is easily seen

to be equal to 9 by the same methods and Theorem C of [Fei 74]. (The situation is very similar to the 5-dimensional case.)

The point of view taken in this chapter can be described by the question: How complete is our knowledge of crystallographic groups in a given (low) dimension? Other views have been taken by R. Schwarzenberger, J.D. Jarrat, N. Broderick and G. Maxwell. Apart from Schwarzenberger's asymtotic estimates [Sch 76], [Sch 80b], there is a series of papers by G. Maxwell e.g. [Max 75], [BrM 77], [Max 80] , who considers space groups with geometrically interesting point groups, e.g. reflection groups and rotation subgroups of reflection groups (not necessarily finite), independent of the dimension. A more complete bibliography up to 78 can be found in [BBNWZ 78].

V. APPENDIX

Theorem (I.2) and Theorem (I.3) are proved in this chapter.

Proof of (I.2):

(i) Clearly $T(R) = T_n \cap R$ is a normal abelian subgroup of R. Let N be any normal abelian subgroup and $\tau_s\varphi \in N$ (for some $s \in \mathbb{R}^n$, $\varphi \in O(\mathbb{R}^n)$). For any $\tau_t \in T(R)$ the elements $\tau_s\varphi$ and $\tau_t(\tau_s\varphi)\tau_t^{-1}$ commute, since both of them lie in N . But this is equivalent to $(id_{\mathbb{R}^n} - \varphi^{-1})^2 (t) = 0$ for all $t \in L(R)$. Since $L(R)$ contains a basis of $\mathbb{R}^n$, one has $(id_{\mathbb{R}^n} - \varphi^{-1})^2 = 0$. Hence all eigenvalues of the orthogonal mapping φ^{-1} are equal to 1 and therefore $\varphi = id_{\mathbb{R}^n}$. This proves $N \subseteq T(R)$, i.e. every normal abelian subgroup of R is contained in $T(R)$.

$C_R(T(R)) = T(R)$ follows immediately from the conjugation rule $(\tau_s\varphi)^{-1} \tau_t(\tau_s\varphi) = \tau_{\varphi(t)}$ for all $\tau_t \in T(R)$, $\tau_s\varphi \in R$ and the fact that $L(R)$ contains an $\mathbb{R}$-basis of $\mathbb{R}^n$.

(ii) The homomorphism $\lambda : R \to O(\mathbb{R}^n) : \tau_s\varphi \mapsto \varphi$ has kernel $T(R)$. Because of the conjugation rule the image of λ operates on $L(R)$ and therefore it is a discrete subgroup of the compact group $O(\mathbb{R}^n)$. Hence $R/T(R)$ is finite.

q.e.d.

Based on the fact that $R/T(R)$ is isomorphic to a finite subgroup of $GL(n,\mathbb{Z})$ further restrictions for $R/T(R)$ can be derived; namely Lemma (II.2) yields rough descriptions of the Sylow subgroups of $R/T(R)$ cf. [AbP 78], and one can easily determine the possible orders of the elements of $R/T(R)$ by using the Euler Φ-function.

Proof of (I.4):

(i) This follows immediately from the proof of (II.1).

(ii) Since the translation subgroup $T(R)$ of a space group R is a characteristic subgroup of R by (I.2), the isomorphism $\alpha : R_1 \to R_2$ induces an isomorphism of $T(R_1)$ onto $T(R_2)$, hence also from $L(R_1)$ onto $L(R_2)$. Therefore there is a unique $\Phi \in GL(\mathbb{R}^n)$ with $\Phi^{-1}\tau_t\Phi = \alpha(\tau_t)$ for all $\tau_t \in T(R_1)$. Furthermore α induces an isomorphism $\bar{\alpha} : R_1/T(R_1) \to R_2/T(R_2)$ such that $T(R_1)$ and $T(R_2)$ become isomorphic $R_1/T(R_1)$-lattices, where $x \in R_1/T(R_1)$ operates on $T(R_1)$ by conjugation and on $T(R_2)$ by conjugation with $\bar{\alpha}(x)$. Hence, after replacing R_1 by $\Phi^{-1}R_1\Phi$, one can assume $T(R_1) = T(R_2)$ and that the groups $P_i = \{\varphi \in GL(\mathbb{R}^n) \mid \text{ex. } \tau_t \in T_n \text{ with } \tau_t\varphi \in R_i\}$ are equal for $i = 1$ and 2.

Let $\{\tau_{t(\varphi)}\varphi \mid \varphi \in P_1\}$ be a set of representatives of the cosets of $T(R_1)$ in R_1 and define $t'(\varphi) \in \mathbb{R}^n$ by $\tau_{t'(\varphi)}\varphi = \alpha(\tau_{t(\varphi)}\varphi)$. (Note that $\alpha : R_1 \to R_2$ now induces the identity on $T(R_1) = T(R_2)$ and on $P_1 = P_2$.) $c : P_1 \to \mathbb{R}^n : \varphi \mapsto t(\varphi) - t'(\varphi)$ is easily checked to be a 1-cocycle. Since P_1 is finite, the 1-cohomology group $H^1(P_1, \mathbb{R}^n)$ vanishes, and hence there exist a $t_o \in \mathbb{R}^n$ (e.g. $t_o = |P_1|^{-1} \sum_{\varphi \in P_1} c(\varphi)$) with $c(\varphi) = t_o - \varphi(t_o)$.

Conjugation with the translation τ_{t_o} induces the isomorphism α , since τ_{t_o} transforms $\tau_{t(\varphi)}\varphi$ into $\tau_{t'(\varphi)}\varphi$ $(\varphi \in P_1)$.

q.e.d.

REFERENCES

[AbP 78] Abold,H., Plesken,W.: Ein Sylowsatz für endliche p-Untergruppen von GL(n$\mathbb{Z}$) . Math. Ann. 232 (1978), 183-186 .

[Bac 80] Backhouse,N.: Clifford theory and its application to the representation theory of crystallographic groups. To appear in [DrN 80].

[BBNWZ 78] Brown,H., Bülow,R., Neubüser,J., Wondratschek,H., Zassenhaus,H.: Crystallographic groups of four-dimensional space. Wiley, New York 1978.

[Beh 62] Behr,H.: Über die endliche Definierbarkeit von Gruppen. Crelles Journal 211 (1962), 116-122.

[Bie 10a] Bieberbach,L.: Über die Bewegungsgruppen des n-dimensionalen euclidischen Raumes mit einem endlichen Fundamentalbereich. Nachr. Königl. Ges. Wiss. Göttingen Math. Phys. Kl. (1910) 75-84.

[Bie 10b] Bieberbach,L.: Über die Bewegungsgruppen der Euklidischen Räume. (Erste Abhandlung) Math. Ann. 70 (1910) 297-336.

[Bie 12] Bieberbach,L.: Über die Bewegungsgruppen der Euklidischen Räume. (Zweite Abhandlung) Die Gruppen mit einem endlichen Fundamentalbereich. Math. Ann. 72 (1912) 400-412.

[Bir 80] Birman,J.L.: Representation theory, selection rules, and physical processes in crystals./ Selection rules and symmetry breaking. To appear in [DrN 80].

[BiS 28] Bieberbach,L., Schur,I.: Über Minkowskische Reduktionstheorie der positiven quadratischen Formen. Sitzungsber. der Preuss. Akad. der Wissensch. (1928), Physic.-Math. Klasse, 510-535.

[BNW 71] Bülow,R., Neubüser,J., Wondratschek,H.: On crystallography in higher dimensions. I. General definitions. II. Procedure of computations in R_4. III. Resulsts in R_4. Acta Crystallogr. A27 (1971), 517-535.

[BNZ 72,73] Brown,H., Neubüser,J., Zassenhaus,H.: On integral groups. I. The reducible case. Numer. Math. 19 (1972), 386-399. II. The irreducible case. Numer. Math. 20 (1972), 22-31. III. Normalizers. Math. Comput. 27 (1973), 167-182.

[BrM 77] Broderick,N., Maxwell,G.: The crystallography of Coxeter groups. II. J. Algebra 44 (1977), 290-318.

[Bro 69] Brown,H.: An algorithm for the determination of space groups. Math. Comput. 23 (1969), 499-514.

[Bue 56] Buerger,M.J.: Elementary crystallography. An introduction to the fundamental geometrical features of crystals. Wiley,New York, 1956. Revised printing, Wiley,New York, 1963.

[Bue 70] Buerger,M.J.: Contemporary crystallography. McGraw-Hill,New York, 1970.

[Bül 67] Bülow,R.: Eine Ableitung der Kristallklassen im R_4 mit Hilfe gruppentheoretischer Programme. University of Kiel, 1967.

[Bül 70] Bülow,R., Neubüser,J.: On some applications of group-theoretical programmes to the derivation of the crystal classes of R_4. pp.131-135 in Leech,J.,Ed.: Computational problems in abstract algebra (Proc. Conf. Oxford,1967). Pergamon Press, Oxford, 1970.

[Bül 73] Bülow,R.: Über Dadegruppen in GL(5,$\mathbb{Z}$). Dissertation, RWTH Aachen, 1973.

[Bur 66] Burckhardt,J.J.: Die Bewegungsgruppen der Kristallographie. 2nd ed., Birkhäuser, Basel, 1966.

[CuR 62] Curtis,C.W., Reiner,I.: Representation theory of finite groups and associative algebras. Interscience, New York, 1962.

[Cra 79] Cracknell,A.P., Davies,B.L., Miller,S.C., Love,W.F.: Kronecker Product Tables I-III. IFI/Plenum, New York,Washington,London 1979.

[Dad 65] Dade,E.C.: The maximal finite groups of 4×4 matrices. III. J.Math.9 (1965), 99-122.

[DrN 80] Dress,A., Neubüser,J. (edit.): Proceedings of the interdisciplinary conference on crystallographic groups (Bielefeld 1978). To appear in MATCH / informal communications in mathematic chemistry (1980), (vol. 9 and 10).

[Fei 74] Feit,W.: On integral representations of finite groups. Proc. London Math. Soc. (3) 29 (1974),633-683.

[Fel 63] Felsch,V., Neubüser,J.: Ein Programm zur Berechnung des Untergruppenverbandes einer endlichen Gruppe. Mitt. Rhein.-Westfäl.Inst.Instrum.Math.Bonn 2 (1963), 39-74.

[FiNP 80] Finken,H., Neubüser,J., Plesken,W.: Space groups and groups of prime power-order. II. Classification of space groups by finite factor groups. Archiv d. Math. Vol. 35 (1980), 203-209.

[FNP 80] Felsch,W., Neubüser,J., Plesken,W.: Space groups and groups of prime-power order. IV. Counterexamples to the class-breadth conjecture. To appear Proceed. London Math. Soc.

[Fro 11] Frobenius,F.G.: Über die unzerlegbaren diskreten Bewegungsgruppen. Sitzungsber. Preuss. Akad. Wiss. Berlin Phys. Math. Kl. (1911), 654-665. Gesammelte Abhandlungen III, Springer, Berlin, 1968, 507-518.

[GrS 80] Grunewald,F., Segal, D.: Some general algorithms. I. Arithmetic groups. II. Nilpotent groups. To appear in Annals of Math.

[Hen 69] Henry,N.F.M., Lonsdale,K.,Eds.: International tables for X-ray crystallography. Vol. I. Symmetry groups. 3rd ed., The Kynoch Press, Birmingham, England, 1969.

[Jan 80] Janner,A.G.M.: Symmetry of incommensurate crystal phases in the superspace group approach. To appear in [DrN 80].

[Jar 80] Jarrat,J.D.: The decomposition of crystal families. To appear in Math. Proc. Cambridge Phil.Soc.

[Jor 80] Jordan,C.: Mémoire sur l'équivalence des formes. J. Ecole Polytech. 48 (1880), 112-150. Oeuvres de C. Jordan, Vol. III, Gauthier-Villars,Paris,1962, 421-460.

[Köh 73] Köhler,K.-J.: Subperiodische kristallographische Bewegungsgruppen. Diplomarbeit, RWTH Aachen, 1973.

[Köh 80] Köhler,K.-J.: On the structure and the determination of n-dimensional partially periodic crystallographic groups. To appear in [DrN 80].

[LeN 80a,b] Leedham-Green,C.R., Newman,M.F.: Space groups and groups of prime-power order. I. Archiv d. Math. Vol. 35 (1980), 193-202. III. in preparation.

[Max 75] Maxwell,G.: The crystallography of Coxeter groups. J.Algebra 35 (1975), 159-177.

[Max 77] Maxwell,G.: Compact Euclidean space forms. J.Algebra 44 (1977), 191-195.

[Max 80] Maxwell,G.: Space groups of Coxeter type. To appear in [DrN 80].

[Mey 79] Meyer,J.: Les Puissances tensorielles de l'ideal augmentation d'un groupe fini et leurs extensions. These de doctorat, Grenoble 1979.

[Min 87] Minkowski,H.: Über den arithmetischen Begriff der Äquivalenz und über die endlichen Gruppen linearer ganzzahliger Substitutionen. Crelles Journal 100 (1887), 449-458. / Zur Theorie der positiven quadratischen Formen, Crelles Journal 101 (1887), 196-202. Gesammelte Abhandlungen I, Teubner, Leipzig 1911, 201-218.

[Min 05] Minkowski,H.: Diskontinuitätsbereich für arithmetische Äquivalenz, J. Reine Angew. Math. 129 (1905), 220-274. Gesammelte Abhandlungen II,Teubner, Leipzig, 1911, 53-100.

[NPW 80] Neubüser,J., Plesken,W., Wondratschek,H.: An emendatory discursion on defining crystal system. To appear in [DrN 80].

[Ple 74] Beiträge zur Bestimmung der endlichen irreduziblen Untergruppen von GL(n,$\mathbb{Z}$). Dissertation RWTH Aachen, 1974.

[Ple 77] Plesken,W.: The Bravais group and the normalizer of a reducible finite subgroup of GL(n,$\mathbb{Z}$). Comm. Algebra 5 (1977), 375-396.

[Ple 78] Plesken,W.: On reducible and decomposable representations of orders. Crelles Journal 297 (1978), 188-210.

[Ple 80a] Plesken,W.: Bravais groups in low dimensions. To appear in [DrN 80].

[Ple 80b] Plesken,W.: Gruppenringe über lokalen Dedekindbereichen. Habilitationsschrift Aachen 1980.

[PlP 77,80] Plesken,W., Pohst,M.: On maximal finite irreducible subgroups of GL(n,ℤ). I. The five and seven dimensional case. II. The six dimensional case. III. The nine dimensional case. IV. Remarks on even dimensions with applications to n = 8. V. The eight dimensional case and a complete description of dimensions less than ten. Math. Comput. 31 (1977),536-577, and Math. Comput. 34, 149 (1980), 245-301.

[Rog 70] Roggenkamp,K.W.: Lattices over orders II. Springer Lecture Notes in Math. 142, Berlin, Heidelberg, New York 1970.

[Rys 72a] Ryskov,S.S.: On maximal finite groups of integer (n×n)-matrices. Dokl.Akad. Nauk SSSR 204 (1972), 561-564. Sov.Math.Dokl. 13 (1972), 720-724.

[Rys 72b] Ryskov,S.S.: Maximal finite groups of integral n×n matrices and full groups of integral automorphisms of positive quadratic forms (Bravais models). Tr.Mat. Inst. Steklov 128 (1972), 183-211. Proc. Steklov Inst. Math. 128 (1972), 217-250.

[Sch 74] Schwarzenberger,R.L.E.: Crystallography in spaces of arbitrary dimension. Proc. Cambridge Philos. Soc. 76 (1974), 23-32.

[Sch 76] Schwarzenberger,R.L.E.: The use of directed graphs in the enumeration of orthogonal space groups. Acta Cryst. (1976), A 32, 556-559.

[Sch 80a] Schwarzenberger,R.L.E.: N-dimensional crystallography. Research Notes in Mathematics 41. Pitman; San Francisco,London, Melbourne 1980.

[Sch 80b] Graphical representation of n-dimensional space groups. To appear in Proc. London Math. Soc.

[Sen 79] Senechal,M.: Color groups. Discrete Appl. Math.1 (1979), 51-73.

[Sie 40] Siegel,C.L.: Einheiten quadratischer Formen. Abh. Math. Sem. Hamburg 13 (1940), 209-239.

[Sie 43] Siegel,C.L.: Discontinuous groups. Ann. Math. (2) 44 (1943), 674-689.

[Wol 67] Wolf,J.A.: Spaces of constant curvature. McGraw-Hill, New York, 1967.

[Zas 38] Zassenhaus,H.: Neuer Beweis der Endlichkeit der Klassenzahl bei unimodularer Äquivalenz endlicher ganzzahliger Substitutionsgruppen. Abh. Math. Sem. Hamburg 12 (1938), 276-288.

[Zas 48] Zassenhaus,H.: Über einen Algorithmus zur Bestimmung der Raumgruppen. Comment. Math. Helv. 21 (1948), 117-141.

[Zas 72] Zassenhaus,H.: On the units of orders. J. Alg.20 (1972), 368-395.

Graham Higman's Thesis "Units in Group Rings"

Robert Sandling
The University
Manchester M13 9PL.

The extent to which Higman's D. Phil. thesis [Hig 2] differs from his published paper [Hig 1] is not widely known. There have been few references to it: a mention in [Coh], an acknowledgement in [Jac 1], a remark by Higman himself [Hig 3]. It is not listed in the standard directory (ASLIB) of British theses.

Higman's paper, entitled "The Units of Group Rings", is widely known. It is important to algebraists for its definitive results on units in integral group rings of finite abelian groups and its characterisation of finite groups which have only trivial units. It is looked to by a broad spectrum of pure mathematicians as one of the papers establishing K-theory because of its connection with the work of Higman's supervisor, J.H.C. Whitehead, on what is now called Whitehead torsion.

The thesis represents the work of another year or more, the paper having been submitted in February 1939, the thesis deposited in the library in December 1940. Most of the paper is included in the thesis but in a completed or altered form; for example, the results on trivial units are derived in the context of the theory of orders. While further classes of infinite groups having only trivial units are introduced, the connection with Whitehead's original motivation in simple homotopy theory is no longer even hinted at. The proofs of what is now referred to as the triviality of the Whitehead group of an infinite cyclic group and of certain finite abelian groups are omitted.

There is more in the thesis, however, than in the paper, much of it still of great interest. It raises many questions still unanswered. More surprisingly, it supplies answers to many questions raised and answered in the intervening years by researchers unaware of Higman's results. Some of these will be apparent in the following summary of the notable acheivements of the thesis.

The thesis marks an early stage in the study of integral group rings as orders and of maximal orders in group rings; it also displays an early use of algebraic methods in the investigation of units in orders. It provides early examples of integral representations, in particular for certain metacyclic groups (namely, the affine group over the field of p elements). These representations have the feature, subsequently seen to be important in other cases, of reducing to triangular form modulo p.

A more complete characterisation is given of group rings having only trivial units. Much is said about non-trivial units. Illustrations are given to exemplify various points: non-conjugacy to trivial units; possibilities for the orders of units of finite order. In the integral case, it is shown that a non-trivial unit of finite order has no central element in its support. There are also results on the order and exponent of a group of normalised units of finite order.

The isomorphism problem for integral group rings of finite groups makes its appearance in two forms: whether the group ring determines the group; whether every subgroup of normalised units of finite order is isomorphic to a subgroup of the group. A positive solution for the second is given for metabelian nilpotent groups and for the subgroups of the metacyclic groups mentioned earlier. His proof for metabelian nilpotent groups contains the main elements of the eventual proof [Whitc.] of the general metabelian case.

It offers an early use of the homomorphism induced on their group rings by the projection of a group onto a factor group. Ideals such as the kernel of this homomorphism are manipulated with facility and the basis for such results as the isomorphism between G/G' and I/I^2 laid down.

Problems appearing in the paper in one form or another receive further discussion and more precise formulation. The existence of one-sided inverses is queried, as is the existence of zero divisors and non-trivial units in group rings of torsion-free groups. For the study of the latter, the one and two unique product conditions are formulated.

Although there are instances in all this material in which Higman did not see a point in as full a fashion as it is presently understood, it is remarkable how complete his insight was into the topics he included and how confident was his treatment of them.

This report provides a detailed summary of the thesis. The Abstract, most of the theorems and certain proofs are quoted in full. There are several reasons for attempting to bring this thesis to a wider audience. It is refreshing to observe how one of the first explorers on the ground saw the terrain, and surprising in hindsight to note how correctly he saw it: raising many of the main questions, spotting many of the main theorems, developing many of the main techniques. For a number of problems, it displays yet another independent researcher reaching the same conclusions. This may indicate the extent of what is true in a given area and so offer a guide to present researches.

It is a challenging exercise in mathematical bibliography to seek out the reappearances (at least in published form) of material in the thesis. Paradoxically I began this exercise before I was aware of Higman's thesis in an effort to correct inadequate attributions I had made in an ad hoc survey of the isomorphism problem at the previous Oberwolfach Tagung (February 1977) in this series organised by Klaus Roggenkamp. I had, in particular, overlooked much of the work of Berman and his collaborators (I have now translated many of their articles and can supply copies). Although no doubt still incomplete, my findings concerning the material in Higman's thesis are included here. Only occasionally do I discuss later developments. There are presently available several comprehensive monographs and surveys on group rings

[Zal-Mih; Bov; Den; Pas 4; Seh 2; Pol; Far]. These indicate the current status of most of the problems Higman considered.

A typewritten manuscript copy of the thesis is kept in the Radcliffe Science Library, Parks Road, Oxford, OX1 3QP. It is unbound and is noted as being "Deposited 20-12-40". It carries the reference number "MS.D.Phil.d.387". Photocopies are available on application to the library. In the United Kingdom, it is available on inter-library loan.

The manuscript consists of:

Acknowledgement: I wish to thank Professor Higman for his permission to include excerpts from his thesis in this article.

Abstract (pp. 1-6)

In the Abstract, Higman describes his results. His interests and emphases are clearly expressed in it and for this reason I include it in its entirety.

It also defines his notation. In my commentary I use a form of notation more common at present. Having parallel notations is clumsy at times, mainly for rings of coefficients. Higman uses K for a general ring or field of coefficients; I have used R for a ring and K for a field. He uses C for a ring of algebraic integers; I have again used R with K being the corresponding quotient field (Higman uses k). My notation for the augmentation ideal of the group ring RG is I(G), with I(H)RG denoting the ideal of RG generated by I(H), H a normal subgroup of G; Higman uses λ_H for this.

"UNITS IN GROUP RINGS

Abstract of Thesis

If G is any group, written multiplicatively, and K is any ring, then the finite formal sums

$$k_1g_1 + k_2g_2 + \dots + k_rg_r, \ k_i \in K, \ g_i \in G, \ (i = 1,\dots,r) ,$$

form a ring, when addition and multiplication is defined in the obvious way. This ring we call the group ring of G over K, and we write it as R(G,K). More precisely, R(G,K) may be defined as a ring which is a linear set over K, and has a basis which is a multiplicative group isomorphic to G. If K has the unity element

1, and if g_0 is the identity of G, then $1.g_0$ is a unity element of R(G,K). We shall identify, when confusion cannot arise thereby, elements k of K with the elements $k.g_0$ of R(G,K). The object of this thesis is to establish certain theorems on the units of R(G,K).

An element e_1 in a ring with unity element 1 will be called a left unit if there exists also an element e_2 such that $e_1e_2 = 1$; and e_2 will then be called a right unit, and a right inverse of e_1. If e_1 is also a right unit, then it has a uniquely defined inverse which we write as e_1^{-1}, and e_1 is then called a unit, simply. We shall deal always with rings R(G,K) in which the coefficient ring K has no right units which are not left units. Whether it is even so possible for R(G,K) to have right units which are not left units, I do not know. Certainly it is not in the most important cases, - for instance, if K can be embedded in a field, and G is of finite order. Any group ring has units of the form e.g, where e is a unit in K, and g is an element of G, for e.g has the inverse $e^{-1}.g^{-1}$. Such a unit we shall call trivial. One of the questions that naturally arises is, for what group rings these are the only units.

This thesis is chiefly concerned with the case in which G is a group of finite order, and K is the ring C of integers in an algebraic field k. We shall then speak of R(G,C) as an integral group ring. R(G,C) is then an order, but not in general a maximal order, of the linear associative algebra R(G,k). Accordingly, after an introductory section, and one on definitions, we discuss in Section 3 the group algebra R(G,k). This section is an exposition of well-known facts concerning the decomposition of the semi-simple algebra R(G,k) into simple components, and is based on the classical theory of representations of finite groups in algebraic fields, as developed by I. Schur.

In Section 4 we pass on to the consideration of integral group rings, and more particularly, to a determination of what such rings have only trivial units. The result is, that R(G,C), where C is the integer ring of the field k, has only trivial units in the following four cases, and (these) only:

(i) G is Abelian, and the orders of its elements all divide two; k is the rational field or an imaginary quadratic extension of it;

(ii) G is Abelian, and the orders of its elements all divide six; k is the rational field, or the extension of it by a complex cube root of unity;

(iii) G is Abelian, and the orders of its elements all divide four; k is the rational field, or the extension of it by a complex fourth root of unity (that is, by $i = \sqrt{-1}$);

(iv) G is the direct product of a quaternion group and an Abelian group the orders of whose elements all divide two; k is the rational field.

To establish this theorem, we prove first a theorem which is a particular case of a result due to O. Schilling, namely, that if k is of finite degree over the rational field the unit group of R(G,C) has a finite index in the unit group of any maximal order of R(G,k) containing it. Secondly, we show that the group of trivial units in R(G,C) is not contained as a proper sub-group of finite index in any group of units of R(G,C). From these two theorems together, it follows that R(G,C) has only trivial units if and only if a maximal order of R(G,k) containing it has a finite unit group. This imposes severe restrictions on the possible simple components of R(G,k), which give rise to our main result.

In the following section, we turn our attention to the units of finite order in R(G,C). We find it convenient to treat normalised units only, - that is to say, units in which the sum of the coefficients is unity. This involves no essential loss, since any unit of finite order is the product of a normalised unit of finite order and a root of unity in C; and any group of units of finite order is isomorphic to the direct product of the corresponding group of normalised units and a group of roots of unity in C. In terms of normalised units, a theorem from section 4 becomes: The group G is not contained as a proper sub-group of finite index in any group of normalised units of R(G,C). In particular this implies that a unit of finite order in the centrum of R(G,C) must be trivial. If G is Abelian, therefore, all the units of finite order in R(G,C) are trivial. This is of course no longer true if G is not Abelian. In fact, a unit of finite order in R(G,C) need not necessarily even be conjugate to a trivial unit. We are able, however, to prove two results on units of finite order. The first states that the elements of any group of normalised units of finite order in R(G,C) are linearly independent, and even linearly independent modulo m, for any integer m, and that therefore the order of such a group cannot exceed the order of G. The second states that the prime factors of the order of a normalised unit of R(G,C) divide the order of G. If we add the assumption that G is soluble, we can show that the order of the unit divides the order of G. Lastly, we show for the very special class of groups whose lower central series terminates in the identity and whose second derived group consists of the identity, that a group of normalised units of finite order in R(G,C) is isomorphic to a subgroup of G. This implies that R(G,C) is not isomorphic to any other group ring R(H,C) unless G is isomorphic to H. These last theorems are proved by methods different from those of the rest of the section, and our chief tools are the two-sided ideals λ_H, where H is a self conjugate subgroup of G, generated by the elements h-1, for all h in H. It should be added, that throughout this section the coefficient ring C is an arbitrary ring of algebraic integers, and may be taken to be ring of all algebraic integers.

In section 6, we apply the theorems we have proved to the detailed investigation of the ring R(G,C) where G is generated by two elements a, b subject to the relations:-

$$a^p = b^{p-1} = 1 , \qquad b^{-1}ab = a^r$$

where p is an odd prime, and r a primitive number modulo p. Here, too, we show that a group of normalised units in R(G,C) is isomorphic to a subgroup of G.

Finally, in section 7, we consider group rings of groups without elements of finite order. Naturally, the theorems proved are of an entirely different character. Notably, they do not depend at all on the coefficient ring K, provided that it has no zero divisors. We show, in fact, that if G satisfy the condition that every subgroup generated by a finite number of elements of G has a homomorphism on the free cyclic group, then if K has no zero divisors neither has R(G,K), and the units of R(G,K) are all trivial. The condition is satisfied by free groups and by free Abelian groups; and generally, by the direct product and the free product of any two groups that satisfy it.

As we have said, section 3 is a repetition of well-known facts; and the first theorems of section 4 are a particular case of a theorem due to Schilling. The rest of the thesis is original, though some of it has been published previously."

1. <u>Introduction and summary of results</u> (pp. 1-5)

The introductory section is much the same as the abstract, although there are differences in details included and excluded. More attempt is made here to place the work in context:

"The study belongs more particularly to the Theory of Groups, because from the algebraic point of view there is nothing distinctive about the particular non-maximal order R(G,C) - indeed it is quite possible for two distinct groups G, G^1 to have isomorphic algebras, even over the rational field, though their group rings over C are not isomorphic." (p.2).

"The group algebra of a finite group is, of course, well known, and affords one of the most convenient examples of a semi-simple algebra...Considerably less attention has hitherto been paid to integral group rings, or to the group rings of infinite groups." (p.4).

Reference is made to work of Reidemeister [Rei] and Franz [Fra 1,2] as dealing with "particular problems in connection with integral group rings" (p.4) but of "no relevance to the problems with which we are concerned" (p.5). It is interesting to note here that the facts about trivial units for cyclic groups were known to Franz [Fra 3]. He cites as "more relevant ... the work on units in maximal orders in semi-simple algebras" (p.5) in [Schi.] and [Eic]. Although he refers to his Theorem 4 as a particular case of a theorem of Schilling, he finds no overlap in the rest of his work on group rings as orders because "in a maximal order of a general semi-simple algebra there is clearly no analogue of a trivial unit" (p.5). Lastly he thanks his "successive supervisors, Mr J.H.C. Whitehead and Mr P. Hall, for many suggestions" (p.5).

2. <u>Definitions and Preliminary Considerations</u> (pp. 6-10)

Aside from setting out basic definitions, this section repeats the question (still unsettled; see [Den]) of whether right units are left units in group rings, a question raised in his paper. He gives an example of an algebra for which they are not, and remarks "there seems nothing to suggest that this is impossible in a group ring" (p.9).

He also sets down the basic relationships between the group ring of a group and that of a subgroup or factor group. In particular, he shows that, if H is a subgroup of G, then an element of RH is a unit in RG if and only if it is a unit in RH. He views the augmentation homomorphism as a particular case of the natural homomorphism from a group ring onto the group ring of a factor group. He introduces normalised units and observes that "the whole group of units of R(G,K) is the direct product of the group of normalised units and the unit group of K" (p.10).

These notions, as with that of the kernel I(H)RG of the homomorphism from RG to RG/H (p.43), are now taken for granted. In the 1930's they were not so familiar and facility in their use developed slowly and as much in connection with topology and cohomology [Rha, Hop, Fre, Lyn, Fox] as with algebra. Algebraists encountered these notions principally in connection with dimension subgroups, and developed methods based on identities in elements of the augmentation ideal (also used by Higman) rather than on structural relationships among ideals.

3. The Group Algebra of a Finite Group (pp. 11-20)

In this section the representation theory of a finite group over a field K which is an algebraic extension of the rationals, is reviewed for subsequent use. It is done from two points of view, that of the theory of algebras for which his references are [Wae] and [Alb], and that of "the classical theory" (p.12), the reference here being [Spe]. He finds the classical theory, with its explicit formulae connecting the basis of group elements of KG with the basis of primitive central idempotents, as more "in keeping with the nature of the subject" (p.12) because, if R denotes the integers of K, RG is "only defined in terms of the special basis (of group elements)" (p.12).

The result he reports in Theorem 1 is the general version of the result about the decomposition of KG given in his paper for the case in which G is abelian (this specialisation is here Theorem 2). Formulae in the matrix entries of the group elements in the irreducible representations are given in detail and play a large role in section 4.

The section concludes with remarks on isomorphisms of groups and group rings. He cites as an "obvious case" (p.19) of non-isomorphic groups having isomorphic group rings, that of two abelian groups of the same order over any field containing roots of unity of that order [Per-Wal]. He goes on to give an example over the rational field, "and therefore" (p.19) over any field of characteristic 0, that of the two non-abelian groups of order p^3, p an odd prime [Ber 3, Pas 1]. Another early example is given in an exercise in [Bou, p.117], that of the two non-abelian groups of order 8 over fields of characteristic not equal to 2 in which -1 is a square.

He remarks that, in both of his examples, the corresponding integral group rings are not isomorphic as shown by results of section 5. In [Col], there is a mention of this for the second example, and, in [Coh-Liv 1], a full proof.

He finishes the section with a formulation of the isomorphism problem for group rings: "Whether it is possible for two non-isomorphic groups to have isomorphic integral group rings I do not know; but the results of section 5 suggest that it is unlikely" (p.20). This tentative conjecture is repeated later in a stronger form. Subsequent work on the problem usually traces its origin to the problem of determining all groups H for which KG is isomorphic to KH for a given field K, a problem which was "proposed by R.M. Thrall at the Michigan Algebra Conference in the summer of 1947" [Per-Wal, p.420]. It took many years for the problem to be formulated as it is known today. Major impetus towards its study was its inclusion in the survey [Bra] and the text [Cur-Rei, p.262].

4. Integral Group Rings of Finite Groups (pp. 21-32).

One of the main results of his published paper is improved in this section. He classifies those group rings RG which have only trivial units with respect to R as well as with respect to G. While the result is little different, the method is very much so. The exposition is entirely in the context of orders, a concept not present in the paper. The fact that RG is an order in KG does not seem to have been the commonplace then that it is now; for example, expository treatments of orders did not cite this as a standard example. Thus the first item in the section is a proof that RG is an order in KG, the main argument of which shows that each of the elements of RG satisfies a monic polynomial equation with coefficients in R.

In the introductory section he discussed the relevance of the theory of orders to this special case and pointed out its distinguishing features. In this section he exploits such features and is able to prove statements about these orders in a wholly algebraic manner; this is in strong contrast with the analytic and geometric style of the theory for more general orders, particularly in the articles of Eichler and Schilling which he cites.

Having noted that RG is not generally a maximal order in KG ("the element $\frac{1}{n}\sum_{i=1}^{n} g^{(i)}$ is an idempotent of the centrum of R(G,k) and is therefore in every maximal order of R(G,k), but it is not in R(G,C)" (p.21) — the $g^{(i)}$ are the n elements of the group G), he then gives a theorem which measures how far removed from being maximal it is:

"Theorem 3. If k is a finite extension of the rational field, and O is any order of R(G,k) containing R(G,C), then R(G,C) contains the set n O of all multiples of elements in O by the order n of G."(p.22).

In the proof he assumes that the irreducible representations are such "that in each representation all the elements of O are represented by matrices whose elements are integers"(p.22). He then needs only remark that, from the classical formulae reviewed in the previous section, each of the standard basis elements E_{ij} of the matrix algebras in the representations corresponds to an element of KG having coefficients which are integers times $1/n$. For the initial assumption he refers to [Spe, Ch. 14, § 65] which, however, applies only to group elements and so to RG, not O (Speiser's methods are those of group matrices and determinants). He goes on to give his own proof:

"We may suppose Γ to be a representation in some finite algebraic field k' containing k. Let O' be the order in R(G,k') consisting of all elements $\sum_i c_i O_i$ where c_i are integers in k', and O_i are elements in O. Let $e_1, \ldots e_s$ be the unit vectors of the vector space in which the matrices of $\Gamma(G)$ act, and consider the vectors

$$(1) \qquad Xe_1 = x_{11}e_1 + x_{12}e_2 + \ldots + x_{1s}e_s,$$

for all matrices X representing elements of O'. Call a vector (1) of length r if $x_{1t} = 0$ for $t > r$ and consider the set I_r of all r-th coefficients x_{1r} of vectors of length r. For each r, I_r is fractional ideal in k' distinct from the zero ideal. For I_r plainly contains the difference of any two of its elements, and the product of any one of them by an integer in k'. Moreover we can find a number m_r such that any element of $m_r I_r$ is an integer. If $r = 1$, we have $Xe_1 = x_{11}e_1$ so that $x - x_{11}$ is a factor of the characteristic function of X. Since O' is an order, this function has integral coefficients, and therefore x_{11} is an integer, so that we may take $m_1 = 1$. Now since Γ is irreducible, we can, by Burnside's Theorem[2] choose an element E_r in R(G,C'), and therefore in O', such that its image in Γ is $m_r E_{1r}$, for some number m_r, where $E_{1r} e_r = e_1$, $E_{1r} e_k = 0$, $k \neq r$.
Then by (1) we have

$$m_r E_{1r} \cdot Xe_1 = m_r x_{ir} e_1,$$

so that $m_r x_{1r}$ must be integral, as required. Thus I_r is an ideal in k'. If it were the zero ideal, there would be less than s linearly independent vectors in (1), contrary to the hypothesis that Γ is irreducible. Since k' is a finite extension of the rational field, we can find an extension k" of k', such that the ideals I_r^1 in k", having the elements of I_r as basis, are principal ideals $(\alpha_r)^1$. Extending the order O' again to O" in R(G,k") and supposing X in (1) to be chosen now from the matrices representing elements of O", we have that there are vectors $f_1, f_2, \ldots, f_s$ in the set (1) such that

$$f_r = f_{r1}e_1 + \ldots + f_{r\,r-1}e_{r-1} + \alpha_r e_r .$$

Then every vector of (1) can be written as $\sum_r y_r f_r$ with integral coefficients y_r. Choosing $f_1, f_2, \ldots, f_s$ as a new basis for the vector space, we obtain an equivalent representation Γ' with the desired properties.

(2) Cf. van der Waerden, op.cit. Ch. XVII, § 121.

(1) See Hilbert, Jahresberichte der Deutschen Mathematiker-Vereinigung, 4, p. 224, (1897)." (pp. 23-24).

With the groundwork completed, he can readily establish the theorem, Theorem 5, which he needs for the classification of group rings with all units trivial. Theorems 4 and 6 are ancillary to it, the latter being proved by a lemma which is restated as Theorem 10 in the subsequent section.

"Theorem 4. If C is the integer ring of a finite extension k of the rational field, then the unit group of R(G,C) has a finite index in the unit group of any order O of R(G,k) containing R(G,C)." (pp. 24-25).

"Theorem 5. A necessary and sufficient condition that all the units of R(G,C) be trivial, is that a maximal order of R(G,k) containing R(G,C) have a unit group of finite order." (p.26).

"Theorem 6. The group of trivial units in R(G,C) is not contained as a proper subgroup of finite index in any group of units of R(G,C)." (p.26).

Having transformed the question into one about maximal orders, he can apply known facts about orders in full matrix rings, fields and division algebras to narrow down the candidates to those of the four cases (as listed in the Abstract). The fourth case requires extra discussion as in the paper. The section ends (there being no Theorem 7) with the statement of the main theorem and its extension to the case of torsion groups:

"Theorem 8. The group ring R(G,C) of a finite group G over a ring of algebraic integers C has only trivial units if and only if G and k, the quotient field of C, have one of the forms I, II, III, IV, listed above." (p. 31).

"Theorem 9. The statement of Theorem 8 remains true if for the words "a finite group G" we substitute "a group G all of whose elements have finite order."" (p.32). For G abelian, Theorem 8 appears in [Coh-Liv 2].

5. Units of Finite Order in Integral Group Rings (pp. 33-51)

This section opens with a stronger form of the isomorphism problem. It and the next are concentrated on providing examples to support this "plausible theorem" (p.33). He states it as follows:

"The theorems that we prove are all partial cases of the plausible theorem:- A group of units of finite order in R(G,C) is isomorphic to a group of trivial units" (p.33).

The "starting point"(p.33) for the results here, a lemma in the previous section, is repeated as Theorem 10:

"Theorem 10. The coefficient of the identity of G in a non-trivial unit of finite order in G is zero." (p.33).

It is a theorem usually attributed to Berman [Ber 1] (see [Tak] also). Berman was the next investigator to study units in group rings systematically and so to rederive results already discovered by Higman. There has been a similar ignorance of the extent of Berman's contributions, partly due to the relative inaccessibility of his publications and to insufficient reviewing. A case in point is his theorem that the integral group ring determines the class sums of the group. This and other Soviet contributions to group ring problems are described in [Seh 2] (see also [Bov] and [Zal-Mih]).

Higman's proof is the same as that which is now considered standard:

"Let E be a unit of finite order in R(G,C), in which the coefficient of the identity is the non-zero integer a. Let $\bar{E}$ be the image of E in the regular representation of G. We have trace $\bar{E}$ = na, where n is the order of G. The characteristic roots of $\bar{E}$ are roots of unity, $w_1, \ldots, w_n$, and therefore

$$(1) \qquad |\text{trace } \bar{E}\,| = |\, w_1 + w_2 + \ldots + w_n| \leq |w_1| + |w_2| + \ldots + |w_n| = n$$

and the same is true for all the conjugates of trace $\bar{E}$. Therefore $|N(a)| \leq 1$, where N(a) stands for the norm of a, and is therefore a non-zero rational integer. Thus equality must hold here, and therefore also in (1). That is to say

$$|w_1 + w_2 + \ldots + w_n| = |w_1| + |w_2| + \ldots + |w_n|,$$

which can only happen if $w_1 = \ldots = w_n$. Then, however, the matrix $\bar{E}$, being of finite order, must be scalar, so that E = w.1, and therefore E is trivial, which proves the Lemma." (p.27).

The next two theorems follow readily from Theorem 10. The first, a result in [Coh-Liv 2], is a "quantitative" (p.33) theorem:

"Theorem 11. <u>The elements of a group of normalised units of finite order in R(G,C) are linearly independent. The order of such a group is at most equal to the order of G.</u>" (p.33).

The second is "qualitative" (p.35):

"Theorem 12. <u>The prime factors of the order of a group of normalised units of finite order in R(G,C) divide the order of G.</u>" (p.35).

For G soluble, this is improved to:

"Theorem 13. <u>If the group G is soluble and of order n, then the order of any normalised unit of finite order in R(G,C) divides n.</u>" (p.37).

The proof, as he records in a Corollary (p.42), shows rather more, namely that the order divides the product of the exponents of the quotients of any series of normal subgroups of G having abelian quotients. The proofs of Theorem 12 and 13 involve matrix and algebraic number calculations as well as an induction argument in the latter. When rediscovered, these results were in a more precise and general form. In [Ber 1], Theorem 13 is established for any finite group. In [Coh-Liv 2], it is shown that the exponent of a group of normalised units divides the exponent of G; by symmetry, the exponent of a group basis equals that of G, a result also in [Pas 1].

Specific non-trivial units have always been difficult to find. The most familiar, a unit of infinite order in the group ring of the cyclic group of order 5, was one of the first to be recorded; it is attributed to Mordell in that part of Whitehead's paper [White., p.284] which makes reference to Higman's yet to be published calculations for certain cyclic groups. Many units of finite order are given in the thesis. To emphasise the significance of the ring of coefficients being one of algebraic integers, he provides a unit of order 3 in the rational group algebra of the quaternion group, namely, $1 - \frac{1}{4}(1 - xyz)(3 + x + y + z)$ where $G = \langle x, y : x^2 = y^2 = z^2 = xyz \rangle$ (p.36). He considers it "a corollary to Theorem 10" (p.36) that a non-trivial unit of finite order has no central elements in its support [Ber 2], and that therefore there are no examples of non-trivial units of finite order to be found in abelian groups, the result in his paper. Berman had stated a stronger form of this corollary in [Ber 1] and used it to draw this same conclusion; note, however, that the results of [Ber 1] are for the ring of rational integral coefficients while many of those in [Ber 2] are stated for rings of algebraic integers.

He gives an infinite family of units of order 2 for the non-abelian group of order 6 (namely, $y + k(x - x^2)(1 + y)$, where k is a rational integer and $G = \langle x, y : x^3 = y^2 = (xy)^2 = 1 \rangle$ (p.36)). At the time, Taussky also worked on units in group rings [Tau 2] and discovered, among other things, various units for this group although she had no occasion to publish them before [Tau 1]. The two units of order 2 which she gives correspond to Higman's units with $k = 1$ and $k = -1$ for a suitable choice of generators. Her method of group matrices, and, in particular, circulants in the case of a cyclic group, reproves Higman's theorem on units for abelian groups in the cyclic case and provides explicit non-trivial units when they exist [New-Tau]. As reported in [Tau 1], Iwasawa constructed many units for the non-abelian group of order 6. His constructions were methodical and yielded Higman's units and many more, for example, $k(x - x^2) + ((1 + k^2) + k^2x - 2k^2x^2)y$. He also noted that his methods gave units for any dihedral group of order at least 6, namely, $\alpha + (1 \pm \alpha)y$ where $\alpha = k(x - x^{-1})$. (My thanks to Professors Taussky-Todd and Iwasawa for their assistance.)

Higman included these examples in the non-abelian group of order 6 to show that units of finite order need not be conjugate to trivial units. This he established by examining their images in an integral representation of the group and by observing that, for odd k, the matrix representing the unit above could not be similar to any of the matrices representing group involutions. It is reported in [Ber 4] that this point was demonstrated in [Ros]; her example is presumably that for the dihedral group of order 8 published in [Ber-Ros] (my thanks to G. Karpilovsky for the details of this reference). According to [Hug-Pea], Zassenhaus also raised this question and it served as one of the motivations for [Hug-Pea]; Hughes and Pearson answer it with a unit which is one of Higman's. Recently much interest has focussed on the question of conjugacy in the larger rational group algebra and of conjugacy of group bases; these are questions that Higman does not mention.

The second half of the section turns to the proof that, for a nilpotent metabelian group, a group of units of finite order is isomorphic to a group of trivial units. His methods are "rather different methods from those used in proving the previous theorem" (p.43). They involve manipulation of the ideals $\lambda_H = I(H)RG$ where H is a normal subgroup of G. Before proving the theorem, Theorem 14, he gathers together some facts about these ideals:

"Lemma 1. If $x_1,\dots,x_r$, of orders $n_1,\dots,n_r$ are a basis of the Abelian group G, then

$$x = \mu_1(x_1 - 1) + \mu_2(x_2 - 1) + \dots + \mu_r(x_r - 1) \equiv 0\ (\lambda_G^2) \quad (\mu_\alpha \text{ integers})$$

if and only if μ_α is divisible by n_α, $\alpha = 1,\dots,r$." (p.43).

"Lemma 2. If ϕ is an element of R(H,C), where H is a self-conjugate subgroup of G, then $\phi \equiv 0\ (\lambda_G\lambda_H)$ if and only if $\phi \equiv 0\ (\lambda_H^2)$." (p.44).

"Corollary to Lemmas 1 and 2. Let G be a group, H a self conjugate subgroup of G. Then if g is an element of G

(i) $g \equiv 1 \mod \lambda_G^2$ if and only if g is in G';

(ii) $g \equiv 1 \mod \lambda_G\lambda_H$ if and only if g is in H';

(iii) $g \equiv 1 \mod (\lambda_G\lambda_H, \lambda_G\lambda_H)$ if and only if g is in (G,H)." (p.46).

Part (iii), in which $(\lambda_G\lambda_H, \lambda_H\lambda_G)$ means I(G)I(H) + I(H)I(G) and (G,H) denotes the subgroup generated by the commutators of elements of G with those of H, is included to accommodate the fact "that multiplication of the ideals λ_H is not in general commutative" (p.45).

Lemma 1 is presently understood as asserting an isomorphism between G/G' and $I(G)/I(G)^2$, G' being the commutator subgroup of an arbitrary group G. This appears in [Hop, pp. 49-50], [Fre, p.291] and [Coh], where it is proved in a basis-free manner. (My thanks to M. Oganjuren for the Hopf reference). He also gives the formula

$$\text{"}\ x_1^{a_1}x_2^{a_2}\dots x_r^{a_r} \equiv 1 + a_1(x_1 - 1) + \dots + a_r(x_r - 1),\ (\lambda_G^2)\ \text{"} \quad (p.45);$$

this would establish the isomorphism between G/G' and $I + I(G)/I(G)^2$ [San 3].

Part (ii), that $H' = G \cap 1 + I(G)I(H)$, is proved for free groups in [Fox; Gru 1] where it is attributed to [Schu. 1]. There is a characterisation of H', based on what are now known as Fox derivatives, in [Schu. 1] and in [Bla; Lyn] to whom Fox also refers the result. But the ideal I(G)I(H) does not enter into these articles. (Schumann's treatment of this point is more clearly and concisely expressed in [Schu.2]. My thanks to R. Griess for providing me with [Bla], an undergraduate thesis which translates [Schu. 1] into the language of Fox derivatives.)

This identity is the key step in establishing the isomorphism between H/H' and $\mathbb{Z}GI(H)/I(G)I(H)$. Although Higman does not state this isomorphism, he effectively uses it in the case H = G' in his proof of Theorem 14; in the case H = G, this is the previous isomorphism. For G free, the isomorphism appears in [Gru 1] and, for the general case, in [Oba 1; Whitc.]. Both groups may be interpreted as right $\mathbb{Z}G$-modules; that the isomorphism respects this structure is discussed in [Gru 1, 2; Oba 2;

San 1, 2; Gru-Ros]. A consequence of this observation is that $H/[H,G]$ is isomorphic to $\mathbb{Z}GI(H)/(I(G)I(H) + I(H)I(G))$, a fact related to part (iii) of Higman's Corollary.

In view of the subsequent history of the isomorphism problem in the metabelian case, it is instructive to quote the proof of Theorem 14 for the case of rational integral coefficients:

"Theorem 14. Let G be a group whose second derived group consists of the identity alone, and the c-th group of whose upper central series is the whole group G. Then a group of normalised units of finite order in R(G,C) is isomorphic to a subgroup of G.

By Theorem 11, a group of normalised units of finite order in R(G,C) is finite; it therefore suffices to prove theorem 14 in case the quotient field of C is a finite extension of the rational field.

Let E be a unit of finite order in R(G,C). The image of E in the natural homomorphism of R(G,C) on R(G/G',C) is a trivial unit in the latter group ring, since G/G' is Abelian. If furthermore E is normalised, then this image is an element of G/G'. Let g_1 be any element of this residue class modulo G'. We have then

$$E = g_1 + \sum_{i=1}^{r} \phi_i(h_i - 1) ;$$

where $h_1,\ldots,h_r$, of orders $n_1,\ldots,n_r$, are a basis of the Abelian group G'. If the sum of the coefficients of the elements in ϕ_i is ρ_i then $\phi_i \equiv \rho_i \mod \lambda_G$, and therefore

(5) $$E \equiv g_1 + \sum_{i=1}^{r} \rho_i(h_i - 1) \mod \lambda_G\lambda_{G'}.$$

Let us for the sake of clarity first suppose that we are dealing with the ring of rational integers as coefficient ring C, and treat the general case afterwards. Then from equation (5) we have,

$$\begin{aligned} E &\equiv g_1 + \sum_{i=1}^{r} \rho_i (h_i - 1) \mod \lambda_G\lambda_{G'} \\ &\equiv g_1 + \sum_{i=1}^{r} (1 + h_i + \ldots + h_i^{\rho_i - 1})(h_i - 1) \mod \lambda_G\lambda_{G'}, \\ &\equiv g_1 + \sum_i g_1 h_1^{\rho_1} \ldots h_{i-1}^{\rho_{i-1}} (h_i^{\rho_i} - 1) \mod \lambda_G\lambda_{G'}, \\ &\equiv g_1 h_1^{\rho_1} \ldots h_r^{\rho_r} \mod \lambda_G\lambda_{G'}. \end{aligned}$$

That is to say, every normalised unit E of finite order in R(G,C) satisfies a congruence

(6) $$E \equiv g \mod \lambda_G\lambda_{G'}.$$

This congruence determines g uniquely. For by the corollary to Lemmas 1 and 2, $g \equiv g' \mod \lambda_G\lambda_{G'}$ implies that gg'^{-1} is in the derived group of G', – that is to say, is the identity. Moreover, if we have also $E_1 \equiv g_1 \mod \lambda_G\lambda_{G'}$, then

$$EE_1 \equiv gg_1 \mod \lambda_G\lambda_{G'} ,$$

so that the correspondence $E \to g$ provides a homomorphism of any group of normalised units of finite order on a subgroup of G. To show that it is an isomorphism, we must show that if

$$E \equiv 1 \mod \lambda_G\lambda_{G'}, \tag{7}$$

and E is a unit of finite order, then E = 1. By hypothesis G has the upper central series $Z_o = 1, Z_1, \ldots, Z_c = G$. We proceed by induction on c; as if c = 1, G' consists of the identity alone, $\lambda_{G'}$ is the ideal (0), and the assertion is therefore trivial. Assume then that the assertion is true of the factor group G/Z_1. Let $\bar{E}$ be the unit of $R(G/Z_1, C)$ corresponding to E. Then (7) implies

$$\bar{E} \equiv 1 \mod \lambda_{G/Z_1} \lambda_{(G/Z_1)'}$$

and so by hypothesis $\bar{E} = 1$, or $E \equiv 1 \mod \lambda_{Z_1}$. It follows that the sum of the coefficients of elements of Z_1 in E is 1, so that not all of these coefficients are zero. Since Z is the centrum of G, it follows from Theorem 10 that E is trivial, say E = z. But we have already shown that (6) determines g uniquely, and therefore by (7), z = 1. Thus Theorem 14 is true if C is the ring of rational integers." (pp.46-48).

In the case of a more general coefficient ring C, a basis $w_o = 1, w_1, \ldots, w_s$ of C over $\mathbb{Z}$ is chosen and the fact that CG is the direct sum of the $\mathbb{Z}Gw_i$ used to allow the proof to proceed along similar lines (pp.49-50).

Jackson, a student of Higman, set out to remove the nilpotency restriction in Theorem 14 in [Jac 1, 2] (his paper and thesis are much the same, the paper having an additional section on nilpotent groups). Jackson follows the lines of the above proof but his concluding steps (reduction argument, use of an integral representation in a special case) are much more involved and are not generally understood.

Recent authors have written on various classes of metabelian groups, often using ideas like those of Jackson, to show that, in the group of normalised units, the subgroup G has a torsion-free normal complement. For a group ring with this property, it is immediate that Higman's strong form of the isomorphism conjecture is correct. The following cases have yielded a positive result: the symmetric group of degree 3 [Den], the alternating group of degree 4 [All-Hob], dihedral groups of odd degree [Miy], groups having a normal abelian subgroup of index 2 [Pas-Smi], groups with a commutator factor group of exponent 2, 3, 4 or 6 [Sek; Cli-Seh-Wei; Rog] or of odd exponent [Cli-Seh-Wei] (this last was reported on at this Tagung).

Higman concludes the section with Corollaries asserting the correctness of the isomorphism conjecture for finite metabelian nilpotent groups. The case of finite metabelian groups was proved in [Whitc.]. All the elements of Whitcomb's proof appear in the above proof of Theorem 14. Whitcomb is able to derive his conclusion using these arguments alone by means of symmetry as he deals only with group bases and not with arbitrary groups of normalised units of finite order (the symmetry argument is spelled out in [Seh 1]). The special case of a nilpotent group of class 2 was resolved in [Pas 1] and in [Sak] for a more general coefficient ring.

6. An Example (pp.52-62)

"In this section we consider the detailed application of the theorems of the last three sections to a particular group ring. We take one of the simplest groups not of prime power order, namely the group G generated by two elements a, b, subject to the relations:-

$$a^p = b^{p-1} = 1; \quad b^{-1}ab = a^r ;$$

where p is an odd prime, and r a primitive number modulo p" (p.52).

The section illustrates certain techniques of Higman in studying representations and provides an early example of an explicit integral representation.

He begins by describing the absolutely irreducible representations of G : $p-1$ linear representations and one of degree $p-1$, which he obtains as a summand of the linear representation corresponding to the doubly transitive permutation representation of G on the cosets of the subgroup generated by b. These provide a decomposition of KG into the direct sum of two algebras : A_1, isomorphic to the group ring KG^* where G^* is the factor group of G with respect to the (normal) subgroup generated by a; and A_2, "the total matrix algebra of index $p-1$" (p.53). K and its ring of integers R are not explicitly defined but K can be taken to be any field of algebraic numbers containing $(p-1)$-th roots of unity. A_1 is obtained by uniting "the components corresponding to representations of degree 1, into a single component" (p.53). This is a technique he had introduced in section 3 and used there in the example of non-isomorphic groups having isomorphic rational group algebras. It allows the decomposition to be effected over smaller fields such as the rational field.

Most of the section is taken up with locating the integral group ring RG in this decomposition. The "first component" (p.54) of an element of RG (that is, its component in A_1) is in RG^*; any member of RG^* is a first component of such an element. Thus, interest centres on the second components. Here the problem is finding a basis for the representation, an integral one, in which the representing matrices are in "the most convenient form" (p.54).

Before giving an explicit basis, he analyses the representation theoretically:

"an element in $R(G,k)$ with first component zero, and second component a matrix whose elements are integers divisible by p is in $R(G,C)$, provided only that our representation has been chosen so that the matrices representing elements of G have integral elements. That is to say, what we have to consider is the form the representation takes modulo p. Now over the Galois field of order p, the group algebra of G ceases to be semi-simple. We have in fact $(a-1)^p \equiv a^p-1 = 0 \mod p$; and since, if g is any element in G, $g(a-1) = (a^s-1)g = (a-1).(a^{s-1}+\ldots+a+1)g$, the ideal $(a-1)$

is nilpotent. As a matter of fact it forms the radical of the group algebra, and therefore the quotient algebra with respect to the radical is isomorphic to the group algebra of a cyclic group of order $p-1$. It is, in particular, commutative, and any representation of G in the Galois field is therefore equivalent to a representation by matrices whose elements above the main diagonal are zero[1].

(1) Cf. B.L. van der Waerden, Moderne Algebra, vol. II, ch. XVII, § 121." (pp.54-55). The radical of such a group algebra was not known in general at the time; it was determined in the case of a normal Sylow p-subgroup in [Bru; Lom].

His first approximation to a convenient basis is that obtained from the permutation representation; its elements are denoted $x_1, x_2, \ldots, x_{p-1}$ and correspond to the cosets of a^i, $i = 1, 2, \ldots, p-1$. By realising each pE_{ij} as representing an element of RG, he concludes that all elements of KG having zero first component and second component of the form pA, A an integral matrix, are in RG.

Next he changes basis

$$\text{"}\ y_\alpha = \sum_{\beta=1}^{p-1} \binom{p-\beta}{a} x_\beta\,, \quad \alpha = 1,\ldots, p-1\ \text{"} \qquad \text{(p.56).}$$

He notes that, by properties of the binomial coefficients, this is effected by "an integral unimodular matrix" (p.57). Using this basis, he finds that, modulo p, the second components of a and b are given respectively by the matrices

$$\text{"}\left\|\begin{matrix} 1 & 0 & & & & 0 \\ 1 & 1 & \cdot & & & \\ & 1 & \cdot & \cdot & & \\ & & \cdot & \cdot & & \\ & & & \cdot & 1 & 0 \\ 0 & & & & 1 & 1 \end{matrix}\right\| \quad \text{and} \quad \left\|\begin{matrix} r^* & 0 & & 0 \\ * & r^{*2} & & \\ & & \cdot & \\ & & & \cdot \\ & & & \cdot\ 0 \\ * & & * & r^{*\,p-1} \end{matrix}\right\|\text{"} \qquad \text{(p.58).}$$

(here r^* is the multiplicative inverse of r modulo p).

Further analysis then leads to the conclusion:

"<u>An element of $R(G,k)$ is in $R(G,C)$ if and only if its component in A_1 is $\phi(b^*)$ in $R(G^*,C)$, and its component in A_2 is a matrix with integer elements, those above the main diagonal being congruent to zero, those in it to $\phi(r^*)$, $\phi(r^{*2})$, ..., $\phi(r^{*\,p-1})$ respectively modulo p</u>" (pp.59-60).

Here b^* is the image of b in the factor group G^* and $\phi(b^*)$ is the expression for an element of A_1 as a function ϕ of b^*, the function ϕ then defining the diagonal entries of the component in A_2 as described. It is easy to see (and Higman observes on p.59) that the projection of $\mathbb{Z}G$ with respect to the primitive central idempotent corresponding to A_2 is the ring of matrices

$$\begin{pmatrix} \mathbb{Z} & p\mathbb{Z} & & & p\mathbb{Z} \\ & \mathbb{Z} & \ddots & & \\ & & \ddots & \ddots & \\ & & & \ddots & p\mathbb{Z} \\ \mathbb{Z} & & & & \mathbb{Z} \end{pmatrix},$$

an important hereditary order.

The group G is metacyclic; the integral representation of the triangular form given here is a special case of that in [Gal-Rei-Ull]. For $p = 3$, Higman has already provided such a representation in section 5, a and b being sent respectively to

$$\begin{pmatrix} 1 & -3 \\ 1 & -2 \end{pmatrix} \quad \text{and} \quad \begin{pmatrix} 1 & 0 \\ 1 & -1 \end{pmatrix} .$$

Up to signs and choice of generators, these same matrices are used in [Al-S] and by Roggenkamp in lectures on [Rog]; on the other hand, the representation in [Hug-Pea] is not triangular.

From the description of RG in the decomposed KG, conclusions are drawn about the units of RG:

"We have then obviously:-

The group U of units in R(G,C) has a self-conjugate subgroup U_o consisting of all units whose component in A_1 is 1. U_o is isomorphic to the group of all matrices with determinant unity and elements in C, those above the main diagonal being congruent to zero and those in it to unity modulo p. The factor group U/U_o is isomorphic to the unit group of $R(G^*,C)$." (p.60).

Next he describes normalised units and explicitly verifies that a group of normalised units of finite order is isomorphic to a subgroup of G. The section ends with a calculation showing that the same is true for the integral group ring of any subgroup of G which contains a and in particular for "the dihedral group of order 2p" (p.62). In [Hug-Pea] the units are described in the case $p = 3$ with rational integer coefficients. In [Col], there is some discussion of the isomorphism problem for a wider class of metacyclic groups than that treated here.

7. Group Rings of Groups without Elements of Finite Order (pp.63-78)

The question of which torsion free groups have trivial Whitehead group is one which draws the active interest of many mathematicians not primarily algebraists. Having trivial Whitehead group implies having only trivial units in the integral group ring. In his paper, Higman had identified a large class of torsion-free groups with the latter property, those which he called "indicable throughout". This term means that every non-trivial subgroup maps homomorphically onto the infinite cyclic group.

The term is not used in the thesis, its place being taken by Condition II which is the same except for the refinement that only finitely generated subgroups need be surveyed (as this is sufficient for the application). In the paper, he showed the class to be closed under extensions and free products and to include such important groups as free and free abelian groups. He also recognised that the nature of the ring of coefficients was unimportant as long as it had no zero divisors and that, in this case, RG had no zero divisors.

The class of groups satisfying Condition II behaves the same way. It is contained in a larger class introduced in the thesis, that described by Condition I (p.65), a class again with the same formal properties and the same consequences for the group ring. These formal properties for the two conditions (closure with respect to extensions, direct products and free products) are the assertions of the final numbered theorems of the thesis, Theorems 17, 18 and 19 respectively. Theorems 15 and 16 state that RG has no zero divisors and no non-trivial units if G satisfies Conditions I and II respectively.

Condition I, given in an equivalent form below, is a complicated one involving what he calls the isolated product set of subsets of the group. He finds it "very unwieldy in application" (p.78) and could not determine whether it held for groups which did not satisfy Condition II (an example of such a group is obtained from $G = \langle x, y : x^3 = y^2 \rangle$ as (p.77) the free product of two copies of G with the subgroup generated by yx^{-1} and $(yx^{-1})^6 x^{-3}$ amalgamated; this is a finitely generated torsion-free perfect group, and this suffices).

The relevance of Condition I to the group ring is to be found in the

"Lemma. If in any two non-vacuous finite subsets X, Y of G, we can find a pair x, y of elements having an isolated product, then R(G,K) has no zero divisors. If furthermore whenever X or Y has more than one element, we can choose two such pairs, then R(G,K) has only trivial units." (p.64).

Here "isolated product" means that the element xy can be expressed as a product of an element from X and one from Y in no other way. At present the term "unique product" is used, and a group satisfying the first hypothesis is called a unique product (or u.p.) group [Rud-Sch]. The second hypothesis defines what is now called a two unique products (or t.u.p.) group; it appears in [Kem] and is applied to group rings in [Sch-Wei]. Higman states (p.65) that the second hypothesis is equivalent to Condition I. He also writes "Whether or not the hypothesis of the second half of the lemma follows from the hypothesis of the first half, I do not know" (p.65). It has recently been established in [Str] that the two are equivalent.

Much of the section is taken up in providing additional illustrations of groups satisfying Condition II. He gives three types. The first (p.72) is a group generated by elements $a_1,\dots,a_n$ subject to relations, for $s = r + 1,\dots,n$ (r fixed), of the form: $a_s^{\lambda_s}$ is a non-trivial word in $a_1,\dots,a_{s-1}$. He cites as "the simplest case of the above type (the group $\langle x, y : x^m = y^n \rangle$ which) if m is prime to n (is) the group of the torus knot" (p.74).

Secondly he takes "another class of knots, - those formed by the process of doubling a simple circuit - their groups are generated by generators a, b, subject to the relation

$$a^{2n+1} = b\,a^{n}\,b^{-2}\,a^{n}\,b \text{" (p.74).}$$

"As a third example, consider the group G generated by $a_1, a_2, \ldots$ ad inf., subject to

$$a_{i-1} = a_i\,a_{i+1}\,a_i^{-1}\,a_{i-1}^{-1}, \quad i = 2, 3, \ldots \text{ ad inf." (p.76).}$$

He shows that G is perfect, locally free and imbeddable in a finitely presented group which also satisfies Condition II.

Towards the end of the chapter comes a conjecture which has proven to be an important focus in the development of the theory of group rings:

"It is obvious that neither condition I nor condition II can hold in a group having elements of finite order. It is a plausible hypothesis, however, that if a group has no elements of finite order then its group ring is without zero divisors and has only trivial units." (p.77).

Acknowledgement

My thanks to Bridget Iveson for her excellent typing.

Bibliography

Citations in the text give the first three (or more) letters of the name(s) of the author(s). Multiple articles are numbered according to the ordering in this bibliography. References are supplied here to reviews from Mathematical Reviews and occasionally from Zentralblatt für Mathematik und ihre Grenzgebiete and Jahrbuch über die Fortschritte der Mathematik.

Albert, A. A., Structure of Algebras, Amer. Math. Soc., New York, 1939. Zbl.23.199.

Allen, P. J.; Hobby, C., A characterization of the units in $Z[A_4]$, J. Algebra, to appear.

Al-Sohebani, A.A., On the units of certain group rings, Ph.D. thesis, Univ. of Birmingham, 1978.

Berman, S. D., On certain properties of integral group rings. (Russian), Dokl. Akad. Nauk SSSR (N. S.)91(1953), 7-9. MR 15-99.

Berman, S. D., On the equation $x^m = 1$ in an integral group ring. (Russian), Ukrain. Mat. Ž. 7(1955), 253-261. MR 17-1048.

Berman, S. D., On certain properties of group rings over the field of rational numbers. (Russian), Užgorod. Gos. Univ. Naučn. Zap. Him. Fiz. Mat. 12(1955), 88-110. MR 20 #3920.

Berman, S. D., Representations of finite groups. (Russian), Algebra 1964, p.83-122, Akad. Nauk SSSR Inst. Naučn. Informacii, Moscow, 1966 = Itogi Nauki - Seriya Matematika. Alg. Top. Geom. 3. MR 25 #2973.

Berman, S. D.; Rossa, A. R., Integral group rings of finite and periodic groups. (Russian), Algebra and Math. Logic : Studies in Algebra. (Russian), p.44-53, Izdat. Kiev. Univ., Kiev, 1966. MR 35 #265.

Blanchfield, R. C., Application of free differential calculus to the theory of groups, Senior thesis, Princeton Univ., 1949.

Bourbaki, N., Algèbra Linéaire, Hermann, Paris, 1947. MR 9-406.

Bovdi, A. A., Group Rings. (Textbook). (Russian), University publishers, Užgorod, 1974. Zbl.339.16004.

Brauer, R., Representations of finite groups, Lectures on Modern Mathematics, Vol. I, pp.133-175, Wiley, New York, 1963. MR 31 #2314. Also in: Richard Brauer : Collected Papers, Vol. I-III, The MIT Press, Cambridge MA, 1980.

Brummund, H., Über Gruppenringe mit einem Koeffizientenkörper der Characteristik p, Dissertation, Münster 1939, Emsdetten, Heinrich und J. Lechte, Münster, 1939. Zbl.22.119. F.d.M.65_{II},1136.

Cliff, G. H; Sehgal, S. K.; Weiss, A. R., Units of integral group rings of metabelian groups, to appear.

Cohn, J. A.; Livingstone, D. On groups of the order p^3, Canad. J. Math. 15(1963), 622-624. MR 27 #3700.

Cohn, J. A.; Livingstone, D., On the structure of group algebras. I., Canad. J. Math. 17(1965), 583-593. MR 31 #3514.

Cohn, P. M., Generalization of a theorem of Magnus, Proc. London Math. Soc. (3)2(1952), 297-310. MR 14-532.

Coleman, D. B., Finite groups with isomorphic group algebras, Trans. Amer. Math. Soc. 105(1962), 1-8. MR 26 #191.

Curtis, C. W.; Reiner, I., Representation theory of finite groups and associative algebras, Interscience - Wiley, New York, 1962. MR 26 #2519.

Dennis, R. K., The structure of the unit group of group rings, Ring theory, II, pp.103-130, Dekker, New York, 1977. MR 56 #3047.

Eichler, M. Über die Einheiten der Divisionsalgebren, Math. Ann. 114(1937), 635-654. Zbl.17.244.

Farkas, D. R., Group rings: an annotated questionnaire, Comm. Algebra 8 (1980), 585-602.

Fox, R. H., Free differential calculus. I. Derivation in the free group ring, Ann. of Math. (2)57(1953), 547-560. MR 14-843.

Franz, W., Überdeckungen topologischer Komplexe mit hyperkomplexen Systemen, J. Reine Angew. Math. 173(1935), 174-184. Zbl.12.127.

Franz, W., Über die Torsion einer Überdeckung, J. Reine Angew. Math. 173(1935), 245-254. Zbl.12.127.

Franz, W., (review of Whitehead, J. H. C., Simplicial spaces, nuclei and m-groups), Zbl. Math. 22(1940), 407-408.

Freudenthal, H., Der Einfluss der Fundamentalgruppe auf die Bettischen Gruppen, Ann. of Math. (2)47(1946), 274-316. MR 8-166.

Galovich, S; Reiner, I.; Ullom, S., Class groups for integral representations of metacyclic groups, Mathematika 19(1972), 105-111. MT 48 #4087.

Gruenberg, K. W., Resolutions by relations, J. London Math. Soc. 35(1960), 481-494. MR 23 #A 3165.

Gruenberg, K. W., Some cohomological topics in group theory, Queen Mary College, London, 1967. MR 41 #3608.

Gruenberg, K. W.; Roseblade, J. E., The augmentation terminals of certain locally finite groups, Canad. J. Math. 24(1972), 221-238. MR 45 #3538.

Higman, G., The units of group-rings, Proc. London Math. Soc. (2)46(1940), 231-248. MR 2-5. Zbl.25.243. F.d.M. 66_I, 104.

Higman, G. Units in group rings, D. Phil. thesis, Oxford Univ., 1940.

Higman, G., (Review of Passman, Isomorphic groups and group rings), Math. Reviews 33(1967), 244.

Hopf, H., Über die Bettischen Gruppen, die zu einer beliebigen Gruppe gehoren, Comment. Math. Helv. 17(1945), 39-79. MR 6-279.

Hughes, I.; Pearson, K. R., The group of units of the integral group ring ZS_3, Canad. Math. Bull. 15(1972), 529-534. MR 48 #4089.

Jackson, D. A., On a problem in the theory of integral group rings, D. Phil. thesis, Oxford Univ., 1967/1968.

Jackson, D. A., The groups of units of the integral group rings of finite metabelian and finite nilpotent groups, Quart. J. Math. Oxford Ser. (2)20(1969), 319-331. MR 40 #2766.

Kemperman, J. H. B., On complexes in a semigroup, Nederl. Akad. Wetensch. Proc. Ser.A. 59 = Indag. Math. 18(1956), 247-254. MR 18-14. MR 19-13.

Lombardo-Radice, L., Intorno alle algebre legate ai gruppi di ordine finito. Nota 2[a], Rend. Sem. Mat. Roma (4)3(1939), 239-256. MR 1-258. Zbl.23.103.

Lyndon, R. C., Cohomology theory of groups with a single defining relation, Ann. of Math. (2) 52 (1950), 650-665. MR 13-819.

Miyata, T., On the units of the integral group ring of a dihedral group, J. Math. Soc. Japan, to appear.

Newman, M.; Taussky, O., On a generalization of the normal basis in abelian algebraic number fields, Comm. Pure Appl. Math. 9(1956), 85-91. MR 17-829.

Obayashi, T., Solvable groups with isomorphic group algebras, J. Math. Soc. Japan 18(1966), 394-397. MR 33 #5750.

Obayashi, T., Integral group rings of finite groups, Osaka J. Math. 7(1970), 253-266. MR 43 #2122.

Passman, D. S., Isomorphic groups and group rings, Pacific J. Math. 15(1965), 561-583. MR 33 #1381.

Passman, D. S., The Algebraic Structure of Group Rings, Wiley - Interscience, New York, 1977. Zbl.368.16003.

Passman, D. S.; Smith, P. F., Units in integral group rings, to appear. Abstract in: Abstracts Amer. Math. Soc. 1(1980), 280.

Perlis, S.; Walker, G., Abelian group algebras of finite order, Trans. Amer. Math. Soc. 68(1950), 420-426. MR 11-638.

Polcino Milies, F. C., Anéis de Grupos, Instituto de Matemática e Estatistica, Univ. de São Paulo, São Paulo, 1976. Zbl.407.16008.

Reidemeister, K., Überdeckungen von Komplexen, J. Reine Angew. Math. 173(1935), 164-173. Zbl.12.126.

Rham, G. de, Sur les complexes avec automorphismes, Comment. Math. Helv. 12(1940), 191-211. Zbl.22.408.

Roggenkamp, K. W., to appear.

Rossa, A. R., On integral group rings of finite groups. (Russian), Tezicy Dokl. Soobšč. Užgorod. Gos. Univ. Ser. Fiz.-Mat. Nauk 1964, 4-5.

Rudin, W.; Schneider, H., Idempotents in group rings, Duke Math. J. 31(1964), 585-602. MR 29 #5119.

Saksonov, A. I., Group rings of finite groups. I, Publ. Math. Debrecen 18(1971), 187-209. MR 46 #5425.

Sandling, R., The modular group rings of p-groups, Ph.D. thesis, Univ. of Chicago, 1969.

Sandling, R., Note on the integral group ring problem, Math. Z. 124(1972), 255-258. MR 45 #3594.

Sandling, R., Group rings of circle and unit groups, Math. Z. 140(1974), 195-202. MR 52 #3217.

Schilling, O., Einheitentheorie in rationalen hyperkomplexen Systemen, J. Reine Angew. Math. 175(1936), 246-251. Zbl.15.2.

Schneider, H.; Weissglass, J., Group rings, semigroup rings and their radicals, J. Algebra 5(1967), 1-15. MR 35 #4317.

Schumann, H.-G., Über Moduln und Gruppenbilder, Math. Ann. 114(1937), 385-413. Zbl.16.294.

Schumann, H.-G., Zum Beweis des Hauptidealsatzes, Abh. Math. Sem. Hansische Univ. 12(1937), 42-47. Zbl.16.103.

Sehgal, S. K., On the isomorphism of integral group rings. II, Canad. J. Math. 21(1969), 1182-1188. MR 42 #392.

Sehgal, S. K., Topics in Group Rings, Marcel Dekker, New York, 1978. Zbl.411.16004.

Sekiguchi, K., Units in integral group rings, Proceedings of the 12th Symposium on Ring Theory (Hokkaido Univ., Sapporo, 1979), pp. 39-50, Okayama Univ., Okayama, 1980.

Speiser, A., Die Theorie der Gruppen von endlicher Ordnung. (Dritte Auflage?), Springer, Berlin, (1937?). Zbl.17.153.

Strojnowski, A., A note on u.p. groups, Comm. Algebra 8(1980), 231-234.

Takahashi, S., Some properties of the group ring over rational integers of a finite group, Notices Amer. Math. Soc 12(1965), 463.

Taussky, O., Matrices of rational integers, Bull. Amer. Math. Soc. 66(1960), 327-345. MR 22 #10994.

Taussky-Todd, O., Olga Taussky-Todd, Number Theory and Algebra, pp. xxxiv - xlvi, Academic Press, New York, 1977. MR 58 #10227.

Waerden, B. L. van der, Moderne Algebra II, Springer, Berlin, 1931. Zbl.2.8.

Whitcomb, A., The group ring problem, Ph.D. thesis, Univ. of Chicago, 1968.

Whitehead, J. H. C., Simplicial spaces, nuclei and m-groups, Proc. London. Math. Soc. (2)45(1939), 243-327. Zbl.22.407. Also in: The mathematical works of J. H. C. Whitehead, Vol. II. Complexes and Manifolds, pp. 99-183, Pergamon Press, Oxford, 1962. MR 30 #4667b.

Zalesskiĭ, A. E.; Mihalev, A. V., Group rings. (Russian), Current problems in mathematics, Vol. 2 (Russian), pp. 5-118, Akad. Nauk SSSR Vsesojuz. Inst. Naučn. i Tehn. Informacii, Moscow, 1973. MR 54 #2723.

On the Supercentre of a Group and its Ring Theoretic Generalization

by Sudarshan K. Sehgal and
Hans Zassenhaus

Introduction

The union of the finite conjugacy classes of a unital ring R under the unit group U(R) of R is a subring FC(R) of R containing the centre of R. It is also said to be the F.C. subring of R.

An element of R belongs to the centre of C(R) if it commutes with any element of R under multiplication. Thus it is unique in its U(R)-conjugacy class and belongs to FC(R).

For any two elements x_1, x_2 of FC(R) the U(R)-conjugacy classes $x_j^{U(R)}$ formed by the U(R)-conjugates

$$x_j^{\epsilon} = \epsilon^{-1} x_j \epsilon \qquad (\epsilon \in U(R))$$

are finite subsets of R (j = 1,2).

Since the mapping

$$\underline{\epsilon}: \ R \to R$$
$$\epsilon(x) = \epsilon^{-1} x \epsilon \qquad (x \in R)$$

constitutes an automorphism of R for each unit ϵ of R (inner automorphism of R) it follows that

$$(x_1 \pm x_2)^{U(R)} \subseteq x_1^{U(R)} \pm x_2^{U(R)}$$
$$(x_1 x_2)^{U(R)} \subseteq x_1^{U(R)} x_2^{U(R)}$$

so that the $U(R)$-conjugacy classes represented by $x_1 \pm x_2$, x_1x_2 are finite. Thus $FC(R)$ is shown to be a subring of R that is invariant under the automorphisms of R and containing the centre of R.

The F.C. subring of $FC(R)$ coincides with $FC(R)$.

The F.C. subring of an arbitrary ring R may be defined as the intersection of R with the F.C. subring of the standard unital overring $\mathbb{Z}+R$ of R:

$$FC(R) = RC(\mathbb{Z} \dot{+} R) \cap R \qquad (1)$$

If R is already unital then (1) defines the same subring of R as the one defined above already.

The <u>supercentre</u> of <u>a group</u> G is defined as the intersection of the F.C. subrings of the group ring of G (over $\mathbb{Z}$) with G:

$$SC(G) = FC(\mathbb{Z}G) \cap G \qquad (2)$$

It is a characteristic subgroup of G containing the centre of G.

The study of the supercentre concept was suggested by a recent paper of Alan Williamson [6] about the following theorem:

"Let G be a periodic group. An element x in G has finite $U(\mathbb{Z}G)$ conjugacy class if and only if either

(I) x is central in G,

or

(II) x has order 4 and

is contained in an abelian subgroup H of index 2 in G where G can be generated by H and an element c of G for which

$$c^2 = x^2,$$

$$h^c = h^{-1}$$

for all h of H."

The theorem of Williamson gives a necessary and sufficient condition for an element of a group to belong to its supercentre. We shall derive it from the following

Theorem 1: Given a $\mathbb{Z}$-order Λ which is to say a unital ring with a finite $\mathbb{Z}$-basis and with semisimple quotient ring

$$A = \mathfrak{Q} \underset{\mathbb{Z}}{\otimes} \Lambda = \mathfrak{Q}\,\Lambda. \tag{3}$$

Then an element x of Λ belongs to the F.C. subring Λ precisely if its residue class modulo any maximal ideal M of A belongs to the F.C. subring of the $\mathbb{Z}$-order Λ/M.

If A is simple then the F.C. subring of Λ is the centre of Λ unless A is a totally definite quaternion algebra in which case Λ is its own F.C. subring.

Let us recall that a quaternion algebra over an arbitrary field F is defined as a simple non commutative hypercomplex system of dimension 4 over F.

A finite extension F of the rational number field $\mathfrak{Q}$ is said to be an algebraic number field of degree $[F:\mathfrak{Q}]$ over $\mathfrak{Q}$.

The quaternion algebra A over an algebraic number field F is said to be definite if among the MacLagan - Wedderburn components of the tensor product algebra of the real number field $\mathfrak{R}$ with A over $\mathfrak{Q}$

there is a Hamilton quaternion algebra $\mathfrak{H}$.

Here $\mathfrak{H}$ is the unital hypercomplex system with the generators i, j and the defining relators

$$i^2 = j^2 = (ij)^2 = -1 \qquad (4)$$

over $\mathfrak{R}$ such that the 4 elements

$$1,\ i,\ j,\ k = ij \qquad (5)$$

form an $\mathfrak{R}$ -basis of $\mathfrak{H}$.

$\mathfrak{H}$ is a division algebra. The only other quaternion algebra over $\mathfrak{R}$ is the ring of matrices of degree 2 over $\mathfrak{R}$, but it is <u>not</u> a division algebra.

Any definite quaternions algebra A of finite dimension over $\mathfrak{Q}$ is a division ring. $\mathfrak{H}$ is said to be <u>totally definite</u> if $A \otimes_{\mathfrak{Q}} \mathfrak{R}$ is the algebraic sum of $[F: \mathfrak{Q}]$ isomorphic copies of the Hamilton quaternion algebra. Equivalently, $A \otimes_{F} \mathfrak{R} = \mathfrak{H}$.

In this case it follows that the $\mathfrak{Q}$-tensor product algebra of $\mathfrak{R}$ with F, the centre of A, is the algebraic sum of $[F: \mathfrak{Q}]$ isomorphic copies of $\mathfrak{R}$.

In other words, there are $[F: \mathfrak{Q}]$ distinct ismorphism of F in $\mathfrak{R}$, F is said to be <u>totally real</u>.

In §1 of this paper some general properties of the F.C. subrings of a ring are investigated and as an application theorem 1 is proven.

In §2 theorem 1 is applied to the group ring of a finite group.

The F.C. subring of a finite group ring is characterized by means of

<u>Theorem 2</u>: Let G a finite group. Then the F.C. subring of $\mathbb{Z}G$ consists of all those elements x of $\mathbb{Z}G$ for which Γx is central in $\Gamma(\mathbb{Z}G)$ for every irreducible representation Γ of $\mathbb{Q}G$ over $\mathbb{Q}$ for which $\Gamma(\mathbb{Q}G)$ is not a totally definite quaternion algebra.

We shall prove again a special part of theorem 1 of A. A. Bovdi [1] viz:

<u>Theorem 3</u>: For any group G the torsion elements of $SC(G)$ coincide up to the sign with the torsion elements of the unit group of the F.C. subring of the group ring which form the characteristic subgroup tor $SC(G)$:

$$\text{tor } U(FC(\mathbb{Z}G)) = \pm \text{ tor } SC(G). \qquad (6)$$

Finally A. Williamson's theorem will be demonstrated anew in §3, first by an application of theorem 1, then by short ad hoc arguments. In this way a deeper insight in the connection between groups and rings will be obtained.

§1. <u>F.C. subrings</u>.

Proposition 1: If

$$\Delta: \; R \to \Lambda \qquad (6a)$$

is a homomorphism of the unital ring R into the unital ring Λ for which the image of the unit group of R is of finite index in the unit group of Λ then the Δ-image of the F.C. subring of R is contained in the F.C. subring of Λ:

$$\Lambda(FC(R)) \subseteq FC(\Lambda) \qquad (6b)$$

Proof: trivial.

In particular the F.C. subring concept depends only on the structure of the ring.

Proposition 2: For any subring S of a ring R we have

$$FC(R) \cap S \subseteq FC(S). \qquad (7)$$

Proof: trivial.

If the unit group of a unital ring is finite then the ring clearly coincides with its F.C. subring.

For example every finite ring coincides with its F.C. subring.

Similarly, if R is unital and the index of $U(R)$ over the unit group of the centre is finite then R is its own F.C. subring.

Proposition 3: The F.C. subring of any infinite division ring is its centre.

Proof: See I. N. Herstein [3].

Proposition 4: Let D a division ring and let $R = D^{f \times f}$ the ring of matrices of finite degree $f > 1$ over D. Then any proper subring S of R which is a $C(D)$-algebra invariant under the inner automorphism group of R, is central in R, excepting the $GF(4)$ contained in $GF(2)^{2 \times 2}$.

Proof: See Herstein [4] page 19.

Remark: Using the same method one can even show that the assumption that S is a $C(D)$-algebra is not necessary. As an application of propositions

3, 4 we obtain

Proposition 5: The F.C. subring of an infinite simple ring is its centre.

As is well known the unit group of an algebraic sum

$$R = \oplus_{i=1}^{s} R_i \tag{8}$$

of a finite number of unital rings $R_1, \ldots, R_s$ is isomorphic to the direct product of the unit group of the component rings inasmuch as

$$U(R) = \oplus_{i=1}^{s} U(R_i) \tag{8a}$$

The centre of (10a) is the algebraic sum of the centres of the component rings:

$$C(R) = \oplus_{i=1}^{s} C(R_i) \tag{8b}$$

Proposition 6: The F.C. subring of the algebraic sum of the F.C. subrings of the component rings:

$$FC(\oplus_{i=1}^{s} R_i) = \oplus_{i=1}^{s} FC(R_i) \tag{8c}$$

<u>Proof</u>: clear.

Propositions 5,6 imply

Proposition 7: The F.C. subring of any semisimple ring without finite minimal ideal coincides with its centre.

Using Proposition 1 we obtain

Proposition 8: If the unit group of the unital subring Λ' of the unital ring Λ is a subgroup of finite index of the unit group of Λ then the F.C. subring of Λ' is the intersection of Λ' with the F.C. subring of Λ.

In order to apply these propositions to $\mathbb{Z}$ - orders Λ with quotient ring A as in (3) we remark that there are only finitely many maximal ideals of A, say $M_1, M_2, \ldots, M_s$. The complementary ideals

$$A_i = \bigcap_{j \neq i} M_j \qquad (1 \leq i \leq s) \tag{9}$$

$$(A_1 = A \text{ if } s = 1)$$

are the minimal ideals of A and there holds the MacLagan - Wedderburn decomposition

$$A = \oplus_{i=1}^{s} A_i \tag{9a}$$

of A into the algebraic sums of simple hypercomplex systems over $\mathbb{Q}$. The first part of theorem 1 is the assertion that an element of Λ belongs to $FC(\Lambda)$ precisely if its components relative to the decomposition (9a) belong to the F.C subring of the component order

$$\Lambda_i = (\Lambda + M_i) \cap A_i \tag{9b}$$

of Λ for $i = 1, 2, \ldots, s$.

As second part of theorem 1 emerges the assertion that in case $s = 1$ the F.C. subring of Λ coincides with the centre of Λ precisely if A is not a totally definite quaternion algebra. Otherwise Λ itself is its own F.C. subring.

We make reference to the well known

Proposition 9: If the $\mathbb{Z}$-order Λ' is contained in the $\mathbb{Z}$-order Λ and is of the same $\mathbb{Z}$-rank then the unit group of Λ' is of finite index in the unit group of Λ (see [8]).

Propositions 8, 9 yield

Proposition 10: If the $\mathbb{Z}$-order Λ' is contained in the $\mathbb{Z}$-order Λ and is of the same $\mathbb{Z}$-rank then the F.C. subring of Λ' is obtained by intersecting the F.C. subring of Λ with Λ'.

The first half of theorem 1 follows by an application of the propositions 6, 10 and the well known

Proposition 11: Any $\mathbb{Z}$-order Λ with central quotient ring as in (3) is contained in the algebraic sum of the component orders Λ_i as a $\mathbb{Z}$-order of the same rank

$$\Lambda \subseteq \oplus_{i=1}^{s} \Lambda_i \tag{9c}$$

as in fact the containment relation (11) is generally true for subdirect sums and the rank equality is implied by the relation (9a) for the quotient rings.

In order to show the second half of theorem 1 we use

Proposition 12: For any simple hypercomplex system A over a field F any proper F-subalgebra S that is invariant under the inner automorphism group of A is central.

<u>Proof</u>: Proposition 4 and a theorem of Cartan-Brauer-Hua ([5] p. 86).

Now we show

Proposition 13: If A is a division ring and S is a subgroup of finite index in the unit group of the $\mathbb{Z}$-order Λ of A such that the division ring $\mathfrak{D}S$ generated by S is properly contained in A then A is a totally definite quaternion algebra with $\mathfrak{D}S$ as its centre.

<u>Proof</u>: For any subfield E of D not contained in $\mathbb{Q}S$ the number of left cosets of the unit group of the $\mathbb{Z}$-order $\Lambda \cap E$ of E over its intersection with S is finite in accordance with the finiteness of $[U(\Lambda):S]$. A fortiori the factor group of $U(\Lambda \cap E)$ over the unit group of the order $\Lambda \cap E \cap \mathbb{Q}S$ of the subfield $E \cap \mathbb{Q}S$ is finite. Note that $\Lambda \cap E \cap \mathbb{Q}S$ is properly contained in $\Lambda \cap E$. Applying Dirichlet's unit theorem to the two commutative $\mathbb{Z}$-orders $\Lambda \cap E$, $\Lambda \cap E \cap \mathbb{Q}S$ it follows that E is a totally complex quadratic extension of the totally real algebraic number field $E \cap \mathbb{Q}S$.

Now suppose F is a subfield of $\mathbb{Q}S$ not contained in the centre $C(D)$ of A then it follows from proposition 12 that there is a non zero element Z of A for which $Z^{-1} F Z$ is a subfield of A not contained in $\mathbb{Q}S$. Hence $Z^{-1} F Z$ is a totally quadratic extension of a totally real subfield. The same applies to F itself. It follows that either $\mathbb{Q}S$ belongs to $C(A)$ or else $\mathbb{Q}S$ is a totally definite quaternion algebra.

If $\mathbb{Q}S$ belongs to $C(A)$ then every non central subfield E of the non commutative division ring A is a totally complex quadratic extension of a totally real subfield E_1. Hence E_1 belongs to $C(D)$. In particular let E be a maximal subfield of A. Then E_1 <u>is</u> the centre of A. Hence $C(A)$ is totally real, $n = 2$, every quadratic extension of $C(A)$ is totally complex. A is a totally definite quaternion algebra. Applying Dirichlet's unit theorem to the subfield $\mathbb{Q}S$ $C(A)$ generated by the subgroup S of <u>finite</u> index in $U(A)$ it follows that $\mathbb{Q}S = C(A)$.

Now let us deal with the alternative that $\mathfrak{Q}S$ is a totally definite quaternion algebra.

There is a maximal subfield E of $\mathfrak{Q}S$. By assumption E is totally complex. There cannot be any subfield of A that is larger than E. Hence E is a maximal subfield of A. The centre of A is contained in $\mathfrak{Q}S$. Since $\mathfrak{Q}S \subset A$ it follows that $C(A) \subset A$. It follows that $C(A) \subset C(\mathfrak{Q}S)$.

It follows from proposition 12 that there is a non zero element Z of A for which $Z^{-1} C(\mathfrak{Q}S) Z$ is a subfield of A not contained in $\mathfrak{Q}S$. But that is impossible because $C(\mathfrak{Q}S)$ is totally real. Hence $C(A) = C(\mathbb{Q}S)$, a contradiction.

Thus proposition 12 is established.

Proposition 13: Let the quotient ring A of the $\mathbb{Z}$-order Λ be a totally definite quaternion algebra over its centre $F = C(A)$. Then the unit group of $\Lambda \cap F$ is of finite index in $U(\Lambda)$.

<u>Proof</u>: Any element ζ of A not contained in F generates a quadratic extension E of F that is a totally complex algebraic number field. Hence the relative norm $N_{E/F}(\zeta)$ is totally positive. Since A is a quaternion algebra over F it follows that the regular norm $N_{A/\Gamma}$ restricts to $N_{E/F}$ on E.

If ζ is a torsion unit of $U(N)$ then $N_{E/F}(\zeta) = N_{A/F}(\zeta)$ is a torsion unit of $U(\Lambda \cap F)$. But F is totally real so that ± 1 are the only torsion units of $U(\Lambda \cap F)$. Hence $N_{A/F}(\zeta) = 1$ for any torsion unit of $U(\Lambda)$. Conversely if $N_{A/F}(\zeta) = 1$ for a unit ζ of $U(\Lambda)$ then either $\zeta \in F$ and hence $\zeta^2 = 1$, $\zeta = \pm 1$ is a torsion unit of $U(\Lambda \cap F)$, or else $\zeta \notin F$, $F(\zeta)$ is a quadratic extensions of F that is a totally complex number field, $N_{E/F}(\zeta) = 1$, all algebraic conjugates of the unit ζ have absolute value 1. Hence ζ is a unit root. The torsion units of $U(\Lambda)$ form the torsion subgroup

$$\mathrm{tor}U(\Lambda) = \{\zeta | \zeta \in U(\Lambda) \ \& \ N_{A/F}(\zeta) = 1\}.$$

Also we see now that there is a monomorphism

$$\iota: \ A \to \mathbb{H}$$

of A into the Hamilton quaternion algebra which maps $\mathrm{tor}U(\Lambda)$ isomorphically on a torsion subgroup of the group $\ker N_{U(\mathbb{H})/U(\mathbb{R})}$ which is known to be isomorphic to the special unitary group $SU(2)$ of degree 2.

Note that the order of the elements of $\mathrm{tor}U(\Lambda)$ is bounded because the degree of their minimal polynomial over $\mathbb{Q}$ is bounded by $2[F:\mathbb{Q}]$. But the torsion subgroups of $SU(2)$ of bounded exponent are known to be finite. Hence $\mathrm{tor}U(\Lambda)$ is a finite normal subgroup of $U(\Lambda)$. The norm mapping $N_{A/F}$ maps the normal subgroup $U(\Lambda \cap f)\ \mathrm{tor}U(\Lambda)$ on the square subgroup of $U(\Lambda \cap F)$.

Because of the total reality of F and Dirichlet's theorem the factor group of $U(\Lambda \cap F)$ over its square subgroup is elementary abelian of exponent 2 and order $2^{[F:\mathbb{Q}]-1}$. Since the norm map $N_{A/F}$ maps the unit group of Λ into the unit group of $\Lambda \cap F$ it follows that the factor group of $U(\Lambda)$ over $U(\Lambda \cap F)\mathrm{tor}U(\Lambda)$ is isomorphic to a subgroup of the square factor group of $U(\Lambda \cap F)$:

$$[U(\Lambda):U(\Lambda \cap F)\ \mathrm{tor}\ U(A)] \mid 2^{[F:\mathbb{Q}]-1}, \qquad (10)$$
$$U(\Lambda)^2 \subseteq U(\Lambda \cap F)\ \mathrm{tor}\ U(A).$$

Thus we have demonstrated the finiteness of the factor group of $U(\Lambda)$ over $U(\Lambda \cap F)$. As a consequence we find that

$$SC(\Lambda) = \Lambda \cap F.$$

Proposition 14: If the quotient ring A of the $\mathbb{Z}$-order Λ is simple, but neither a totally complex algebraic number field nor a totally definite quaternion algebra over its centre then the subring $\mathbb{Z}S$ of Λ generated by any subgroup S of finite index in $U(\Lambda)$ is a $\mathbb{Z}$-order of A of maximal rank.

<u>Proof</u>: A is a ring of matrices of finite degree f over a division ring D of finite dimension over $\mathbb{Q}$.

By assumption the intersection of the $U(\Lambda)$-conjugates of S is a normal subgroup of $U(\Lambda)$ of index dividing $[U(\Lambda):S]!$ so that by Fermat's theorem of group theory

$$\epsilon^{[U(\Lambda):S]!} \in S \quad (\epsilon \in U(\Lambda)) \tag{11}$$

If $f > 1$ then we apply (11) to the transvection units

$$\epsilon = I_f + \mu\lambda e_{hj}$$

of Λ where

$$1 \leq h \leq f,\ 1 \leq j \leq f,\ h \neq j$$
$$e_{hj} = (\delta_{ih}\delta_{kj}),$$
$$\lambda \in \Lambda \cap D,$$

and μ is a natural number for which

$$\mu e_{ik} \in \Lambda \quad (i,k = 1,2,\ldots,f). \tag{12}$$

It follows from (11) that

$$\mu\lambda[U(\Lambda):S]!\, e_{hj} \in \mathbb{Z}S \quad (h,j = 1,2,\ldots,f; h \neq j) \tag{13}$$

where the elements on the left of (13) generate A over $\mathbb{Q}$.

Hence, $\mathbb{Z}S$ is a submodule of Λ of maximal $\mathbb{Z}$-rank.

If A is a division ring then the quotient ring of $\mathbb{Z}S$ is a division subring $\mathfrak{Q}S$ of A.

Any element a of A that is not contained in $\mathfrak{Q}S$ generates an algebraic number field $\mathfrak{Q}(a)$ such that the index of the unit group of the $\mathbb{Z}$-order $\mathbb{Z}S \cap \mathfrak{Q}(a)$ is finite in the unit group of the $\mathbb{Z}$-order $\mathbb{Z}S \cap \mathfrak{Q}(a)$.

Since by assumption $\mathfrak{Q}(a) \supset \mathfrak{Q}S \cap \mathfrak{Q}(a)$ it follows from Dirichlet's unit theorem that a is totally complex and that $\mathfrak{Q}S \cap \mathfrak{Q}(a)$ is totally real with relative degree 2. But a non unital element f of $C(A)$ must be totally complex, because in accordance with proposition 12 there exists a $U(A)$-conjugate of f which does not belong to $\mathfrak{Q}S$. In other words the totally real elements of $\mathfrak{Q}S$ belong to the centre of A. They generate a totally real subfield B of $C(A)$.

If A is a field then it follows by assumption that A is not totally complex. Hence it follows from Dirichlet's unit theorem that $\mathfrak{Q}S = A$, $\mathbb{Z}S$ is of maximal rank.

Now let A be noncommutative. It follows that any maximal subfield $\mathfrak{Q}(a)$ of A contains $C(A)$ properly. In accordance with proposition 12 there exists a $U(A)$-conjugate of a not belonging to S.

It satisfies a quadratic equation over B. Hence $B(a)$ is a quadratic extension of B, hence $B = C(A)$, $B(a) = B(b)$, where $b^2 = \beta$ is a totally negative element of B, A is a quaternion algebra over B.

There is a nonzero element c of A for which $cbc^{-1} = -b$,

$c^2 = j \in B$, $c \notin B$. Since there is a $U(A)$-conjugate of c that does not belong to S it follows that c is totally complex, j is a totally negative element of B. Hence A is a totally definite quaternion algebra over its centre B. Thus proposition 14 is demonstrated.

Corollary of Proposition 14: For any $\mathbb{Z}$-order Λ either the index of the unit group of the centre of Λ in the unit group of Λ is finite or the unit group of Λ contains a noncyclic free subgroup.

Proof: If for every maximal ideal M of the semisimple hypercomplex system $A = Q \otimes_{\mathbb{Z}} \Lambda$ the projection order $\Lambda/\Lambda \cap M$ is abelian or totally definite quaternion order then for each of those orders the index of the unit group of the centre in the full unit group is finite. But Λ is isomorphic to a $\mathbb{Z}$-order contained in the algebraic sum $\Lambda' = \bigoplus_M \Lambda/\Lambda \cap M$ over the finitely many maximal ideals M of A. Hence also Λ' and Λ have the property that the unit group of the centre is of finite index in the full unit group of the order.

If A is simple noncommutative and if A is not a totally definite quaternion algebra then according to proposition 14 the order $\mathbb{Z}S$ is of finite index in Λ for any subgroup of S that is of finite index in $U(\Lambda)$. Since A is noncommutative it follows that $U(\Lambda)$ is infinite.

For any solvable subgroup S of finite index in $U(\Lambda)$ the order $\mathbb{Z}S$ is of finite index in Λ. We have $\mathfrak{Q}S = A$ and there is a faithful irreducible representation $\Delta: S \to \mathfrak{D}^{f \times f}$ of S of finite degree f over $\mathfrak{D}$. It follows that the centre of S is of finite index in S. Hence $C(S)$ is of finite index in $U(\Lambda)$, hence $\mathfrak{Q}\,C(S) = A$ is commutative, a contradiction.

Thus it follows that $U(\Lambda)$ is infinite and contains no solvable subgroup of finite index. By a theorem of Tits the group $U(\Lambda)$ contains a noncyclic free subgroup.

Finally let $A = \mathbb{Q}\Lambda$ be not simple such that for some maximal ideal M of A the projection order $\Lambda/M \cap \Lambda$ is noncommutative and not a totally definite quaternion order.

For the minimal ideal $\hat{M}$ of A not contained in M (complementary ideal of M) the intersection order $\hat{M} \cap \Lambda$ is isomorphic to a suborder of $\Lambda/\Lambda \cap M$ of finite index. Hence also $\hat{M} \cap \Lambda$ is neither commutative nor a totally definite quaternion order so that its unit group contains a noncyclic free group S. But the order $\hat{M} \cap \Lambda \oplus M \cap \Lambda$ is of finite index in Λ and its unit group contains the noncyclic free group $S \oplus 1_M$. The same is true for $U(\Lambda)$.

Now we prove the second part of theorem 1 as follows.

Let A a simple noncommutative hypercomplex system over $\mathbb{Q}$ which is not a totally definite quaternion algebra. Let Λ a $\mathbb{Z}$-order of maximal rank of A. We must show that the F.C. subring of Λ coincides with the centre of Λ.

If there is a noncentral element x of $FC(\Lambda)$ then the centralizer $C_\Lambda(x)$ of x in Λ is a subring which is <u>not</u> of maximal $\mathbb{Z}$-rank. On the other hand the centralizer S of x in $U(A)$ is a subgroup of finite index. According to proposition 14 we know that $\mathbb{Z}S$ is of maximal rank in Λ. But $\mathbb{Z}S$ is contained in $C_\Lambda(x)$, a contradiction.

§2. <u>Application of Theorem 1</u>

In this section we study the application of theorem 1 to the group ring of a finite group.

For any homomorphism Δ of a $\mathbb{Z}$-order R into a $\mathbb{Z}$-order Λ the images form a $\mathbb{Z}$-suborder of Λ, possibly of lesser $\mathbb{Z}$-rank.

The unit group of R is homomorphically mapped by Δ into the unit group of Λ. If Δ is an epimorphism then the index of $\Delta U(R)$ in $U(\Lambda)$ is finite. This is because $\mathbb{Q}R$ is semisimple so that there holds an algebraic decomposition

$$\mathbb{Q}R = \mathbb{Q}\ker \Delta \oplus B$$

and the component orders of R are

$$R_1 = (R + B) \cap \mathbb{Q}\ker \Delta, \; R_2 = (R + \mathbb{Q}\ker \Delta) \cap B$$

such that R is of finite index in $R_1 \oplus R_2$ and Δ extends uniquely to an epimorphism $\bar{\Delta}$ of $R_1 \oplus R_2 = R_1$ on Λ with R_1 as kernel such that $R_2 \simeq \Lambda$. On the other hand Λ contains the $\mathbb{Z}$-order $(\Lambda \cap R_1) \oplus (\Lambda \cap R_2)$ as a module of finite index such that the unit groups $U(\Lambda \cap R_1)$, $U(\Lambda \cap R_2)$, $U((\Lambda \cap R_1)) \oplus U(\Lambda \cap R_2)$ are subgroups of finite index in $U(R_1)$, $U(R_2)$, $U(\Lambda)$ respectively. It follows that

$$U(\Lambda \cap R_2) \simeq \Delta U(\Lambda \cap R_2) \subseteq \Delta(U(\Lambda)) \subseteq \bar{\Delta}\, U(R_2) = U(\Lambda).$$

Now let G a finite group.

Consider an epimorphism

$$\Delta : \mathbb{Z}G \to \Lambda \qquad (13a)$$

of the group ring of G on a totally definite quaternion order Λ. Thus the tensor product algebra of the totally definite algebra $A = \mathbb{Q}\Lambda$ and the real number field $\mathbb{R}$ over the centre of A is the Hamilton quaternion algebra

$$\mathbb{H} = A \otimes_{C(A)} \mathbb{R},$$

$$H \otimes_{\mathbb{R}} \mathbb{C} = \mathbb{C}^{2 \times 2} = A \otimes_{C(A)} \mathbb{C}$$

defines a representation

$$\Gamma: G \to \mathbb{C}^{2\times 2} \tag{13b}$$

of G of degree 2 with the same kernel as Δ such that the ring $\mathbb{Z}\Gamma G$ generated by ΓG is isomorphis to $\Delta(\mathbb{Z}G) = \Lambda$. Because of the definiteness of A over C(A) the reduced relative norm is totally positive which means that the determinant of Γ is 1 so that Γ can be transformed into special unitary form

$$\Gamma: G \to SU(2). \tag{13c}$$

Because of the well known locally isomorphic epimorphism

$$\eta: SU(2) \to SO(3) \tag{13d}$$

with

$$\ker \eta = < -I_2 > \tag{13e}$$

the epimorphic images $\eta\Gamma G$ are among the finite subgroups of SO(3) which also are well known (s.e.g. Zassenhaus [9]).

Set $\bar{G} = \Gamma G$ and assume $\tilde{\Delta}$ to be the rationally irreducible representation of G with kernel ker Δ.

Proposition 15: With the previous notations we obtain the following totally definite quaternion orders generated by finite groups:

I

$$\bar{G} = < a, b >$$

$$a = \begin{pmatrix} & -1 \\ 1 & \end{pmatrix}, \quad b = \begin{pmatrix} \zeta_{2m} & \\ & \zeta_{2m}^{-1} \end{pmatrix}$$

$$(2 \le m \in \mathbb{Z}^{>0})$$

where we set generally

$$\zeta_m = \exp(2\pi i/m) \quad (\eta \in \mathbb{Z}^{>0}).$$

Here the order of $\bar{G}$ is 4m with defining relators

$$a^2 = b^m = (ab)^2 \qquad (14)$$

between the generators a, b, the totally definite quaternion order $\mathbb{Z}\bar{G}$ is of $\mathbb{Z}$-rank $2\varphi(2m)$ and the epimorphic image $\eta\bar{G}$ is a dihedral group D_{2m}. The degree of $\tilde{\Delta}$ is $2\varphi(2m)$.

The centre of A is isomorphic to the maximal real subfield $\mathfrak{Q}(\zeta_{2m} + \zeta_{2m}^{-1})$ of the 2m-th cyclotomic field.

II $\qquad \bar{G} = \langle a, b \rangle,$

$$a = \begin{pmatrix} & -1 \\ 1 & \end{pmatrix}, \quad b = \frac{1}{2}\begin{pmatrix} 1+i & 1+i \\ -1+i & 1-i \end{pmatrix},$$

$$i = \sqrt{-1},$$

of order 24 with defining relators

$$a^2 = b^3 = (a\,b)^3, \qquad (14a)$$

the <u>tetrahedral</u> rotation group as η-image:

$$\eta\,\bar{G} = T \simeq \mathfrak{A}_4,$$

4 as degree of $\tilde{\Delta}$ and

$$C(A) = \mathfrak{Q},$$

$$\dim \mathfrak{Q}\,\bar{G} = 4.$$

III $\qquad \bar{G} = \langle a,b \rangle,$

$$a = \begin{pmatrix} & -1 \\ 1 & \end{pmatrix}, \quad b = \begin{pmatrix} 1/2 & \sqrt{3}/2 \\ -\sqrt{3}/2 & 1/2 \end{pmatrix}$$

of order 48 with defining relators

$$a^2 = b^3 = (ab)^4, \qquad (14b)$$

the <u>octahedral rotation group</u> as η-image

$$\eta\,\bar{G} = O \simeq \mathfrak{S}_4,$$

4 as degree of $\tilde{\Delta}$ and

$$C(A) = \mathfrak{Q},$$
$$\dim_{\mathfrak{Q}} \bar{G} = 4.$$

IV $\bar{G} = < a, b >$,

$$a = \begin{pmatrix} & -1 \\ 1 & \end{pmatrix}, \quad b = \begin{pmatrix} 1/2 & \kappa + \kappa' i \\ -\kappa + \kappa' i & 1/2 \end{pmatrix}$$

$$i = \sqrt{-1}, \quad \kappa = (-1 + \sqrt{5})/4, \quad \kappa' = (-1 - \sqrt{5})/4$$

of order 60 with defining relators

$$a^2 = b^3 = (ab)^5, \tag{14c}$$

the icosahedral rotation group as η-image:

$$\eta \bar{G} = I \simeq \mathfrak{A}_5,$$

8 as degree of $\tilde{\Delta}$ and

$$C(A) = \mathfrak{Q}(\sqrt{5}),$$
$$\dim_{\mathfrak{Q}} C(A) = 2 .$$

Proposition 16: There is a rationally irreducible representation $\tilde{\tilde{\Delta}}$ of G with the same kernel as $\eta\Gamma$ that is unique up to rational equivalence such that $\mathfrak{Q}\tilde{\tilde{\Delta}}G$ is not a totally definite quaternion algebra over its centre, except for $I : m = 2$, in which case there are four distinct representations of degree 1 in $\mathfrak{Q}$ with kernels intersecting in $\ker \eta\Gamma$.

The degree of $\tilde{\tilde{\Delta}}$ is as follows:

case I, $m \geq 3$: $\varphi(m)$

case II : 3

case III : 3

case IV : 6.

Applying theorem 1 we obtain

Proposition 17: The largest ideal of the group ring of a finite group G that is contained in its F.C. subring, is the ideal $X_{FC(G)}$ of $\mathbb{Z}G$ formed as the intersection of the kernels of all irreducible representations of $\mathbb{Z}G$ (over $\mathfrak{Q}$, or over any fixed extension of $\mathfrak{Q}$) for which the image group of G is neither cyclic nor isomorphic to any one of the groups I-IV enumerated in proposition 15.

The factor ring of $FC(\mathbb{Z}G)$ over $X_{FC(G)}$ is the centre of the factor ring of the group ring over $X_{FC(G)}$:

$$FC(\mathbb{Z}G)/X_{FC(G)} = C(\mathbb{Z}G/X_{FC(G)}) \tag{15}$$

For the proof it suffices to remark that for each of the groups enumerated in proposition 16 there is only <u>one</u> rationally irreducible faithful representation up to rational equivalence.

Also we point out that the corresponding representation orders are nonisomorphic.

Moreover, any irreducible representation Ψ of G over a field of zero characteristic for which the image ΨG is not cyclic determines a noncommutative representation order $\mathbb{Z}\Psi G$.

Note that proposition 17 constitutes a sharpened version of theorem 2.

Prior to proving theorem 3 let us recall that any element of the supercentre of an arbitrary group G belongs to a <u>finite</u> G-conjugacy class, as follows from the definition of SC(G) and restriction from $U(\mathbb{Z}G)$ to the subgroup G:

$$C(G) \subseteq SC(G) \subseteq \varphi(G) \tag{16}$$

where $\varphi(G)$ denotes the characteristic subgroup of G that is formed as the union of the finite G-conjugacy classes.

If G is the quaternion group then $C(G) \subset G = SC(G)$. If G is the noncommutative group of 6 elements then $SC(G) = 1 \subset G = \varphi(G)$.

On the other hand each of the functors C, SC, φ are idempotent:

$$\begin{aligned} C(C(G)) &= C(G), \\ SC(SC(G)) &= SC(G), \\ \varphi(\varphi(G)) &= \varphi(G). \end{aligned} \tag{17}$$

This is obvious for C, φ. It follows for SC from

Proposition 18: For any subgroup H of a group G we have

$$SC(H) \supseteq SC(G) \cap H = SC(\mathbb{Z}G) \cap H \tag{18}$$

which is analogous to proposition 2 and proved as easily.

The set theoretical relation (17) can be sharpened to

$$C(\mathbb{Z}G) \subseteq FC(\mathbb{Z}G) \subseteq \mathbb{Z}\varphi(G) \tag{19}$$

as follows by the finiteness of $\epsilon^{U(\mathbb{Z}G)}$ for ϵ of $FC(\mathbb{Z}G)$ and from the relation

$$\epsilon^{U(\mathbb{Z}G)} = \epsilon$$

for ϵ of $C(\mathbb{Z}G)$.

Turning to the proof of theorem 3 we remark that the torsion elements of $\varphi(G)$ form a locally finite characteristic subgroup $\mathrm{tor}\varphi(G)$ of G. Hence also $\mathrm{tor}SC(G)$ is a locally finite subgroup of $SC(G)$.

Supposing $\varepsilon = \Sigma\, \varepsilon(g),\ \varepsilon(g) \in \mathbb{Z}$ is a torsion element of prime power order q. Then by ([6], p. 177) there is an element g_o of G for which $g_o = 1$ and $\varepsilon(g_o) \neq 0$. Since $< \varepsilon^{U(FC(\mathbb{Z}G))} >$ is finite and ε is torsion it follows that

$$N = < \varepsilon^{U} >$$

is a finite normal subgroup of $U(FC(\mathbb{Z}G))$. Hence $< g_o, N >$ is finite. It contains the element

$$g_o^{-1}\ \varepsilon = \alpha = \Sigma\alpha(g)g,\ \alpha(g_o) \neq 0$$

Thus we have by [6] p. 45

$$\alpha = \alpha(1) = \pm 1,\ \varepsilon = \pm g_o.$$

§3. Order Theory and Group Ring Theory

In order to apply the previous theory to the proof of A. Williamson's theorem mentioned in the introduction let us restate it as follows: "Let G a periodic group. An element x of G belongs to the supercentre of G if and only if either

(I) x is central in G,

or

(II) x has order 4 with G-centralizer H of index 2 such that the square of every element of G not in H is x^2.

Note that the equation $c^2 = c'^2 = x^2$ implies that $c(cc')c^{-1} = c^2c'^2c'^{-1}c^{-1} = x^4\,(cc')^{-1} = (cc')^{-1}$ so that we see at once the equivalence of the new version with the other one.

Note that any element x of $SC(G)$ represents a finite G-conjugacy class x^G. Since x is a torsion element of G it follows that the subgroup $< x^G >$ of G generated by x^G is finite normal.

If x is not central in G then there exists an element j of G not commuting with x. The subgroup generated by y and $< x^G >$ is finite. Now an application of theorem 1 and the analysis of the F.C. subring of the group ring of a finite group done in §2 and applied to $G = < y, x^G >$ shows indeed that x is of order 4 and that the centralizer of x in G is of index 2 in G, such that $c^2 = x^2$ for every element c of G not belonging to $C_G(x)$.

Hence all elements of G normalize $< x >$ and $C_G(x)$ is a subgroup of G of index 2 such that $c^2 = x^2$ for every element c of G not belonging to $C_G(x)$.

The following purely group ring theoretic argument will lead us to the same goal directly. Let us restate A. Williamson's theorem as follows: "Let G a torsion group.

Then

1) $SC(C) = C(G)$

or

2) $SC(G) = G$ if G is a Hamiltonian 2-group

or

3) $SC(G) = < C(G), x >$, $C(G)$ is an elementary abelian 2-group, in case G has an element x of order 4 with abelian G-centralizer $H = C_G(x)$ such that

$$G = < H,y \mid y^2 = x^2 \& \forall h\ (h \in H \Rightarrow h^y = h^{-1}) >$$

is neither abelian nor a Hamiltonian 2-group."

Lemma 1. For any finitely generated subgroup A of $SC(G)$ there is a fixed natural number n such that all n-th powers of units of $\mathbb{Z}G$ centralize A.

Proof: Let $A = \langle g_1,\ldots,g_t \rangle$, $t \in \mathbb{Z}^{>0}$, $g_i \in SC(G)$ $(1 \leq i \leq t)$, hence $C_i = g_i^{U(\mathbb{Z}G)}$ is finite and for ϵ of $U(\mathbb{Z}G)$ the permutation representation

$$\epsilon \to \pi(\epsilon) = \begin{pmatrix} x \\ x^{\epsilon} \end{pmatrix} \Big| \; x \in \bigcup_{i=1}^{t} C_i$$

extends uniquely to an automorphism of $\langle A^{U(\mathbb{Z}G)} \rangle$. Now let n be the exponent of the finite permutation group $\pi(U(\mathbb{Z}G))$. It follows that ϵ^n centralizes A for all ϵ of $U(\mathbb{Z}G)$.

Let $\mathfrak{T} = \text{tor}\, SC(G)$.

Lemma 2. If x is in $\mathfrak{T}$, z in tor G and y in G then $(x,y) \in \langle z \rangle$ or $(y,z) \in \langle z \rangle$.

Proof: Consider $u = 1 + \hat{z}\, y(1-z)$ where $\hat{z} = \sum_{i=0}^{|z|-1} z^i$. Then $u^n x = x u^n$ by Lemma 1. Hence

$$\hat{z}\, y(1-z)\, x = x\, \hat{z}\, y(1-z).$$

Thus

$$xy = \begin{cases} x z^i y z \Rightarrow (y,z) \in \langle z \rangle \\ \langle x,y \rangle \in \langle z \rangle . \end{cases}$$

Taking $z = x$ in the last Lemma, we get the

Corollary of Lemma 2: Every subgroup of $\mathfrak{T}$ is normal in G.

Lemma 3. If $x \in \mathfrak{T}$, $y \in$ tor G then x normalizes $\langle y \rangle$.

Proof: Consider $u = 1 + \hat{y}\; x\,(1-y) \in U(\mathbb{Z}G)$

then

$$x\, \hat{y}\, x\,(1-y) = \hat{y}\, x\,(1-y)\, x$$

(in accordance with Lemma 1) give us

$$(1-y)\, x\, \hat{y}\, x\, (1-y) = 0$$

This give us

$$x^2 = x\, y^i\, x\, y \quad \text{or} \quad y\, x\, y^i\, x$$

for some i and $y^x \in \langle y \rangle$.

Lemma 4. $\mathfrak{J}$ is abelian or a Hamiltonian 2-group and for every abelian subgroup A of $\mathfrak{J}$, $\in G$, $x \notin C_G(A)$ we have $a^x = a^{-1}$ for all a of A. The centralizer of A in G is of index at most 2 and $C(G)$ is of exponent 2.

<u>Proof</u>. Consider $H = \langle a, x \rangle$ and $a^x \neq a$, $a \in A$. We claim that every subgroup of H is normal in H. By Lemma 2, for any integers l, m we have either $(a^m, x) \in \langle x^l a^m \rangle$ or

$$(x^l a^m, x) \in \langle x^l a^m \rangle .$$

But $(a^m, x) = (x^l a^m, x)$.

Thus x normalizes $\langle x^l a^m \rangle$ and by the last lemma a normalizes $\langle x^l a^m \rangle$. It follows that $\langle x^l a^m \rangle$ is normal in H. Thus H and also $\mathfrak{J}$ are abelian or Hamiltonian. We have H a Hamiltonian group. Write $H = K_8 \times B$ where $K_8 = \langle i, j \mid i^2 = j^2 = t,\ t^2 = 1,\ ji = ijt \rangle$ is the quaternion group of order 8 and B is an abelian group. We wish to prove that B is of exponent 2. We write $a = ib$, $b \in B$. Then i has a finite number of $U(\mathbb{Z}H)$-conjugates. Thus also j has only a finite number of $U(\mathbb{Z}H)$-conjugates. Consequently, $U(\mathbb{Z}H)$ is an F.C. group. It follows from [6], p. 209 that B is of exponent 2. Clearly then $a^x = a^{-1}$.

From this conclusion of the lemma follows easily the <u>Proof of the theorem</u>.

a) First suppose $\mathfrak{J}$ is abelian, but not central. Then $C(G) \subseteq \mathfrak{J}$ and it follows by the last lemma that $C(G)^2 = 1$. Pick $x \in \mathfrak{J}$, $y \in G$ such that $(x,y) \neq 1$. Consider $<\mathfrak{J}, y> = G_1$. Then y^2 centralizes $\mathfrak{J}$ and $y^2 \in C(G_1)$ which is contained in $SC(G_1)$. It follows that $y^4 = 1$. Thus $y^x = y^3$. Now, $x^{-1} = y^{-1} x y = y^2 x$

$$\Rightarrow x^2 = y^2 \Rightarrow x^4 = 1.$$

We have proved that any element in $\mathfrak{J}$ not commuting with y has order 4 whereas any element in $\mathfrak{J}$ commuting with y has order 2. Write

$$\mathfrak{J} = E \times C_1 \times \ldots \times C_r$$

$$E^2 = 1, \ | C_i | = 4 \quad (1 \leq i \leq r),$$

a direct product of cyclic groups. Since y^2 is a fixed element we conclude that $\mathfrak{J} = E \times C_1$. Let H be the centralizer of $\mathfrak{J}$ in G. Then considering $y\,b$, $b \in H$ instead of y we obtain $(y\,b)^2 = y^2$ which implies that $y\,b\,y^{-1} = b^{-1}$. Thus H is abelian and of index 2. Conversely, suppose we have

$$G = <H, y \mid y^2 = x^2 \in H \;\&\; \forall h\ (h \in H \Rightarrow h^y = h^{-1})>$$

which is not an abelian or a Hamiltonian 2-group. Clearly H is an abelian (normal) subgroup of index 2 in G. Due to [6], p. 209, we know that y does not belong to $\mathfrak{J}$. Moreover, if $y\,h \in \mathfrak{J}$ for some h of H then $(y\,h)^y = y\,h^{-1} = y\,h$ due to Lemma 4. This implies that $h^2 = 1$ and $h \in C(G)$, $y \in \mathfrak{J}$. Also it follows by Lemma 3 that if $h \in H \cap \mathfrak{J}$ then $h \in <x\ , C(G)>$. We have therefore only to prove that x is in $\mathfrak{J}$. This has been proved by Bovdi [2].

b) Now let $\mathfrak{J}$ be nonabelian. Therefore, by Lemma 4, $\mathfrak{J} = E \times K_8$, $E^2 = 1$. Let C_1 and C_2 be the centralizer in G of i and j respectively. Then $(G:C_1) = 2 = (G:C_2)$. It follows that $G = C_1 \cup C_2 \cup C_{1j}$ and $(G:C_1 \cap C_2) \leq 4$. Since $C_1 \cap C_2$ commutes elementwise with C_1 and j it is central and $(G:C(G)) \leq 4$. But $C(G)$ is of exponent 2 and contained in $\mathfrak{J}$, hence $G = \mathfrak{J}$.

References

1. Bovdi, A. A.: The periodic normal divisors of the multiplicative group ring; Sibirski Matem. Zh., 9 (1968), 495-498.

2. Bovdi, A. A.: The periodic normal divisors of the multiplicative group ring II; Sibirski Matem. Zh., 11 (1970), 495-511.

3. Herstein, I. N.: Conjugates in Division Rings; Proc. Amer. Math. Soc., 7 (1956), 1021-1022.

4. Herstein, I. N.: Topics in Ring Theory; Univ. of Chicago Press, Chicago (1972).

5. Jacobson, N.: Structure of Rings; Amer. Math. Soc., Providence, R.I. (1968).

6. Sehgal, S. K.: Topics in Group Rings; Manuel Dekker, N.Y. (1978).

7. Williamson, A.: On the conjugacy classes in a group ring; Canad. Math. Bull., 21 (1978), 491-496.

8. Zassenhaus, Hans: „Neuer Beweis der Endlichkeit der Klassenzahl bei unimodularer Äquivalenz ganzzahliger Substitutionsgruppen; Hamb. Abh. 12 (1938), 276-288.

9. Zassenhaus, Hans: The Theory of Groups, second edition; Chelsea Publishing Company, New York (1958).

SOME FACTS CONCERNING INTEGRAL REPRESENTATIONS OF IDEALS IN AN ALGEBRAIC NUMBER FIELD

Olga Taussky
California Institute of Technology
Pasadena, CA 91125/USA

Introduction.

The ring $Z^{n\times n}$ of $n \times n$ integral matrices (integral is to mean rational integral), has a number of very attractive features. It is a principal ideal domain and it is a maximal order in $Q^{n\times n}$. The latter algebra is simple and all its automorphisms are inner. The units in $Z^{n\times n}$ are the unimodular $n \times n$ matrices.

The ring $Z^{n\times n}$ contains a great many items. Anything in that ring seems interesting and interactions between any two elements produces some results. The fact that it is non commutative gives new structure to objects from commutative systems when their representations by elements in $Z^{n\times n}$ is considered.

I start with some examples to illustrate the last claim.

(1) The ring $Z^{n\times n}$ contains representations of orders from algebraic number fields and algebras over Q. It also contains representations of groups of order n by integral $n \times n$ matrices, like permutations, among these are the Galois groups of the fields if they are normal fields. In this case the action of the Galois group on a generator α of the field leads to a polynomial $p(\alpha)$. If α is a zero of an irreducible polynomial $f(x)$ then all $n \times n$ integral matrices A with $f(A) = 0$ are similar via rational matrices. Hence $p(A) = S^{-1}AS$ for S an integral matrix. Hence A and $S^{-1}AS$ must commute. Hence normality is replaced by commutativity facts. This approach may help to understand the polynomials $p(\alpha)$ whose structure is not yet known as well as one may wish.

In general A and $S^{-1}AS$ do not commute and their commutators may have interesting properties. On the other hand, the representation hides the fact that $f(x)$ has exactly n zeros which are algebraic numbers.

(2) Since $Z^{n\times n}$ contains the representation of all algebraic numbers of degree n relations between different fields can now be considered. As an example a theorem

by Taussky will be mentioned:

Let A,B be 2×2 rational matrices and at least one of them, say A, with irrational eigenvalues α, α'. Then $\det(AB - BA)$ is a norm from $Q(\alpha)$. (The converse of this theorem is also true. A result concerning cases where the norm can be obtained via integral commutators will be mentioned later.

(3) Fields in which a given p splits into principal ideals given by eigenvalues of an integral matrix of $\det \pm p$: this is particularly useful for $n = 2$ [It was pointed out by Estes and P. Morton that there exist infinitely many quadratic fields in which a given prime p splits into principal ideals]. There is a connection there with ideal matrices for prime ideals of norm p. This will become evident later.

(4) The fields in which the eigenvalues of sums or products of integral matrices lie are of interest.

(5) Characterization of elements in the SL(2,Z) which are commutators in the SL(2,Z) are of interest (see Taussky, Problem.)

The main issue of this account concerns facts about the ideals in an order of a fixed algebraic number field and the role of integral matrices connected with them. In this connection two canonical forms of integral matrices play an important role, the Smith normal form and the Hermite normal form.

Another important concept is that of the greatest common divisors of $n \times n$ integral matrices. The existence of such divisors implies the fact that $Z^{n \times n}$ is a principal ideal ring so that every ideal is a left or right principal ideal. The definition of a greatest common right (left) divisor is based on the definition of right divisor and left multiple (left divisor and right multiple): if

$$A = CD, \quad A, C, D \in Z^{n \times n}$$

then D is a right divisor of A and A is left multiple of D. (C is a left divisor and D a right multiple). A greatest common right divisor g c r d of A and B is a common right divisor which is a left multiple of every common right divisor of A and B, (and analogously for g c ℓ d).

There are routine methods for obtaining the common divisors.

These common divisors are unique apart from multiplication by unimodular matrices on the appropriate side.

1. *Ideal matrices.*

This is a concept that goes back to Poincaré, was taken up by MacDuffee, later by Taussky and generalized by G. B. Wagner, Bhandari, Nanda to abstract situations.

In what follows F is an algebraic number field of degree n; its maximal order $\mathfrak{o}$ has basis $\omega_1, \dots, \omega_n$; the algebraic integer α is zero of an integral monic polynomial $f(x)$ and generates F; the algebraic integers $\alpha_1, \dots, \alpha_n$ form a Z-basis for an ideal $\mathfrak{a}$ in an order of F; the matrix C is the companion matrix of $f(x)$.

Definition of ideal matrix.

The integral matrix $X_{\mathfrak{a}}$ for which

$$X_{\mathfrak{a}} \begin{pmatrix} \omega_1 \\ \vdots \\ \omega_n \end{pmatrix} = \begin{pmatrix} \alpha_1 \\ \vdots \\ \alpha_n \end{pmatrix}$$

is called an ideal matrix for $\mathfrak{a}$ with respect to the bases mentioned. Hence all ideal matrices for a given ideal are of the form $U X_{\mathfrak{a}} V$ where U,V are unimodular integral matrices.

An important fact concerning ideal matrices for maximal order is *the* $A \cup B$ *theorem* (Taussky).

Theorem 1. Let A,B be ideal matrices for the ideals $\mathfrak{a}, \mathfrak{b}$ in $\mathfrak{o}$. Then there exists a unimodular matrix U such that the product $\mathfrak{a}\mathfrak{b}$ has $A \cup B$ as ideal matrix.

Characterization of ideal matrices .

Theorem 2. (MacDuffee) Let

$$\omega_i \begin{pmatrix} \omega_1 \\ \vdots \\ \omega_n \end{pmatrix} = X_i \begin{pmatrix} \omega_1 \\ \vdots \\ \omega_n \end{pmatrix} \qquad i = 1, \dots, n$$

and $\mathfrak{X}$ the space over Z with the X_i as basis. Then a matrix X is an ideal matrix with respect to an ideal in an order in F if and only if $X \mathfrak{X} X^{-1}$ is again a ring of integral matrices.

If $\mathcal{O} = Z[\alpha]$ then the set $\mathfrak{X}$ can be replaced by the companion matrix C. For Z[C] is then an isomorphic map of $Z[\alpha]$.

A commutator theorem involving ideal matrices .

Theorem 3. (Hanlon) Let $-n$ be the norm of an algebraic integer in $Q(\sqrt{m})$, m a square free integer. Then there exists an ideal matrix G attached to $Q(\sqrt{m})$ which satisfies

$$-\det(AG - GA) = n$$

where A is the companion matrix of the polynomial of which $\sqrt{m}$, respectively $\frac{1+\sqrt{m}}{2}$ is a zero, depending on whether $m \equiv 2, 3$ respectively $\equiv 1(4)$.

If $-n$ can be put in the form $-n = \text{norm } (s + t\sqrt{m})$, respectively norm $\left(s + t \cdot \frac{1+\sqrt{m}}{2}\right)$ then G can be chosen an ideal matrix for an ideal of norm $(t + 1)$. [This commutator involves integral matrices in contrast to the theorem by Taussky mentioned in the introduction].

Proof. *Case 1.* $m \equiv 1(4)$. Suppose

$$-n = \text{Norm}(s + t(\tfrac{-1}{2} + \tfrac{\sqrt{m}}{2})) = + (s^2 - ts - (\tfrac{m-1}{4})t^2)$$

Let $$G = \begin{pmatrix} t+1 & 0 \\ s & 1 \end{pmatrix}$$

In the present case $$A = \begin{pmatrix} 0 & 1 \\ \frac{m-1}{4} & 1 \end{pmatrix}$$

$$AG - GA = \begin{pmatrix} s & -t \\ t(\frac{m-1}{4})+s & -s \end{pmatrix}$$

$$\det(AG - GA) = -s^z + t^2(\frac{m-1}{4}) + st = -n$$

Assume $(t+1) \mid (\frac{m-1}{4} - s^2 - s)$. We show that G is an ideal matrix or equivalently that GAG^{-1} is an integral matrix.

$$G^{-1} = \begin{pmatrix} \frac{1}{t+1} & 0 \\ \frac{-s}{t+1} & 1 \end{pmatrix}$$

$$GAG^{-1} = \begin{pmatrix} -s & t+1 \\ \frac{(\frac{m-1}{4} - s^2 - s)}{t+1} & s+1 \end{pmatrix}$$

Case 2: $m \equiv 2, 3 \pmod 4$ Suppose $-n = \text{norm}\,(s + t\sqrt{m})$

$$\text{Let } G = \begin{pmatrix} t+1 & 0 \\ s & 1 \end{pmatrix}$$

$$\text{This time } A = \begin{pmatrix} 0 & 1 \\ m & 0 \end{pmatrix}$$

$$AG - GA = \begin{pmatrix} s & -t \\ tm & -s \end{pmatrix}$$

$$\det(AG = GA) = -s^2 + t^2 m = +n.$$

Assume $(t+1) \mid (m - s^2)$.

$$G^{-1} = \begin{pmatrix} \frac{1}{t+1} & 0 \\ \frac{-s}{t+1} & 1 \end{pmatrix}$$

$$GAG^{-1} = \begin{pmatrix} -s & t+1 \\ \frac{m-s^2}{t+1} & s \end{pmatrix}$$

The role of ideal matrices for finding a Z-basis of an ideal .

Theorem 4 (MacDuffee). Let $\alpha_1, \ldots, \alpha_n$ be a Z-basis for the ideal $\mathfrak{a}$ in $\mathfrak{o}$ generated by the numbers $\beta_1, \ldots, \beta_k$ and let $A = (a_{ik})$ be such that

$$\alpha_i = \sum a_{ij}\, \omega_j$$

i.e. A is an ideal matrix for $\mathfrak{a}$. Let $M(\mathfrak{o})$ be a regular representation of $\mathfrak{o}$ into $Z^{n\times n}$. Then A is a g c r d of the matrices $M(\beta_1), \ldots, M(\beta_k)$.

Proof. Since the presentation by MacDuffee is slightly cumbersome a briefer matrix account is given here. The notation is also changed. By definition of the representation

$$\beta_i \begin{pmatrix} \omega_1 \\ \vdots \\ \omega_n \end{pmatrix} = Q_i\, A \begin{pmatrix} \omega_1 \\ \vdots \\ \omega_n \end{pmatrix} = M(\beta_i) \begin{pmatrix} \omega_1 \\ \vdots \\ \omega_n \end{pmatrix}$$

Here Q_i is the n x n matrix (q_{ris}) for which

$$\omega_r\, \beta_i = \sum q_{rij}\, \alpha_j$$

This implies

$$M(\beta_i) = Q_i\, A$$

Hence A is common right divisor of the matrices $M(\beta_i)$.
It will now be shown that every common right divisor of all $M(\beta_j)$ is a right divisor of A, hence A is a g c r d of the $M(\beta_j)$. For this purpose express the α_r in terms of the β_j:

$$\alpha_r = \sum_j \pi_{rj}\, \beta_j$$

where $\pi_{rj} \in \mathcal{O}$. Let

$$\pi_{rj} = \sum_h p_{rjh} \omega_h$$

For each $j = 1, \ldots, k$ let P_j be the matrix whose r,h entry is p_{rjh}. Then

$$\sum P_j M(\beta_j) \begin{pmatrix} \omega_1 \\ \vdots \\ \omega_n \end{pmatrix} = \sum P_j \beta_j \begin{pmatrix} \omega_1 \\ \vdots \\ \omega_n \end{pmatrix} = \sum \beta_j \left(P_j \begin{pmatrix} \omega_1 \\ \vdots \\ \omega_m \end{pmatrix} \right) =$$

$$= \sum \beta_j \begin{pmatrix} \pi_{1j} \\ \vdots \\ \pi_{nj} \end{pmatrix} = \begin{pmatrix} \sum_j \beta_j \pi_{1j} \\ \vdots \\ \sum_j \beta_j \pi_{nj} \end{pmatrix} = \begin{pmatrix} \alpha_1 \\ \vdots \\ \alpha_n \end{pmatrix} = A \begin{pmatrix} \omega_1 \\ \vdots \\ \omega_n \end{pmatrix}$$

Hence

$$A = \sum_j P_j M(\beta_j)$$

Theorem 4 plays an important role for the *representation of ideals* into $Z^{n\times n}$ via ideal matrices.

Here *the fact that an ideal is non principal in* $\mathcal{O}$ *is mapped onto a non commutativity statement in* $Z^{n\times n}$.

The following fact holds for the mapping of $\mathcal{O}$ into $Z^{n\times n}$ mentioned in Theorem 4:

Theorem 5 (Taussky) A principal ideal $\mathfrak{a}$ in $\mathcal{O}$ is mapped onto a set $M(\mathfrak{a})$ in $Z^{n\times n}$ and generates an ideal there. An ideal which is not principal is mapped onto a principal ideal, generated by an element outside $M(\mathfrak{a})$ which does not commute with all the elements of $M(\mathfrak{a})$. [In virtue of Theorem 4 this generator is an ideal matrix of $\mathfrak{a}$].

In case of a principal ideal the generator of the map has to be chosen with a suitable unimodular factor, this factor must ensure that the generator commutes with all of $M(\mathfrak{a})$.

An example is now given to demonstrate the last mentioned fact. It concerns the principal ideal with Z-basis $(9, 2+\sqrt{58})$ and generator $61 + 8\sqrt{58}$ in $Z[\sqrt{58}]$. The ring $Z[\sqrt{58}]$ has basis $1, \sqrt{58}$ and can be represented by the matrices I and $\begin{pmatrix} 0 & 1 \\ 58 & 0 \end{pmatrix}$. An ideal matrix for our ideal can be chosen as $\begin{pmatrix} 9 & 0 \\ 2 & 1 \end{pmatrix}$. But it does not commute with $\begin{pmatrix} 0 & 1 \\ 58 & 0 \end{pmatrix}$. However, the matrix $\begin{pmatrix} 5 & 8 \\ 38 & 61 \end{pmatrix}\begin{pmatrix} 9 & 0 \\ 2 & 1 \end{pmatrix} = \begin{pmatrix} 61 & 8 \\ 464 & 61 \end{pmatrix}$ does commute with $\begin{pmatrix} 0 & 1 \\ 58 & 0 \end{pmatrix}$, being $61\ I + 8 \begin{pmatrix} 0 & 1 \\ 58 & 0 \end{pmatrix}$, the map of the generator. The unimodular matrix $\begin{pmatrix} 5 & 8 \\ 38 & 61 \end{pmatrix}$ is found by solving the linear system obtained from the commutativity requirement. This system of four equations has rank 2. Hence the resulting matrix has as elements linear forms in two parameters. Its determinant is a quadratic form, in our case $9x^2 + 2xy - 6y^2$. It represents 1 for $x = 1$, $y = -1$. This is only possible for a principal ideal.

An example of a non principal ideal is $(3, 8 + \sqrt{79})$ which belongs to an ideal class of order 3 in $Z[\sqrt{79}]$. As ideal matrix we can choose $\begin{pmatrix} 3 & 0 \\ 8 & 1 \end{pmatrix}$. Every ideal matrix corresponding to this ideal will have $\det = \pm 3$, hence 3 is either an eigenvalue of the ideal matrix or 3 factorizes in the field of the eigenvalues of any ideal matrix into principal ideals.

The role of ideal matrices in the similarity of integral matrices.

Theorem 5 (Taussky) Let A, B be integral $n \times n$ matrices satisfying $f(A) = f(B) = 0$. Then the similarity S between A and B can be expressed in terms of the ideal matrices $X_{\mathfrak{a}}$, $X_{\mathfrak{b}}$ of two ideals $\mathfrak{a}, \mathfrak{b}$:

$$(X_{\mathfrak{a}} X_{\mathfrak{b}}^{-1})^{-1} A X_{\mathfrak{a}} X_{\mathfrak{b}}^{-1} = B$$

This fact was used recently to obtain a matrix method for composition of binary quadratic forms.

Similarity of integral matrices plays a role in the theorem of Latimer and MacDuffee to be studied in 2. It concerns the concept of *matrix class :* Let A be an $n \times n$ integral matrix. Then the set $\{S^{-1} AS\}$, S integral unimodular, is called a matrix class.

2. *The theorem of Latimer and MacDuffee.*

This theorem is meaningful for all orders $Z[\alpha]$, not only for the maximal order.

Theorem 6. Let $\{A\}$ be the set of all $n \times n$ integral matrices which satisfy the equation $f(A) = 0$. There is a 1-1 correspondence between the ideal classes in $Z[\alpha]$ and the matrix classes obtained from the set $\{A\}$. The correspondence is derived from the equation

$$A \begin{pmatrix} \alpha_1 \\ \vdots \\ \alpha_n \end{pmatrix} = \alpha \begin{pmatrix} \alpha_1 \\ \vdots \\ \alpha_n \end{pmatrix}$$

(For more details see Taussky).

The matrix classes are bound to the polynomial $f(x)$. If another polynomial is chosen to generate the same $Z[\alpha]$ another set of matrix classes is obtained. e.g. let $n = 2$ and A a 2×2 matrix. Then $-A$ defines, in general, another set of matrix classes.

If $Z[\alpha]$ is the maximal order then its ideals have inverses. It was shown by Taussky that the class of the inverse of an ideal $\mathfrak{a}$ corresponds to the matrix class of the transpose of the matrix A which corresponds to the class of $\mathfrak{a}$.

This leads to results concerning ideal classes of order 2 and the corresponding matrix classes.

For $n = 2$ an explicit rule for multiplication of matrix classes was obtained by Taussky. This does not use the correspondence with ideals. It is a by-product of the matrix method for composition of binary quadratic forms.

Applications of the theorem of Latimer and MacDuffee have been made in group theory, via the polynomial $f(x) = x^n - 1$ (this is not irrecucible, but the theorem of Latimer and MacDuffee holds also for algebras determined by polynomials with distinct zeros), see, e.g. Magnus, Plesken.

It is also useful in the study of conjugacy classes of the SL(n,Z).

It is important to point out that the ideal class group in the narrow sense is not in correspondence with matrix classes defined via unimodular matrices of det $=\pm 1$. This becomes clear from observing that the equation

$$S^{-1}AS \left(S^{-1} \begin{pmatrix} \alpha_1 \\ \vdots \\ \alpha_n \end{pmatrix}\right) = S^{-1} \begin{pmatrix} \alpha_1 \\ \vdots \\ \alpha_n \end{pmatrix}$$

shows that the matrix $S^{-1}AS$ corresponds to the same ideal, only its basis is changed. Conversely, the above equation is not changed if the α_i are multiplied by an element in $Q(\alpha)$ (as long as the products remain integral) irrespective of the sign of the norm of this element).

The situation is different for quadratic form classes connected with ideal classes. There the correspondence is only a homomorphism. If we study the connection with matrix classes the following emerges:

Let $A = \begin{pmatrix} a_{11} & a_{12} \\ a_{21} & a_{22} \end{pmatrix}$ be an integral matrix as studied previously. Consider the quadratic form

$$f(x,y) = a_{21}\, x^2 + (a_{22} - a_{11})\, xy - a_{12}\, y^2$$

It has the same discriminant as A. Let $S = \begin{pmatrix} s_{11} & s_{12} \\ s_{21} & s_{22} \end{pmatrix}$ be a unimodular integral matrix with $\det S = +1$. Then the transformation

$$\begin{pmatrix} x \\ y \end{pmatrix} \to S \begin{pmatrix} x \\ y \end{pmatrix}$$

transforms S into a form in the same class as f. If $\det S = -1$ we can express S as $\begin{pmatrix} 0 & 1 \\ 1 & 0 \end{pmatrix} S_1$ where $\det S_1 = +1$. The matrix $\begin{pmatrix} 0 & 1 \\ 1 & 0 \end{pmatrix}$ transforms A into $\begin{pmatrix} a_{22} & a_{21} \\ a_{12} & a_{11} \end{pmatrix}$ and the corresponding form is then $-f$.

If two matrices A, B do not lie in the same class, it is of interest to ask whether they are similar via an integral matrix in an extension field of Q, with det a unit. This led to the Guralnick-Jacobinski-Taussky theorem. While Taussky studied precisely the question asked above it was later realized that it can be obtained from a more abstract situation; Guralnick generalized Jacobinski's work. In an unpublished paper by Dade the result of Taussky was generalized to include reducible characteristic polynomials with non repeated zeros and invertible ideals in non maximal orders.

3. *Connections with integral quaternions.*

The ring of Hurwitz quaternions is the maximal order in the algebra of rational quaternions. It is a principal ideal domain. It was pointed out by Rehm that the concept of ideal matrix has an analog in this ring. Rehm was stimulated by a treatment of Venkov who used quaternions in the study of Gauss' result concerning sums of three squares. Instead of matrix classes one studies sets - called bundles - $\{\varepsilon\mu\varepsilon^{-1}\}$ where $\mu^2 = -m$ (m a fixed rational integer; μ a quaternion without scalar part, ε a unit in the Hurwitz ring.

Hanlon's thesis studies further facts which are analogs in these two situations. He further uses Hurwitz' quaternions for the study of ideal classes in imaginary quadratic fields. [The proof of Theorem 4 and the numerical example were developed in collaboration with him].

Bibliography

Baumert, L.:
Query 153.
Notices Amer. Math. Soc. 25, 252 (1978).

Bhandari, S. K.:
Ideal matrices for Dedekind domains.
J. Indian Math. So. 42, 109-126 (1978).

Bhandari, S. K. and Nanda, V. C.:
Ideal matrices for relative extensions.
Abh. Math. Sem. Univ. Hamburg 49, 3-17 (1977).

Dade, E. C.:
Invertible ideals capitulate sooner or later.
Manuscript.

Estes, D.:
Determinants of Galois autormorphisms of maximal commutative rings of 2 x 2 matrices.
Linear Alg. and Appl. 21 (1979), 225 - 243.

Foster, L.:
On the characteristic roots of the product of certain rational integral matrices of order two.
California Institute of Technology thesis, 1964.

Gustafson, W. H.:
Remarks on the history and applications of integral representations.
These Springer Lecture Notes

Guralnich, R.:
Isomorphism of modules under ground ring extension.
Lin. Alg. and Appl. to appear.

Jacobinski, H.:
Uber Geschlechter von Ordnungen.
J. reine und angew. Math. 230, 29-39 (1968).

Kruglzak, S. A.:
Precise ideals of integer matrix rings of the second order.
Ukrain. Mat. Z. 18, 58-64 (1966).

Levy, L. S.:
Almost diagonal matrices over Dedekind domains.
Math. Zeitschr. 124, 89-99 (1972).

MacDuffee, C. C.:
The theory of matrices.
Ergebnisse der Mathematik, Springer 1933.

MacDuffee, C. C.:
An introduction to the theory of ideals in linear associative rings.
Trans. Amer. Math. Soc. 31, 71-90 (1928).

MacDuffee, C. C.:
A method for determining the canonical basis of an ideal in an algebraic field.
Math. Ann. 105, 663-665(1931.

Magnus, W.:
Non euclidean tesselations and their groups.
Academic Press 1974.

Mahler, K.:
Inequalities for ideal bases in algebraic number fields.
J. Austral. Math. Soc. 4, 425-448 (1964).

Malysev, A. V. and Paceo, U. V.:
On the arithmetic of second order matrices.
Zap. Nauc. Ses., Moscow 93, 41-86 (1980).

Mann, H. and Yamamoto, K.:
On canonical bases of ideals.
J. of Combinatorial Theory 2, 71-76 (1967).

Newman, M.:
Integral matrices.
Acad. Press 1972.

Plesken, W.:
Beiträge zur Bestimmung der endlichen irreduziblen Untergruppen von GL(n,Z) und ihrer ganzzahligen Darstellungen.
Ph.D. Thesis, Aachen, 1-69 (1976).

Plesken, W. and Pohst, M.:
On maximal finite irreducible subgroups of GL(n,Z) II, The six dimensional case.
Mathematics of Computation 31, 552-573 (1977).

Rademacher, H.:
Zur Theorie der dedekindschen Summen.
Math. Z. 63, 445-463 (1955)

Rehm, H. P.:
On Ochoa's special matrices in matrix classes.
Linear Algebra and Appl. 17, 181-188 (1977).

Rehm, H. P.:
On a theorem of Gausz concerning the number of integral solutions of the equation
$x^2 + y^2 + z^2 = m$.
Seminar Notes on ternary forms and norms, to appear, Dekker.

Reiner, I., Roggenkamp, K. W.:
Integral representations.
Lecture Notes in Mathematics, 744, Springer 1979

Roggenkamp, K. W. and Huber-Dyson ,V.:
Lattices over orders.
Lecture Notes in Mathematics 115, Springer 1970.

Rosenbrock, H. H.:
State-Space and Multivariable Theory.
J. Wiley, 1970.

Schur, I.:
Über Ringbereiche im Gebiet der ganzzahligen linearen Substitutionen.
Sitzg. Ber. Preuss. Akad. Wiss. (1922), 145 - 168.

Siegel, C. L.:
Über die analytische Theorie der quadratischen Formen III.
Ann. Math. 38, 212-291 (1937).

Taussky, O.:
On a theorem of Latimer and MacDuffee.
Canadian Journal Mathematics 1, 300-302 (1949).

Taussky, O.:
On matrix classes corresponding to an ideal and its inverse.
Illinois Journal Mathematics 1, 103-113 (1957).

Taussky, O.:
Ideal matrices, I.
Archiv der Mathematik 13, 275-282 (1962)

Taussky, O.:
Ideal matrices, II.
Math. Ann. 150, 218-225 (1963).

Taussky, O.:
On the similarity transformation between an integral matrix with irreducible characteristic polynomial and its transpose.
Math. Ann. 166, 60-63 (1966).

Taussky, O.:
Research Problem 10.
Bull. Amer. Math. Soc. 64, 124 (1958).

Taussky, O.:
Additive commutators of rational 2 x 2 matrices.
Linear Alg. and Appl. 12 (1975), 1 - 6.

Taussky, O.:
Connections between algebraic number theory and integral matrices.
Appendix to H. Cohn, A classical invitation to algebraic numbers and class fields, Springer (1978).

Taussky, O.:
A diophantine problem arising out of similarity classes of integral matrices.
J. of Number Theory 11, 472-475 (1979).

Taussky, O.:
Some facts concerning integral representations of the ideals in an algebraic number field.
Linear Alg. and Appl. 31, 245-248 (1980).

Taussky, O.:
Composition of binary integral quadratic forms and composition of matrix classes.
To appear in Lin. and Multilin Algebra.

Venkow, V.:
On the arithmetic of quaternions.
Bull. de l'académie des Sciences de l'URSS, 205-246; 489-504; 535-562; 607-622 (1922, 1929).

Wagner, G. B.:
Ideal matrices and ideal vectors.
Math. Ann. 183, 241-249 (1969).

ZETA-FUNCTIONS OF ORDERS

Colin J. Bushnell & Irving Reiner (*)

Introduction

The Dedekind zeta-function $\zeta_R(s)$ is a generalisation of the ordinary Riemann zeta-function $\zeta(s)$, and is defined as follows. Let R be the ring of all algebraic integers in a number field K, and let $\underline{\underline{a}}$ range over all ideals of R of finite index $(R:\underline{\underline{a}})$. We then define

$$\zeta_R(s) = \sum_{\underline{\underline{a}}} (R:\underline{\underline{a}})^{-s}, \qquad \mathrm{Re}(s) > 1.$$

Hecke proved that $\zeta_R(s)$ has an analytic continuation to the whole complex s-plane, and that $\zeta_R(s)$ has a simple pole at $s = 1$. Further, there is a functional equation connecting $\zeta_R(s)$ and $\zeta_R(1-s)$. The Dedekind zeta-function plays a fundamental role in analytic number theory, in studying the distribution of prime ideals of R, and in calculating the ideal class number of R.

This zeta-function was generalised further, by Hey, to the case of maximal orders in simple algebras. More recently, Louis Solomon [3] introduced a still more general zeta-function defined thus. Let Λ be a $\mathbb{Z}$-order in a semisimple $\mathbb{Q}$-algebra A, and let L range over all left ideals of Λ of finite index $(\Lambda:L)$. The Solomon zeta-function is defined by

$$\zeta_\Lambda(s) = \sum_{L} (\Lambda:L)^{-s},$$

which converges and gives an analytic function for Re(s) sufficiently large.

(*) Research of the second author was supported by the National Science Foundation.

This is, in turn, a special case of a more general situation. Let N be a fixed Λ-lattice, and let

$$\zeta_\Lambda(N;s) = \sum_{L \subseteq N} (N:L)^{-s},$$

the sum extending over all sublattices L of N of finite index (N:L). This series converges to an analytic function of s for Re(s) sufficiently large. Solomon introduced this function $\zeta_\Lambda(N;s)$ with the aim of computing the number of isomorphism classes of such Λ-lattices L in a fixed Λ-lattice N. Unfortunately, the theory needed for such applications is still in the process of being developed, but meanwhile one must establish the basic properties of such zeta-functions. Here, we shall describe some of Solomon's results, as well as the recent work by Bushnell and Reiner. For the sake of simplicity, we shall restrict our attention to $\zeta_\Lambda(s)$, rather than the more general $\zeta_\Lambda(N;s)$ just defined.

§1 Euler Product

Let R be the ring of all algebraic integers in a number field K, and let P range over all prime ideals of R. We set NP = (R:P), the absolute norm of P. Since every ideal of R is a power product of prime ideals, one obtains an Euler product formula

$$\zeta_R(s) = \prod_P \{1 + NP^{-s} + NP^{-2s} + \dots\} = \prod_P \{1 - NP^{-s}\}^{-1},$$

valid for Re(s) > 1. Furthermore, if R_P denotes the P-adic completion of R, then the factor $(1 - NP^{-s})^{-1}$ is precisely the zeta-function of the valuation ring R_P (with the obvious definitions), so we have

$$\zeta_R(s) = \prod_P \zeta_{R_P}(s), \qquad \mathrm{Re}(s) > 1.$$

Suppose now that Λ is a $\mathbb{Z}$-order in a finite-dimensional semisimple $\mathbb{Q}$-algebra A, and define $\zeta_\Lambda(s)$ as above. For each rational prime p, let the subscript p indicate p-adic completion. Then Λ_p is a $\mathbb{Z}_p$-order in the semi-simpl $\mathbb{Q}_p$-algebra A_p, and we may form its zeta-function $\zeta_{\Lambda_p}(s)$. For each left ideal L of finite index in Λ, we have

$$(\Lambda:L) = \prod_p (\Lambda_p : L_p),$$

and almost all factors are 1. Using this, Solomon established the Euler product formula

$$\zeta_\Lambda(s) = \prod_p \zeta_{\Lambda_p}(s).$$

Thus, we can compute $\zeta_\Lambda(s)$ by working locally.

For almost all primes p, the completion A_p is a direct sum of full matrix algebras over fields, and Λ_p is a maximal $\mathbb{Z}_p$-order in A_p. Such primes p will be called "good" primes, and all other primes "bad", so there are only a finite number of bad primes. At each good prime, Λ_p is isomorphic to a direct sum of full matrix rings over complete discrete valuation rings. In this situation, we can describe all of the left ideals of Λ_p, and can calculate $\zeta_{\Lambda_p}(s)$ explicitly. In particular, if $\Lambda = M_n(R)$, where R is a discrete valuation ring, it turns out that

$$\zeta_\Lambda(s) = \prod_{j=0}^{n-1} \zeta_R(ns - j) = \zeta_R(ns)\,\zeta_R(ns-1)\ldots\zeta_R(ns-(n-1)).$$

It is convenient to introduce a "global" zeta-function $\zeta_A(s)$, depending on the $\mathbb{Q}$-algebra A rather than on Λ, such that

(*) $$\zeta_{A_p}(s) = \zeta_{\Lambda_p}(s) \qquad \text{for each good prime } p.$$

To define $\zeta_A(s)$, first write $A = \Pi A_i$ (Wedderburn components), and set $\zeta_A(s) = \Pi \zeta_{A_i}(s)$, so it suffices to give the definition for the case of a simple algebra. Let K_i be the centre of A_i, n_i^2 the dimension of A_i over K_i, and R_i the ring of algebraic integers in K_i. Then define

$$\zeta_{A_i}(s) = \prod_{j=0}^{n_i-1} \zeta_{R_i}(n_i s - j),$$

a product of Dedekind zeta-functions. Using this definition, Solomon showed that (*) holds. Therefore we may write

$$\zeta_\Lambda(s) = \zeta_A(s) \prod_{\text{bad } p} \phi_p(s), \qquad \text{where } \phi_p(s) = \zeta_{\Lambda_p}(s)/\zeta_{A_p}(s).$$

Thus, $\zeta_\Lambda(s)$ is expressed in terms of Dedekind zeta-functions, apart from a finite number of correction factors $\phi_p(s)$ at the bad primes p.

§2 Maximal Orders

When Λ is a maximal $\mathbb{Z}$-order, K.Hey determined the correction factors $\phi_p(s)$ explicitly, as we now describe. For each p, Λ_p is a maximal $\mathbb{Z}_p$-order in A_p, and so Λ_p decomposes according to the Wedderburn decomposition of A_p. Changing notation, let A be a simple $\mathbb{Q}_p$-algebra with centre F, and let

$$A = M_k(D), \quad \dim_F(D) = e^2, \quad n = ke,$$

where D is a skewfield with centre F. Let R be the valuation ring in F, P its prime ideal, and let Δ be the integral closure of R in D. Then $\Lambda \cong M_k(\Delta)$, and Δ is a non-commutative discrete valuation ring. Using the fact that all left ideals of Λ are principal, Hey showed that

$$\zeta_\Lambda(s) = \Pi\ \zeta_R(ns - j),$$

where the product is taken over $j \equiv 0 \pmod e$, $0 \leq j < n$. Since

$$\zeta_A(s) = \prod_{j=0}^{n-1} \zeta_R(ns - j),$$

in this case, we obtain

$$\phi_p(s) = \Pi\ (1 - NP^{-(ns-j)}),$$

with the product over $j \not\equiv 0 \pmod e$, $0 \leq j < n$. Therefore $\phi_p(s) \in \mathbf{Z}[p^{-s}]$, since NP is a power of p. Note further that $\phi_p(s) = 1$ whenever $e = 1$, which proves our earlier statement about the behaviour of $\zeta_\Lambda(s)$ at good primes.

§3 Solomon's Conjecture

Now let Λ be an arbitrary $\mathbf{Z}$-order in a $\mathbf{Q}$-algebra A, and define

$$\phi_p(s) = \zeta_{\Lambda_p}(s)/\zeta_{A_p}(s)$$

as in §1. Hey's formula showed that $\phi_p(s) \in \mathbf{Z}[p^{-s}]$ whenever Λ is a maximal order. Solomon conjectured that the same holds true for <u>every</u> order Λ. Using an ingenious combinatorial argument based on Möbius functions defined on the lattice of left ideals of Λ, Solomon proved that $\phi_p(s)$ is a rational function of p^{-s}. Below, we shall sketch our proof that $\phi_p(s)$ is always a polynomial in p^{-s}, with rational integral coefficients.

It follows as an immediate consequence that $\zeta_\Lambda(s)$ has an analytic continuation to the whole complex s-plane, and it is easy to find the order of its pole at $s = 1$. Furthermore, when Λ is a self-dual order such as an integral group ring RG, where G is a finite group and R is a P-adic ring of integers, we can establish a functional equation.

§4 Calculations and Conjectures

We have seen how to calculate $\zeta_\Lambda(s)$ when Λ is maximal. Similar techniques work when Λ is a hereditary order. On the other hand, when Λ is an integral group ring, the calculation of $\zeta_\Lambda(s)$ can be rather difficult. Since $\zeta_\Lambda(s)$ can be expressed in terms of Dedekind zeta-functions, we need only calculate the correction factors $\phi_p(s)$. In practice, this local problem is best handled by expressing Λ as a fibre product:

$$\begin{array}{ccc} \Lambda & \longrightarrow & \Lambda_1 \\ \downarrow & & \downarrow \\ \Lambda_2 & \longrightarrow & \overline{\Lambda} \end{array}$$

where $\overline{\Lambda}$ is a finite ring. Each left ideal L of Λ gives rise to a pair of ideals L_1, L_2 of Λ_1, Λ_2 respectively. If the ideal theory of Λ_1 and Λ_2 is sufficiently simple, one can then determine all ideals L corresponding to a given pair L_1, L_2. This technique has been applied to the cases where $\Lambda = \mathbb{Z}G$, with G a cyclic p-group or a dihedral group of order 2p.

There are many exciting directions for future research, and we may list a few:

i) Find systematic procedures for calculating $\zeta_\Lambda(s)$.

ii) Does the zeta-function defined by using right ideals of Λ coincide with the zeta-function defined via left ideals? (yes, if Λ is a group ring!)

iii) How can these zeta-functions, and in particular their behaviour at the pole at s = 1, be used to determine the number of isomorphism classes and the number of genera of left ideals of Λ ?

iv) What information can we obtain about the distribution of maximal left ideals of Λ, or more generally, about the maximal sublattices of a given Λ-lattice ?

v) Let M range over the representatives of the isomorphism classes of left ideals of Λ of finite index. Then

$$\zeta_\Lambda(s) = \sum_M Z_\Lambda(M;s),$$

where

$$Z_\Lambda(M;s) = \sum_{\substack{L \subseteq \Lambda \\ L \simeq M}} (\Lambda:L)^{-s}.$$

How does this "partial zeta-function" vary with M ?

vi) What is the appropriate theory of L-series for such zeta-functions ?

§5 Local Analytic Theory

Local zeta-functions of fields first achieved prominence in Tate's thesis [4], in an attempt to supersede Hecke's global methods of treating Dedekind zeta-functions and the L-functions attached to "grössencharacters". Tate's ideas were subsequently extended to cover the much more complicated case of local simple algebras, as in [2]. The zeta-functions of orders may be viewed, with some advantage, as a straightforward special case of this general theory. See [1] for a full treatment of the material of this section.

5.1 We adopt the following notations: A is a finite-dimensional semisimple $\mathbb{Q}_p$-algebra, Λ is a $\mathbb{Z}_p$-order in A, and M is a fixed full Λ-lattice in A (that is, M spans A over $\mathbb{Q}_p$). As before, we have

$$\zeta_\Lambda(s) \quad = \quad \sum_{L \subseteq \Lambda} (\Lambda:L)^{-s},$$

and the partial zeta-function

$$Z_\Lambda(M;s) \quad = \quad \sum_{\substack{L \subseteq \Lambda \\ L \simeq M}} (\Lambda:L)^{-s}.$$

Both these series converge, and define analytic functions, in some right-hand complex half-plane. They may be regarded as elements of the ring $\mathbb{Z}[[p^{-s}]]$of formal power series in p^{-s} over $\mathbb{Z}$. Some more notation: if M_1, M_2 are full left Λ-lattices in A, we put

$$\{M_1,M_2\} \quad = \quad \{x \in A \mid M_1 x \subseteq M_2\}.$$

This is a full $\mathbb{Z}_p$-lattice in A, and hence a compact open subgroup of A, in its natural topology. If $M_1 = M_2$, it is an order.

Clearly, a left Λ-lattice $L \subseteq \Lambda$ is Λ-isomorphic to M if and only if there exists $x \in A^\times \cap \{M,\Lambda\}$ such that $L = Mx$. Moreover, $Mx = My$ if and only if $xy^{-1} \in \{M,M\}^\times$. (The notation $^\times$ means the group of units of a ring.) Thus

$$Z_\Lambda(M;s) \quad = \quad \sum (\Lambda:Mx)^{-s},$$

where the summation extends over $x \in \{M,M\}^\times \backslash (A^\times \cap \{M,\Lambda\})$. We may transform this further by first introducing the "generalised group index"

$$(M_1:M_2) = \frac{(M_1:M_1 \cap M_2)}{(M_2:M_1 \cap M_2)}.$$

The index $(M:Mx)$ depends only on x, not on M, so we denote it by

$$\|x\|^{-1} = \|x\|_A^{-1} = (M:Mx).$$

This is a continuous homomorphism from $A^\times$ to the group of positive real numbers, and its values are integral powers of p. We let $\Phi = \Phi_{M,\Lambda}$ denote the characteristic function of the set $\{M,\Lambda\}$ in A. That is, Φ takes the value one on the lattice $\{M,\Lambda\}$, and zero elsewhere on A. Then we have

$$Z_\Lambda(M;s) = (\Lambda:M)^{-s} \sum_x \Phi(x) \|x\|^s,$$

where the sum is now taken over $x \in \{M,M\}^\times \backslash A^\times$. This range of summation is a principal homogeneous space over the group $A^\times$, and its natural topology is discrete. Therefore this summation process represents invariant integration. So, we choose a Haar measure $d^\times x$ on $A^\times$, such being necessarily bi-invariant; then $Z_\Lambda(M;s)$ takes the form

$$Z_\Lambda(M;s) = (\Lambda:M)^{-s} \mu^\times(\{M,M\}^\times)^{-1} \int_{A^\times} \Phi(x) \|x\|^s d^\times x.$$

Here, $\mu^\times$ denotes the measure of a set with respect to $d^\times x$. As a special case, if Λ' is a maximal order in A, we obtain

$$\zeta_{\Lambda'}(s) = \mu^\times(\Lambda'^\times)^{-1} \int_{A^\times} \Phi(x) \|x\|^s d^\times x,$$

where Φ is now the characteristic function of Λ' in A.

5.2 To place this expression for $Z_\Lambda(M;s)$ in its correct formal context, we let S(A) denote the vector space of complex-valued functions on A which are locally constant and of compact support. Then, for $\Phi \in S(A)$, we define

$$Z(\Phi;s) \quad = \quad \int_{A^\times} \Phi(x)\, \|x\|^s \, d^\times x.$$

This expression is valid for Re(s) sufficiently large. In fact, Re(s) > 1 will do here. When A is a field, these are precisely the functions considered in [4]. Apart from a harmless exponential factor, $Z_\Lambda(M;s)$ is of this form.

By choosing special Φ, Tate was able to recover the local Dedekind zeta-function, and, in a more general context, the L-series. The classical functions play an important role here, abstracted in [2] in the guise of the "Euler factor". In our situation, the Euler factor is a function $L_A(s)$ which satisfies

i) $L_A(s) = f(p^{-s})^{-1}$, for some $f(X) \in \mathbb{C}[X]$ with $f(0) = 1$;

ii) there exists $\Phi \in S(A)$ such that $L_A(s) = Z(\Phi;s)$;

iii) for any $\Phi \in S(A)$, $L_A(s)^{-1}Z(\Phi;s) \in \mathbb{C}[p^s, p^{-s}]$.

These conditions determine $L_A(s)$ uniquely, provided it exists.

The module theory provides us with a candidate for $L_A(s)$, namely $\zeta_{\Lambda'}(s)$, for any maximal order Λ' in A. This is independent of the choice of Λ', and is therefore canonically associated with A. Moreover, it is of the form $f(p^{-s})^{-1}$, where $f(X) \in \mathbb{Z}[X]$ has constant term 1. This is in §2 when A is simple. In general, $\zeta_{\Lambda'}$ is the product of the corresponding functions of the Wedderburn components of A. We have just seen that it satisfies ii), and indeed:

THEOREM: $\zeta_{\Lambda'}(s) = L_A(s)$.

To prove this, it is enough to check that $\zeta_{\Lambda'}(s)^{-1} Z(\Phi;s)$ lies in $\mathbb{C}[p^s,p^{-s}]$ for a set of functions Φ which spans the vector space S(A). However, for $y \in A^\times$, we have

$$\int_{A^\times} \Phi(xy) \|x\|^s d^\times x = \|y\|^{-s} \int_{A^\times} \Phi(x) \|x\|^s d^\times x,$$

so we need only do it for a set whose multiplicative translates span S(A). A convenient choice here is the set of characteristic functions of spheres, i.e.,sets of the form $\alpha + p^f\Lambda'$, $\alpha \in \Lambda'$, $f \in \mathbb{Z}$, $f > 0$. This leads to a rather complicated calculation, but one can reduce to the case of A simple, and work explicitly with matrices. When Φ is the characteristic function of a sphere, $Z(\Phi;s)$ turns out to be the product of an exponential factor and some of the factors from the product expression for $\zeta_{\Lambda'}$ given in §2.

Corollary 1: $\zeta_{\Lambda'}(s)^{-1} Z_\Lambda(M;s) \in \mathbb{Z}[p^{-s}]$.

For, this quotient lies in both $\mathbb{C}[p^s,p^{-s}]$ and $\mathbb{Z}[[p^{-s}]]$, and thus in $\mathbb{Z}[p^{-s}]$. Solomon's Conjecture follows.

Corollary 2: <u>For any $\Phi \in S(A)$, the function $Z(\Phi;s)$ admits analytic continuation to a meromorphic function of s, whose singularities are among those of $\zeta_{\Lambda'}(s)$.</u>

5.3 Another aspect of Tate's thesis concerns functional equations. We outline a general result in this area. Let F be a subfield of the centre of A, and let $\lambda: A \times A \to F$ be a nondegenerate symmetric F-bilinear form. For example, if A were F-central simple, $\lambda(x,y)$ could be the reduced trace of xy. Let ψ be a non-trivial continuous character of the additive group of the field F. Then the pairing $(x,y) \mapsto \psi(\lambda(x,y))$ is nondegenerate, and induces an isomorphism (of topological groups) between A and its Pontrjagin dual.

For a certain Haar measure dx on A, we can now define the Fourier transform $\hat{\Phi}$ of a function $\Phi \in S(A)$ by

$$\hat{\Phi}(y) = \int_A \Phi(x)\ \psi(\lambda(x,y))\ dx, \qquad y \in A.$$

Then $\hat{\Phi}$ again lies in S(A). The choice of dx here is not arbitrary; it is the unique Haar measure on A for which the Fourier Inversion Formula $\hat{\hat{\Phi}}(x) = \Phi(-x)$, $x \in A$, $\Phi \in S(A)$, holds. It does, of course, depend on ψ and λ.

FUNCTIONAL EQUATION (Tate, Godement, Jacquet):

<u>For any</u> $\Phi, \Psi \in S(A)$, <u>we have</u>

$$Z(\Phi;s)\ Z(\hat{\Psi};1-s) = Z(\hat{\Phi};1-s)\ Z(\Psi;s),$$

<u>in the sense of analytic continuation.</u>

The hard part of this result is the case of A simple. The transition to the semisimple case is then easy. When A is a field, it is due to Tate. Tate's proof works equally well for division algebras. The general case is due to Godement and Jacquet [2], where it is proved as a special case of a very much harder and more general theorem. It is possible to give a straightforward proof of the case to hand, as in the Appendix to [1].

Now let us specialise to the case of group rings: G is a finite group,

$A = FG$, where $F/\mathbb{Q}_p$ is a finite field extension, $\Lambda = RG$, where R is the valuation ring in F. We let λ be the pairing $(x,y) \mapsto t(xy)$, where $t: FG \to F$ is the F-linear map given, for $g \in G$, by

$$t(g) = \begin{cases} 1 & \text{if } g = 1_G, \\ 0 & \text{otherwise.} \end{cases}$$

The character ψ is chosen to be trivial on R, but non-trivial on $\pi^{-1}R$, where π denotes a prime of R. For these choices, if Φ is the characteristic function of a lattice M, then $\hat{\Phi}$ is the product of a positive constant and the characteristic function of $\{M,RG\}$. We also need the F-linear involution * of FG given by $g^* = g^{-1}$, $g \in G$. Let us write

$$\tilde{M} = \{M,RG\}^*.$$

THEOREM: <u>Let Λ' be a maximal order in</u> FG <u>containing</u> RG. <u>If</u> M <u>is any full left</u> RG-<u>lattice in</u> FG, <u>we have</u>

$$\frac{Z_{RG}(M;s)}{Z_{RG}(\tilde{M};1-s)} = (\Lambda':RG)^{1-2s} \frac{\zeta_{\Lambda'}(s)}{\zeta_{\Lambda'}(1-s)}$$

$$= \frac{\zeta_{RG}(s)}{\zeta_{RG}(1-s)}.$$

One proves this by taking Φ to be the characteristic function of $\tilde{M}$ and Ψ the characteristic function of Λ' in the general functional equation.

It is perhaps more enlightening to write

$$\phi(s) = \frac{\zeta_{RG}(s)}{\zeta_{\Lambda'}(s)}.$$

Then $\phi(s) \in \mathbb{Z}[p^{-s}]$, and has constant term 1. The functional equation reads

$$\phi(s) \quad = \quad (\Lambda':RG)^{1-2s}\ \phi(1-s).$$

If we let $(\Lambda':RG) = p^n$, and $\phi(s) = f(p^{-s})$, $f(X) \in \mathbb{Z}[X]$, this says

$$f(X) \quad = \quad p^n X^{2n}\ f(1/pX).$$

So the polynomial f has highest term $p^n X^{2n}$, and there is a symmetry among its coefficients:

$$f(X) \quad = \quad a_0 + a_1 X + a_2 X^2 + \ldots + a_{2n} X^{2n},$$

$$p^{n-i} a_i \quad = \quad a_{2n-i}, \qquad 0 \leq i \leq 2n.$$

This is consistent with a "Riemann hypothesis" for $\phi(s)$, which is certainly not true in this generality. However, the location of the zeros of $\phi(s)$ will be important in later applications.

REFERENCES

1 Bushnell, C.J. & Reiner, I. Zeta-functions of arithmetic orders and Solomon's Conjectures. Mathematische Zeitschrift 173 (1980) 135-161.

2 Godement, R. & Jacquet, H. Zeta-functions of simple algebras. Lecture Notes in Mathematics 260. Berlin-Heidelberg-New York: Springer 1972.

3 Solomon, L. Zeta-functions and integral representation theory. Advances in Math. 26, 306-326 (1977).

4 Tate, J.T. Fourier analysis in local fields and Hecke's zeta-functions. Thesis, Princeton University 1950 (= Cassels J.W.S. & Fröhlich A. (eds.):

Algebraic Number Theory, 305-347. London: Academic Press, 1967).

University of London King's College,
Department of Mathematics,
Strand, London WC2R 2LS.

University of Illinois,
Department of Mathematics
Urbana, Illinois 61801.

The Class Group à la Fröhlich

Jürgen Ritter (Heidelberg)

§ 1. Introduction

In my lecture I'm going to talk about the description of the locally free class group of an arithmetic order spanning a semisimple algebra, as it was given by Fröhlich in terms of certain homomorphisms which are defined on the Grothendieck group belonging to the corresponding semisimple algebra. Main references are [F1, F2, A]. This description of the class group has got many nice properties; so far example it doesn't refer to how the algebra splits into its simple components and thus reproduces the functorial behaviour of the class group in very clear way. As a main example which reflects the worth of the homomorphism language in this context very nicely, we should like to interpret the ring o_L of integers of L, which is a finite Galois extension of a number field K with corresponding Galois group G, as an element of the locally free class group of $o_K[G]$, provided the extension L/K is only tamely ramified; and for this effect we have to introduce Fröhlich's generalized concept of the Lagrange resolvent (which was only defined for cyclic extensions).

(1.1) To begin with let us fix some notation. A is a semisimple algebra over K which is the quotient field of a Dedekind domain o, and a is an o-order in A. We call o (or a) local or global if K is a finite extension of either the p-adic number field $\underline{Q}_p$ or of the field $\underline{Q}$ of rational numbers, respectively. Now, what we are mainly interested in is the category of isomorphism classes of a-lattices $L(a)$, or to be somewhat more modest, its subcategory $P(a)$ of isomorphism classes of projective a-lattices. At this stage it is perhaps a good idea to recall some basic facts on orders and their lattices, all of which can be looked up in Reiner's book [R].

(1.1.a) In case o is local the Krull-Schmidt theorem holds for a; this is no longer true if o is global.

(1.1.b) For a maximal order a the decomposition of A into its simple components A_i leads to a decomposition of a itself, namely a is the direct sum of the projections a_i of a in A_i, which by themselves are maximal orders in A_i with respect to o as well as to the integral closure of o in the centre of A_i. In case of a maximal order one therefore can always restrict oneself to the central simple situation.

(1.1.c) In the global case a Hasse global-local-principle is valid for various notations in connection with orders and their lattices: so for the notation of the maximality of an order, so for the notation of the projectivity of a lattice, and for the notation of exact sequences, kernels and cokernels, with respect to lattices. One should also keep in mind that two full lattices in a semisimple algebra A differ only at finitely many places p of K, since each of them has a finite index in their sum; especially an order is locally almost everywhere a maximal order. Finally we should mention that, as follows from the approximation theorem, a full lattice in A may be locally defined in the following way: up to finitely many exceptions it has to coincide with a given full lattice in A and at the exceptional places p it may be any full lattice in the localized algebra A_p.

(1.1.d) Assume now a to be maximal and local, and A to be central simple. Then, essentially by Hensel's lemma, one gets: the lattices are all free and hence fully determined by their rank.

(1.1.e) If a is maximal and global (and again A central simple) then by (1.1.c,d) all lattices are still projective. To see now what the category $L(a)$ in this case will be we have to look at the Grothendieck group $K_o(a)$ [R, § 36] on the one side and, on the other side, at cancellation properties:

An a-lattice M is locally free and thus has a local rank n

$$M \otimes_o K_p \simeq A_p^n \qquad \text{(some } p\text{)} .$$

By the Noether-Deuring theorem [CR, 29.11] also $M \otimes_o K \simeq A^n$. Hence M has a global rank, rk M = n, and this gives rise to the exact sequence:

$$0 \to cl(a) \to K_o(a) \to \underline{Z} \to 0 \quad ,$$

by which the class group $cl(a)$ of a is defined. Using Eichler's beautiful approximation theorem for the group $A^{\times'}$ of units in A having reduced norm 1 Swan [Sw2] showed that the reduced norm induces an isomorphism $cl(a) \simeq cl_A(o)$ of the class group of a with the so-called A-classgroup of o, which is the factor group of the group of all fractional ideals of K divided by the subgroup of the principle ideals (a) with a > 0 at each real place that doesn't split A [R, p.309]. If

moreover A satisfies the Eichler condition, which means that there is at least one infinite place p of K such that the localization A_p of A at p is not isomorphic to the division ring $\underline{H}$ of quaternions over $\underline{R}$, one has also the cancellation property, i.e. equality in $K_o(a)$ implies equality in $L(a)$.
I would like to mention here that in the following we won't need this result, in fact it will be contained in our description of the locally free class group of a not necessarily maximal global order. Nevertheless I wouldn't feel very happy if I just passed by the results in the maximal case which were the origin of all that followed [e.g. J and F1].

(1.1.f) Let now A = KG be the group algebra of a finite group G over a number field K, and $a = o[G]$. Then Swan [Sw1, p.571] showed that for any projective a-lattice M, K ⊗ M is a free KG-module, from which it follows, first, that we have a rank again, and second, that M is locally free; the latter since the Cartan matrix belonging to G with respect to the residue field o/p is non-degenerate [S, III 1,2,3], which implies that the map $K_o(a_p) \overset{e}{\to} K_o(A)$ is injective, and also since by (1.1.a) two elements of $P(a_p)$ are isomorphic if and only if they induce the same element in $K_o(a_p)$; here p has been any prime ideal of o.

(1.2) From now on we always assume o to be global and take the observation just made as a motive for studying the category $L\mathfrak{f}(a)$ of isomorphism classes of locally free a-lattice instead of studying $P(a)$, but we would like to repeat that for group algebras $L\mathfrak{f}(a) = P(a)$. To that end, as in (1.1.e), we shall look at $K_o(a)$, which now by definition is the Grothendieck group belonging to $L\mathfrak{f}(a)$, and again it is enough here to look at the "locally free" class group cl(a) which has been defined as the kernel of the rank map; and then we have to think about cancellation properties. But before going into the details let us come back to our announced example:
L/K is a tamely ramified Galois extension of number fields with Galois group G, A = KG, $a = o_K[G]$. How does o_L sit in $K_o(a)$? Obviously, here one first has to make sure that o_L in fact is projetive over a. As L/K is tamely ramified the trace map $tr_{L/K}$ from L to K maps o_L onto o_K [CF, p.21], so there is an element $e \in o_L$ having trace 1. Take now any surjection $X \twoheadrightarrow o_L$, where X is some a-lattice, split it by an o_K-map φ, which is possible since each a-lattice is projetive over o_K, and alter this φ to $\tilde{\Phi}$:

$$\tilde{\varphi}(a) = \sum_{x\in G} x^{-1}\,\varphi(xa\cdot e) \qquad (a \in o_L) ;$$

then $\tilde{\varphi}$ is an a-splitting map for $X \twoheadrightarrow o_L$, i.e. o_L is projective. Conversely, let us show that the extension L/K can only be tamely ramified if o_L is $o_K[G]$-projective. The localized module $o_{L,p} = \bigoplus_{P|p} o_P$ is then projective as a $o_p[G]$-module; here p is any prime ideal of o_K, the $P|p$ are the primes of o_L lying above p, and we have abbreviated $(o_K)_p$ with o_p and $(o_L)_P$ with o_P. For the equality above see for example [R, 11.6]. Denote now by H the decomposition group of some P lying above p. As $o_{L,p}$ is the induced module $o_p[G] \underset{o_p[H]}{\otimes} o_P$, it follows that o_P is $o_p[H]$-projective: for an epimorphism of $o_p[H]$-modules $Y \twoheadrightarrow o_P$ induces an epimorphism of the induced $o_p[G]$-modules $G \otimes_H Y \twoheadrightarrow o_{L,p}$ which splits by some G-map ψ since $o_{L,p}$ is G-projective; now $\pi\psi$, with π denoting the H-projection from $G \otimes Y$ to Y, is a H-splitting map for $Y \twoheadrightarrow o_P$. By the same argument as in the end of (1.1.f) we conclude that o_P is free over $o_p[H]$, that is $o_P \simeq o_p[H]$. Hence there is an $a \in o_P$ such that $\{xa : x \in H\}$ is a o_p-basis of o_P. Now write $1 = \sum_{x\in H} t_x xa$ with $t_x \in o_p$. By applying the elements of H to this equation one sees $t_x = t_1$ for all $x \in H$, that is $\sum_{x\in H} xa = tr_{P|p}(a)$ is a unit, and therefore the trace $tr_{P|p} : o_P \to o_p$ is surjective.

§ 2. Idel groups

(2.1) We need some more notation. $J(A) = \{\alpha = (\ldots, \alpha_p, \ldots)\}$ is the idel group of A; here the components α_p of α lie in the unit group $A_p^\times$ of the algebra A_p and almost everywhere even in the unit group $a_p^\times$ of the order a_p (where $a_p = A_p$ at an infinite p). Observe that, in a way, J(A) is independent of our ground field K: we might as well consider A to be defined over $\underline{Q}$ since

$$\bigoplus_{p|\mathrm{p}} A_p = A \otimes_K (\bigoplus_{p|\mathrm{p}} K_p) = A \otimes_K (K \otimes_{\underline{Q}} \underline{Q}_\mathrm{p}) = A \otimes_{\underline{Q}} \underline{Q}_\mathrm{p} \quad ;$$

the same argument applies to a, too, as $o \otimes_{\underline{Z}} \underline{Z}_\mathrm{p} = \bigoplus_{p|\mathrm{p}} o_p$.

We have two distinguished subgroups in J(A), the unit group $A^\times$ of A and $U(a) = \prod_p a_p^\times$. The latter carries the canonic product topology; call

now a subgroup of J(A) open in J(A) if its intersection with $U(\mathfrak{a})$ is open in $U(\mathfrak{a})$. The corresponding topology of J(A) is independent of the special order $\mathfrak{a}$ in A which was used to define it, since we have $U(\mathfrak{a} \cap \mathfrak{b}) \subset U(\mathfrak{a})$ for two orders $\mathfrak{a}$ and $\mathfrak{b}$ in A and since the canonic topology of $U(\mathfrak{a})$ induces the canonic topology on $U(\mathfrak{a} \cap \mathfrak{b})$.
We should finally mention that the decomposition of A into its simple components A_i gives rise to a corresponding direct decomposition of the idel group of A. Hence when looking at J(A) it sometimes might help to first split J(A) into its direct factors $J(A_i)$ and to consider the group $J(A_i)$ as defined over the centre of A_i. Especially when working with the reduced norm on the idel level we should keep this in mind.

One more notation: $R^{m\times m}$ is the ring of m×m-matrices with entries in some ring R and Gl(m, R) is its unit group.

(2.2) Take now a locally free $\mathfrak{a}$-lattice M of rank m. Then as was already pointed out in (1.1.e) $M \otimes K \cong A^m$ so that we may consider M as imbedded in A^m as a full lattice. Hence by (1.1.c) outside of a finite set S of places we have $M_p = \mathfrak{a}_p^m$; at the exceptional places $p \in S$ we still have an isomorphism $\beta_p : \mathfrak{a}_p^m \to M_p$ since $M \in \mathcal{Lf}_m(\mathfrak{a})$, the latter being the category of isomorphism classes of locally free $\mathfrak{a}$-lattices of rank m. We now consider β_p as an element in $Gl(m, A_p)$; recall here that M_p is a full lattice in A_p. Taking $\beta_p = 1$ outside of S we then get an idel $\beta \in J(A^{m\times m})$ to which we attach a full lattice $\beta\mathfrak{a}^m$ in A^m by local dates (compare again (1.1.c)): $(\beta\mathfrak{a}^m)_p = \mathfrak{a}_p^m$ for $p \notin S$ and $(\beta\mathfrak{a}^m)_p = \beta_p\mathfrak{a}_p^m$ for $p \in S$. This way we achieve $M \cong \beta\mathfrak{a}^m$. When is $\beta\mathfrak{a}^m \cong \gamma\mathfrak{a}^m$ for $\beta, \gamma \in J(A^{m\times m})$? Call this isomorphism α; then $\alpha \in Gl(m, A) \subset J(A^{m\times m})$ and $\beta\mathfrak{a}^m = \alpha\gamma\mathfrak{a}^m$, hence locally $\beta_p\mathfrak{a}_p^m = (\alpha\gamma)_p\mathfrak{a}_p^m$, which means that $\beta \equiv \alpha\gamma \bmod U(\mathfrak{a}^{m\times m})$. So what we have shown is the following double coset description:

$$\mathcal{Lf}_m(\mathfrak{a}) \cong Gl(m, A) \setminus J(A^{m\times m})/U(\mathfrak{a}^{m\times m}) .$$

We would like to replace this set of double cosets by a group, and this we can do most easily by factoring out the subgroup of $J(A^{m\times m})$ spanned by Gl(m, A), $U(\mathfrak{a}^{m\times m})$, and by the commutator subgroup $J(A^{m\times m})'$ of $J(A^{m\times m})$. For reasons of simplicity we would rather like to define $J(A^{m\times m})'$ as the kernel of the reduced norm

$$nr : J(A^{m\times m}) \to J(Z) \quad , \quad Z = \text{centre of } A ;$$

where, as was already said, the usual reduced norm is extended to the idel group the natural way. What do we lose when passing from $Gl(m, A)\backslash J(A^{m\times m})/U(\mathfrak{a}^{m\times m})$ to $J(A^{m\times m})/J(A^{m\times m})'Gl(m, A)U(\mathfrak{a}^{m\times m})$?

(2.3) At this stage I'd like to recall Eichler's approximation theorem. For the moment let A be central simple over K. Then, provided A satisfies the Eichler condition (see (1.1.e)), the following holds: to a given idel $\beta \in J(A)'$ and any finite set S of places including all infinite ones there exists an $\alpha \in A^{\times}$ such that $nr(\alpha) = 1$, α_p is close to β_p at $p \in S$, and α_p belongs to $\mathfrak{a}_p^{\times}$ outside of S. For the proof compare [K]. Observe that in this statement the order $\mathfrak{a}$ may be replaced again by any order $\tilde{\mathfrak{a}}$ in A; this is obvious if $\tilde{\mathfrak{a}}$ contains $\mathfrak{a}$ and can be deduced in general by passing from $\tilde{\mathfrak{a}}$ to $\mathfrak{b} = \mathfrak{a} \cap \tilde{\mathfrak{a}}$: Call S_o the finite set of primes at which $\mathfrak{a}$ and $\mathfrak{b}$ differ locally. To solve now for a given β the approximation task with respect to the new order $\mathfrak{b}$ and some set S we have to apply our approximation property with respect to $\mathfrak{a}$ twice. First we find an $\alpha \in A^{\times}$, $nr\ \alpha = 1$, α_p close to β_p at S, $\alpha_p \in \mathfrak{a}_p^{\times}$ outside of S, then we find to the idel γ which is defined by $\gamma_p = 1$ outside of $S_o \setminus S$ and $\gamma_p = \alpha_p$ for $p \in S_o \setminus S$, an $\varepsilon \in A^{\times}$ such that $nr\ \varepsilon = 1$, ε_p is sufficiently close to γ_p for $p \in S_o \cup S$, and $\varepsilon_p \in \mathfrak{a}_p^{\times}$ for $p \notin S_o \cup S$. Then $\alpha\varepsilon^{-1}$ is a solution of our new approximation task.
Now, given β, choose S that big that it contains all places p for which β_p doesn't belong to $\mathfrak{a}_p^{\times}$, choose then the approximation at S that good that $\alpha_p^{-1}\beta_p \in \mathfrak{a}_p^{\times}$, then automatically $\alpha_p^{-1}\beta_p \in \mathfrak{a}_p^{\times}$ also outside of S. That means

$$J(A)' = A^{\times'} \cdot V$$

where V is any open subgroup of $J(A)'$ contained in $U(\mathfrak{a})$. This equality extends at once to the case of a semisimple algebra A when $\mathfrak{a} = \mathfrak{m}$ is a maximal order (cf. (1.1.b)). Since for an arbitrary order $\mathfrak{a}$ lying in a maximal order $\mathfrak{m}$ the group $U(\mathfrak{a})$ is open in $U(\mathfrak{m})$ the result also extends to semisimple algebras and non-maximal orders. Obviously, here each simple component of A is assumed to satisfy the Eichler condition.

As a corollary we get $J(A^{m\times m})' \subset Gl(m, A)\cdot U(\mathfrak{a}^{m\times m})$ if $m \geq 2$ (which implies the Eichler condition), thereby answering our question at the end of (2.2). Hence:

if $m \geq 2$, $Lf_m(\mathfrak{a}) \simeq J(A^{m\times m})/J(A^{m\times m})'Gl(m, A)U(\mathfrak{a}^{m\times m})$;

this being also true for m = 1 provided A satisfies the Eichler condition (which means that each simple component of A satisfies this condition).

(2.4) Now the nice point is that for all $m \geq 1$ the reduced norm yields an isomorphism

$$J(A^{m\times m})/J(A^{m\times m})'Gl(m, A)U(\mathfrak{a}^{m\times m}) \simeq J(Z)/Z^{\times}.\, nr\, U(\mathfrak{a}) ;$$

here again Z denotes the centre of A. This follows from (1) to (5) below, where for reasons of simplicity in the first four statements we have restricted ourselves to simple algebras.

(1) From [R, 33.1] we have $nr\, A^{m\times m} = nr\, A$ and $nr\, A_p^{m\times m} = nr\, A_p$.

(2) For a finite p we have $nr\, A_p = Z_p$, as follows from the fact that a local p-adic division ring always contains both, maximal subfields which are unramified and which are totally ramified over the centre [H]; hence, on the one side, one gets all units of the centre as norms and, on the other side, one gets, possibly up to a sign, some prime element of the centre as a norm.

(3) For an infinite prime p we have $nr\, A_p = Z_p$ unless A is ramified at p, i.e. p is real and $A_p \simeq \underline{H}^{r\times r}$ (some r), in which case $nr\, A_p = \{t \in \underline{R},\, t \geq 0\}$.

(4) $nr\, A = Z_+ = \{\alpha \in Z : \alpha \geq 0$ at each ramified infinite place of A$\}$; this is the Hasse-Schilling-Maaß theorem [R, 33.15].

(5) $nr(\mathfrak{a}_p^{m\times m})^{\times} = nr\, \mathfrak{a}_p^{\times}$ since for a finite p the order $\mathfrak{a}_p$ modulo its Jacobson radical is finite hence semisimple, and therefore each matrix in $(\mathfrak{a}_p^{m\times m})^{\times}$ can be written as a product of elementary matrices in $(\mathfrak{a}_p^{m\times m})^{\times}$, which obviously have reduced norm 1, and a diagonal matrix of the form $\begin{pmatrix} \gamma_1 & & 0 \\ & \ddots & \\ 0 & & 1 \end{pmatrix}$ with $\gamma \in \mathfrak{a}_p^{\times}$, which has reduced norm $nr\, \gamma$; compare for this [Sw3, 11.8; and Sw4, 8.5].

Observe in addition to the five points above, that the infinite primes at which some simple component of $A^{m\times m}$ is ramified occur in the numerator as well as in the denominator of our factor group $J(A^{m\times m})/J(A^{m\times m})'Gl(m, A)U(\mathfrak{a}^{m\times m})$, hence when looking at the quotient group $J(Z)/Z^{\times}\, nr\, U(\mathfrak{a})$ we can forget about the restrictions they induce on the image of nr.

(2.5) Next we want to study how the summation of locally free lattices is reflected in $J(Z)/Z^{\times}_{\cdot}\, nr\, U(\mathfrak{a})$. To that effect we again pick our $M \simeq \beta\mathfrak{a}^m$ with $\beta \in J(A^{m\times m})$ and try to modify the idelmatrix β modulo $J(A^{m\times m})'Gl(m, A)U(\mathfrak{a}^{m\times m})$ to a diagonal idelmatrix in $J(A^{m\times m})$. Namely, denote by S the finite set of places at which $\mathfrak{a}_p$ is not maximal, and choose at each $p \in S$ a surjection $g_p : M_p \to \mathfrak{a}_p$. Because of $\mathrm{Hom}_{\mathfrak{a}_p}(M_p, \mathfrak{a}_p) = (\mathrm{Hom}_{\mathfrak{a}}(M, \mathfrak{a}))_p$ we may approximate the finitely many g_p by some $g \in \mathrm{Hom}_{\mathfrak{a}}(M, \mathfrak{a})$, say $g \equiv g_p \bmod p$; then by Nakayama's lemma also g is surjective at each p. If $p \notin S$ perhaps gM_p is not all of $\mathfrak{a}_p$ but it is isomorphic to $\mathfrak{a}_p$ by (1.1.d). Hence the sequence $0 \to \ker g \to M \to \mathrm{im}\, g \to 0$ splits locally everywhere, which by (1.1.a) implies that ker g is locally free of rank m-1, and by (1.1.c) that $M \simeq \ker g \oplus \mathrm{im}\, g$. Since im g is locally free of rank 1 an induction argument now yields a diagonal idelmatrix belonging to M.

Assume next $M \simeq \beta_1\mathfrak{a} + \ldots + \beta_m\mathfrak{a} \simeq \gamma_1\mathfrak{a} + \ldots + \gamma_m\mathfrak{a}$ with $\beta_i, \gamma_i \in J(A)$. We could as well put this in the form $M \simeq \beta\mathfrak{a}^m \simeq \gamma\mathfrak{a}^m$, where β and γ are the diagonal matrices in $J(A^{m\times m})$ with entries $\beta_1, \ldots, \beta_m$ and $\gamma_1, \ldots, \gamma_m$, respectively. By (2.3) for $m \geq 2$ this is equivalent to $\beta \equiv \gamma \bmod J(A^{m\times m})'\mathrm{Gl}(m, A)U(\mathfrak{a}^{m\times m})$, and by (2.4) to $\mathrm{nr}\,\beta \equiv \mathrm{nr}\,\gamma \bmod Z^\times\,\mathrm{nr}\,U(\mathfrak{a})$. But as for example is shown in the proof of [R, 33.1], $\mathrm{nr}\,\beta = \prod \mathrm{nr}\,\beta_i = \mathrm{nr}(\prod \beta_i)$, where the last two reduced norms are defined on J(A) and the first one on $J(A^{m\times m})$.

From this we get our first main result (cf. [F1]):

<u>Theorem</u>. $K_o(\mathfrak{a}) = \underline{Z} \oplus J(Z)/Z^\times.\mathrm{nr}\,U(\mathfrak{a})$ <u>and</u> $\mathrm{cl}(\mathfrak{a}) = J(Z)/Z^\times.\mathrm{nr}\,U(\mathfrak{a})$; <u>moreover, in</u> $L\mathfrak{f}(\mathfrak{a})$ <u>the cancellation property is true for lattices of rank</u> ≥ 2, <u>and also for rank</u> 1-<u>lattices if</u> A <u>satisfies the Eichler condition</u>.

Namely, attach to $M \in L\mathfrak{f}(\mathfrak{a})$ its rank m and the reduced norm $\mathrm{nr}(\prod \beta_i) \bmod Z^\times\,\mathrm{nr}\,U(\mathfrak{a}) \in J(Z)/Z^\times.\mathrm{nr}\,U(\mathfrak{a})$; here up to an isomorphism M is again written as $\beta_1\mathfrak{a} + \ldots + \beta_m\mathfrak{a}$. By what we have seen above this leads to a surjective group homomorphism $K_o(\mathfrak{a}) \to \underline{Z} \oplus J(Z)/Z^\times\,\mathrm{nr}\,U(\mathfrak{a})$. Let now the two lattices M and N have the same image. Then they have the same rank m and, when $m \geq 2$, they have to be isomorphic by (2.3) and (2.4). If m = 1, add $\mathfrak{a}$ to M and to N; then we get $M \oplus \mathfrak{a} \simeq N \oplus \mathfrak{a}$, and again M = N in $K_o(\mathfrak{a})$.

I should like to mention here that by passing over from the idel language to the ideal language the obtained description of $\mathrm{cl}(\mathfrak{a})$ will give Jacobinski's class group formula [J; also F1, p.119].

(2.6) Assume now $\mathfrak{a} = \mathfrak{m}$ to be a maximal order. Then $\mathrm{nr}\,U(\mathfrak{m}) = U(\mathcal{O})_+$, where $\mathcal{O}$ is the maximal order in Z and the +-sign indicates again that the local components at the infinite ramified places of some simple component of A have to be positive. Namely for maximal local orders we can first restrict our attention to the central simple case, next in view of (2.4 (5)) to the division ring case, where there is a unique maximal order consisting of all elements that have integral reduced norm [H], and therefore by the argument given in (2.4 (2)) we indeed get all integral elements as reduced norms.

This observation now yields the following description of the kernel group $D(\mathfrak{a})$, which is defined as the kernel of the natural map of class groups, $cl(\mathfrak{a}) \to cl(\mathfrak{m})$, corresponding to an embedding of $\mathfrak{a}$ into a maximal order $\mathfrak{m}$:

$$\begin{array}{ccccccccc} 0 \to & D(\mathfrak{a}) & \to & cl(\mathfrak{a}) & \to & cl(\mathfrak{m}) & \to 0 \\ & \| & & \| & & \| & \\ 0 \to & U(\mathcal{O})_+/\mathcal{O}_+^\times\, nr\, U(\mathfrak{a}) & \to & J(Z)/Z^\times . nr\, U(\mathfrak{a}) & \to & J(Z)/Z^\times . U(\mathcal{O})_+ & \to 0 \end{array} ,$$

where at the left bottom corner we have used some appropriate isomorphism theorem. Observe that $D(\mathfrak{a})$ is independent of the special choice of m; observe too that the scalar extension map $cl(\mathfrak{a}) \to cl(m)$ is surjective. Let us mention here that for the same reason the extension map $cl(\mathfrak{a}) \to cl(\mathfrak{a}_1)$ is surjective provided the homomorphism from $\mathfrak{a}$ to $\mathfrak{a}_1$, which now may be orders sitting in two semisimple algebras, induces an epimorphism on the algebra level; for then the corresponding maps on the centre level and hence on the idel group level are surjective.

By the way, the equality $cl(\mathfrak{m}) = J(Z)/Z^\times . U(\mathcal{O})_+$ is nothing else than the ideltheoretic version of Swan's result stated in (1.1.e).

§ 3. The homomorphism language

In this section we restrict ourselves to the case that A is the group algebra of a finite group G over a number field K, and $\mathfrak{a} = \mathcal{o}[G]$. However, it is to remark here that in a very natural way the formula for $cl(\mathcal{o}G)$, which we are going to derive from our theorem in (2.5), can be generalized to the class group of an arbitrary global order; compare [J3, I].

(3.1) Pursuing section (2.5) we want to separate the components of the centre Z of KG. By [Hu, p.544] these are isomorphic to the fields $K(\chi) = K(\chi(x) : x \in G)$, where χ runs through the set Φ of absolutely irreducible characters of G. At this stage let us choose once for all a splitting field E for G, of which we assume, first, that it is a finite Galois extension of K with corresponding Galois group $G = G_{E/K}$, and second, that it is big enough to contain certain number fields that will be defined later on. Now, as all absolutely irreducible representations of G are realizable over E, we get an action of G on Φ by $\chi^{\omega}(x) = \chi(x)^{\omega}$, where again $\chi \in \Phi$, $x \in G$, and where $\omega \in G$; especially we have $K(\chi) = K(\chi^{\omega})$. By Galois theory the G-orbit of χ, as a G-set, can be identified with the Galois group G_{χ} of $K(\chi)/K$ which is a subextension of E/K. Thus Φ itself, as a G-set can be identified with the disjoint union $\bigcup_{\Phi \bmod G} G_{\chi}$. Because of $Z = \bigoplus_{\Phi \bmod G} K(\chi)$, we finally get an identification of Φ with the set Σ of all non-zero homomorphisms from Z to E, on which G operates by composition: $\sigma^{\omega} = \sigma\omega : Z \overset{\sigma}{\to} E \overset{\omega}{\to} E$ $(\sigma \in \Sigma,\ \omega \in G)$.

(3.2) This being said we concentrate ourselves on the set Σ. Each $\sigma \in \Sigma$ induces maps on the various group levels $Z^{\times} \to E$, $\mathcal{O}^{\times} \to \mathcal{o}_E^{\times}$, $J(Z) \to J(E)$, and $U(\mathcal{O}) \to U(\mathcal{o}_E)$. Writing T for either Z, E, or a field $K(\chi)$, and $\mathcal{o}_T$ for its maximal order, we define $F_1(T) = T^{\times}$, $F_2(T) = \mathcal{o}_T^{\times}$, $F_3(T) = J(T)$, and $F_4(T) = U(\mathcal{o}_T)$, and denote by $\sigma_j : F_j(Z) \to F_j(E)$ the induced maps of above. Hence, for each j, we get a pairing $\Sigma \times F_j(Z) \to F_j(E)$ which attaches to a pair (σ, α_j) the element $\langle\sigma, \alpha_j\rangle = \alpha_j^{\sigma_j} \in F_j(E)$. It is obviously multiplicative. To save space let us drop the index j in what follows. Considering the group $G = G_{E/K}$ as an automorphism group of $F(E)$ in the natural way we see at once $\langle\sigma^{\omega}, \alpha\rangle = \langle\sigma, \alpha\rangle^{\omega}$. Next we see that $\langle\sigma, \alpha\rangle = 1$ for all σ implies $\alpha = 1$. Namely, $F(Z) = \bigoplus_{\Phi \bmod G} F(K(\chi))$, and $G_{\chi} \in \Sigma$ acts as a group of automorphisms on $F(K(\chi))$. Putting these observations together we receive an injective homomorphism $F(Z) \to \mathrm{Hom}_G(\Sigma, F(E))$, which sends $\alpha \in F(Z)$ to the map f_{α} defined by $f_{\alpha}(\sigma) = \langle\sigma, \alpha\rangle$. That this homomorphism in fact is an isomorphism follows very easily: Any G-invariant $f \in \mathrm{Hom}(\Sigma, F(E))$ is fully determined by the images $f(1_{\chi})$, where 1_{χ} is the unit element in G_{χ}, and where χ runs through Φ mod G. As 1_{χ} is invariant under the action of the Galois group of $E/K(\chi)$, which is a subgroup of G, also $f(1_{\chi}) \in F(E)$ has to be invariant under the action of this group and hence belongs to $F(K(\chi))$. Thus $f = f_{\alpha}$, where $\alpha \in F(Z) = \bigoplus_{\Phi \bmod G} F(K(\chi))$ has the components $f(1_{\chi})$.

By (3.1) Σ is G-isomorphic to Φ and therefore we can write

$$F(Z) \simeq \mathrm{Hom}_G(\Phi, F(E)) = \mathrm{Hom}_G(RG, F(E)) \quad ,$$

where in the last equality we have replaced the set Φ by the free abelian group RG generated by Φ; RG is also known to be the character ring of G [S, p.II-37]. It results: $J(Z) \simeq \mathrm{Hom}_G(RG, J(E))$ and, obviously compatible with this isomorphism, $Z^\times \simeq \mathrm{Hom}_G(RG, E^\times)$. How can we translate the subgroup nr $U(\mathfrak{a})$ of $J(Z)$ into this homomorphism language? To that end let us look at the diagram

$$\begin{array}{ccc} J(KG) & & \\ \downarrow \mathrm{nr} & \searrow & \\ J(Z) & \xrightarrow{\simeq} & \mathrm{Hom}_G(RG, J(E)) \end{array}$$

and try to understand the diagonal map.

Pick an $\tilde{\alpha} \in J(KG)$, denote by $\alpha \in J(Z)$ its reduced norm, by α_χ the component of α in $J(K(\chi))$, and by $f_\alpha \in \mathrm{Hom}_G(RG, J(E))$ the corresponding homomorphism to α: then $f_\alpha(\chi)$ is the image of α_χ under the embedding $J(K(\chi)) \subset J(E)$ (which depends not only on $K(\chi)$ but on χ itself).

Now start again with the idel $\tilde{\alpha}$ and denote by $\tilde{\alpha}_\chi$ its component in $J(A_\chi)$, where A_χ is the simple part of KG that belongs to χ. Keeping in mind that E is a splitting field of A_χ we compute the reduced norm of $\tilde{\alpha}_\chi$ in the following way: we identify $\tilde{\alpha}_\chi$ with the element $\tilde{\alpha}_\chi \otimes 1$ of $J(A_\chi \otimes_{K(\chi)} E)$, interpret this element as a matrix, which is possible as $A_\chi \otimes_{K(\chi)} E = E^{r\times r}$ with $r = \chi(1)$, and then take the determinant. Obviously, from the way the reduced norm is extended on the idel group level, we must have $\mathrm{nr}(\tilde{\alpha}_\chi) = f_\alpha(\chi)$, and hence $f_\alpha(\chi) = \det(\tilde{\alpha}_\chi \otimes 1)$.

From representation theory it follows that the map $\tilde{\alpha} \to \tilde{\alpha}_\chi \otimes 1 \in J(E^{r\times r})$ induces the irreducible representation T_χ of G with character χ on the J-level. Thus $f_\alpha(\chi) = \det_\chi(\tilde{\alpha})$, where $\det_\chi$ denotes that 1-dimensional linear representation of G that is defined as the determinant of T_χ, i.e. $\det_\chi(\tilde{\alpha}) = \det(T_\chi(\tilde{\alpha}))$. Hence our diagonal map $J(KG) \to \mathrm{Hom}_G(RG, JE)$ can best be denoted by Det, where for some idel $\tilde{\alpha} \in J(KG)$, Det $(\tilde{\alpha})$ means the homomorphism in $\mathrm{Hom}_G(RG, JE)$ which sends χ to $\det_\chi(\tilde{\alpha})$.

The same way we get the diagonal map (compare (2.4)):

$$\begin{array}{ccc} J((KG)^{m\times m}) & & \\ \downarrow \mathrm{nr} & \searrow & \\ J(Z) & \xrightarrow{\simeq} & \mathrm{Hom}_G(RG, JE) \end{array} \quad .$$

Namely, lift up the representation $T_\chi : KG \to E^{r\times r}$ to the m-rank level

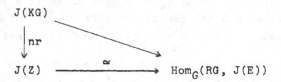

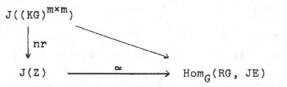

$(KG)^{m\times m} \to E^{rm\times rm}$, take the induced idel representation $T_\chi^m : J((KG)^{m\times m}) \to J(E^{rm\times rm})$, and define $\mathrm{Det}_\chi(\tilde{\alpha}) = \det(T_\chi^m(\tilde{\alpha}))$. We have proved:

<u>Theorem</u> [F2, A] $\mathrm{cl}(\mathfrak{o}G) \simeq \mathrm{Hom}_G(RG, JE)/\mathrm{Hom}_G(RG, E^\times)\ \mathrm{Det}\ U(\mathfrak{o}G)$. <u>On the level of rank</u> m-<u>lattices this isomorphism can be explicitely described by</u> $\tilde{\alpha} \to \mathrm{Det}(\tilde{\alpha})$, <u>where, as in (2.2), the idel</u> $\tilde{\alpha} \in J((KG)^{m\times m})$ <u>represents the rank</u> m-<u>lattice</u> $\tilde{\alpha}\mathfrak{o}[G]^m$.
<u>Moreover</u>: $D(\mathfrak{o}[G]) \simeq \mathrm{Hom}_G^+(RG, U(\mathfrak{o}_E))/\mathrm{Hom}_G^+(RG, \mathfrak{o}_E^\times)\ \mathrm{Det}\ U(\mathfrak{o}G)$.

What is the meaning of the +-sign in this formula for the kernel subgroup? $\mathrm{Hom}_G^+(RG, U(\mathfrak{o}_E))$ has its origin in $U(\mathcal{O})_+$, and here by definition the sign indicates the restriction for a unit idel $\alpha \in U(\mathcal{O})$ to belong to $U(\mathcal{O})_+$: the χ-components α_χ of α, that come from the decomposition $J(KG) = \bigoplus_{\Phi \bmod G} J(A_\chi)$, have to have positive coordinates $(\alpha_\chi)_p$ for all infinite p at which A_χ is ramified. Let us translate this into our new language. If $\alpha \in U(\mathcal{O})$ corresponds to $f \in \mathrm{Hom}_G(RG, J(\mathfrak{o}_E))$, then the χ-component α_χ corresponds to $f(\chi)$, and $(\alpha_\chi)_p$ to the $(f(\chi))_P$, where P runs over all primes of E lying above p. Now, as A_χ is ramified at the real place p of K if and only if χ is a realvalued character having Schur index 2 over <u>R</u> (i.e. χ is <u>symplectic</u>) we can describe $\mathrm{Hom}_G^+(RG, U(\mathfrak{o}_E))$ in the following way: it is the subgroup of $\mathrm{Hom}_G(RG, U(\mathfrak{o}_E))$ consisting of all homomorphisms f such that, for any symplectic χ, $f(\chi)$ is real and positive at each infinite place of E lying above some real place of K. We may as well express this condition by saying that f when restricted to R^sG is real and positive at each such place of E; here R^sG is the subgroup of RG which is spanned by the characters of all representations of G over E that leave some non-singular alternating form invariant; observe that R^sG is freely generated by the symplectic characters χ and by the sums $\chi + \overline{\chi}$ for the non-symplectic χ ($\overline{\chi}$ being the complex conjugate character to χ) and that always $f(\chi + \overline{\chi}) = f(\chi)f(\overline{\chi})$ is real and positive.

(3.3) Before finishing this paragraph we would like at least to mention the most important functorial properties of the class group $\mathrm{cl}(\mathfrak{o}_K G)$ with respect to maps concerning the change of fields and the change of groups; and we are going to express its functorial behaviour in terms of lattices as well as in terms of homomorphisms. However, to avoid some technical details, in all cases dealt with we will omit the proofs showing that the two given expressions are indeed compatible with our isomorphism in (3.2). The proofs are not too hard since the isomorphism in the theorem is given explicitly, but they are also not quite trivial: in (2) and (4) below one has somewhat to worry about the determinant; the main argument to be used here is the one that was used already in (2.4 (5)); compare [F2, A; F3, IV, V].

(1) Let $G \to H$ be a homomorphism of groups. Then by tensoring an $\mathfrak{o}[G]$-lattice with $\mathfrak{o}[H]$ we get a map $cl(\mathfrak{o}G) \to cl(\mathfrak{o}H)$. The corresponding map in the homomorphism language is induced by the natural map of the character rings $RH \to RG$ which comes from the original homomorphism $G \to H$; notice that $Hom_{\Omega}(-, JE)$ is a contravariant functor.

(2) Let U be a subgroup of G. As $\mathfrak{o}[G]$ is free over $\mathfrak{o}[U]$, we can consider any locally free G-lattice also as a locally free H-lattice and thereby get a map $cl(\mathfrak{o}G) \to cl(\mathfrak{o}U)$. The corresponding map in the homomorphism language is described as follows: $f \in Hom_{\Omega}(RG, JE)$ goes to $\tilde{f} \in Hom_{\Omega}(RU, JE)$, where $\tilde{f}$ is defined by $\tilde{f}(\psi) = f(ind_U^G \psi)$; here ψ is any element in RU, and ind_U^G denotes the induction map from RU to RG.

(3) Let K_1 be an extension field of K with ring of integers $\mathfrak{o}_1$. Then for obvious reasons the extension map $cl(\mathfrak{o}G) \to cl(\mathfrak{o}_1G)$ reflects itself in the homomorphism language as the canonic injection $Hom_{\Omega}(RG, JE) \to Hom_{\Omega_1}(RG, JE)$; of course we have to assume here that E contains K_1; Ω_1 is the corresponding Galois group.

(4) Let K_1 be a subfield of K. Then, as $\mathfrak{o}$ is locally free over $\mathfrak{o}_1$, the restriction of scalars induces a map $K_o(\mathfrak{o}G) \to K_o(\mathfrak{o}_1G)$; in case $\mathfrak{o}$ is actually free over $\mathfrak{o}_1$ it even induces a map $cl(\mathfrak{o}G) \to cl(\mathfrak{o}_1G)$. On the other side, now assuming E to be a Galois extension of K_1 we get a map

$$N_{K/K_1} : Hom_{\Omega}(RG, JE) \to Hom_{\Omega_1}(RG, JE)$$

which is defined by $(N_{K/K_1}f)(\chi) = \prod_{\tau} f(\chi^{\tau^{-1}})^{\tau}$, where τ runs through some representative set of Ω in Ω_1.

§ 4. The main example

(4.1) In connection with section (1.2) we take our tamely ramified Galois extension L/K and choose first an $a \in L$ which generates a normal basis of L over K, and second, elements $b_p \in \mathfrak{o}_{L,p}$ such that $\mathfrak{o}_{L,p} = b_p\mathfrak{o}_p[G]$; the latter being possible as $\mathfrak{o}_L$ is locally free over $\mathfrak{o}_K[G]$. By extending scalars we arrive at the equality $aK_pG = b_pK_pG$ and thus have got elements $\lambda_p \in (K_pG)^{\times}$ with $a\lambda_p = b_p$ (all p). Since $a\mathfrak{o}_K[G]$ and $\mathfrak{o}_L$ both are full lattices in L, λ_p lies in $\mathfrak{o}_p[G]^{\times}$ almost everywhere

(1.1.c); hence the λ_p give rise to an idel λ such that $a\lambda o_K[G] \simeq o_L$. Therefore, by (3.2), via the isomorphism $cl(o_K G) \simeq Hom_G(RG, JE)/Hom_G(RG, E^\times)\,Det\,U(o_K G)$ the element o_L is represented by the map $Det(\lambda)$ (since $Det(a) \in Hom_G(RG, E^\times)$).

(4.2) We interrupt the discussion of our example here and introduce the generalized Lagrange resolvents of Fröhlich. Namely we define

$$(c|\psi) = \det_\psi\Big(\sum_{x\in G} c^x x^{-1}\Big) \in E \quad ;$$

here c is any element of L, ψ is the character of some representation of G over E, and $\det_\psi$ is again the determinant representation belonging to ψ; $\det_\psi$ may be interpreted as a homomorphism from EG to E. Observe that in our definition we have implicitly assumed that $L \subset E$.

We list some simple properties of the resolvent (compare [F4]).

(1) $(c|\psi_1 + \psi_2) = (c|\psi_1)\cdot(c|\psi_2)$.

(2) $(c\lambda|\psi) = (c|\psi)\,Det_\psi(\lambda)$ for $\lambda = \sum_x r_x x \in KG$.

(3) If ρ is the character of the regular representation of G, then $(c|\rho) = \det(c^{xy^{-1}})_{x,y}$ = square of the discriminant of $\{c^x : x \in G\}$.

Putting c = a in (3) we see that $(a|\rho) \neq 0$, and hence by (1) and the fact that ρ contains any irreducible character of G, that also $(a|\psi) \neq 0$. Thus we can understand $(a|-)$ as a homomorphism in $Hom(RG, E^\times)$ (it is not compatible with the G-action).

Now we repeat all this for the local case, i.e. $c \in L_p = L \otimes_K K_p$, and we get $(b_p|\psi) \in E_p^\times$. Since we may assume that the matrices $T_\psi(x)$ all have integral entries [Hu, p.528] and since $b_p \in o_{L,p}$, we also have $(b_p|\psi) \in o_p$. From this we deduce $(b_p|\psi) \in o_p^\times$ almost everywhere, as by additivity, for an irreducible character χ, $(b_p|\chi)$ divides the square $(b_p|\rho)$ of the discriminant of $o_{L,p}/o_p$ which is the p-part of the discriminant of o_L/o_K. Thus we arrive at an idel $(b|-) \in Hom(RG, JE)$, given by $(b|\psi)_p = (b_p|\psi)$.

(4.3) Putting (4.1) and (4.2) together we get the

<u>Theorem</u>. <u>In</u> $cl(o_K G) = Hom_G(RG, JE)/Hom_G(RG, E^\times)\,Det\,U(o_K G)$ <u>the ring</u> o_L <u>of integers of</u> L <u>is represented by the homomorphism</u>

$$\psi \to \frac{(b|\chi)}{(a|\chi)} = Det_\chi(\lambda) \quad .$$

I would like here to add a remark. By rather deep number theoretical

methods it can be shown (compare (3.3 (4))) that $N_{K/\underline{Q}}(o_L)$ actually belongs to the kernel subgroup $D(\underline{Z}G)$, cf. [F2; F5]. Moreover it is conjectured that the only obstructions to the vanishing of o_L in $D(\underline{Z}G)$ are the signs of the Artin root numbers of symplectic characters; for this compare also M. Taylor's lecture [T].
To get some more insight in what $N_{K/\underline{Q}}(o_L) \in D(\underline{Z}G)$ might be, Fröhlich [F2, A] defined certain maps h_p (and h) on $D(\underline{Z}G)$ and computed the image of $N_{K/\underline{Q}}(o_L)$ under these maps. Let me finish my talk by describing what these maps h_p are, so that everybody who is interested can go right away into the number theoretical part of the integral normal basis business.

(4.4) Things becomes a bit easier if in

$$D(\underline{Z}G) = \mathrm{Hom}_G^+(RG, U(o_E))/\mathrm{Hom}_G^+(RG, o_E^\times)\mathrm{Det}\, U(\underline{Z}G)$$

we "reduce" modulo some prime number p:
To that effect we look at the surjection $U(o_E) \twoheadrightarrow (o_E/P)^\times$, where P denotes the product of all primes of E lying above p, and remember that, by Hensel's lemma, it splits. Thus it induces an surjection on the Hom-level

$$\mathrm{Hom}_G(RG, U(o_E)) \twoheadrightarrow \mathrm{Hom}_G(RG, (o_E/P)^\times) .$$

As $(o_E/P)^\times$ is without p-torsion, in the right-hand homomorphism group we may restrict RG to any subgroup z of RG that has p-power index in RG. Now recall that the Cartan matrix c_p of G with respect to $\underline{Z}/p$ has a determinant which is a power of p [S, p.III-13]. Hence, with d_p denoting the decomposition map of G with respect to p [S, III], we can take for example $z = d_p^{-1}$ (im c_p). Now, since the kernel of d_p sits in d_p^{-1} (im c_p) as a direct summand [S, p.III-13], we may restrict ourselves further from z to ker d_p and arrive at a surjection
$r_p : \mathrm{Hom}_G(RG, U(o_E)) \twoheadrightarrow \mathrm{Hom}_G(\ker d_p, (o_E/P)^\times)$. Clearly r_p annihilates Det $U(\underline{Z}G)$, as the determinant of some representation T_p of G in characteristic p doesn't depend on T_p itself but only on its composition factors. Define now h_p to be the induced map

$$h_p : D(\underline{Z}G) \to E_p(G) := \mathrm{Hom}_G(\ker d_p, (o_E/P)^\times)/r_p\, \mathrm{Hom}_G(RG, o_E^\times) .$$

If $p \nmid |G|$ then ker $d_p = 0$, hence $h_p = 0$.
Similarly, one can define a map
$h : D(\underline{Z}G) \to \prod_{p||G|} \mathrm{Hom}_G(\ker d_p, (o_E/P)^\times)/(\prod_{p||G|} r_p)\mathrm{Hom}_G(RG, o_E^\times)$. For an

interpretation of the kernels of h_p and h as the kernel subgroups belonging to the embedding of $\underline{Z}G$ into certain orders $\Lambda(p)$ and Λ, respectively, see [C-N, Appendice I].

References

[CF] Cassels a. Fröhlich, "Algebraic Number Theory". Academic Press, 1967

[C-N] Cassou-Noguès, Quelques Théorèmes de Base Normale d'Entiers. Annales de l'institut Fourier de l'université de Grenoble 28 (1978), 1-33

[CR] Curtis a. Reiner, "Representation Theory of Finite Groups and Associative Algebras". Interscience Publ. 1962

[F1] Fröhlich, Locally free modules over arithmetic orders. Crelle J. 274/275 (1975), 112-124

[F2] Fröhlich, Arithmetic and Galois module structure for tame extensions. Crelle J. 286/287 (1976), 380-440

[F3] Fröhlich, Class groups, in particular Hermitian Class groups. To appear

[F4] Fröhlich, Resolvents and trace form. Proc. Camb. Phil. Soc. 78 (1975), 185-210

[F5] Fröhlich, Galois module structure. In: Fröhlich, "Algebraic Number Fields". Academic Press 1977

[H] Hasse, Über p-adische Schiefkörper und ihre Bedeutung für die Arithmetik hyperkomplexer Zahlsysteme. Math. Ann. 104 (1931), 495-534

[Hu] Huppert, "Endliche Gruppen I". Springer 1967

[J] Jacobinski, Genera and decompositions of lattices over orders. Acta Math. 121 (1968), 1-29

[K] Kneser, Starke Approximation in algebraischen Gruppen I. Crelle J. 218 (1965), 190-203

[R] Reiner, "Maximal Orders". Academic Press 1975

[S] Serre, "Représentations Linéaires des Groupes Finis". Hermann, 1967

[Sw1] Swan, Induced representations and projective modules. Ann. of Math. 71 (1960), 552-578

[Sw2] Swan, Projective modules over group rings and maximal orders. Ann. of Math. 76 (1962), 55-61

[Sw3] Swan, Algebraic K-Theory. Springer LNM 76 (1968)

[Sw4] Swan a. Evans, K-theory of finite groups and orders. Springer LNM 149 (1970)

[T] Taylor's lecture at this conference.

THE HERMITIAN CLASSGROUP

A. Fröhlich

1. Motivation

Throughout we shall use the following notations. F is a number field (with $[F : \mathbb{Q}] < \infty$) or a local field (with $[F : \mathbb{Q}_p] < \infty$, p a rational prime - we shall not discuss Archimedean local fields, where the problems we have to consider are relatively trivial). Γ is a finite group. $N \supset K \supset F$ is a tower of field extensions, with N/K Galois with Galois group Γ and tame (at most tamely ramified). o_N, o_K, $o_F = o$ (without subscript) are the respective rings of integers. E is the algebraic closure of F, supposed to contain N and K, and $\Omega = \Omega_F = \mathrm{Gal}(E/F)$. $t_{N/F}$ is the trace $N \to F$. $o\Gamma$, $F\Gamma$ are group rings.

The motivation for the work to be described here comes from the study of the pair - to be called a <u>Hermitian</u> <u>$o\Gamma$-module</u> - consisting of the $o\Gamma$-module o_N together with the Γ-invariant form $x,y \mapsto t_{N/F}(xy)$ on N over F. Note that, because of tame ramification, o_N is now locally free over $o\Gamma$ - no doubt a "Queyrut-type" generalisation to the wild case is possible. It should also be kept in mind that the form $x,y \mapsto t_{N/F}(xy)$ as a form on o_N is not unimodular, i.e. is in general singular in the sense the word is used in ordinary Hermitian K-theory, and thus falls outside the framework of that theory. Our aim here will be to indicate briefly some basis aspects of the algebraic theory of Hermitian modules, which has been developed with a view to arithmetic applications.

These applications have their origin in the theory of Galois module structure of rings of algebraic integers which has come into being in the last ten years (cf. [F2], [F6]). We now take F, K, N to be number fields. As in the tame case o_N is locally free over $\mathbb{Z}\Gamma$, it defines an element $(o_N) = U_{N/K}$ of the locally free class group $\mathrm{Cl}(\mathbb{Z}\Gamma)$ of $\mathbb{Z}\Gamma$. As had been conjectured, M. Taylor (cf. [Ty]) has shown that $U_{N/K}$ is determined by the Artin root numbers $W(N/K, \chi)$ (cf. [M]) for symplectic characters χ. It is however not true that $U_{N/K}$ in turn determines the $W(N/K, \chi)$. One thus looks for additional elements of structure, to get additional information on these symplectic root numbers, and indeed the Hermitian structure provides this. One can however go a good deal beyond that. The question of determining the Artin root number can be reduced to a local one, as it is in fact the product of the local "Langlands constants" (cf. [Tt]. So now let F etc. be local fields. The $o\Gamma$-module o_N is now free, i.e. its structure problem is entirely trivial. But this is not so for the Hermitian $o\Gamma$-module consisting of o_N and the trace form $x,y \mapsto t_{N/F}(xy)$ on N. The latter does yield non-trivial information about the values of the Langland constants for symplectic characters and one can conjecture that these values are completely determined by the Hermitian structure. One thus has a theory of local Hermitian Galois module structure, in parallel to the theory of global Galois module structure, and in fact the theory of global Hermitian Galois module structure is a common source for both of these. Here we shall however not go into the arithmetic applications, apart from a briefest hint.

Our aim is to give an outline of some of the steps involved in creating the appropriate algebraic tools. (Now, the role of Γ to be a Galois group is irrelevant of course.) For a systematic and detailed account see the forthcoming notes [F5]. For results on local Hermitian Galois module structure see [F3] and also [F4].

2. The basic problem

Following our example, we are considering pairs (X,β) where

$$(2.1)\quad \begin{cases} X \text{ is a } \begin{Bmatrix} \text{free} \\ \text{locally free} \end{Bmatrix} \mathcal{O}\Gamma\text{-module} \begin{cases} \text{local case,} \\ \text{global case,} \end{cases} \\ \beta \text{ a non degenerate symmetric bilinear form on } X \otimes_{\mathcal{O}} F = XF \\ \text{over } F \text{, which is } \Gamma \text{ invariant, i.e. } \beta(v_1\gamma, v_2\gamma) = \beta(v_1, v_2) \\ \text{for all } v_1, v_2 \in XF, \text{ all } \gamma \in \Gamma \end{cases}$$

Recall again that β is <u>not</u> assumed to be unimodular on X. Our aim is to generalise the notion of the discriminant of a quadratic lattice (the case $\Gamma = 1$). Discriminants are to be elements in a suitable group $H\,Cl(\mathcal{O}\Gamma)$, the Hermitian class group. More precisely, if henceforth $K_0H(\mathcal{O}\Gamma)$ denotes the Grothendieck group of pairs (X,β) under orthogonal sum, then the discriminant is a homomorphism

$$(2.2)\qquad d\colon K_0H(\mathcal{O}\Gamma) \to H\,Cl(\mathcal{O}\Gamma).$$

Classically the discriminant of a quadratic lattice was an ideal. But this definition is too weak. To see this, consider the local case, when the discriminant as ideal is an element of $F^*/\mathcal{O}^*$ (* denotes invertible elements), whereas the "proper" discriminant is an element of $F^*/\mathcal{O}^{*2}$. (For the global analogue see [F1]). Indeed let $\{x_i\}$ $(1 \leq i \leq q)$ be a free $\mathcal{O}$-basis of X, giving a discriminant matrix

$$(2.3)\qquad (\beta(x_i,x_\ell))$$

Its determinant is then uniquely determined by X, modulo squares of determinants of $GL_q(\mathcal{O})$, and so the discriminant is

$$(2.4)\qquad d((X,\beta)) = \mathrm{Det}(\beta(x_i,x_\ell))\,\mathcal{O}^{*2} \in F^*/\mathcal{O}^{*2}.$$

We shall have to reformulate our basic concept. For this we need the notion of an involution j of a ring R. This is an automorphism of the additive group of R, an antiautomorphism of its multiplicative monoid (i.e. $(a\,b)^j = b^j\,a^j$) and is of period 2, i.e. $j^2 = 1$. Such an involution j has a <u>matrix extension,</u> again denoted by j, say to the ring $M_m(R)$ of m by m square matrices over R: we apply j to the entries and transpose. The group ring of Γ over a commutative ring R has the standard involution

$^-$, given by

$$(2.5) \qquad \overline{\sum_{\gamma\in\Gamma} a_\gamma \gamma} = \sum_{\gamma\in\Gamma} a_\gamma \gamma^{-1} \quad (a_\gamma \in R) .$$

A <u>Hermitian</u> $\mathcal{O}\Gamma$-<u>module</u> is a pair (X,h) where

$$(2.6) \quad \begin{cases} X \text{ is a } \begin{Bmatrix} \text{free} \\ \text{locally free} \end{Bmatrix} \mathcal{O}\Gamma\text{-module} \begin{cases} \text{local case,} \\ \text{global case,} \end{cases} \\ \\ h \text{ a non degenerate Hermitian form} \\ XF \times XF \to F\Gamma \end{cases}$$

(with respect to $^-$). To fix ideas, $h(v_1,v_2)$ is $F\Gamma$-linear in v_2, and $h(v_2,v_1) = \overline{h(v_1,v_2)}$. The pairs (X,β) of (2.1) correspond biuniquely to the Hermitian modules (X,h), if we define

$$h(v_1,v_2) = \sum_\gamma \beta(v_1,v_2\gamma^{-1})\gamma \ ,$$

(and conversely β the coefficient of 1 in the expansion of h) (cf. [FM]). Thus for our arithmetic example of §1, we now consider the pair $(\mathcal{O}_N, h_{N/K/F})$ where

$$(2.7) \qquad h_{N/K/F}(v_1,v_2) = \sum_\gamma t_{N/F}(v_1.v_2^\gamma)\gamma^{-1} .$$

We are looking for a definition of the discriminant in the "Hom-language", used so successfully elsewhere (cf. [F2], [F6], [R]). With each representation $T : \Gamma \to GL_n(E)$ (E the algebraic closure of F) we associate its character χ, viewed as a function $\Gamma \to E$. The ring R_Γ of <u>virtual characters</u> is then - as additive group - generated under pointwise addition by the characters χ. Equivalently R_Γ is the Grothendieck group of $E\Gamma$. One can in fact - with the obvious analogue to R_Γ as a Grothendieck group - treat quite analogously any semisimple F-algebra A with involution, with an order $\mathcal{A}$ spanning A and stable under the involution. We shall not give the precise formulations in this general case, but we will use such orders $\mathcal{A}$ for purposes of illustration.

The approach to the generalisation of the discriminant of a quadratic lattice, which suggests itself, is via the generalisation of the concept of a determinant (see [R], or [F2] Appendix), of a groupring (or algebra). To recall the definition let $T : \Gamma \to GL_n(E)$ be a representation. Extend this, for any m, to an algebra homomorphism $M_m(F\Gamma) \to M_{mn}(E)$, yielding a homomorphism $a \mapsto \mathrm{Det}_\chi(a)$

$$GL_m(F\Gamma) \to GL_{mn}(E) \xrightarrow{\mathrm{Det}} E^* ,$$

which indeed only depends on the character χ of Γ. The map

$$\mathrm{Det}(a) : \chi \mapsto \mathrm{Det}_\chi(a)$$

extends to R_Γ. If now (X,h) is a Hermitian module, let us say in the local case, then analogously to (2.4) one might wish to define its discriminant in terms of the determinant $\mathrm{Det}(h(x_i,x_\ell))$ of the discriminant matrix $(h(x_i,x_\ell))$, i.e. in terms of the map

$$\chi \mapsto \mathrm{Det}_\chi(h(x_i,x_\ell)).$$

This approach will often work, but not always.

To describe the difficulty which arises here, and to prepare the ground for our later solution, we need the notion of a <u>symplectic</u> <u>involution</u> of a matrix ring $M_n(L)$, L any field of characteristic $\neq 2$. We view the matrices as acting on the n-dimensional space L^n from the right. If b is a non degenerate symmetric or alternating form on L^n then the <u>adjoint</u> <u>involution</u> j of b is given by

$$b(x_1P,x_2) = b(x_1,x_2P^j), \quad \forall x_1,x_2 \in L^n, \quad \forall P \in M_n(L).$$

j is <u>orthogonal</u>, or <u>symplectic</u>, according to whether b is symmetric, or alternating - this does in fact not depend on the particular choice of b for the given j. Every involution of $M_n(L)$ is orthogonal or symplectic (and not both), if it fixes L.

A representation $T : \Gamma \to GL_n(E)$ is <u>orthogonal</u> or <u>symplectic</u> if $T(\Gamma)$ leaves a non-degenerate symmetric or alternating form, respectively, invariant, and we then call the associated character orthogonal, or symplectic. We can restate this in terms of the algebra homomorphism $T : F\Gamma \to M_n(E)$ extending the group representation above: we have

$$(2.8) \qquad T(\bar{a}) = T(a)^j \qquad \forall\ a \in F\Gamma$$

for some orthogonal, or symplectic j, respectively. In this form the definition extends to representations of an F-algebra A with involution $^-$. Moreover (2.8) implies the same for the matrix extensions of $^-$ and j, and the extension of the representation T to $M_m(F\Gamma) \to M_{mn}(E)$. Now - as will be seen below - the determinant $\mathrm{Det}_\chi(a)$ for a invertible symmetric element $a = \bar{a}$ and a symplectic character χ is always a square in the field $F(\chi)$ of values of χ over F, and one wishes to find a canonical square root. Moreover this should be done in such a way that it leads to a unified definition, and not one "by cases" .

<u>Example</u> We consider $M_2(L)$ with the involution $^\times$ given by

$$(2.9)\qquad \begin{pmatrix} a & b \\ c & d \end{pmatrix}^{\times} = \begin{pmatrix} d & -b \\ -c & a \end{pmatrix} .$$

The symmetric elements are the scalar matrices aI (which proves that $^{\times}$ is symplectic), and $\mathrm{Det}(aI) = a^2$, and this has the "canonical square root" a, as a function on $^{\times}$ symmetric matrices. Any generalised quaternion group Γ has a two dimensional irreducible representation T with $T(\bar{a}) = T(a)^{\times}$.

There is another way to look at this problem. If A is a simple F-algebra with involution $^{-}$ which leaves F elementwise fixed then the irreducible representation (unique to within equivalence) $T : A \subset A \otimes_F E \cong M_n(E)$ is either symplectic or orthogonal, and not both; accordingly we call A <u>symplectic</u> or <u>orthogonal</u>. If A, as an algebra with involution, is indecomposable (and semisimple) then it is as above, or $^{-}$ moves the centre cent(A), and in the latter case it is called <u>unitary</u>. Now for indecomposable unitary or orthogonal A the definition of discriminant "via Det" is satisfactory, but for symplectic A - as in our example above (cf. (2.9)) - it is not. Our aim is thus to generalise the definition of the "canonical square root" given above. Distinct definitions "by cases" would be useless, and so our procedure must be entirely general and include all three types of indecomposable algebras with involution, and must do so in a unified manner, applicable e.g. to grouprings which may have components of all types. This will be outlined in the next section.

3. Pfaffians

With L as in §2, let j be a symplectic involution of $M_n(L)$. If $Q \in GL_n(L)$, $QQ^j = I$, the identity matrix, then

$$(3.1)\qquad \mathrm{Det}(Q) = 1 .$$

On the other hand if $S = S^j \in GL_n(L)$ is j-symmetric then $\exists\ P \in GL_n(L)$ with

$$S = PP^j .$$

By (3.1), the function

$$\mathrm{Det}(P) = \mathrm{Pf}^j(S)$$

only depends on S. It is the <u>Pfaffian</u> of S. (Classically the Pfaffian is defined for skew symmetric matrices with respect to transposition, i.e. to an orthogonal involution - here we define it for symmetric matrices with respect to a symplectic involution).

Example (2.9); $j = {}^{\times}$. Then

$$\begin{pmatrix} a & 0 \\ 0 & a \end{pmatrix} = \begin{pmatrix} 0 & 1 \\ -a & 0 \end{pmatrix} \begin{pmatrix} 0 & 1 \\ -a & 0 \end{pmatrix}^{\times}$$

and so indeed

$$\mathrm{Pf}(aI) = a.$$

<u>3.1. Proposition.</u> <u>Let</u> T_1 <u>and</u> T_2 <u>be equivalent representations</u> $\Gamma \to GL_n(E)$, <u>extended to homomorphisms</u> $F\Gamma \to M_n(E)$, <u>so that for all</u> $a \in F\Gamma$

$$T_1(\bar{a}) = T_1(a)^{j_1}, \quad T_2(\bar{a}) = T_2(a)^{j_2},$$

<u>with</u> j_1, j_2 <u>symplectic involutions</u>. <u>Then if</u> $a = \bar{a}$ <u>is symmetric and invertible</u>,

$$\mathrm{Pf}^{j_1}(T_1(a)) = \mathrm{Pf}^{j_2}(T_2(a)).$$

<u>Similarly with</u> $F\Gamma$ <u>replaced by</u> $M_m(F\Gamma)$.

For proofs of this and all following results see [F5] (or [F3], [F4]).

If χ is the character of T_1 and T_2, $a = \bar{a} \in (F\Gamma)^*$, or more generally $a = \bar{a} \in GL_n(F\Gamma)$, we shall write

(3.2) $$\mathrm{Pf}_\chi(a) = \mathrm{Pf}^{j_1}(T_1(a)) .$$

As we have seen $\mathrm{Pf}_\chi(a)$ is uniquely defined for any invertible, $^-$ symmetric a and any symplectic χ. Let now R_Γ^s be the additive subgroup of R_Γ generated by the symplectic χ. Any actual character in R_Γ^s is again symplectic.

<u>3.2. Proposition.</u> <u>If</u> χ, ψ <u>are symplectic then</u> $\mathrm{Pf}_{\chi+\psi}(a) = \mathrm{Pf}_\chi(a)\ \mathrm{Pf}_\psi(a)$. <u>Hence</u>

$$\mathrm{Pf}(a) : \chi \mapsto \mathrm{Pf}_\chi(a) .$$

<u>extends to a homomorphism</u> $R_\Gamma^s \to E^*$, <u>and in fact</u> $\mathrm{Pf}(a) \in \mathrm{Hom}_{\Omega_F}(R_\Gamma^s, E^*)$.

Recall that $\Omega_F = \mathrm{Gal}(E/F)$.

We also have

<u>3.3. Proposition.</u> <u>If</u> $a = \bar{a} \in GL_m(F\Gamma)$, $b \in GL_m(F\Gamma)$ <u>then</u>

$$\mathrm{Pf}_\chi(\bar{b}\ a\ b) = \mathrm{Det}_\chi(b)\ \mathrm{Pf}_\chi(a) \ \underline{\text{for}}\ \chi \in R_\Gamma^s .$$

<u>In particular</u>

$$Pf_\chi(\bar{b}\, b) = Det_\chi(b)$$

If now χ is a character, let $\bar{\chi}$ be its contragredient (classically complex conjugate). Then $\chi + \bar{\chi}$ is symplectic. Thus if $\chi \in R_\Gamma$ then $\chi + \bar{\chi} \in R_\Gamma^s$.

3.4. Proposition. If $a = \bar{a} \in GL_m(F)$ and $\chi \in R_\Gamma$ then

$$Pf_{\chi + \bar{\chi}}(a) = Det_\chi(a).$$

In particular if $\chi \in R_\Gamma^s$ then

$$Det_\chi(a) = Pf_\chi(a)^2.$$

The first of these equations tells us that determinants of symmetric elements are determined by their Pfaffians, the second one that determinants of symmetric elements at symplectic characters are squares (in $F(\chi)$).

3.5. Proposition. If $a = \bar{a} \in GL_m(F\Gamma)$, $b = \bar{b} \in GL_q(F\Gamma)$; so that now

$$\begin{pmatrix} a & 0 \\ 0 & b \end{pmatrix} \in GL_{m+q}(F\Gamma) \quad,$$

then

$$Pf_\chi \begin{pmatrix} a & 0 \\ 0 & b \end{pmatrix} = Pf_\chi(a)\ Pf_\chi(b).$$

4. Discriminants

We shall give separate, distinct definitions of the discriminant for the local, and the global case, respectively. There is a unified, general formulation which however provides less insight than the ones given here.

First we take F to be a local field (as stated in §1). Let (X,h) be a Hermitian $o\Gamma$-module, $\{x_i\}$ $(1 \leqslant i \leqslant q)$ a basis of X over $o\Gamma$. Then, by Proposition 3.2., the Pfaffian $Pf(h(x_i,x_\ell))$ of the discriminant matrix lies in $Hom_{\Omega_F}(R_\Gamma^s, E^*)$. The discriminant matrix $(h(x_i,x_\ell))$ is unique modulo the substitution by $\bar{b}(h(x_i,x_\ell))b$, with $b \in GL_q(o\Gamma)$. By Proposition 3.3, its Pfaffian is unique modulo multiplication by the map $Det^s(b): \chi \mapsto Det_\chi(b)$ $(\chi \in R_\Gamma^s)$. (Here the superscript s indicates restriction to R_Γ^s). These maps form a group $Det^s(GL_q(o\Gamma))$ which coincides with $Det^s(o\Gamma)^*$. We thus arrive at the definition of the Hermitian classgroup

$$(4.1) \qquad H\,Cl(o\Gamma) = Hom_{\Omega_F}(R_\Gamma^s,\ E^*)/Det^s(o\Gamma)^* \qquad,$$

and of the discriminant

(4.2) $$d(X,h) = \mathrm{Pf}(h(x_i,x_\ell)) \text{ modulo } \mathrm{Det}^s(o\Gamma)^* .$$

By Proposition 3.5, d is multiplicative with respect to orthogonal sums, in other words we have a homomorphism

(4.3) $$d : K_oH(o\Gamma) \to H\,Cl(o\Gamma) \quad ,$$

where $K_oH(o\Gamma)$ is the Grothendieck group of Hermitian $o\Gamma$-modules. With the appropriate changes, these definitions go over to an involution invariant order A in a semisimple F-algebra A with involution $\bar{\ }$.

Examples. (all local) 1. $\Gamma = 1$, i.e. $A = o$, $F\Gamma = A = F$, $\bar{\ }$ is the identity. This is the simplest type of a simple orthogonal algebra. The Grothendieck group which now appears as R_Γ, is generated by the embedding $\chi : F \to E$, and $R_\Gamma^s = (2\chi)\mathbb{Z}$. The evaluation map $f \mapsto f(2\chi)$ yields an isomorphism $\mathrm{Hom}_{\Omega_F}(R_\Gamma^s, E^*) \cong F^*$, and gives rise to an isomorphism $\mathrm{Det}^s(o^*) \cong o^{*2}$. (For indeed

$$\mathrm{Det}\begin{pmatrix} a & 0 \\ 0 & a \end{pmatrix} = a^2).$$ Thus now

(4.4) $$H\,Cl(o) \cong F^*/o^{*2} \quad ,$$

as we expected.

2. A is a quadratic extension of F, $\bar{\ }$ its non-trivial automorphism over F, A the integral closure of o in A. This is the simplest unitary case. Now the Grothendieck group R_Γ is generated by the two embeddings χ and $\bar{\chi} : A \to E$ over F, and $R_\Gamma^s = (\chi + \bar{\chi})\mathbb{Z}$. Via the evaluation map we get an isomorphism

(4.5) $$H\,Cl(A) \cong F^*/\mathrm{norm}_{A/F}\, A^* .$$

3. $A = M_2(F)$, $A = M_2(o)$ and the involution is that in (2.9). This is the simplest example of a symplectic algebra. Here $R_\Gamma = R_\Gamma^s = \mathbb{Z}$, and the evaluation map yields

(4.6) $$H\,Cl(A) \cong F^*/o^* .$$

The three examples (4.4) - (4.6) illustrate nicely how our definition in terms of the "Hom language" can reflect and unify seemingly quite different situations.

We now turn to the global case, i.e. F is now a numberfield, o its ring of algebraic integers. If we put the local groups together (including those at infinity which we have not defined here) via an idelic, restricted direct product procedure, we do get a group, which we shall call the adelic Hermitian classgroup of $o\Gamma$ - given by

$$(4.7)\qquad \mathrm{Ad\,H\,Cl}(o\Gamma) = \mathrm{Hom}_{\Omega_F}(R_\Gamma^s, JE)/\mathrm{Det}^s Uo\Gamma.$$

Here JE is the idele group of E, i.e. the direct limit (in fact union) of the idele groups JL, L running over the finite extensions of $\mathbb{Q}$ in E. Also

$$Uo\Gamma = \prod_p (o_p\Gamma)^*$$

(product over all prime divisors p of F). o_p denotes the completion of o (respectively of F) at p if p is finite (respectively infinite). The classification via the group (4.7) is however not fine enough, and in particular cannot reflect genuinely global properties. The Hermitian classgroup will in fact be defined by

$$(4.8)\qquad \begin{cases} \mathrm{H\,Cl}(o\Gamma) = \mathrm{Cok}\,\Delta \\[2mm] \Delta\ :\ \mathrm{Hom}_{\Omega_F}(R_\Gamma, E^*) \to \left(\dfrac{\mathrm{Hom}_{\Omega_F}(R_\Gamma, JE)}{\mathrm{Det}\,Uo\Gamma} \times \mathrm{Hom}_{\Omega_F}(R_\Gamma^s, E^*)\right). \end{cases}$$

We have to explain the symbols and show how this definition gives us a good discriminant. $Uo\Gamma$ was defined above, Det $Uo\Gamma$ is the group of maps Det(a) : $\chi \mapsto \mathrm{Det}_\chi(a)$, all $\chi \in R_\Gamma$, with $a \in Uo\Gamma$. Also

$$\Delta(g) = (g^{-1}\,\mathrm{Det}\,Uo\Gamma,\ g^s),$$

where g^s is the restriction of g to $R_\Gamma^s \subset R_\Gamma$.

Let then (X,h) be a Hermitian $o\Gamma$-module. Choose a free basis $\{v_i\}$ $(1 \leq i \leq q)$ of XF over FΓ. For each prime divisor p of F there is an element $a_p \in GL_q(F_p\Gamma)$ which transforms $\{v_i\}$ into a free basis of $X_p = X \otimes_o o_p$ over $o_p\Gamma$. (Subscripts denote completions, except that for infinite primes we define o_p to be F_p). The discriminant is then

$$(4.9)\qquad d(X,h) = \text{class of }(\mathrm{Det}(a)\ \mathrm{mod}\ \mathrm{Det}\,Uo\Gamma,\ \ \mathrm{Pf}(h(v_i,v_\ell))).$$

Here Det(a) is the map with local components $\chi \mapsto \mathrm{Det}_\chi(a_p)$. The element in (4.9) is independent of choices. Indeed with another choice of the a_p we still get the same coset Det(a) mod Det $Uo\Gamma$. On the other hand, by Proposition 3.3., if we change the basis $\{v_i\}$ via an element $b \in GL_q(F\Gamma)$, then we have to multiply $\mathrm{Pf}(h(v_i,v_\ell))$ by $\mathrm{Det}^s(b)$, and Det(a) by $\mathrm{Det}(b)^{-1}$.

Next observe that the map

$$(f \bmod \mathrm{Det}\, \mathfrak{U}\mathfrak{o}\Gamma,\ g) \mapsto f^s g \bmod \mathrm{Det}^s\, \mathfrak{U}\mathfrak{o}\Gamma$$

(f^s the restriction of f to R_Γ^s) yields a homomorphism

$$(4.10) \qquad \underline{P} : \mathrm{H\,Cl}(\mathfrak{o}\Gamma) \to \mathrm{Ad\,H\,Cl}(\mathfrak{o}\Gamma) \quad .$$

Indeed with the notation leading up to (4.9), and denoting by $\{x_{p,i}\}$ the free basis of X_p over $\mathfrak{o}_p\Gamma$, obtained via a_p from $\{v_i\}$, we see from Proposition 3.3, that

$$\mathrm{Pf}(h(v_i,v_\ell))\ \mathrm{Det}(a_p) = \mathrm{Pf}(h(x_{p,i},\ x_{p,\ell})).$$

Thus the p-component of $\underline{P}d$ is the local discriminant:

$$(\underline{P}d(X,h))_p = d(X_p,h_p).$$

The map $\underline{P}$ is thus essentially simultaneous arithmetic localisation at all p. Its kernel and its cokernel, neither of which vanish in general, are genuinely global objects, measuring the deviation from the Hasse principle.

<u>Example</u>: 1. $\Gamma = 1$. Then $\underline{P}$ is injective and via the evaluation map we get an isomorphism

$$\mathrm{Cok}\ \underline{P} \cong JF/(JF)^2\, F^* .$$

This equation implies in particular an old result on the discriminant ideal of a quadratic lattice: its ideal class is always a square (Hecke's theorem). Thus in general the group Cok $\underline{P}$ yields a generalisation of this theorem.

2. The preceding theory applies quite generally in the context of orders, and involutions on algebras. In the simplest type of unitary algebra we consider again a quadratic extension A of the numberfield F, $\bar{\ }$ its non trivial automorphism over F, $\mathfrak{A}$ the ring of integers of A. Then

$$\mathrm{Cok}\ \underline{P}_{\mathfrak{A}} \cong JF/(\mathrm{norm}_{A/F}\, JA) \cdot F^* \quad .$$

5. The role of the Hermitian classgroup

One reason for introducing the Hermitian classgroup is that it provides a useful "approximation" to the Grothendieck group $K_0\, H(\mathfrak{o}\Gamma)$. The latter group is very difficult to get hold of in any explicit way, whereas H Cl($\mathfrak{o}\Gamma$) is relatively easier to compute. The functorial behaviour on change of group or basering becomes much more transparent in the Hom description of $H\,Cl(\mathfrak{o}\Gamma)$. Also there are a number of basic maps from $K_oH(\mathfrak{o}\Gamma)$ into certain other groups (i.e. functions from Hermitian $\mathfrak{o}\Gamma$-modules which are multiplicative with respect to orthogonal sums), which factorise through $H\,Cl(\mathfrak{o}\Gamma)$, and can thus be computed. We shall give a number of examples. For proofs and details see [F5].

I. Here we consider a local field F, the corresponding global maps really coming from the local ones. We consider $\mathfrak{o}\Gamma$-modules of form X/Y, where X and Y are free $\mathfrak{o}\Gamma$-modules spanning the same $F\Gamma$-module - thus X/Y is $\mathfrak{o}$-torsion, in fact finite. The Grothendieck group of such modules and exact sequences, will be denoted by $K_oT(\mathfrak{o}\Gamma)$. It is generated by symbols [X/Y] with X,Y as above. If X,Y are free $\mathfrak{o}\Gamma$-modules of finite rank, spanning the same $F\Gamma$-module, then we can define, in more generality, uniquely a symbol $[X/Y] \in K_oT(\mathfrak{o}\Gamma)$ by $[X/Y] = [X/U] - [Y/U]$ where $U \subset X \cap Y$, and U is free and spans XK = YK.

Now let (X,h) be a Hermitian $\mathfrak{o}\Gamma$-module. Write

$$\hat{X}_h = [y \in XF, \quad h(y,X) \subset \mathfrak{o}\Gamma]$$

for the h-dual of X. Then we get a map

$$(5.1) \qquad d' : K_oH(\mathfrak{o}\Gamma) \to K_oT(\mathfrak{o}\Gamma), \quad (X,h) \mapsto [\hat{X}_h/X].$$

On the other hand the map $Nr : R_\Gamma \to R_\Gamma^s$

$$Nr(\phi) = \phi + \bar{\phi} \qquad (\bar{\phi} \text{ the contragredient})$$

yields a homomorphism

$$(5.2) \qquad \hat{Nr} : Hom_{\Omega_F}(R^s, E^*) \to Hom_{\Omega_F}(R_\Gamma, E^*),$$

and we know (cf. [F5]) that

$$K_oT(\mathfrak{o}\Gamma) = Hom_{\Omega_F}(R_\Gamma, E^*)/Det(\mathfrak{o}\Gamma)^*.$$

Now we get

<u>5.1. Proposition.</u> $\hat{Nr}$ <u>gives rise to the homomorphism</u> ν <u>in the diagram</u>

$$\begin{array}{ccc} K_oH(\mathfrak{o}\Gamma) & \xrightarrow{d'} & K_oT(\mathfrak{o}\Gamma) \\ & {\scriptstyle d}\searrow \quad \nearrow{\scriptstyle \nu} & \\ & H\,Cl(\mathfrak{o}\Gamma) & \end{array}$$

<u>and this diagram commutes</u>.

II. Here F is a numberfield, $\mathcal{o}$ its ring of integers. The forgetful map

$$(X,h) \mapsto (X)$$

defines a homomorphism (we keep the notation of [F5])

$$\text{(5.3)} \qquad \delta' : K_o H(\mathcal{o}\Gamma) \to \mathrm{Cl}(\mathcal{o}\Gamma)$$

into the locally free classgroup of $\mathcal{o}\Gamma$. Recall now the Hom description of $\mathrm{Cl}(\mathcal{o}\Gamma)$ (cf. [F2] (Appendix), [R], [F6]),

$$\mathrm{Cl}(\mathcal{o}\Gamma) = \mathrm{Hom}_{\Omega_F}(R_\Gamma, JE)/(\mathrm{Det}\, U\mathcal{o}\Gamma) \cdot \mathrm{Hom}_{\Omega_F}(R_\Gamma, E^*).$$

We thus get a quotient map

$$\text{(5.4)} \qquad (\mathrm{Hom}_{\Omega_F}(R_\Gamma, JE)/\mathrm{Det}\, U\mathcal{o}\Gamma) \times \mathrm{Hom}_{\Omega_F}(R_\Gamma^s, E^*) \to \mathrm{Cl}(\mathcal{o}\Gamma),$$

which takes (f mod $\mathrm{Det}\, U\mathcal{o}\Gamma$, g) into the class of f in $\mathrm{Cl}(\mathcal{o}\Gamma)$.

<u>5.2. Proposition</u>. <u>The map</u> (5.4) <u>gives rise to the homomorphism</u> δ'' <u>in the diagram</u>

$$\begin{array}{ccc} K_o H(\mathcal{o}\Gamma) & \xrightarrow{\delta'} & \mathrm{Cl}(\mathcal{o}\Gamma) \\ & {\scriptstyle d}\searrow \quad \nearrow{\scriptstyle \delta''} & \\ & H\,\mathrm{Cl}(\mathcal{o}\Gamma) & \end{array}$$

<u>and this diagram commutes</u>.

<u>Remark</u>. If one replaces $H\,\mathrm{Cl}$ by $\mathrm{Ad}\, H\,\mathrm{Cl}$ such a map δ'' can in general not be defined. The Proposition shows that the proper global discriminant fixes the module class of X.

III. From now on we treat the global and the local case together. Suppose we are given a homomorphism $\theta : \Gamma \to \Sigma$ of groups. This yields homomorphisms $R_\Sigma \to R_\Gamma$, $R_\Sigma^s \to R_\Gamma^s$ (if Σ is a quotient of Γ this is inflation, if Γ is a subgroup of Σ it is restriction). Applying the functor Hom_{Ω_F}, we get a contravariant homomorphism

$$\text{(5.5)} \qquad \theta^* : \mathrm{Hom}_{\Omega_F}(R_\Gamma^s, \quad) \to \mathrm{Hom}_{\Omega_F}(R_\Sigma^s, \quad)$$

(and similarly with R_Γ, R_Σ in place). On the other hand, extension of scalars gives rise to a homomorphism

(5.6) $$K_oH(o\theta) : K_oH(o\Gamma) \to K_oH(o\Sigma).$$

<u>5.3. Proposition</u>. <u>The maps</u> θ^* <u>give rise to the homomorphism</u> $H\,Cl(o\theta)$ <u>in the diagram</u>

$$\begin{array}{ccc} K_oH(o\Gamma) & \xrightarrow{K_oH(o\theta)} & K_oH(o\Sigma) \\ \downarrow d & & \downarrow d \\ H\,Cl(o\Gamma) & \xrightarrow{H\,Cl(o\theta)} & H\,Cl(o\Sigma) \end{array} ,$$

<u>and this diagram commutes</u>.

IV. Now let Δ be a subgroup of Γ. Induction of characters is a map $R_\Delta \to R_\Gamma$, restricting to $R_\Delta^s \to R_\Gamma^s$, and gives rise to contravariant homomorphisms

(5.7) $$\lambda : \mathrm{Hom}_{\Omega_F}(R_\Gamma^s, \quad) \to \mathrm{Hom}_{\Omega_F}(R_\Delta^s, \quad),$$

(and similarly with R_Γ, R_Δ).

Next any free (locally free) $o\Gamma$-module retains this property when considered over $o\Delta$. Let $\ell : F\Gamma \to F\Delta$ be the F-linear map, with

$$\ell(\gamma) = \begin{cases} \gamma & , \gamma \in \Delta \\ 0 & , \gamma \in \Gamma \quad , \gamma \notin \Delta \end{cases} .$$

If h is a Hermitian form $V \times V \to F\Gamma$, then we get a Hermitian form $\ell h : V \times V \to F\Delta$, by setting $\ell h(v_1,v_2) = \ell(h(v_1,v_2))$. The functor $(X,h)_\Gamma \mapsto (X,\ell h)_\Delta$ yields a homomorphism (abuse of notation)

(5.8) $$K_oH\lambda : K_oH(o\Gamma) \to K_oH(o\Delta).$$

<u>5.4. Proposition</u>. <u>The maps</u> λ <u>give rise to the homomorphism</u> $H\,Cl\,\lambda$ <u>in the diagram</u>

$$\begin{array}{ccc} K_oH(o\Gamma) & \xrightarrow{K_oH\Lambda} & K_oH(o\Delta) \\ \Big\downarrow d & & \Big\downarrow d \\ H\,Cl(o\Gamma) & \xrightarrow{H\,Cl\,\Lambda} & H\,Cl(o\Delta), \end{array}$$

<u>and this diagram commutes</u>.

V. We now consider "orthonormal $o\Gamma$-modules". These are pairs (M,b) where

$$(5.9)\qquad \begin{cases} M \text{ is an } o\text{-torsion free, finitely generated } o\Gamma\text{-module} \\ b : M \times M \to o \text{ is a symmetric } \Gamma\text{-invariant bilinear form} \\ M \text{ has an } o\text{-basis } \{m_i\} \text{ with } b(m_i,m_\ell) = \delta_{i\ell}. \text{ (Kronecker symbol)} \end{cases}$$

These modules, with orthogonal sum, define a Grothendieck group $G^{on}(o\Gamma)$, which has the structure of a commutative ring (via $M, N \mapsto M \otimes_o N$, with diagonal action of Γ). Associating with M the character of the $E\Gamma$-module $M \otimes_o E$, we get a homomorphism

$$(5.10)\qquad s : G^{on}(o\Gamma) \to R_\Gamma$$

of rings, whose image we denote by $R_\Gamma^{on}(o)$.

Let (X,h) be a Hermitian $o\Gamma$-module, (M,b) an orthonormal $o\Gamma$-module. We define a new Hermitian $o\Gamma$-module

$$(5.11)\qquad (X,h)\cdot(M,b) = (X \otimes_o M,\ h\cdot b),$$

where Γ acts on $X \otimes_o M$ diagonally and $h\cdot b$ is given by

$$(5.12)\qquad \begin{cases} h\cdot b(x_1 \otimes m_1,\ x_2 \otimes m_2) = \sum\limits_{\gamma\in\Gamma} h_\gamma(x_1,x_2)\ b(m_1,m_2\gamma^{-1})\gamma \\ h_\gamma \text{ given by } h(x_1,x_2) = \sum\limits_{\gamma\in\Gamma} h_\gamma(x_1,x_2)\gamma, \quad h_\gamma(x_1,x_2) \in F, \end{cases}$$

($\forall\ x_1, x_2 \in XF$, $\forall\ m_1, m_2 \in MF$). The product (5.11) is easily seen to define on $K_oH(o\Gamma)$ the structure of a $G^{on}(o\Gamma)$-module.

On the other hand, we can define on $\mathrm{Hom}_{\Omega_F}(R_\Gamma,\ \ )$ and $\mathrm{Hom}_{\Omega_F}(R_\Gamma^s,\ \ )$ the structure of $R_\Gamma^{on}(o)$-modules, by setting

$$(5.13)\qquad \phi f(\chi) = f(\phi\chi), \quad \phi \in R_\Gamma^{on}(o), \quad (\chi \in R_\Gamma^s \text{ or } \chi \in R_\Gamma).$$

5.5. *Proposition*. *The product* (5.13) *induces the structure of a* $R_\Gamma^{on}(o)$-*module on* $H\,Cl(o\Gamma)$, *and the diagram*

$$\begin{array}{ccc} G^{on}(o\Gamma) \times K_o H(o\Gamma) & \xrightarrow{\ s \times d\ } & R_\Gamma^{on}(o) \times H\,Cl(o\Gamma) \\ \downarrow & & \downarrow \\ K_o H(o\Gamma) & \xrightarrow{\ d\ } & H\,Cl(o\Gamma) \end{array}$$

commutes, where the vertical maps come from the given module structures.

Using the results outlined under III - V, one can set up the usual apparatus of Frobenius modules over Frobenious functors. To use the induction techniques which thus become available, a more detailed knowledge of $R_\Gamma^{on}(o)$ will be needed.

Remark. Results analogous to those discussed under III - V are also available on algebraic change of basefield, both going up and going down.

6. Arithmetic applications

Just as in the theory of global Galois module structure, the generalised resolvents provide the link between the algebraic invariants and the arithmetic ones (conductors, Gauss sums etc). We return to the situation considered in §1. The problem at this stage is to compute Pfaffians for the Hermitian form $h_{N/K} = h_{N/K/K} : N \times N \to K\Gamma$, defined via the trace (cf. (2.7)). If, say, a is a free generator of N over $K\Gamma$ (generates a *normal basis* of N/K) we have to compute

$$\mathrm{Pf}(\sum_\gamma t_{N/K}(a \cdot a^\gamma)\gamma^{-1}) = \mathrm{Pf}(h_{N/K}(a,a)).$$

Recall on the other hand the definition of the resolvent:

$$(a|\chi) = \mathrm{Det}_\chi(\sum a^\gamma \gamma^{-1}) \quad (\chi \in R_\Gamma) \quad .$$

We then have

Theorem. *For all* $\chi \in R_\Gamma^s$,

$$\mathrm{Pf}_\chi(h_{N/K}(a,a)) = (a|\chi).$$

Proof. Use Proposition 3.3 and the equation

$$\sum_\gamma t_{N/K}(a \cdot a^\gamma)\gamma^{-1} = \overline{\sum a^\gamma \gamma^{-1}} \cdot \sum a^\gamma \gamma^{-1}.$$

On the basis of this theorem, the theory now proceeds via the connections between resolvents on the one hand and conductors and Galois Gauss sums on the other. This lies outside the scope of the present survey.

Literature

[F1] A. Fröhlich, Discriminants of algebraic number fields. Math.Z. 74 (1960), 18-28.

[F2] A. Fröhlich, Arithmetic and Galois module structure for tame extensions. Crelle 286/287, (1976), 380-440.

[F3] A. Fröhlich, Symplectic local constants and Hermitian Galois module structure, Proc. Internat. Symposium Tyoto 1976, (Ed. S. Iyanaga), Japan Soc. for the Promotion of Science, Tokyo 1977, 25-42.

[F4] A. Fröhlich, Local Hermitian group modules, Proc. Conference on quad. forms, Queen's Univ., Kingston, Ontario, 1977.

[F5] A. Fröhlich, Classgroups, in particular Hermitian classgroups, Lecture Notes to be published.

[F6] A. Fröhlich, Galois module structure of rings of integers, Springer Ergoluisse Berich t , Forthcoming.

[FM] A. Fröhlich and A. McEvett, Forms over rings with involution, J. of Alg. 12 (1969) 79-104.

[M] J. Martinet, Character theory and Artin L-functions, Durham Proceedings, 1977, 1-88.

[R] J. Ritter, see present volume.

[Tt] J. Tate, Local constants, Durham Proceedings 1977, 89-131.

[Ty] M.J. Taylor, On Fröhlich's conjecture for rings of integers of tame extensions, to be published in Invent. Math.

FRÖHLICH'S CONJECTURE, LOGARITHMIC METHODS AND SWAN MODULES

by M. J. TAYLOR*

1. Let E/M be a Galois extension of number fields. We write $\Gamma = \mathrm{Gal}(E/M)$ and we view the ring of integers of E, $\mathcal{O}_E$, as a (right) $\mathbb{Z}\Gamma$-module in the natural way.

A prime ideal $\mathfrak{p}$ of M is said to be tamely ramified in E if the ramification index of $\mathfrak{p}$ in E/M is prime to p. It is known that $\mathcal{O}_E$ is locally free over $\mathcal{O}_M\Gamma$ (resp. $\mathbb{Z}\Gamma$) at a prime of M (resp. of $\mathbb{Z}$) if, and only if, that prime is (resp. the primes of M above that prime are) tamely ramified in E/M. The first of these results is usually ascribed to E. Noether for the result in [N]; however, for a completely general proof see page 21 of [CF].

The principal aim in the study of rings of integers as Galois modules is to discover whether $\mathcal{O}_E$ is a free $\mathbb{Z}\Gamma$-module or not. By the local considerations described above, it will be sufficient henceforth to assume E/M to be tame.

Let $K_o(\mathbb{Z}\Gamma)$ be the Grothendieck group of finitely generated, locally free $\mathbb{Z}\Gamma$-modules. We denote by $\mathrm{Cl}(\mathbb{Z}\Gamma)$ the kernel of the rank homomorphism $K_o(\mathbb{Z}\Gamma) \to \mathbb{Z}$. For a locally free $\mathbb{Z}\Gamma$-module of rank r, M say, we let (M) denote the difference in $K_o(\mathbb{Z}\Gamma)$ of the class of M and r copies of $\mathbb{Z}\Gamma$. In the sequel we shall always write the group operation in $\mathrm{Cl}(\mathbb{Z}\Gamma)$ multiplicatively.

---

* Part of the work contained in this paper was done while I received financial support from the C. N. R. S. and the kind hospitality of the University of Besançon.

It is natural to ask whether or not it is always true that $(\mathcal{O}_E) = 1$. In [M1] J. Martinet showed that for the case when $M = \mathbb{Q}$ there exist extensions E with Γ isomorphic to H_8, the quaternion group of order eight, such that $(\mathcal{O}_E) \neq 1$. Subsequently, in [F1], A. Fröhlich gave a beautiful interpretation of this result in terms of the constant in the functional equation of the Artin L-function of the irreducible non-abelian character of Γ.

To make this more precise we need to introduce further notation. Let R_Γ be the group of virtual characters of Γ. For $\chi \in R_\Gamma$ we let $\Lambda(s,\chi)$ denote the "extended" Artin L-function associated to χ. It is well-known that $\Lambda(s,\chi)$ satisfies a functional equation

$$\Lambda(s,\chi) = W(\chi)\,\Lambda(1-s,\bar{\chi})$$

where $\bar{\chi}$ is the complex conjugate of χ (cf. [M2]). The constant $W(\chi)$ is called the Artin root number of χ. Using the fact that the $\Lambda(s,\chi)$ are not identically zero we deduce from the functional equation that $W(\chi+\bar{\chi}) = 1$, and so, in the case when χ is real-valued, $W(\chi) = \pm 1$.

Again we take Γ to be isomorphic to H_8 and $M = \mathbb{Q}$. We let ψ be the irreducible non-abelian character of Γ. A. Fröhlich showed (loc.cit)

<u>THEOREM 1</u>:

<u>$\mathcal{O}_E$ is a free $\mathbb{Z}\Gamma$-module if, and only if, $W(\psi) = +1$.</u>

We now return to the general situation where Γ is arbitrary and where M is not necessarily the rationals.

The main aim of this article is to describe the algebraic methods used in the proof of Theorem 2 (below) which was first conjectured by Fröhlich (cf. [T1] for details). In the final section we show how these methods can be applied to describing the classes of Swan modules.

THEOREM 2 :

$(\mathcal{O}_E)^2 = 1$ and further, if the Artin root numbers of the symplectic characters of Γ are all $+1$, then $(\mathcal{O}_E) = 1$, i. e. the Artin root numbers of such characters are the only obstructions to the vanishing of $(\mathcal{O}_E)$.

Remark 1 : For a more precise description of the class of $(\mathcal{O}_E)$ - using Ph. Cassou-Noguès class $t(W)$ - see [T1].

Remark 2 : Here our point of view is that we wish to describe $(\mathcal{O}_E)$. One can adopt the other point of view that one wishes to describe Artin root numbers in terms of a (more general) module class of $\mathcal{O}_E$. For this approach see [F2].

2. The classgroup of Fröhlich : We let $D(\mathbb{Z}\Gamma)$ be the sub-group of $Cl(\mathbb{Z}\Gamma)$ of classes which become trivial under extension by a maximal order containing $\mathbb{Z}\Gamma$. (Jacobinski has shown that the sub-group $D(\mathbb{Z}\Gamma)$ is independent of the choice of maximal order). The reason for our introducing $D(\mathbb{Z}\Gamma)$ is that A. Fröhlich, following a conjecture of J. Martinet, showed that $(\mathcal{O}_E) \in D(\mathbb{Z}\Gamma)$ (cf. § 9 of [F3]).

For the remainder of this article, unless stated to the contrary, Γ will always be a group of ℓ-power order, ℓ^n say, for some prime number ℓ. We impose this restriction on Γ in order that the main ideas be as clear as possible. We now give Fröhlich's description of $D(\mathbb{Z}\Gamma)$. (For his description for an arbitrary finite group Γ see Appendix II of [F3]).

Let F be the cyclotomic field obtained by adjoining the ℓ^nth roots of unity to $\mathbb{Q}$. We put $\Omega = \mathrm{Gal}(F/\mathbb{Q})$, $\mathcal{O}_{F_\ell} = \mathcal{O}_F \otimes_{\mathbb{Z}} \mathbb{Z}_\ell$, $U = \mathcal{O}_{F_\ell}^*$.

Both U and R_Γ are Ω-modules (Ω acts on R_Γ value-wise). We let $\mathrm{Hom}_\Omega(R_\Gamma, U)$ denote the group of homomorphisms from R_Γ to U which commute with Ω-action.

Similarly we let $\mathrm{Hom}_{\Omega^+}\left(R_\Gamma, \mathcal{O}_F^*\right)$ be the group of Ω-homomorphisms from R_Γ to $\mathcal{O}_F^*$ which are totally positive on all symplectic characters of Γ. We view $\mathrm{Hom}_{\Omega^+}\left(R_\Gamma, \mathcal{O}_F^*\right)$ as a subgroup of $\mathrm{Hom}_\Omega\left(R_\Gamma, U\right)$ via the natural embedding $\mathcal{O}_F^* \hookrightarrow U$.

To each $z \in \mathbb{Z}_\ell \Gamma^*$ we wish to associate a homomorphism $\mathrm{Det}(z) : R_\Gamma \to U$. By $\mathbb{Z}$-linearity it is enough to define the value of Det(z) on a character of Γ. So let χ be a character of Γ which is afforded by a representation $T : \Gamma \to GL_m(F)$. We extend T to an algebra homomorphism $T : \mathcal{O}_\ell \Gamma \to M_m\left(F \otimes \mathcal{O}_\ell\right)$ and we define

$$\mathrm{Det}(z)(\chi) = \det(T(z)).$$

It is easily checked (cf. Appendix I of [F3]) that Det(z) is an Ω-homomorphism. We denote the sub-group of such homomorphisms by $\mathrm{Det}\left(\mathbb{Z}_\ell \Gamma^*\right)$.

Fröhlich described an isomorphism (loc. cit)

$$D(\mathbb{Z}\Gamma) \cong \frac{\mathrm{Hom}_\Omega\left(R_\Gamma, U\right)}{\mathrm{Det}\left(\mathbb{Z}_\ell \Gamma^*\right) \mathrm{Hom}_{\Omega^+}\left(R_\Gamma, \mathcal{O}_F^*\right)} \tag{2.1}$$

For the person seeing this description for the first time, it is worthwhile pointing out that a module class is represented by an Ω-homomorphism from R_Γ to U, and, furthermore, if we wish to determine whether or not such a class is trivial, we have to try and develop methods for deciding whether or not such a homomorphism lies in the denominator.

Example (Swan, Ullom):

Let r be an integer prime to ℓ and let (r, Σ) be the two sided $\mathbb{Z}\Gamma$ ideal $r\mathbb{Z}\Gamma + \mathbb{Z} \sum_{\gamma \in \Gamma} \gamma$. Swan showed that such modules are locally free (cf. [S]). Steve Ullom (cf. [U]) showed that their classes form a sub-group $T(\mathbb{Z}\Gamma)$ of $D(\mathbb{Z}\Gamma)$, and, moreover, that the class of the module (r, Σ) is represented under (2.1) by the homomorphism :

(2. 2) $$\chi \mapsto r^{(\chi,\, \varepsilon)}$$

for $\chi \in R_\Gamma$. Here ε is the identity character of Γ and $(\,,\,)$ is the standard inner product of character theory.

In the next section we give a new logarithmic description of $\mathrm{Det}(\mathbb{Z}_\ell\Gamma^*)$. Then, in section 4, we explain how this description is used to prove Fröhlich's conjecture. Lastly, in section 5, we use the same technique to describe the group of Swan classes $T(\mathbb{Z}\Gamma)$.

3. Integral logarithms : The proofs of all results in this section, together with the appropriate extensions to arbitrary Γ are to be found in [T2].

We denote the Jacobson radical of $\mathbb{Z}_\ell\Gamma$ by $\mathfrak{r}$, and we define a map $\mathcal{L}_1 : 1+\mathfrak{r} \to \mathbb{Q}_\ell\Gamma$ by stipulating that for $r \in \mathfrak{r}$

$$\mathcal{L}_1(1-r) = -\ell \sum_{n=1}^{\infty} \frac{r^n}{n} + \sum_{n=1}^{\infty} \frac{\Psi^\ell(r^n)}{n}$$

where $\Psi^\ell : \mathbb{Q}_\ell\Gamma \to \mathbb{Q}_\ell\Gamma$ is the $\mathbb{Q}_\ell$-linear map induced by $\gamma \mapsto \gamma^\ell$, for $\gamma \in \Gamma$. (This series converges because the $\mathfrak{r}$-adic and ℓ-adic topologies coincide).

In order to obtain a homomorphism we let $\mathfrak{C}$ be the set of conjugacy classes of Γ, and we let $\phi : \mathbb{Q}_\ell\Gamma \to \mathbb{Q}_\ell\mathfrak{C}$ be the $\mathbb{Q}_\ell$-linear map induced by mapping γ to its conjugacy class $C(\gamma)$. Then we obtain that the composite map $\mathcal{L} = \phi \circ \mathcal{L}_1$ is a group homomorphism. Moreover, $\mathcal{L}$ is in fact an integral logarithm in the sense that $\mathcal{L}(1+\mathfrak{r}) \subseteq \ell\mathbb{Z}_\ell\mathfrak{C}$. (This is the underlying reason for our adjustment of the usual logarithm in $\mathcal{L}_1$ by $\ell - \Psi^\ell$).

Because $\left(\mathbb{Z}_\ell\Gamma^* : 1+\mathfrak{r}\right) = \ell - 1$ and because the image space of $\mathcal{L}$ is uniquely divisible, $\mathcal{L}$ extends in a unique manner to $\mathbb{Z}_\ell\Gamma^*$. Further, as the image space is a pro-ℓ-group,

(3. 1) $$\mathcal{L}(\mathbb{Z}_\ell\Gamma^*) \subseteq \ell\mathbb{Z}_\ell\mathfrak{C}.$$

Next we choose $\chi \in R_\Gamma$ and we view $\chi : \mathbb{Z}_\ell \mathfrak{C} \to \mathcal{O}_{F_\ell}$ by $\mathbb{Z}_\ell$-linearity and by stipulating that $\chi(C(\gamma)) = \chi(\gamma)$. It can then be shown that for any $z \in \mathbb{Z}_\ell \Gamma^*$

(3. 2) $$\chi(\mathcal{L}(\mathrm{Det}(z))) = \log(\mathrm{Det}(z)(\ell\chi - \psi^\ell\chi))$$

where $\psi^\ell : R_\Gamma \to R_\Gamma$ is the ℓth Adams operation i. e. for $\gamma \in \Gamma$, $\psi^\ell\chi(\gamma) = \chi(\gamma^\ell)$.

Using (3. 2) we can show that $\mathcal{L}$-factors through the homomorphism Det. That is to say, there exists a unique group homomorphism $\nu : \mathrm{Det}\left(\mathbb{Z}_\ell\Gamma^*\right) \to \ell\mathbb{Z}_\ell\mathfrak{C}$ such that the following diagram commutes

$$\begin{array}{ccc} \mathbb{Z}_\ell\Gamma^* & \xrightarrow{\mathcal{L}} & \ell\mathbb{Z}_\ell\mathfrak{C} \\ & {}_{\mathrm{Det}}\searrow & \uparrow \nu \\ & & \mathrm{Det}\,(\mathbb{Z}_\ell\Gamma^*) \end{array}$$

As the image space of ν is torsion free, $\mathrm{Ker}(\nu)$ contains all the torsion of $\mathrm{Det}\left(\mathbb{Z}_\ell\Gamma^*\right)$. Thus we find, that in order to obtain good results from ν, it is best to restrict to determinants coming from the commutator ideal $\mathfrak{a} = \mathrm{Ker}\left(\mathbb{Z}_\ell\Gamma \to \mathbb{Z}_\ell\Gamma^{ab}\right)$. Then the restriction of ν yields an isomorphism

(3. 3) $$\nu : \mathrm{Det}(1+\mathfrak{a}) \xrightarrow{\sim} \ell\phi(\mathfrak{a}).$$

(The author has recently learnt that the result that $\mathrm{Det}(1+\mathfrak{a})$ is torsion free was first shown by C. T. C. Wall in [W]).

There then remains the question of describing $\mathrm{Det}\left(\mathbb{Z}_\ell\Delta^*\right)$ when Δ is an abelian group. This, it turns out, is trivial and we have the isomorphism

(3. 4) $$\mathbb{Z}_\ell\Delta^* \xrightarrow{\sim} \mathrm{Det}\left(\mathbb{Z}_\ell\Delta^*\right).$$

<u>Proof</u> : If $z \in \mathbb{Z}_\ell\Delta^*$ is such that $\mathrm{Det}(z) = 1$, then $\mathrm{Det}(z)(\chi) = 1$ for each abelian character χ of Γ i. e. $\chi(z) = 1$ for each such χ. Consequently z has image 1 at each component of the group algebra $\Omega_\ell\Delta$ and hence $z = 1$.

Using (3. 4) it is easy to verify that the sequence

$$1 \to \mathrm{Det}(1+\mathfrak{a}) \to \mathrm{Det}\left(\mathbb{Z}_\ell\Gamma^*\right) \to \mathrm{Det}\left(\mathbb{Z}_\ell\Gamma^{ab^*}\right) \to 1$$

is exact. Therefore, by (3. 3) and (3. 4), we obtain an exact sequence

$$(3.5) \qquad 0 \to \ell\,\phi(\mathfrak{a}) \to \mathrm{Det}(\mathbb{Z}_\ell \Gamma^*) \to \mathbb{Z}_\ell \Gamma^{ab*} \to 1.$$

4. <u>The fixed point theorem</u> : Let $K/\mathbb{Q}_\ell$ be a finite Galois extension which is non-ramified and let $\Delta = \mathrm{Gal}(K/\mathbb{Q}_\ell)$. Then $\mathrm{Det}(\mathcal{O}_K\Gamma^*)$ is a Δ-module by the rule that for $\delta \in \Delta$, $x \in \mathcal{O}_K\Gamma^*$,

$$\mathrm{Det}(x).\,\delta = \mathrm{Det}(x^\delta).$$

In [T2] it is shown that

<u>THEOREM 3</u> :

<u>For any finite group Γ (i.e. not necessarily an ℓ-group)</u>

$$\mathrm{Det}(\mathcal{O}_K\Gamma^*)^\Delta = \mathrm{Det}(\mathcal{O}_K\Gamma^{*\Delta}) \ (=\mathrm{Det}(\mathbb{Z}_\ell\Gamma^*))$$

<u>where, as usual, the superscript Δ denotes the group of points fixed by Δ.</u>

<u>Remark</u> : In fact one can very easily extend this result to the case when K is a tame extension of $\mathbb{Q}_\ell$. The reader is referred to Theorem 6 of [T1] to see how the above non-ramified fixed point theorem implies the corresponding result in the tame situation.

Next, we outline the main ideas in the proof of Theorem 3, and then we describe how this theorem is applied to prove Fröhlich's conjecture.

<u>Main steps in the proof of Theorem 3</u> :

(1) First we reduce to considering $\mathbb{Q}$-elementary groups by Brauer induction, and by Ullom's Frobenius module structure theorem for Det-groups.

(2) Then we deal with the case of $\mathbb{Q}$-p-elementary groups with $p \neq \ell$, by use of techniques originating from Fröhlich's homomorphisms on classgroups associated with the kernel of Brauer's decomposition map mod (ℓ).

(3) Next we pass from the case of $\mathbb{Q}$-ℓ-elementary groups to ℓ-groups by methods similar to those used in the theory of Galois rings. This leads us to consider in detail the case when Γ is an ℓ-group.

Using the fact that $K/\mathbb{Q}_\ell$ is non-ramified we can extend the exact sequence

(3. 5) in a natural way to

$$(4.1)\qquad 0 \to \ell\phi(\mathfrak{a}) \otimes \mathcal{O}_K \to \mathrm{Det}\left(\mathcal{O}_K \Gamma^*\right) \to \mathcal{O}_K \Gamma^{ab*} \to 1 .$$

We then show that the Δ-fixed points satisfy a corresponding exact sequence ; this yields a commutative diagram

$$\begin{array}{ccccccccc} 0 \to & \left(\ell\phi(\mathfrak{a}) \otimes \mathcal{O}_K\right)^\Delta & \to & \mathrm{Det}\left(\mathcal{O}_K \Gamma^*\right)^\Delta & \to & \left(\mathcal{O}_K \Gamma^{ab*}\right)^\Delta & \to 1 \\ & \uparrow a & & \uparrow b & & \uparrow c & \\ 0 \to & \ell\phi(\mathfrak{a}) & \to & \mathrm{Det}\left(\mathbb{Z}_\ell \Gamma^*\right) & \to & \mathbb{Z}_\ell \Gamma^{ab*} & \to 1 \end{array}$$

It is immediate that the inclusion maps a and c are surjective and whence b also is, as we require.

<u>Application to Theorem 2</u> :

As previous Γ is an ℓ-group (although the ideas used extend very easily to general Γ). We can now describe the strategy for proving Theorem 2. First, by § 9 of [F3] we have an Ω-homomorphism v, which, under (2. 1), represents the class $\left(\mathcal{O}_E\right)$ modified by the obstructions of the signs of the root numbers of symplectic characters. So to prove Theorem 1, it is sufficient to show that v lies in the denominator of (2. 1).

By use of the exact sequence (4. 1) we show that for some Galois extension E/Ω, $v = \mathrm{Det}(z) = \prod_{\mathfrak{L}} \mathrm{Det}\left(z_{\mathfrak{L}}\right)$, with $z_{\mathfrak{L}} \in \mathcal{O}^t_{\mathfrak{L}} \Gamma^*$. Here the direct product extends over the primes $\mathfrak{L}$ of E above ℓ and $\mathcal{O}^t_{\mathfrak{L}}$ is the ring of integers of the maximal tame sub-extension of $E_{\mathfrak{L}}$. However, as v is an Ω-homomorphism it is trivially a Gal (EF/Ω) (= Λ say) homomorphism, where Λ acts on R_Γ and U in the natural way. Consequently for any $\chi \in R_\Gamma$, $\lambda \in \Lambda$, $v\left(\chi^\lambda\right) = v(\chi)^\lambda$ i. e. $\mathrm{Det}(z)\left(\chi^\lambda\right) = \left(\mathrm{Det}(z)(\chi)\right)^\lambda = \mathrm{Det}\left(z^\lambda\right)\left(\chi^\lambda\right)$.

Thus as Det(z) is Λ-fixed, we deduce that each $\mathrm{Det}\left(z_{\mathfrak{L}}\right)$ is fixed by the decomposition group of $\mathfrak{L}$, and so by the generalisation of Theorem 3 to tame extensions $\mathrm{Det}\left(z_{\mathfrak{L}}\right) = \mathrm{Det}(x)$ for some $x \in \mathbb{Z}_\ell \Gamma^*$. Moreover, because Λ transitively permutes the primes $\mathfrak{L}$ and because Det(z) is Λ-fixed, we see that

$\mathrm{Det}(z) = \prod_{\mathfrak{L}} \mathrm{Det}(x) \in \mathrm{Det}\left(\mathbb{Z}_\ell \Gamma^*\right)$, and hence v lies in the denominator of (2. 1),

as we required.

5. Swan modules : In this section we apply the logarithmic techniques of § 3 to calculate the group of Swan classes $T(\mathbb{Z}\Gamma)$ when Γ is an ℓ-group. For details see [T3].

THEOREM 4 (conjectured by S. Ullom) :

(i) If $\ell \neq 2$ and Γ is not cyclic, then $T(\mathbb{Z}\Gamma)$ is cyclic of order $|\Gamma|\ \ell^{-1}$.

(ii) If $\ell = 2$ and Γ is neither cyclic, dihedral, semi-dihedral nor quaternion, then $T(\mathbb{Z}\Gamma)$ is cyclic of order $|\Gamma|/4$.

Remark : If Γ is cyclic then $T(\mathbb{Z}\Gamma) = \{1\}$; while if Γ is quaternion (resp. dihedral, resp. semi-dihedral) then $T(\mathbb{Z}\Gamma)$ has order 2 (resp. 1, resp. 2) (cf. [U] for details of these exceptional cases).

Proof : Swan (cf. § 6 of [S]) showed that $T(\mathbb{Z}\Gamma)$ is a quotient of the group $\left(\mathbb{Z}/_{|\Gamma|}\mathbb{Z}\right)^*$ factored out by the image of $\langle \pm 1 \rangle$, and as such it is clearly cyclic. Ullom (loc. cit) showed that $T(\mathbb{Z}\Gamma)$ is annihilated by the Artin exponent, and so, because the Artin exponent divides $|\Gamma|$ (cf. [L]), we deduce

$$|T(\mathbb{Z}\Gamma)| \leq \begin{cases} |\Gamma|\ \ell^{-1} & \text{if } \ell \neq 2, \\ |\Gamma|/4 & \text{if } \ell = 2. \end{cases}$$

(5. 1) Consequently to prove the theorem it is enough to show that the class of Swan module $(1+\ell, \Sigma)$ has order greater than or equal to $|\Gamma|\ \ell^{-1-\delta}$, where $\delta = 0$ if $\ell \neq 2$ and $\delta = 1$ if $\ell = 2$.

Let ρ denote the regular character of Γ. We define a homomorphism e,

$$e : \mathrm{Hom}_\Omega\left(R_\Gamma, U\right) \to \mathbb{Z}_\ell \bmod (\ell\,|\Gamma|)$$

by stipulating that for $f \in \mathrm{Hom}_\Omega\left(R_\Gamma, U\right)$

$$e(f) = \log\left(f\left(\ell\rho - \psi^\ell \rho\right)\right) \bmod (\ell\,|\Gamma|).$$

Next we will show that e vanishes on the denominator of (2. 1). Firstly, by (3. 2), we know that for $z \in \mathbb{Z}_\ell \Gamma^*$

$$e(\mathrm{Det}(z)) = \rho(\mathfrak{L}(z)) \in \rho\left(\ell \mathbb{Z}_\ell \mathfrak{C}\right) \subseteq \ell\,|\Gamma|\,\mathbb{Z}_\ell$$

since ρ is zero outside the identity element and $|\Gamma|$ on the identity element. Secondly, if $g \in \mathrm{Hom}_\Omega\left(R_\Gamma, \mathcal{O}_F^*\right)$, then, as $\ell\rho - \psi^\ell\rho$ is rational valued, $g\left(\ell\rho - \psi^\ell\rho\right)$ is a rational unit, i. e. it is ± 1, and so $e(g) = 0$. Hence e induces a homomorphism

$$\bar{e} : D(\mathbb{Z}\Gamma) \to \mathbb{Z}_\ell \bmod (\ell\,|\Gamma|\,).$$

Now we calculate $\bar{e}(1+\ell, \Sigma)$. From (2. 2) the class of $(1+\ell, \Sigma)$ is represented under (2. 1) by the homomorphism h where $h(\chi) = (1+\ell)^{(\chi, \epsilon)}$, so that

$$\bar{e}(1+\ell, \Sigma) = e(h) = \log\left(h\left(\ell\rho - \psi^\ell\rho\right)\right) \bmod (\ell\,|\Gamma|\,)$$

(5. 2) $$= \left(\ell\rho - \psi^\ell\rho, \epsilon\right)\log(1+\ell) \bmod (\ell\,|\Gamma|\,).$$

By an elementary calculation $\log(1+\ell) = u\,\ell^{1+\delta}$ for some $u \in \mathbb{Z}_\ell^*$. Also it is immediate that $(\ell\rho, \epsilon) = \ell$ and

$$\left(\psi^\ell\rho, \epsilon\right) = \frac{1}{|\Gamma|}\sum_{\gamma \in \Gamma} \rho\left(\gamma^\ell\right) = \sum_{\gamma^\ell = 1} 1.$$

<u>THEOREM 5</u> :

(i) <u>(Kulakoff cf. [Z]). If $\ell \neq 2$ and if Γ is non cyclic, then</u> $\sum_{\gamma^\ell = 1} 1$ <u>is divisible by ℓ^2</u>.

(ii) <u>(Alperin, Feit, Thompson cf. [L]). If $\ell = 2$ and Γ is neither cyclic, dihedral, semi-dihedral nor quaternion, then</u> $\sum_{\gamma^2 = 1}$ <u>1 is divisible by 4</u>.

Thus we see that for all Γ as stated in Theorem 4, $\left(\ell\rho - \psi^\ell\rho, \epsilon\right) = \ell m$ for some integer m prime to ℓ. Hence by (5. 2) we have now shown

$\bar{e}(1+\ell, \Sigma) = \ell^{2+\delta} m.u \bmod (\ell\,|\Gamma|\,)$, which establishes (5. 1).

BIBLIOGRAPHIE

[F1] A. FRÖHLICH - Artin root numbers and normal integral bases for quaternion fields, Invent. Math. 17 (1972), p. 143-166.

[F2] A. FRÖHLICH - Hermitian classgroups, same volume.

[F3] A. FRÖHLICH - Arithmetic and Galois module structure for tame extensions, J. reine angew. Math. 286/7 (1976), p. 380-440.

[L] T-Y. LAM - Artin exponents of finite groups, J. Algebra 9 (1968), p. 94-119.

[M1] J. MARTINET - Modules sur l'algèbre du groupe quaternonien, Ann. Sci. Ecole Norm. Sup. 4 (1971), p. 229-308.

[M2] J. MARTINET - Character theory and Artin L-functions, Algebraic Number Fields (Proc. Durham Symposium), Academic Press (1977), London.

[S] R. SWAN - Periodic resolutions for finite groups, Ann. Math. 72 (1960), p. 267-291.

[T1] M. J. TAYLOR - On Fröhlich's conjecture for rings of integers of tame extensions, to appear.

[T2] M. J. TAYLOR - A logarithmic approach to classgroups of integral group rings, to appear in J. Algebra.

[T3] M. J. TAYLOR - Locally free classgroups of groups of prime power order, J. Algebra 50 (1978), p. 463-487.

[U] S. V. ULLOM - Non-trivial bounds for classgroups, Illinois J. Math. 20 (1976), p. 361-371.

[W] C. T. C. WALL - Norms of units in group rings, Proc. L. M. S. (3) 29 (1974), p. 593-632.

[Z] H. ZASSENHAUS - "Theory of groups", Chelsea, New-York, 1958.

M. J. TAYLOR
Queen Mary College, London,
and
E. R. A. C. N. R. S. n° 070654,
25030 BESANCON CEDEX

S-GROUPES DE GROTHENDIECK ET STRUCTURE GALOISIENNE DES ANNEAUX D'ENTIERS

par Jacques QUEYRUT (Bordeaux)*

Soit $\mathbb{Z}$ l'anneau des entiers et soit $\mathbb{Q}$ son corps des fractions. On fixe une clôture algébrique $\overline{\mathbb{Q}}$ de $\mathbb{Q}$ et l'on note $\overline{\mathbb{Z}}$ la clôture intégrale de $\mathbb{Z}$ dans $\overline{\mathbb{Q}}$. Le groupe de Galois $G_{\mathbb{Q}}$ de $\overline{\mathbb{Q}}$ sur $\mathbb{Q}$ est un groupe profini. Les corps de nombres N, i.e. les extensions de $\mathbb{Q}$ de degré fini incluses dans $\overline{\mathbb{Q}}$, sont en bijections avec les sous-groupes ouverts G_N de $G_{\mathbb{Q}}$.

Tout groupe d'automorphismes Γ de N opère sur la clôture intégrale $\mathbb{Z}_N$ de $\mathbb{Z}$ dans N. On se propose d'étudier la structure de $\mathbb{Z}_N$ en tant que $\mathbb{Z}[\Gamma]$-module.

Soit $\mathcal{P}(\mathbb{Z})$ l'ensemble des nombres premiers de $\mathbb{Z}$. Un automorphisme γ de N est dit ramifié (resp. sauvagement ramifié) en $p \in \mathcal{P}(\mathbb{Z})$ s'il existe un idéal premier $\mathfrak{P}$ de $\mathbb{Z}_N$ au-dessus de p et un entier i supérieur ou égal à 1 (resp. à 2) tel que $\gamma(a) - a \in \mathfrak{P}^i$, $\forall a \in \mathbb{Z}_N$. On note S, ou $S(\Gamma)$, une partie de $\mathcal{P}(\mathbb{Z})$ contenant l'ensemble des nombres premiers p tel que Γ contienne un automorphisme sauvagement ramifié en p.

Le premier résultat est dû à Hilbert qui montre dans son rapport (1913) que pour une extension N abélienne de $\mathbb{Q}$ de groupe de Galois Γ, $\mathbb{Z}_N$ est un $\mathbb{Z}[\Gamma]$-module libre si Γ ne contient pas d'automorphisme ramifié aux diviseurs premiers de l'ordre de Γ ([Hi]).

E. Noether a montré ensuite (1932) que $\mathbb{Z}_N$ est un $\mathbb{Z}_K[\Gamma]$-module projectif si Γ ne contient pas d'automorphisme ramifié aux diviseurs premiers de l'ordre de Γ. En fait, on peut montrer plus précisément que $\mathbb{Z}_N$ est un $\mathbb{Z}[\Gamma]$-module localement projectif en p si et seulement si Γ ne contient pas d'automorphisme sauvagement ramifié en p ([N] et [M 1]).

---

* Laboratoire associé au C.N.R.S. n° 226

Le résultat de Hilbert a été généralisé par Leopoldt (1959). Si N est une extension abélienne de $\mathbb{Q}$ de groupe de Galois Γ, alors $\mathbb{Z}_N$ est un $\mathfrak{O}$-module libre où $\mathfrak{O}$ est l'ensemble du $\lambda \in \mathbb{Q}[\Gamma]$ tels que $\lambda \mathbb{Z}_N \subset \mathbb{Z}_N$ (appelé ordre associé). Il a construit une base de $\mathbb{Z}_N$ sur $\mathfrak{O}$ à l'aide de sommes de Gauss ([L]).

Le sujet a été relancé par Martinet tout d'abord en 1969 en montrant que pour N extension diédrale de degré 2p (p premier) de $\mathbb{Q}$ de groupe de Galois Γ, $\mathbb{Z}_N$ est un $\mathbb{Z}[\Gamma]$-module libre si Γ ne contient pas d'automorphisme sauvagement ramifié, puis en 1971 en étudiant les extensions quaternioniennes de degré 8 de $\mathbb{Q}$ ([M 1]).

Il a montré que pour les groupes Γ quaternioniens d'ordre 8, il n'y avait que deux classes d'isomorphismes de $\mathbb{Z}[\Gamma]$-modules projectifs, toutes les deux représentées par des anneaux d'entiers d'extensions quaternioniennes de $\mathbb{Q}$ ([M 2]). Fröhlich a alors montré que $\mathbb{Z}_N$ appartient à la classe des $\mathbb{Z}[\Gamma]$-modules libres si et seulement si la constante $W(\chi)$ vaut +1 pour le caractère irréductible de degré 2 de Γ (qui est à valeurs réelles non rationnel sur le corps des réels) ([F 1]).

J'ai montré qu'il en était de même pour les extensions quaternioniennes de $\mathbb{Q}$ de degré 12 ([Q 1]).

Sous l'impulsion de Fröhlich, le lien entre la structure des anneaux d'entiers de N et la constante de W s'est précisé sous l'hypothèse que Γ ne contienne pas d'automorphismes sauvagement ramifiés, dite hypothèse de ramification modérée, et une théorie très complète s'est développée. Cela a débouché sur le théorème et la conjecture suivants ([F 2]).

Soit $\mathfrak{M}$ un ordre maximal de $\mathbb{Z}$ dans $\mathbb{Q}[\Gamma]$ contenant $\mathbb{Z}[\Gamma]$, c'est-à-dire un sous-anneau unitaire de $\mathbb{Q}[\Gamma]$ qui est un $\mathbb{Z}$-module de type fini, contenant Γ et maximal pour l'inclusion. Fröhlich a montré le résultat suivant conjecturé par Martinet : sous l'hypothèse de ramification modérée, alors $\mathfrak{M}\, \mathbb{Z}_N$ est un $\mathfrak{M}$-module stablement libre (i. e. $\mathfrak{M}\, \mathbb{Z}_N \oplus \mathfrak{M}$ est isomorphe à $\mathfrak{M}^r \oplus \mathfrak{M}$ où r est le degré de K sur $\mathbb{Q}$). Puis il a fait la conjecture suivante justifiée par un grand nombre de nouveaux cas particuliers obtenus par Cassou-Noguès, Cougnard, lui-même et Taylor : sous l'hypothèse de ramification modérée, $\mathbb{Z}_N \oplus \mathbb{Z}_N$ est un $\mathbb{Z}[\Gamma]$-module libre et de plus si $W(\chi) = +1$ pour les caractères à valeurs réelles non réalisables sur $\mathbb{R}$, alors $\mathbb{Z}_N$ est un $\mathbb{Z}[\Gamma]$-module stablement libre.

L'étude de la structure d'un $\mathbb{Z}[\Gamma]$-module projectif (à isomorphisme stable près) revient à déterminer sa classe dans le groupe des classes des $\mathbb{Z}[\Gamma]$-modules projectifs de rang 1. Fröhlich a donné une nouvelle description de ce groupe et a décrit la classe de $\mathbb{Z}_N$ à l'aide d'une généralisation de la notion de résolvante de Lagrange. Ces résolvantes sont engendrées par les sommes de Gauss. Tout cela utilise de façon fondamentale le fait que $\mathbb{Z}_N$ est un $\mathbb{Z}[\Gamma]$-module localement libre ([F 2]).

Je montre ici comment se généralisent tous les résultats précédents sans hypothèse de ramification modérée.

La première difficulté est que l'on connaît très mal la structure locale des anneaux d'entiers pour les idéaux premiers sauvagement ramifiés. A.-M. Bergé ([B]) a montré qu'il existe des contraintes très fortes sur la ramification pour que les anneaux d'entiers soient libres ou seulement projectifs sur leur ordre associé. Une façon de travailler avec des modules projectifs sans avoir besoin d'hypothèse de ramification consiste à étendre les scalaires à un ordre maximal contenant $\mathbb{Z}[\Gamma]$. Quelques cas particuliers ont été étudiés (essentiellement par Cougnard) et ont donné des résultats positifs. Toutefois, Cougnard et Wilson ont construit un contre-exemple montrant que le théorème de Fröhlich ne se généralise pas au cas où Γ contient des automorphismes sauvagement ramifiés, tout au moins sous la même forme ([C 1], [Wi]).

L'idée de départ de ce travail a été de considérer la classe de l'anneau des entiers dans de nouveaux groupes de Grothendieck. Les idéaux premiers sauvagement ramifiés jouant un rôle particulier, ces groupes dépendent de l'ensemble S des nombres premiers p tels que Γ contienne un automorphisme sauvagement ramifié en p.

1. - S-groupes de Grothendieck

Soit R un anneau de Dedekind de corps des fractions K, et soit S une partie de l'ensemble $\mathcal{P}(R)$ des idéaux premiers de R. L'indexation par un élément $\mathfrak{p} \in \mathcal{P}(R)$ désignera la complétion en $\mathfrak{p}$.

Soit $\mathfrak{O}$ un odre de R dans une K-algèbre A semi-simple, de dimension finie sur K et séparable.

A toute catégorie $\mathcal{C}$ de $\mathfrak{O}$-modules de type fini, on associe le groupe abélien libre $\mathcal{G}(\mathcal{C})$ engendré par les classes d'isomorphismes (M) des modules M de $\mathcal{C}$ et le sous-groupe $\mathcal{G}'_S(\mathcal{C})$ de $\mathcal{G}(\mathcal{C})$ engendré par les éléments de la forme $(M)-(M')-(M'')$, où M, M' et M'' sont des objets de $\mathcal{C}$ vérifiant : il existe une suite exacte $0 \longrightarrow M' \xrightarrow{\varphi} M \xrightarrow{\psi} M'' \longrightarrow 0$ telle que, pour tout $\mathfrak{p} \in \mathcal{P}(R)-S$, la suite $0 \longrightarrow M'_{\mathfrak{p}} \xrightarrow{\varphi_{\mathfrak{p}}} M_{\mathfrak{p}} \xrightarrow{\psi_{\mathfrak{p}}} M'' \longrightarrow 0$ est scindée.

DÉFINITION 1.1. - <u>On appelle</u> S-<u>groupe de Grothendieck de</u> $\mathcal{C}$ <u>le groupe quotient</u> $\mathcal{G}(\mathcal{C})/\mathcal{G}'_S(\mathcal{C})$. <u>On le note</u> $\mathcal{K}_o^S(\mathcal{C})$.

On note $[M]$, la classe dans $\mathcal{G}(\mathcal{C})/\mathcal{G}'_S(\mathcal{C})$ d'un module M de $\mathcal{C}$.

On s'intéresse en particulier aux deux catégories suivantes :

$\mathcal{C}(\mathfrak{O})$, la catégorie des $\mathfrak{O}$-modules de type fini sans R-torsion ;

$\mathcal{C}_{1p}^S(\mathfrak{O})$, la catégorie des $\mathfrak{O}$-modules M de type fini sans R-torsion et tels que $M_{\mathfrak{p}}$ est un $\mathfrak{O}_{\mathfrak{p}}$-module projectif pour tout $\mathfrak{p} \in \mathcal{P}(R)-S$.

On note $\mathcal{G}_{\oplus}^S(\mathfrak{O})$ (resp. $\mathcal{K}_o^S(\mathfrak{O})$) le S-groupe de Grothendieck de $\mathcal{C}(\mathfrak{O})$ (resp. $\mathcal{C}_{1p}^S(\mathfrak{O})$).

Pour faire le lien avec les groupes de Grothendieck déjà connus, on choisit quelques ensembles S particuliers.

Tout d'abord si S est vide, on note plus simplement $\mathcal{K}_o(\mathcal{C})$ le groupe $\mathcal{K}_o^S(\mathcal{C})$. Ainsi $\mathcal{K}_o(\mathfrak{O})$ est l'habituel groupe de Grothendieck de la catégorie des $\mathfrak{O}$-modules projectifs et $\mathcal{G}_{\oplus}(\mathfrak{O})$ est le quotient de $\mathcal{G}(\mathcal{C}(\mathfrak{O}))$ par le sous-groupe engendré par les éléments de la forme $(M' \oplus M'')-(M')-(M'')$.

Ensuite si S est gros, plus précisément si S contient l'ensemble $\mathfrak{I}(\mathfrak{O})$ des idéaux premiers $\mathfrak{p}$ de R tels que $\mathfrak{O}_{\mathfrak{p}}$ n'est pas un ordre héréditaire de $R_{\mathfrak{p}}$ dans $A_{\mathfrak{p}}$, alors $\mathcal{G}_{\oplus}^S(\mathfrak{O})$ est égal à $\mathcal{K}_o^S(\mathfrak{O})$ et coïncide avec le groupe de Grothendieck noté habituellement $\mathcal{G}_o(\mathfrak{O})$.

On a de plus $\mathcal{G}_{\oplus}^S(\mathfrak{O}) = \mathcal{G}_{\oplus}^{S \cap \mathfrak{I}(\mathfrak{O})}(\mathfrak{O})$ et $\mathcal{K}_o^S(\mathfrak{O}) = \mathcal{K}_o^{S \cap \mathfrak{I}(\mathfrak{O})}(\mathfrak{O})$.

<u>On supposera pour la suite que</u> $S \subset \mathfrak{I}(\mathfrak{O})$.

Enfin soit S' une partie de $\mathcal{P}(R)$ contenant S. Le foncteur identité définit un homomorphisme surjectif de $\mathcal{G}_{\oplus}^S(\mathfrak{O})$ sur $\mathcal{G}_{\oplus}^{S'}(\mathfrak{O})$, car $\mathcal{G}'_S(\mathcal{C}(\mathfrak{O})) \subset \mathcal{G}'_{S'}(\mathcal{C}(\mathfrak{O}))$. L'inclusion $\mathcal{C}_{1p}^S(\mathfrak{O}) \subset \mathcal{C}_{1p}^{S'}(\mathfrak{O})$ permet de plonger $\mathcal{G}(\mathcal{C}_{1p}^S(\mathfrak{O}))$ dans $\mathcal{G}(\mathcal{C}_{1p}^{S'}(\mathfrak{O}))$. Comme $\mathcal{G}'_S(\mathcal{C}_{1p}^S(\mathfrak{O}))$ est alors inclus dans $\mathcal{G}'_{S'}(\mathcal{C}_{1p}^{S'}(\mathfrak{O}))$, on en déduit un homomorphisme de $\mathcal{K}_o^S(\mathfrak{O})$ dans $\mathcal{K}_o^{S'}(\mathfrak{O})$.

PROPOSITION 1.2. - Soient M et N deux $\mathfrak{O}$-modules de $\mathcal{C}(\mathfrak{O})$ (resp. $\mathcal{C}^S_{1p}(\mathfrak{O})$). Alors $[M]=[N]$ dans $\mathcal{G}^S_{\oplus}(\mathfrak{O})$ (resp. $\mathcal{K}^S_o(\mathfrak{O})$) si et seulement s'il existe trois $\mathfrak{O}$-modules U, V et W de $\mathcal{C}(\mathfrak{O})$ (resp. $\mathcal{C}^S_{1p}(\mathfrak{O})$) et deux suites exactes :

$$0 \longrightarrow V \longrightarrow M\oplus U \longrightarrow W \longrightarrow 0$$
$$0 \longrightarrow V \longrightarrow N\oplus U \longrightarrow W \longrightarrow 0$$

scindées localement pour tout idéal $\mathfrak{p} \in \mathcal{P}(R)-S$.

Démonstration. - Voir A. Heller, lemme 2.1 ([He]).

Soit $\mathfrak{p} \in \mathcal{P}(R)$ et soit $\mathcal{C}$ une sous-catégorie de $\mathcal{C}(\mathfrak{O})$, on note $\mathcal{C}_{\mathfrak{p}}$ l'image de $\mathcal{C}$ par complétion en $\mathfrak{p}$ et on pose :

$$\mathcal{K}^S_o(\mathcal{C}_{\mathfrak{p}})=\mathcal{K}_o(\mathcal{C}_{\mathfrak{p}}) \quad \text{si} \quad \mathfrak{p}\notin S \quad \text{et} \quad \mathcal{K}^S_o(\mathcal{C}_{\mathfrak{p}})=\mathcal{K}_o^{\{\mathfrak{p}R_{\mathfrak{p}}\}}(\mathcal{C}_{\mathfrak{p}}) \quad \text{si} \quad \mathfrak{p}\in S .$$

Ainsi

$$\mathcal{G}^S_{\oplus}(\mathfrak{O}_{\mathfrak{p}})=\mathcal{K}^S_o(\mathfrak{O}_{\mathfrak{p}})=\mathcal{G}_o(\mathfrak{O}_{\mathfrak{p}}) \quad \text{si } \mathfrak{p}\in S \quad \text{et} \quad \mathcal{K}^S_o(\mathfrak{O}_{\mathfrak{p}})=\mathcal{K}_o(\mathfrak{O}_{\mathfrak{p}}),\ \mathcal{G}^S_{\oplus}(\mathfrak{O}_{\mathfrak{p}})=\mathcal{G}_{\oplus}(\mathfrak{O}_{\mathfrak{p}}) \quad \text{si } \mathfrak{p}\notin S .$$

DÉFINITION 1.3. - On appelle S-groupe des classes de $\mathfrak{O}$ (resp. des classes projectives de $\mathfrak{O}$) le noyau de l'homomorphisme de $\mathcal{G}^S_{\oplus}(\mathfrak{O})$ dans $\prod_{\mathfrak{p}\in\mathcal{P}(R)} \mathcal{G}^S_{\oplus}(\mathfrak{O}_{\mathfrak{p}})$ (resp. de $\mathcal{K}^S_o(\mathfrak{O})$ dans $\prod_{\mathfrak{p}\in\mathcal{P}(R)} \mathcal{K}^S_o(\mathfrak{O}_{\mathfrak{p}})$) obtenu à partir des foncteurs d'extension des scalaires à $\mathfrak{O}_{\mathfrak{p}}$. On note ce groupe $\widetilde{\mathcal{G}}^S_{\oplus}(\mathfrak{O})$ (resp. $\widetilde{\mathcal{K}}^S_o(\mathfrak{O})$).

On associe à $\mathcal{C}(\mathfrak{O})$ la catégorie suivante, notée $\mathcal{C}(\mathfrak{O})'$:

les objets de $\mathcal{C}(\mathfrak{O})'$ sont les triplets (M, α, N) où M et N sont des objets de $\mathcal{C}(\mathfrak{O})$ et α est un A-isomorphisme de $K\otimes_R M$ sur $K\otimes_R N$,

un morphisme de (M, α, N) dans (M', α', N') est un couple (f, g) de $\mathfrak{O}$-homomorphismes de M dans M' et de N dans N' respectivement et tels que $\alpha'\circ f_K = g_K \circ \alpha$ où f_K et g_K proviennent de f et g par extensions des scalaires à K.

Une suite exacte d'objets et de morphismes de cette catégorie est donc donnée par un diagramme commutatif

$$\begin{array}{ccccccccc}
0 & \longrightarrow & K\otimes_R M' & \xrightarrow{f'_K} & K\otimes_R M & \xrightarrow{f''_K} & K\otimes_R M'' & \longrightarrow & 0 \\
 & & \downarrow{\scriptstyle \alpha'} & & \downarrow{\scriptstyle \alpha} & & \downarrow{\scriptstyle \alpha''} & & \\
0 & \longrightarrow & K\otimes_R N' & \xrightarrow{g'_K} & K\otimes_R N & \xrightarrow{g''_K} & K\otimes_R N'' & \longrightarrow & 0 .
\end{array}$$

Cette suite est dite scindée s'il existe deux sections s et t de f'' et g'' respectivement telles que $\alpha \circ s_K = t_K \circ \alpha''$. Par complétion on définit de même une suite localement scindée en $\mathfrak{p} \in \mathcal{P}(R)$.

Soit $\mathcal{C}'$ une sous-catégorie de $\mathcal{C}(\mathfrak{O})'$, on note $\mathcal{K}_{\#}^{S}(\mathcal{C}')$ le quotient du groupe abélien libre engendré par les classes d'isomorphismes, encore notées (M, α, N), des objets (M, α, N) de $\mathcal{C}'$ par le sous-groupe engendré par les éléments de la forme $(M, \alpha, N) - (M', \alpha', N') - (M'', \alpha'', N'')$ où ces trois objets sont liés par une suite exacte localement scindée en dehors de S et les éléments de la forme $(M, \alpha\beta, P) - (M, \alpha, N) - (N, \beta, P)$.

On note $[M, \alpha, N]$ la classe de (M, α, N) dans ce groupe.

DÉFINITION 1.4. - <u>Soit</u> $\mathcal{C}$ <u>une sous-catégorie de</u> $\mathcal{C}(\mathfrak{O})$; <u>on appelle</u> S-<u>groupe de Grothendieck relatif de</u> $\mathcal{C}$, <u>le groupe</u> $\mathcal{K}_{\#}^{S}(\mathcal{C}')$ <u>où</u> $\mathcal{C}'$ <u>est la sous-catégorie de</u> $\mathcal{C}(\mathfrak{O})'$ <u>dont les objets sont les triplets</u> (M, α, N) <u>avec</u> M, N <u>appartenant à</u> $\mathcal{C}$.

On note $\mathcal{K}_{o, rel}^{S}(\mathcal{C})$ ce groupe et en particulier pour $\mathcal{C} = \mathcal{C}(\mathfrak{O})$ (resp. $\mathcal{C} = \mathcal{C}_{lp}^{S}(\mathfrak{O})$), on le note $\mathcal{G}_{\oplus, rel}^{S}(\mathfrak{O})$ (resp. $\mathcal{K}_{o, rel}^{S}(\mathfrak{O})$).

Le groupe $\mathcal{K}_{o, rel}^{S}(\mathfrak{O})$ s'identifie au groupe de Grothendieck de la catégorie des $\mathfrak{O}$-modules de type fini et de R torsion qui sont quotients de deux modules localement projectifs pour tout $\mathfrak{p} \in \mathcal{P}(R) - S$.

Si S contient $\mathfrak{F}(\mathfrak{O})$, $\mathcal{K}_{o, rel}^{S}(\mathfrak{O})$ est égal à $\mathcal{G}_{\oplus, rel}^{S}(\mathfrak{O})$ et ces deux groupes s'identifient au groupe de Grothendieck de la catégorie des $\mathfrak{O}$-modules de type fini et de R-torsion, noté habituellement $\mathcal{G}_{o}^{t}(\mathfrak{O})$.

L'application qui à (M, α, N) associe $[N] - [M]$ se factorise en un homomorphisme, noté ν, de $\mathcal{K}_{o, rel}^{S}(\mathcal{C})$ dans $\mathcal{K}_{o}^{S}(\mathcal{C})$.

DÉFINITION 1.5. - <u>Soit</u> $\mathcal{C}$ <u>une sous-catégorie de</u> $\mathcal{C}(\mathfrak{O})$; <u>on appelle</u> S-<u>groupe de Whitehead de</u> $\mathcal{C}$, <u>le groupe</u> $\mathcal{K}_{\#}^{S}(\mathcal{C}'')$ <u>où</u> $\mathcal{C}''$ <u>est la sous-catégorie de</u> $\mathcal{C}(\mathfrak{O})'$ <u>dont les objets sont les triplets</u> (M, α_K, M) <u>où</u> M <u>appartient à</u> $\mathcal{C}$ <u>et</u> α_K <u>provient par extension des scalaires à</u> A <u>d'un</u> $\mathfrak{O}$-<u>automorphisme</u> α <u>de</u> M (<u>on notera plus simplement</u> (M, α) <u>les objets de</u> $\mathcal{C}''$).

On note $\mathcal{K}_{1}^{S}(\mathcal{C})$ ce groupe et en particulier pour $\mathcal{C} = \mathcal{C}_{lp}^{S}(\mathfrak{O})$ on le note $\mathcal{K}_{1}^{S}(\mathfrak{O})$. Pour S vide ce groupe est l'habituel groupe de Whitehead $\mathcal{K}_{1}(\mathfrak{O})$ de la catégorie des $\mathfrak{O}$-modules projectifs. Si S contient $\mathfrak{F}(\mathfrak{O})$, $\mathcal{K}_{1}^{S}(\mathcal{C}_{lp}^{S}(\mathfrak{O}))$ est égal à $\mathcal{K}_{1}^{S}(\mathcal{C}(\mathfrak{O}))$ et ces deux groupes sont notés $\mathcal{G}_{1}(\mathfrak{O})$.

Comme précédemment, on définit pour $S \subset \mathcal{P}(R)$ et $\mathfrak{p} \in \mathcal{P}(R)$ les groupes $\mathcal{K}^S_{o,rel}(\mathcal{C}_\mathfrak{p})$ et $\mathcal{K}^S_1(\mathcal{C}_\mathfrak{p})$.

Pour $\mathcal{C} = \mathcal{C}(\mathfrak{O})$ ou $\mathcal{C} = \mathcal{C}^S_{lp}(\mathfrak{O})$ et pour tout A-module V, il existe M dans $\mathcal{C}$ et W un A-module tels que $K \otimes_R M = V \oplus W$. L'application qui à $[V, \alpha] \in \mathcal{K}_1(A)$ associe $[M, \alpha \oplus 1_W, M]$ de $\mathcal{K}^S_{o,rel}(\mathcal{C})$ ne dépend pas du vhoix de W et de M. Elle se factorise en un homomorphisme, noté δ, de $\mathcal{K}_1(A)$ dans $\mathcal{K}^S_{o,rel}(\mathcal{C})$.

THÉORÈME 1.6. - <u>Pour</u> $\mathcal{C} = \mathcal{C}^S_{lp}(\mathfrak{O})$ <u>ou</u> $\mathcal{C} = \mathcal{C}(\mathfrak{O})$, <u>la suite suivante est exacte</u> :

$$\mathcal{K}_1(A) \xrightarrow{\delta} \mathcal{K}^S_{o,rel}(\mathcal{C}) \xrightarrow{\nu} \mathcal{K}^S_o(\mathcal{C}) \xrightarrow{\varepsilon} \mathcal{K}_o(A)$$

<u>où</u> ε <u>désigne l'homomorphisme obtenu à partir du foncteur d'extension des scalaires de</u> R <u>à</u> K.

<u>Le noyau de</u> δ <u>contient l'image de</u> $\mathcal{K}^S_1(\mathcal{C})$ <u>par l'homomorphisme obtenu à partir du foncteur d'extension des scalaires de</u> R <u>à</u> K.

<u>Si</u> S <u>est vide, l'image de</u> $\mathcal{K}_1(\mathcal{C})$ <u>est égale au noyau de</u> δ.

<u>Démonstration</u>. - Voir A. Heller, propositions 4.1, 5.1 et 5.2 ([He]).

<u>Notation</u>. - <u>On note</u> $\mathcal{H}^S(\mathcal{C})$ <u>le noyau de</u> δ.

PROPOSITION 1.7. - <u>Pour</u> $\mathcal{C} = \mathcal{C}(\mathfrak{O})$ <u>ou</u> $\mathcal{C} = \mathcal{C}^S_{lp}(\mathfrak{O})$, <u>l'application</u> $\lambda_{o,rel}$ <u>de</u> $\mathcal{K}^S_{o,rel}(\mathcal{C})$ <u>dans</u> $\bigoplus_{\mathfrak{p} \in \mathcal{P}(R)} \mathcal{K}^S_{o,rel}(\mathcal{C}_\mathfrak{p})$ <u>obtenue à partir des foncteurs d'extension des scalaires de</u> R <u>à</u> $R_\mathfrak{p}$ <u>est un isomorphisme</u>.

<u>Démonstration</u>. - Les groupes de Grothendieck relatifs sont engendrés par les éléments de la forme $[M, 1, N]$ où M et N sont des $\mathfrak{O}$-modules d'un même A-module et tels que M est inclus dans N. La seule difficulté pour définir un homomorphisme réciproque est de montrer que si $\alpha \in J(M_n(A))$ et M et N sont des réseaux de A^n alors $[M, 1, N] = [\alpha M, 1, \alpha N]$ ou en d'autres termes que $[\alpha M \oplus N, 1, M \oplus \alpha N] = 0$.

Mais comme $\begin{pmatrix} \alpha^{-1} & 0 \\ 0 & \alpha \end{pmatrix}$ est le produit de matrices élémentaires, le lemme d'approximation donne le résultat.

Pour $\mathcal{C} = \mathcal{C}(\mathfrak{O})$ ou $\mathcal{C} = \mathcal{C}^S_{lp}(\mathfrak{O})$, on en déduit un diagramme commutatif où les applications λ sont obtenues à partir des foncteurs d'extensions des scalaires aux complétés

$$\begin{array}{ccccccccccc}
0 & \longrightarrow & \mathcal{H}^S(\mathcal{C}) & \longrightarrow & \mathcal{K}_1(A) & \xrightarrow{\ \delta\ } & \mathcal{K}^S_{o,rel}(\mathcal{C}) & \xrightarrow{\ \nu\ } & \mathcal{K}^S_o(\mathcal{C}) & \xrightarrow{\ \varepsilon\ } & \mathcal{K}_o(A) \\
 & & \downarrow{\scriptstyle \lambda_1} & & \downarrow{\scriptstyle \lambda_{1,K}} & & \downarrow{\scriptstyle \lambda_{o,rel}} & & \downarrow{\scriptstyle \lambda_o} & & \downarrow{\scriptstyle \lambda_{o,K}} \\
0 & \longrightarrow & \prod_{\mathfrak{p}\in\mathcal{P}(R)} \mathcal{H}^S(\mathcal{C}_{\mathfrak{p}}) & \longrightarrow & \prod_{\mathfrak{p}\in\mathcal{P}(R)} \mathcal{K}_1(A_{\mathfrak{p}}) & \longrightarrow & \prod_{\mathfrak{p}\in\mathcal{P}(R)} \mathcal{K}_{o,rel}(\mathcal{C}_{\mathfrak{p}}) & \longrightarrow & \prod_{\mathfrak{p}\in\mathcal{P}(R)} \mathcal{K}^S_o(\mathcal{C}_{\mathfrak{p}}) & \longrightarrow & \prod_{\mathfrak{p}\in\mathcal{P}(R)} \mathcal{K}_o(A_{\mathfrak{p}})
\end{array}$$

Soit $J\mathcal{K}_1(A)$ le sous-groupe de $\prod_{\mathfrak{p}\in\mathcal{P}(R)} \mathcal{K}_1(A_{\mathfrak{p}})$ formé des éléments $(x_{\mathfrak{p}})_{\mathfrak{p}\in\mathcal{P}(R)}$ tels que $x_{\mathfrak{p}}$ appartienne à $\mathcal{H}(\mathcal{C}_{\mathfrak{p}}(\mathfrak{D}))$ pour presque tout $\mathfrak{p}\in\mathcal{P}(R)$.

La commutativité du diagramme précédent et la proposition 1.7 montrent que l'image de $\lambda_{1,K}$ est incluse dans $J\mathcal{K}_1(A_{\mathfrak{p}})$ et que $\prod_{\mathfrak{p}\in\mathcal{P}(R)} \mathcal{H}^S(\mathcal{C}_{\mathfrak{p}})$ est inclus dans $J\mathcal{K}_1(A_{\mathfrak{p}})$. On note $U^S(\mathcal{C})$ le groupe $\prod_{\mathfrak{p}\in\mathcal{P}(R)} \mathcal{H}^S(\mathcal{C}_{\mathfrak{p}})$ et en particulier pour $\mathcal{C} = \mathcal{C}^S_{1p}(\mathfrak{D})$, on pose $U^S(\mathcal{C}) = U^S(\mathfrak{D})$.

Le lemme du serpent appliqué au diagramme précédent, donne donc le théorème suivant :

THÉORÈME 1.8. - <u>Pour</u> $\mathcal{C} = \mathcal{C}^S_{1p}(\mathfrak{D})$ <u>ou</u> $\mathcal{C} = \mathcal{C}(\mathfrak{D})$, <u>il existe une suite exacte</u>

$$\mathcal{K}_1(A) \xrightarrow{\ \lambda_{1,K}\ } \frac{J\mathcal{K}_1(A)}{U^S(\mathcal{C})} \xrightarrow{\ \partial\ } \tilde{\mathcal{K}}^S_o(\mathcal{C}) \longrightarrow 0 .$$

2. - S-<u>groupes de Grothendieck relatifs</u>

On étend tout d'abord aux algèbres quelconques la définition de la fonction Det introduite par A. Fröhlich pour les algèbres de groupes.

Soit M un A module de type fini et soit α un A-automorphisme de M, l'application $f \mapsto \alpha \circ f$ est un endomorphisme de K-espace vectoriel de $\mathrm{Hom}_A(P, M)$, encore noté α ; on obtient une fonction bilinéaire de P et de (M, α) dans $\mathcal{K}_1(K)$; d'où en identifiant $\mathcal{K}_1(K)$ et K^* au moyen de l'application déterminant sur K, on obtient une fonction bilinéaire de $\mathcal{K}_o(A) \times \mathcal{K}_1(A)$ dans K^*. On la note $\det_A$.

Le corps K est dit assez gros, relativement à A, si pour tout A-module simple S, le commutant de S est isomorphe à K. Une extension $\overline{K}$ de K est dite assez grosse, relativement à A, si les commutants des $\overline{K} \otimes_K A$-modules simples sont des corps isomorphes à $\overline{K}$.

PROPOSITION 2.1. - Soit Det l'application de $\mathcal{K}_1(A)$ dans $\mathrm{Hom}(\mathcal{K}_o(A), K^*)$ définie par : pour tout $x \in \mathcal{K}_1(A)$, $\mathrm{Det}(x)$ est l'homomorphisme qui à ρ dans $\mathcal{K}_o(A)$ associe $\det_A(\rho, x) = \mathrm{Det}_\rho(x)$. Si K est assez gros, l'application Det est un isomorphisme de $\mathcal{K}_1(A)$ dans $\mathrm{Hom}(\mathcal{K}_o(A), K^*)$.

Démonstration. - On utilise les descriptions $\mathcal{K}_o(A)$ et $\mathcal{K}_1(A)$ données par H. Bass ([Ba]) au moyen des A-modules simples.

Soit $\overline{K}$ une extension galoisienne de K de groupe de Galois G. On pose $\overline{A} = \overline{K} \otimes_K A$. Soit M un $\overline{A}$-module de type fini et α un $\overline{A}$-automorphisme de M. Pour tout $g \in G$, on définit sur le groupe additif M une nouvelle structure de $\overline{A}$-module en posant :

$$\forall \lambda \in \overline{K},\ \forall a \in A,\ \forall m \in M \quad (k \otimes a) \cdot m = (g^{-1}(k) \otimes a) m .$$

On note M^g le $\overline{A}$-module ainsi défini. L'application $\overline{A}$ linéaire α définit une application, encore notée α, sur M^g. Le groupe G opère donc sur les groupes $\mathcal{K}_o(\overline{A})$ et $\mathcal{K}_1(\overline{A})$. L'extension des scalaires de K à $\overline{K}$ définit un homomorphisme, noté $\mathrm{Ext}_K^{\overline{K}}$, de $\mathcal{K}_o(A)$ dans $\mathcal{K}_o(\overline{A})$ et de $\mathcal{K}_1(A)$ dans $\mathcal{K}_1(\overline{A})$. Il est clair que l'image de $\mathrm{Ext}_K^{\overline{K}}$ est incluse dans le sous-groupe formé des éléments fixes par G.

Pour la suite on suppose que $\overline{K}$ est une extension assez grosse de K.

L'homomorphisme d'extension des scalaires de K à $\overline{K}$ n'est pas en général un homomorphisme injectif de $\mathcal{K}_1(A)$ dans $\mathcal{K}_1(\overline{A})$. En effet, le problème se ramène au cas où A est un corps gauche D. Supposons que K est le centre de D, $\mathcal{K}_1(A)$ est isomorphe à $D^*/[D^*, D^*]$ et $\mathcal{K}_1(\overline{A})$ est isomorphe à $\overline{K}^*$. L'application $\mathrm{Ext}_K^{\overline{K}}$ devient l'homomorphisme de norme réduite de $D^*/[D^*, D^*]$ dans K^*. Dans ce cas, l'application $\mathrm{Ext}_K^{\overline{K}}$ sera injective si et seulement si le noyau de la norme réduite est inclus dans $[D^*, D^*]$.

PROPOSITION 2.2. - L'application $\mathrm{Ext}_K^{\overline{K}}$ est un homomorphisme injectif de $\mathcal{K}_1(A)$ dans $\mathcal{K}_2(\overline{A})$ dans les trois cas suivants :

- la K-algèbre A est décomposée (i.e. les commutants des A-modules simples sont des corps commutatifs)

- K est un corps de nombres

- K est localement compact.

Démonstration. - Dans le premier cas, on utilise le fait que $\mathcal{K}_1(A)$ est isomorphe au produit des groupes multiplicatifs des commutants des A-modules simples. Si K est un corps de nombres c'est le théorème de Wang ([Wa]), si K est localement compact, c'est le théorème de Nakayama-Matsushima ([NM]).

On notera encore Det le composé de l'homomorphisme $\mathrm{Ext}_K^{\overline{K}}$ et de l'homomorphisme Det de $\mathcal{K}_1(\overline{A})$ dans $\mathrm{Hom}(\mathcal{K}_o(A), \overline{K}^*)$. C'est donc un homomorphisme de $\mathcal{K}_1(A)$ dans $\mathrm{Hom}_G(\mathcal{K}_o(\overline{A}), \overline{K}^*)$.

L'application $A^* \longmapsto \mathcal{K}_1(A)$ est surjective ([Ba], page 366). L'image de $\mathcal{K}_1(A)$ par l'homomorphisme Det est égal à l'image de A^* ; on notera encore Det(a) l'élément Det([A, a]). Explicitement cela se traduit de la façon suivante :

PROPOSITION 2.3. - Soit $a \in A^*$ et M un $\overline{A}$-module de type fini, $\mathrm{Det}_{[M]}(a)$ est le déterminant du $\overline{K}$-endomorphisme de M défini par a .

Démonstration. - On se ramène au cas où A est simple, $\overline{K} = K$ et M est un $\overline{A}$-module simple. L'algèbre A est alors isomorphe à $M_n(\overline{K})$. L'image de a dans $\mathcal{K}_1(A)$ est représenté par [A, a]. Les $\overline{K}$-espaces vectoriels $\mathrm{Hom}_A(M, A)$ et M sont isomorphes à $\overline{K}^n$, et les $\overline{K}$-endomorphismes respectifs définis par a sont identiques sur $\overline{K}^n$.

Remarque. - On retrouve ainsi, dans le cas où A est une algèbre de groupe $K[\Gamma]$, la définition de l'application Det introduite par A. Fröhlich ([F 2]).

On note $\overline{R}$ la clôture intégrale de R dans $\overline{K}$, et $\overline{\mathfrak{D}}$ l'ordre $\overline{R} \otimes_R \mathfrak{D}$.

PROPOSITION 2.4. - On a :

$$\mathrm{Det}(\mathcal{H}(\mathcal{C}(\mathfrak{D}))) = \mathrm{Det}(\mathcal{K}_1(A)) \cap \mathrm{Hom}_G(\mathcal{K}_o(\overline{A}), \overline{R}^*) .$$

Démonstration. - On démontre tout d'abord le résultat lorsque R est complet pour la valuation discrète associée à un idéal premier $\mathfrak{p}$. Il est clair que $\mathrm{Det}(\mathcal{H}(\mathcal{C}(\mathfrak{D})))$ est inclus dans $\mathrm{Det}(\mathcal{K}_1(A)) \cap \mathrm{Hom}_G(\mathcal{K}_o(\overline{A}), \overline{R}^*)$, le diagramme suivant étant commutatif :

$$\begin{array}{ccc}
\mathcal{K}_1(\mathcal{C}(\mathfrak{D})) & \xrightarrow{\mathrm{Ext}_R^K} & \mathcal{K}_1(A) \\
\Big\downarrow {\scriptstyle \mathrm{Ext}_R^{\overline{R}}} & & \Big\downarrow {\scriptstyle \mathrm{Ext}_K^{\overline{K}}} \\
\mathcal{K}_1(\mathcal{C}(\overline{\mathfrak{D}})) & \xrightarrow{\mathrm{Ext}_{\overline{R}}^{\overline{K}}} & \mathcal{K}_1(\overline{A})
\end{array} .$$

Il est aussi clair que le résultat est vrai si K est assez gros.

Le groupe des éléments $x \in \mathcal{K}_1(A)$ tels que $\mathrm{Det}(x) \in \mathrm{Hom}_G(\mathcal{K}_o(\bar{A}), \bar{R}^*)$ est engendré par les éléments de la forme $[S,\alpha]$ où S est un A-module simple. On a $[\bar{K}\otimes_K S] = [\bigoplus_{\omega \in G/G_\chi} (\bar{S}^n)^\omega]$ où $\bar{S}$ est un $\bar{A}$-module simple, n est l'indice de Schur de $[\bar{S}] = \chi$ et G_χ est le sous-groupe d'isotropie de χ dans G. On choisit $\bar{M}$ un sous-$\bar{\mathfrak{O}}$-module de $\bar{S}^n$ et soit M le sous-$\mathfrak{O}$-module de $\bigoplus_\omega \bar{M}^\omega$ formée des éléments m_ω tels que $m_\omega = m_{\omega'}$, $\forall \omega, \omega'$. On a $\bar{R} \otimes_R M = \bar{M}$. Le A-automorphisme α de S se décompose sous la forme $\bigoplus_{\omega \in G/G_\chi} \alpha_\omega$ où $\alpha_\omega \in \mathrm{Aut}_A((\bar{S}^n)^\omega)$. On a $M_n(K) \simeq \mathrm{Aut}_A(\bar{S}^n)$ et $\det(\alpha_\omega)$ est un entier ; donc α_ω est un automorphisme de M^ω et par suite α est un $\mathfrak{O}$-automorphisme de M. Ce qui montre que $[S,\alpha]$ appartient à $\mathcal{K}(\mathcal{C}(\mathfrak{O}))$.

Lorsque R n'est plus complet, le résultat découle de la commutativité du diagramme précédent du théorème 1.8.

COROLLAIRE 2.5. - Soit $S\mathcal{K}_1(A)$ le noyau de l'application Det de $\mathcal{K}_1(A)$ dans $\mathrm{Hom}_G(\mathcal{K}_o(A), K^*)$, alors quel que soit $S \subset \mathcal{P}(R)$, $S\mathcal{K}_1(A)$ est inclus dans $\mathcal{K}^S(\mathcal{C}(\mathfrak{O}))$.

Démonstration. - On a $S\mathcal{K}_1(A) \subset \mathcal{K}(\mathcal{C}(\mathfrak{O})) \subset \mathcal{K}^S(\mathcal{C}(\mathfrak{O}))$.

On note $\mathcal{P}(K)$ l'ensemble des places de K. Soit $\mathfrak{p} \in \mathcal{P}(K)$ et $\mathfrak{P}$ une place de K au-dessus de $\mathfrak{p}$. Si $D_\mathfrak{P}$ désigne le groupe de décomposition de $\mathfrak{P}$, les autres places de K au-dessus de $\mathfrak{p}$ sont de la forme $\mathfrak{P}^\sigma$ pour σ parcourant un système de représentants de G modulo $D_\mathfrak{P}$. On identifie les trois algèbres commutatives $\bigoplus_{\sigma \in G/D_\mathfrak{P}} \bar{K}_{\mathfrak{P}^\sigma}$, $\mathbb{Z}[G] \otimes_{\mathbb{Z}[D_\mathfrak{P}]} \bar{K}_\mathfrak{P}$ et $K_\mathfrak{p} \otimes_K \bar{K}$.

Soit M un $\bar{K}\otimes_K A$-module le $K_\mathfrak{p}$-module ; $K_\mathfrak{p} \otimes_K M$ s'identifie à $\bigoplus_{\sigma \in G/G_\mathfrak{P}} (\bar{K}_{\mathfrak{P}^\sigma} \otimes_{\bar{K}} M)$. On définit ainsi un isomorphisme, noté $\mathrm{Ext}_K^{K_\mathfrak{p}}$, de $\mathcal{K}_o(\bar{A})$ dans $\mathbb{Z}[G] \otimes_{\mathbb{Z}[D_\mathfrak{P}]} \mathcal{K}_o(\bar{K}_\mathfrak{P} \otimes_K A)$ par $\mathrm{Ext}_K^{K_\mathfrak{p}}([M]) = \sum_{\sigma \in G/D_\mathfrak{P}} \sigma \otimes [\bar{K}_\mathfrak{P} \otimes_K M]$. Comme $\mathbb{Z}[G]$ est un $\mathbb{Z}[D_\mathfrak{P}]$-module libre, on en déduit un isomorphisme de $\mathrm{Hom}_{D_\mathfrak{P}}(\mathcal{K}_o(\bar{K}_\mathfrak{P} \otimes_K A), \bar{K}_\mathfrak{P}^*)$ sur $\mathrm{Hom}_G(\mathcal{K}_o(\bar{K}\otimes_K A), \bigoplus_{\sigma \in G/D_\mathfrak{P}} \bar{K}^*_{\mathfrak{P}^\sigma})$ donné par le produit tensoriel par $\mathbb{Z}[G]$ sur $\mathbb{Z}[D_\mathfrak{P}]$. Cet homomorphisme commute avec les applications transposées de l'extension des scalaires. On peut donc par passage à la limite inductive définir un isomorphisme de $\mathrm{Hom}_{G_{K_\mathfrak{p}}}(\mathcal{K}_o(\bar{K}_\mathfrak{P} \otimes_K A), \bar{K}_\mathfrak{P}^*)$

sur $\mathrm{Hom}_{G_K}(\mathcal{K}_o(\bar{A}), (K_\mathfrak{p} \otimes_K \bar{\mathfrak{Q}})^*)$ où cette fois $\bar{K}_\mathfrak{P}$ (resp. $\bar{K}$) désigne une clôture algébrique de $K_\mathfrak{p}$ (resp. K) de groupe de Galois $G_{K_\mathfrak{p}}$ (resp. G_K).

On déduit des remarques précédentes et du corollaire 2.11 la proposition suivante :

PROPOSITION 2.6. - *L'application* Det *se prolonge en une application, encore notée* Det , *de* $J\mathcal{K}_1(A)$ *dans* $\mathrm{Hom}_{G_K}(\mathcal{K}_o(\bar{A}), J(\bar{K}))$. *C'est un isomorphisme si et seulement si pour tout* $\mathfrak{p} \in \mathcal{P}(R)$, *l'application* Det *de* $\mathcal{K}_1(A_\mathfrak{p})$ *dans* $\mathrm{Hom}_{D_\mathfrak{P}}(\mathcal{K}_o(\bar{A}_\mathfrak{P}), \bar{K}^*_\mathfrak{P})$ *est un isomorphisme*.

On se propose maintenant de décrire les groupes $\mathcal{G}^S_{\oplus,\,\mathrm{rel}}(\mathfrak{D})$ et $\mathcal{K}^S_{o,\,\mathrm{rel}}(\mathfrak{D})$. La proposition 1.7 permet de se ramener au cas local et complet. Plus précisément on fait pour la suite de ce paragraphe les hypothèses suivantes : on suppose que K *est un corps local* muni d'une valuation discrète v_K associée à un idéal premier $\mathfrak{p}$, que R est l'anneau de valuation de K. On suppose que K est complet pour la topologie $\mathfrak{p}$-adique définie par v_K. On note k le corps résiduel $R/\mathfrak{p}$. On suppose de plus que k *est parfait*. Cette dernière hypothèse entraîne que K est localement compact.

L'hypothèse de séparabilité faite sur A entraîne qu'il existe une extension galoisienne $\bar{K}$ de K assez grosse relativement à A. On peut en particulier prendre pour $\bar{K}$ une clôture galoisienne de K. On note G_K le groupe de Galois de $\bar{K}$ sur K, $\bar{R}$ la clôture intégrale de R dans $\bar{K}$, $\bar{A}$ la $\bar{K}$-algèbre $\bar{K} \otimes_K A$, $\bar{\mathfrak{D}}$ l'ordre $\bar{R} \otimes_R \mathfrak{D}$ de $\bar{R}$ dans $\bar{A}$.

Il suffit donc de décrire $\mathcal{G}_{\oplus,\,\mathrm{rel}}(\mathfrak{D})$, $\mathcal{K}_{o,\,\mathrm{rel}}(\mathfrak{D})$ et $\mathcal{G}^\mathfrak{p}_{\oplus,\,\mathrm{rel}}(\mathfrak{D}) = \mathcal{K}^\mathfrak{p}_{o,\,\mathrm{rel}}(\mathfrak{D})$ et plus précisément les sous-groupes de $\mathcal{K}_1(A)$: $\mathcal{H}(\mathfrak{D})$, $\mathcal{H}(\mathcal{C}_{1p}(\mathfrak{D}))$ et $\mathcal{H}^\mathfrak{p}(\mathcal{C}(\mathfrak{D})) = \mathcal{H}^\mathfrak{p}(\mathcal{C}^\mathfrak{p}_{1p}(\mathfrak{D}))$.

PROPOSITION 2.7. - *Soit* $\mathcal{C} = \mathcal{C}(\mathfrak{D})$ *ou* $\mathcal{C} = \mathcal{C}_{1p}(\mathfrak{D})$. *L'application* δ *de* $\mathcal{K}_1(A)$ *dans* $\mathcal{K}^S_{o,\,\mathrm{rel}}(\mathcal{C})$ *se factorise en un unique homomorphisme injectif* Δ *rendant le diagramment suivant commutatif*

$$\begin{array}{ccc} \mathrm{Det}(\mathcal{K}_1(A))/\mathrm{Det}(\mathcal{H}^S(\mathcal{C})) & \xrightarrow{\ \Delta\ } & \mathcal{K}^S_{o,\,\mathrm{rel}}(\mathcal{C}) \\ {\scriptstyle \mathrm{Det}} \nwarrow & & \nearrow {\scriptstyle \delta} \\ & \mathcal{K}_1(A) & \end{array}$$

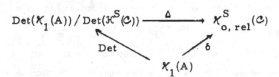

<u>si et seulement si pour</u> S <u>vide et</u> $\mathcal{C} = \mathcal{C}_{1p}(\mathfrak{D})$, $S\mathcal{K}_1(A)$ <u>est inclus dans</u> $\mathcal{K}(\mathcal{C}_{1p}(\mathfrak{D}))$. <u>L'application</u> Δ <u>est surjective si et seulement si l'application</u> ε <u>de</u> $\mathcal{K}_o^S(\mathcal{C})$ <u>dans</u> $\mathcal{K}_o(A)$ <u>est injective</u>.

<u>Démonstration</u>. - Par définition, le noyau de δ est $\mathcal{K}^S(\mathcal{C})$. Le noyau de l'application Det de $\mathrm{Det}(\mathcal{K}_1(A)) / \mathrm{Det}(\mathcal{K}^S(\mathcal{C}))$ est $\mathcal{K}^S(\mathcal{C}).S\mathcal{K}_1(A)$. Le résultat découle du théorème de factorisation, du théorème 1.8 et du corollaire 2.5.

<u>Notation</u>. - Pour $\mathcal{C}$ sous-catégorie de $\mathcal{C}(\mathfrak{D})$, on pose

$$H^S(\mathcal{C}) = \mathrm{Det}(\mathcal{K}^S(\mathcal{C})) \subset \mathrm{Hom}_{G_K}(\mathcal{K}_o(\bar{A}), \bar{K}^*).$$

Comme R est un anneau local, l'application qui à $a \in \mathfrak{D}^*$ associe l'élément $[\mathfrak{D}, a]$ de $\mathcal{K}_1(\mathfrak{D})$ est un homomorphisme surjectif. En utilisant l'abus de notation précédent, on a :

PROPOSITION 2.8. - $H(\mathcal{C}_{1p}(\mathfrak{D})) = \mathrm{Det}(\mathfrak{D}^*)$.

D'après la proposition 2.4., on a $H(\mathcal{C}(\mathfrak{D})) = \mathrm{Det}(\mathcal{K}_1(A)) \cap \mathrm{Hom}_{G_K}(\mathcal{K}_o(\bar{A}), \bar{R}^*)$ il reste à déterminer $H^{\mathfrak{p}}(\mathcal{C}_{1p}^{\mathfrak{p}}(\mathfrak{D})) = H^{\mathfrak{p}}(\mathcal{C}(\mathfrak{D}))$. <u>Pour cela on note</u> e <u>l'homomorphisme de</u> $\mathcal{K}_o(\mathfrak{D})$ <u>dans</u> $\mathcal{K}_o(A)$ <u>obtenu à partir de l'extension des scalaires de</u> R <u>à</u> K (voir [Se 2]).

Le théorème 1.6 appliqué à R donne une suite exacte

$$R^* \longrightarrow K^* \xrightarrow{\delta} \mathcal{K}_{o,\,rel}(R) \xrightarrow{\nu} \mathcal{K}_o(R) \longrightarrow \mathcal{K}_o(K) \longrightarrow 1$$

où $\mathcal{K}_{o,\,rel}(R)$ s'identifie au groupe des idéaux de R (ici isomorphe à K^*/R^*), $\mathcal{K}_o(K)$ s'identifie à $\mathbb{Z}$ par l'application de dimension sur K et $\mathcal{K}_o(R)$ s'identifie au produit de $\mathbb{Z}$ et du groupe des classes d'idéaux de R .

Comme précédemment on peut définir une dualité entre $\mathcal{K}_{o,\,rel}^{\mathfrak{p}}(\mathfrak{D})$ et $\mathcal{K}_o(\mathfrak{D})$. Au couple formé d'un élément [P] dans $\mathcal{K}_o(\mathfrak{D})$ et d'un élément $[M, \alpha, N]$ de $\mathcal{K}_{o,\,rel}^{\mathfrak{p}}(\mathfrak{D})$ on associe l'élément $[\mathrm{Hom}_{\mathfrak{D}}(P, M), \alpha, \mathrm{Hom}_{\mathfrak{D}}(P, N)]$ de $\mathcal{K}_{o,\,rel}(R)$ identifié à K^*/R^*. Grâce au fait que P est supposé projectif on obtient ainsi une application bilinéaire de $\mathcal{K}_o(\mathfrak{D}) \times \mathcal{K}_{o,\,rel}^{\mathfrak{p}}(\mathfrak{D})$ dans K^*/R^*, on la note $\det_{\mathfrak{D}}$.

On note $\overline{\mathrm{Det}}$ l'application qui à $[M, \alpha, N] \in \mathcal{K}_{o,\,rel}(\mathfrak{D})$ associe l'homomorphisme de $\mathrm{Hom}(\mathcal{K}_o(\mathfrak{D}), K^*/R^*)$ donné par la dualité précédente.

PROPOSITION 2.9. - On a la propriété suivante :

$$\forall x \in \mathcal{K}_o(\mathfrak{O}),\ \forall y \in \mathcal{K}_1(A),\quad \det_{\mathfrak{O}}(x, \delta(y)) = \det_A(e(x), y)\, R^* .$$

Le résultat est clair et se traduit par la commutativité du diagramme suivant :

$$\begin{array}{ccc} \mathcal{K}_1(A) & \xrightarrow{\ \mathrm{Det}\ } & \mathrm{Hom}(\mathcal{K}_o(A), K^*) \\ \delta \downarrow & & \downarrow {}^t e \\ \mathcal{K}_{o,\,rel}(\mathfrak{O}) & \xrightarrow{\ \overline{\mathrm{Det}}\ } & \mathrm{Hom}(\mathcal{K}_o(\mathfrak{O}), K^*/R^*) \end{array}$$

où ${}^t e$ est l'application transposée de l'application e de $\mathcal{K}_o(\mathfrak{O})$ dans $\mathcal{K}_o(A)$.

On notera encore $\overline{\mathrm{Det}}$ le composé de l'application d'extension des scalaires de $\mathcal{K}^{\mathfrak{p}}_{o,rel}(\mathfrak{O})$ dans $\mathcal{K}^{\mathfrak{p}}_{o,rel}(\bar{\mathfrak{O}})$ et de l'application $\overline{\mathrm{Det}}$ de $\mathcal{K}^{\mathfrak{p}}_{o,rel}(\bar{\mathfrak{O}})$ dans $\mathrm{Hom}(\mathcal{K}_o(\bar{\mathfrak{O}}), \bar{K}^*/\bar{R}^*)$. C'est donc un homomorphisme de $\mathcal{K}^{\mathfrak{p}}_{o,rel}(\mathfrak{O})$ dans $\mathrm{Hom}_{G_K}(\mathcal{K}_o(\bar{\mathfrak{O}}), \bar{K}^*/\bar{R}^*)$.

PROPOSITION 2.10. - L'application $\overline{\mathrm{Det}}$ de $\mathcal{K}_{o,\,rel}(\mathfrak{O})$ dans $\mathrm{Hom}_{G_K}(\mathcal{K}_o(\bar{\mathfrak{O}}), \bar{K}^*/\bar{R}^*)$ est un homomorphisme injectif. Cet homomorphisme est surjectif si et seulement si la k-algèbre $k \otimes_R \mathfrak{O}$ est décomposée.

Démonstration. - Il est clair qu'il suffit de supposer que $\bar{K}$ est une extension assez grosse de degré fini sur K. Le corps $\bar{K}$ est alors un corps local complet. Soit $\mathfrak{P}$ son idéal de valuation. Comme $\mathfrak{P}$ est inclus dans le radical de Jacobson de $\mathfrak{O}$, l'application de réduction modulo $\mathfrak{P}$ est un isomorphisme de $\mathcal{K}_o(\bar{\mathfrak{O}})$ sur $\mathcal{K}_o(\bar{\mathfrak{O}}/\mathfrak{P}\bar{\mathfrak{O}})$ ([Ba], page 449) ; $\bar{\mathfrak{O}}/\mathfrak{P}\bar{\mathfrak{O}}$ est une $\bar{k}$-algèbre de dimension finie avec $\bar{k} = \bar{R}/\mathfrak{P}$. On a vu (proposition 1.7) que $\mathcal{K}^{\mathfrak{p}}_{o,\,rel}(\mathfrak{O})$ est engendré par les éléments de la forme $[M, 1, N]$ où M et N sont deux $\mathfrak{O}$-réseaux d'un même A-module et $M \subset N$. On en déduit donc un isomorphisme de $\mathcal{K}_{o,\,rel}(\mathfrak{O})$ sur $\mathcal{G}^t_o(\mathfrak{O})$ le groupe de Grothendieck des $\mathfrak{O}$-modules de torsion. Ce dernier s'identifie à $\mathcal{G}_o(\mathfrak{O}/\mathfrak{p}\,\mathfrak{O})$, car tout $\mathfrak{O}$ module simple est un $\mathfrak{O}/\mathfrak{p}\mathfrak{O}$-module. L'application de valuation identifie $\bar{K}^*/\bar{R}^*$ à $\mathbb{Z}$, (cette application se traduit sur $\mathcal{K}_{o,\,rel}(\bar{R})$ en prenant la longueur des suites de Jordan Hölder des $\bar{R}$-modules de torsion). Il est clair que l'on retrouve ainsi l'homomorphisme de $\mathcal{G}_o(\mathfrak{O}/\mathfrak{p}\,\mathfrak{O})$ dans $\mathrm{Hom}_{G_K}(\mathcal{K}_o(\bar{\mathfrak{O}}/\mathfrak{p}\,\bar{\mathfrak{O}}), \mathbb{Z})$ décrit au paragraphe 2 appliqué à la k-algèbre $\mathfrak{O}/\mathfrak{p}\mathfrak{O}$. Le résultat découle alors du théorème 2.7.

L'application e commute avec l'extension des scalaires. On note $\bar{e}$ l'homomorphisme correspondant de $\mathcal{K}_o(\bar{\mathfrak{O}})$ dans $\mathcal{K}_o(\bar{A})$. On déduit que l'image de $\bar{e}$, notée $\mathrm{Im}\,\bar{e}$, ne dépend pas du choix de l'extension assez grosse $\bar{K}$.

PROPOSITION 2.11. - $H^{\mathfrak{p}}(\mathcal{C}(\mathfrak{O})) = \mathrm{Det}(\mathcal{K}_1(A)) \cap \{f / f(\mathrm{Im}\,\bar{e}) \subset \bar{R}^*\}$.

Démonstration. - On utilise la commutativité du diagramme :

$$\begin{array}{ccc} \mathcal{K}_1(A) & \xrightarrow{\mathrm{Det}} & \mathrm{Hom}_{G_K}(\mathcal{K}_o(\bar{A}), \bar{K}^*) \\ \delta \downarrow & & \downarrow {}^t\bar{e} \\ \mathcal{K}_{o,\mathrm{rel}}(\mathfrak{O}) & \xrightarrow{\overline{\mathrm{Det}}} & \mathrm{Hom}_{G_K}(\mathcal{K}_o(\mathfrak{O}), \bar{K}^*/\bar{R}^*) \end{array}$$

et le fait que l'application $\overline{\mathrm{Det}}$ est injective (proposition 2.10).

3. - S-groupes des classes d'un ordre arithmétique

Soit K une extension de degré fini de $\mathbb{Q}$, $\mathbb{Z}_K$ la clôture intégrale de $\mathbb{Z}$ dans K. Soit $\bar{\mathbb{Q}}$ une clôture galoisienne de $\mathbb{Q}$ contenant K, de groupe de Galois G_K, et $\bar{\mathbb{Z}}$ la clôture intégrale de $\mathbb{Z}$ dans $\bar{\mathbb{Q}}$.

Soit A une K-algèbre semi-simple, de dimension finie. On note $\bar{A}$ la $\bar{\mathbb{Q}}$-algèbre semi-simple $\bar{\mathbb{Q}} \otimes_{\mathbb{Q}} A$.

Soit $\mathfrak{O}$ un ordre de $\mathbb{Z}_K$ dans A, on note $\bar{\mathfrak{O}}$ l'ordre $\bar{\mathbb{Z}} \otimes_{\mathbb{Z}} \mathfrak{O}$ de $\mathbb{Z}$ dans $\bar{A}$.

On note $\mathcal{P}(K)$ l'ensemble des places de K et on identifie $\mathcal{P}(\mathbb{Z}_K)$ et l'ensemble des places non archimédiennes de K. On a $\mathcal{P}(K) = \mathcal{P}(\mathbb{Z}_K) \cup \mathcal{P}_\infty(K)$ où $\mathcal{P}_\infty(K)$ est l'ensemble des places archimédiennes de K. L'indexation par un élément $\mathfrak{p} \in \mathcal{P}(K)$ désigne la complétion en $\mathfrak{p}$. On note J(A) (resp. $J(\bar{\mathbb{Q}})$) le groupe des idèles de A (resp. $\bar{\mathbb{Q}}$) et $U(\mathfrak{O})$ le sous-groupe de J(A) formé des idèles $(x_\mathfrak{p})_{\mathfrak{p} \in \mathcal{P}(K)}$ tels que $x_\mathfrak{p} \in \mathfrak{O}_\mathfrak{p}^*$ pour tout $\mathfrak{p} \in \mathcal{P}(\mathbb{Z}_K)$. On pose $\bar{\mathbb{Q}}_\mathfrak{p} = K_\mathfrak{p} \otimes_K \bar{\mathbb{Q}}$ et $\bar{\mathbb{Z}}_\mathfrak{p} = \mathbb{Z}_\mathfrak{p} \otimes_{\mathbb{Z}} \bar{\mathbb{Q}}$. L'application Det se prolonge en un homomorphisme surjectif de J(A) dans $\mathrm{Hom}_{G_K}(\mathcal{K}_o(\bar{A}), J(\bar{\mathbb{Q}}))$ (voir proposition 2.6).

Soit $\mathfrak{p} \in \mathcal{P}(\mathbb{Z}_K)$, on note $P_\mathfrak{p}(\mathfrak{O})$ le sous-groupe de $\mathcal{K}_o(\bar{A})$ engendré par les classes [M] des $\bar{A}$-modules M tels que pour tout $\mathfrak{P}$ au-dessus de $\mathfrak{p}$, $[M_\mathfrak{P}]$ appartient à l'image de $\bar{e}_\mathfrak{P}$.

Soit $\mathfrak{p} \in \mathcal{P}_\infty(K)$, et soit $\mathfrak{P}$ une place de $\bar{K}$ au-dessus de $\mathfrak{p}$; par extension des scalaires de $K_\mathfrak{p}$ à $\bar{K}_\mathfrak{P}$, le groupe $\mathcal{K}_o(A_\mathfrak{p})$ s'identifie à un facteur direct de $\mathcal{K}_o(\bar{A}_\mathfrak{P})$; soit $\mathcal{K}'_o(\bar{A}_\mathfrak{P})$ le sous-groupe de $\mathcal{K}_o(\bar{A}_\mathfrak{P})$ engendré par les $\bar{A}_\mathfrak{P}$-modules simples n'appartenant pas à $\mathcal{K}_o(\bar{A}_\mathfrak{p})$. On note $P_\mathfrak{p}(\mathfrak{O})$ le sous-groupe de

$\mathcal{K}_0(A)$ engendré par les classes $[M]$ des A-modules M tels que pour tout $\mathfrak{P}$-au-dessus de $\mathfrak{p}$, $[M_{\mathfrak{P}}]$ appartienne à $\mathcal{K}'_0(A_{\mathfrak{P}})$.

Soit $H^S(\mathcal{C}(\mathfrak{D}))$ (resp. $H^S(\mathcal{C}^S_{1p}(\mathfrak{D})))$ le sous-groupe de $\mathrm{Hom}_{G_K}(\mathcal{K}_0(\bar{A}), J(\bar{\mathbb{Q}}))$ formé des homomorphismes f tels que

$$\forall \mathfrak{p} \in S, \qquad f(P_{\mathfrak{p}}(\mathfrak{D}))_{\mathfrak{p}} \subset \mathbb{Z}^*_{\mathfrak{p}} \ ;$$

$$\forall \mathfrak{p} \in \mathcal{P}(\mathbb{Z}_K)-S, \quad f(\mathcal{K}_0(A))_{\mathfrak{p}} \subset \bar{\mathbb{Z}}^*_{\mathfrak{p}} \quad \text{(resp. il existe } \alpha \in \mathfrak{D}^*_{\mathfrak{p}} \text{ tel que } f(\rho)_{\mathfrak{p}} = \mathrm{Det}_{\rho}(\alpha) \text{, pour tout } \rho \text{ dans } \mathcal{K}_0(\bar{A})) \ ;$$

$$\forall \mathfrak{p} \in \mathcal{P}_\infty(K), \qquad f(P_{\mathfrak{p}}(\mathfrak{D}))_{\mathfrak{p}} \quad \text{est réel et positif.}$$

THÉORÈME 3.1. - <u>Soit</u> $\alpha \in J(A)$, <u>on note</u> $\mathfrak{D}\alpha$ <u>l'unique</u> $\mathfrak{D}$-<u>module tel que</u> $(\mathfrak{D}\alpha)_{\mathfrak{p}} = \mathfrak{D}_{\mathfrak{p}}\alpha_{\mathfrak{p}}$. <u>Pour</u> $\mathcal{C} = \mathcal{C}(\mathfrak{D})$ <u>ou</u> $\mathcal{C} = \mathcal{C}^S_{1p}(\mathfrak{D})$, <u>l'application</u> $\alpha \longrightarrow [\mathfrak{D}] - [\mathfrak{D}\alpha]$ <u>de</u> $J(A)$ <u>dans</u> $\mathcal{K}^S_0(\mathcal{C})$ <u>se factorise en un unique isomorphisme</u> $\eta_{\mathfrak{D}}$ <u>rendant le diagramme suivant commutatif</u>

$$\begin{array}{ccc} \mathrm{Hom}_{G_K}(\mathcal{K}_0(A), J(\mathbb{Q})) / \mathrm{Hom}_{G_K}(\mathcal{K}_0(A), \mathbb{Q}^*)\, H^S(\mathcal{C}) & \xrightarrow{\ \eta_{\mathfrak{D}}\ } & \mathcal{K}^S_0(\mathcal{C}) \ . \\ \mathrm{Det} \nwarrow & & \nearrow \\ & J(A) & \end{array}$$

<u>Démonstration</u>. - Le résultat se déduit du théorème 1.6 des propositions 2.4, 2.6, 2.8 et du fait que $\mathrm{Det}(J(A))$ est égal à l'ensemble des homomorphismes f de $\mathrm{Hom}_{G_K}(\mathcal{K}_0(\bar{A}), J(\bar{\mathbb{Q}}))$ tel que $f(P_{\mathfrak{p}}(\mathfrak{D}))_{\mathfrak{p}}$ soit réel et positif pour tout $\mathfrak{p} \in \mathcal{P}_\infty(K)$ (théorème de Eichler, voir [We], proposition 3).

Soit $\mathrm{Hom}^+_{G_K}(R_\Gamma, J(\mathbb{Q})) = \{ f \in \mathrm{Hom}_{G_K}(R_\Gamma, J(\mathbb{Q})),\ \forall \mathfrak{p} \in \mathcal{P}_\infty(K),\ f(P_{\mathfrak{p}}(\mathfrak{D}))_{\mathfrak{p}}$ est réel et positif$\}$.

THÉORÈME 3.2. - <u>Avec les notations du théorème 3.1</u>, <u>pour</u> $\mathcal{C} = \mathcal{C}(\mathfrak{D})$ <u>ou</u> $\mathcal{C}^S_{1p}(\mathfrak{D})$ <u>l'application</u> $\alpha \longrightarrow [\mathfrak{D}, 1, \mathfrak{D}\alpha]$ <u>de</u> $J(A)$ <u>dans</u> $\mathcal{K}^S_{0,\mathrm{rel}}(\mathcal{C})$ <u>se factorise en un unique homomorphisme</u> $\Delta_{\mathfrak{D}}$ <u>rendant le diagramme suivant commutatif</u> :

$$\begin{array}{ccc} \mathrm{Hom}^+(\mathcal{K}_0(A), J(\mathbb{Q}))/H^S(\mathcal{C}) & \xrightarrow{\ \Delta_{\mathfrak{D}}\ } & \mathcal{K}^S_{0,\mathrm{rel}}(\mathcal{C}) \ . \\ \mathrm{Det} \nwarrow & & \nearrow \\ & J(A) & \end{array}$$

L'homomorphisme $\Delta_{\mathfrak{D}}$ *est injectif. L'homomorphisme* $\Delta_{\mathfrak{D}}$ *est surjectif si*
$\forall \mathfrak{p} \in \mathcal{P}(R)-S$, *l'homomorphisme* $e_{\mathfrak{p}}$ *est injectif ;*
$\forall \mathfrak{p} \in S$, *l'homomorphisme* $e_{\mathfrak{p}}$ *est une injection directe.*

Démonstration. - Ce théorème découle immédiatement des propositions du paragraphe précédent.

4. - Application à la structure galoisienne des anneaux d'entiers

On généralise, dans ce paragraphe, au cas où S est non vide, les résultats de Fröhlich ([F 2]), Taylor ([T]) et Cougnard ([C 2] et [C 3]) obtenus pour les groupes Γ ne contenant pas d'éléments sauvagement ramifiés.

THÉORÈME 4.1. - *Soit* Γ *un groupe d'automorphismes d'un corps de nombres* N. *Soit* S *un ensemble de nombres premiers contenant les nombres premiers* p *tels qu'il existe un automorphisme de* Γ *sauvagement ramifié en* p. *Alors* $[\mathbb{Z}_N] - r[\mathbb{Z}[\Gamma]] = 0$ *dans* $Q_{\oplus}^{S}(\mathbb{Z}[\Gamma])$ *où* $r = [N^{\Gamma} : \mathbb{Q}]$ *est le rang de* $\mathbb{Z}_N$.

Si Γ ne contient pas d'automorphisme sauvagement ramifié, on peut prendre S vide et l'on retrouve le résultat suivant, dû à Fröhlich ([F 2], théorème 11) :

COROLLAIRE 4.2. - *Soit* $\mathfrak{M}$ *un ordre maximal de* $\mathbb{Z}$ *dans* $\mathbb{Q}[\Gamma]$ *contenant* $\mathbb{Z}[\Gamma]$. *On suppose que* Γ *ne contient pas d'automorphisme sauvagement ramifié ; alors* $\mathfrak{M} \otimes_{\mathbb{Z}[\Gamma]} \mathbb{Z}_N$ *et* $\mathfrak{M}\mathbb{Z}_N$ *sont des* $\mathfrak{M}$*-modules stablement libres.*

Cela se déduit du résultat bien connu suivant (voir [Q 2], proposition 1.7) : soient deux $\mathbb{Z}[\Gamma]$-modules M et M' localement libres ; M et M' deviennent par extension des scalaires des $\mathfrak{M}$-modules stablement isomorphes si et seulement s'il existe un $\mathbb{Z}[\Gamma]$-module X de type fini sans $\mathbb{Z}$-torsion et un isomorphisme de $M \oplus X$ sur $M' \oplus X$.

Quelques résultats ont été démontrés par Fröhlich et Cougnard dans le cas où Γ contient des automoprhismes sauvagement ramifiés. En utilisant les exemples de [Q 2], corollaire 6.4, on retrouve le résultat suivant, dû à Cougnard dans le cas où $N^{\Gamma} = \mathbb{Q}$ ([C 2] et à Fröhlich dans le cas général ([F 3]) :

COROLLAIRE 4.3. - Soit $\mathfrak{M}$ un ordre maximal de $\mathbb{Z}$ dans $\mathbb{Q}[\Gamma]$ contenant $\mathbb{Z}[\Gamma]$. Si Γ est un p-groupe, $\mathfrak{M}\,\mathbb{Z}_N$ est un $\mathfrak{M}$-module stablement libre.

J. Cougnard ([C 3]), a montré que $\mathfrak{M} \otimes_{\mathbb{Z}[\Gamma]} \mathbb{Z}_N$ est isomorphe à $\mathfrak{M}\,\mathbb{Z}_N \oplus T$ où T est un groupe fini dont les facteurs premiers de l'ordre sont les nombres premiers p tels que Γ contienne un automorphisme sauvagement ramifié en p et $\mathfrak{M}$ est toujours un ordre maximal de $\mathbb{Z}$ dans $\mathbb{Q}[\Gamma]$ contenant $\mathbb{Z}[\Gamma]$. On écrit T comme quotient d'un module libre ; on a donc une suite

$$0 \longrightarrow M \longrightarrow \mathfrak{M}^n \longrightarrow T \longrightarrow 0 .$$

En utilisant les exemples de [Q 2], on retrouve les résultats suivants, dus à Cougnard ([C 3]) :

COROLLAIRE 4.4. - Si Γ est un groupe métacyclique d'ordre pq ($q/p-1$, p et q premiers), un groupe quaternionien d'ordre $4p^m$ ou diédral d'ordre $2p^m$ (p premier), alors $[\mathfrak{M}\,\mathbb{Z}_N] - [\mathfrak{M}^r] = [\mathfrak{M}^r] - [M]$ dans le groupe $\mathcal{Cl}(\mathfrak{M})$ des classes des $\mathfrak{M}$-modules de type fini sans torsion.

Le choix de S fait dans le théorème 4.1 entraîne que $\mathbb{Z}_N$ est un $\mathbb{Z}[\Gamma]$-module localement projectif pour tout p n'appartenant pas à S. On peut donc considérer la classe de $\mathbb{Z}_N$ dans le groupe de Grothendieck $\mathcal{K}_o^S(\mathbb{Z}[\Gamma])$ de la catégorie $\mathcal{C}_{lp}^S(\mathbb{Z}[\Gamma])$ des $\mathbb{Z}[\Gamma]$-modules de type fini sans $\mathbb{Z}$-torsion, localement libres pour tout p n'appartenant pas à S (voir paragraphe 1).

CONJECTURE. - L'élément $[\mathbb{Z}_N] - r[\mathbb{Z}[\Gamma]]$ est d'ordre 2 dans $\mathcal{K}_o^S(\mathbb{Z}[\Gamma])$. Il est même trivial si les constantes de l'équation fonctionnelle des fonctions L d'Artin valent +1 pour les caractères symplectiques de Γ.

Cette conjecture généralise celle donnée par A. Fröhlich dans [F 2]. On trouvera une formulation plus précise dans [C N, Q]. Le théorème suivant et plus précisément son corollaire montrent que cette conjecture est vérifiée pour un groupe Γ abélien. Ce résultat permet de démontrer la conjecture précédente pour une large classe de groupes (voir [CN, Q]).

Pour tout idéal premier $\mathfrak{P}$ de N, on note $\Gamma(\mathfrak{P})$ le groupe de décomposition de $\mathfrak{P}$ dans Γ (i.e. le sous-groupe de Γ formé des automorphismes γ tels que $\gamma(\mathfrak{P}) = \mathfrak{P}$).

THÉORÈME 4.5. - <u>Si</u> Γ <u>ne possède pas de caractère symplectique irréductible et si pour tout idéal premier</u> $\mathfrak{P}$ <u>de</u> $\mathbb{Z}_N$, $\Gamma(\mathfrak{P})$ <u>est abélien, alors</u> $[\mathbb{Z}_N] - r[\mathbb{Z}[\Gamma]] = 0$ <u>dans</u> $\mathcal{K}_o^S(\mathbb{Z}[\Gamma])$.

COROLLAIRE. - <u>Si</u> Γ <u>est abélien, alors</u> $[\mathbb{Z}_N]$ <u>est égal à</u> $r[\mathbb{Z}[\Gamma]]$ <u>dans</u> $\mathcal{K}_o^S(\mathbb{Z}[\Gamma])$.

Ce résultat a été démontré par M. J. Taylor ([T]) dans le cas où Γ ne contient pas d'automorphisme sauvagement ramifié.

Soit $U^S_{N/K}$ l'élément de $\mathrm{Hom}_{G_{\mathbb{Q}}}(R_\Gamma, J(\overline{\mathbb{Q}}))/\mathrm{Hom}_{G_{\mathbb{Q}}}(R_\Gamma, \overline{\mathbb{Q}}^*)H^S(\mathcal{C}^S_{1p}(\mathbb{Z}[\Gamma]))$ dont l'image dans $\widetilde{\mathcal{K}}_o^S(\mathbb{Z}[\Gamma])$ par $\eta_{\mathbb{Z}[\Gamma]}$ est $[\mathbb{Z}_N]-r[\mathbb{Z}[\Gamma]]$ (où R_Γ désigne le groupe des caractères virtuels de Γ ; ce groupe s'identifie à $\mathcal{K}_o(\overline{\mathbb{Q}}[\Gamma])$).

Pour tout χ de R_Γ, on note $W_{N/K}(\chi)$ la constante de l'équation fonctionnelle des séries L d'Artin associées au caractère χ. On note $W'_{N/K}$ l'élément de $\mathrm{Hom}_{G_{\mathbb{Q}}}(R(G), J(\overline{\mathbb{Q}}))$ construit de la façon suivante :

- pour tout caractère χ irréductible et non symplectique de G on pose $W'_{N/K}(\chi)=1$
- pour tout caractère χ irréductible et symplectique de G, on définit les composantes locales de $W'_{N/K}(\chi)$ par :

$$W'_{N/K}(\chi)_p = 1 \quad \text{pour tout nombre premier } p$$

$$W'_{N/K}(\chi)_{p_\infty^\omega} = W_{N/K}(\chi^{\omega^{-1}}), \quad \text{pour tout } \omega \in G_{\mathbb{Q}}.$$

On désigne par $t^S_{\mathbb{Z}[\Gamma]}(W_{N/K})$ la classe de $W'_{N/K}$ modulo $\mathrm{Hom}_{G_{\mathbb{Q}}}(R_\Gamma, \overline{\mathbb{Q}}^*)H^S(\mathcal{C}^S_{1p}(\mathbb{Z}[\Gamma]))$. On pose $V^S_{N/K} = U^S_{N/K}/t^S_{\mathbb{Z}[\Gamma]}(W_{N/K})$.

La conjecture de Fröhlich se généralise de la façon suivante :

$$U^S_{N/K} = t^S_{\mathbb{Z}[\Gamma]}(W_{N/K}).$$

Pour démontrer le théorème 4.1, on montre que $U^S_{N/K}$ est représenté par l'élément f de $\mathrm{Hom}_{G_{\mathbb{Q}}}(R_\Gamma, J(\overline{\mathbb{Q}}))$ suivant :

pour tout $\chi \in R_\Gamma$, la p composante de $f(\chi)$ est égale à :

$$f(\chi)_p = \eta_{K/\mathbb{Q}}(a_p/\chi)\,\tau_K(\chi)^{-1}\,W'_{N/K}(\chi)$$

où $\tau_K(\chi)$ est la somme de Gauss galoisienne associée à χ (voir [M 3])

a_p est une base de $\mathbb{Z}_{N_p}$ sur $\mathbb{Z}_{K_p}[\Gamma]$

$\eta_{K/\mathbb{Q}}(a_p/\chi)$ est la résolvante de a_p (définie dans [F 2] ou [Q 2]).

On démontre alors que les applications $\chi \longrightarrow W'_{N/K}(\chi)$ et $\chi \longrightarrow \eta_{K/\mathbb{Q}}(a_p, \chi)\,\tau_K(\chi)^{-1}$ appartiennent à $\mathrm{Hom}_{G_\mathbb{Q}}(R_\Gamma, \bar{\mathbb{Q}}^*)\, H^S(\mathcal{C}^S_{1p}(\mathbb{Z}[\Gamma]))$ (voir [Q 3]).

RÉFÉRENCES

[B] A.-M. BERGÉ, Anneau d'entiers et ordres associés, Thèse, Université de Bordeaux I, 1979.

[Ba] H. BASS, Algebraic K-theory, New York, 1968.

[C 1] J. COUGNARD, Un contre-exemple à une conjecture de J. Martinet, in Algebraic Number Fields, Proc. Sympos. Univ. Durham, Fröhlich ed., Academic Press, London, 1977.

[C 2] J. COUGNARD, Entiers d'une p-extension, Compos. Math. 33 (1976), 303-336.

[C 3] J. COUGNARD, Une propriété de l'anneau des entiers des extensions galoisiennes non abéliennes de degré pq des rationnels, Pub. Math. Fac. des Sciences de Besançon, 1976-1977.

[CN, Q] Ph. CASSOU-NOGUÈS et J. QUEYRUT, Structure galoisienne des anneaux d'entiers d'extensions sauvagement ramifiées, II, (à paraître).

[F 1] A. FRÖHLICH, Artin root numbers and normal integral bases for quaternion fields, Invent. Math. 17 (1972), 143-166.

[F 2] A. FRÖHLICH, Arithmetic and Galois module structure for tame extensions, J. reine angew. Math. 286-287 (1976), 380-439.

[F 3] A. FRÖHLICH, Some problems of Galois module structure for wild extensions, Proc. London Math. Soc. 37 (1978), 193-212.

[He] A. HELLER, Some exact sequences in algebraic K-theory, Topology, 3 (1969), 389-408.

[Hi] D. HILBERT, Gesammelte Abhandlungen, Band. 1.

[L] H. W. LEOPOLD, Über die Hauptordnung des ganzen Elementen eines abelschen Zahlkörpers, J. reine angew. Math. 201 (1959), 119-149.

[M 1] J. MARTINET, Sur l'arithmétique des extensions galoisiennes à groupe de Galois diédral d'ordre 2p , Ann. Inst. Fourier, 19 (1969), 1-80.

[M 2] J. MARTINET, Modules sur l'algèbre du groupe quaternionien, Ann. Scient. Ec. Norm. Sup. 4e série, t. 4 (1971), 399-408.

[N] E. NOETHER, Normal basis bei Körpern ohne höhere Verzweigung, J. reine angew. Math. 167 (1932), 147-152.

[N M] T. NAKAYAMA und Y. MATSUSHIMA, Über die multiplicative group einer p-adischen Division algebra, Proc. Imp. Acad. Tokyo 19 (1943), 622-628.

[Q 1] J. QUEYRUT, Extensions quaternioniennes généralisées et constante de l'équation fonctionnelle des séries L d'Artin, Publ. Math. Univ. de Bordeaux I, 4 (1972-1973), 91-113.

[Q 2] J. QUEYRUT, S-groupes des classes d'un ordre arithmétique, (à paraître).

[Q 3] J. QUEYRUT, Structure galoisienne des anneaux d'entiers d'extensions sauvagement ramifiées, I, (à paraître).

[Se] J.-P. SERRE, Représentations linéaires des groupes finis, 2e édition, Hermann, Paris, 1971.

[T] M. J. TAYLOR, Galois module structure of integers of relative abelian extensions, J. reine angew. Math. 303-304 (1978), 97-101.

[Wa] S. WANG, On the commutator group of a simple algebra, Am. J. of Math. 72 (1950), 323-334.

[We] A. WEIL, Basic number theory, Springer Verlag, 1974.

[Wi] S. WILSON, Some counter-examples in the theory of the Galois module structure of wild extensions, Ann. Inst. Fourier, 30, 3 (1980), 1-9.

U.E.R. de Mathématiques
et d'Informatique
Université de Bordeaux I
F 33405 TALENCE CEDEX

Ratios of Rings of Integers as Galois Modules

by
Stephen V. Ullom*

§1. Introduction.

Let F be an algebraic number field with ring of integers $\operatorname{int} F$ and N/F a normal tamely ramified extension field with Galois group $\Gamma = \operatorname{Gal}(N/F)$. Let E be an intermediate field, $F \subset E \subset N$, fixed by a subgroup Δ of Γ. Int N is a locally free module over the group rings $\operatorname{int} k \cdot \Gamma$, k subfield of F. Fröhlich's conjecture states roughly that the Artin root numbers of the symplectic characters of Γ are the only obstruction to $\operatorname{int} N$ being stably free as a $\mathbb{Z}\Gamma$-module. Recently this conjecture has been proved by M. J. Taylor under slightly stronger hypotheses [T].

The situation is less clear when $\mathbb{Z}\Gamma$ is replaced by $\mathcal{O}\Delta$ where $\mathcal{O} = \operatorname{int} E$. Here we take $\Gamma = H_{4n}$, the quaternion group of order $4n$, $n > 1$, and Δ a cyclic subgroup of order $2n$. The subgroup generated by the element of order 2 of H_{4n} fixes the subfield say K of N. Consider normal tamely ramified extensions N' of F containing K with $\operatorname{Gal}(N'/F) = H_{4n}$. Let $C\ell(\mathcal{O}\Delta)$ be the classgroup of locally free $\mathcal{O}\Delta$-modules. We investigate the ratio of modules

$$(1.1) \qquad (\operatorname{int} N')(\operatorname{int} N)^{-1} \in C\ell(\mathcal{O}\Delta)$$

as N' varies.

On the other hand Fröhlich [F1] has calculated the ratio of Artin root numbers of quaternion characters. Our main result (3.6) is the calculation - under some additional hypotheses - of the ratio in (1.1) and the construction of a homomorphism t from the subgroup of ratios in $C\ell(\mathcal{O}\Delta)$ to ± 1 which determines the ratio of the corresponding root numbers.

This report contains an outline of several of the results of [U] and some further comments in §3.

§2. Invariants for the ratio.

Given F and K as above, let $\bar{F}$ be an algebraic closure

* This research was partially supported by a National Science Foundation Grant.

of F and $\Omega_F = \mathrm{Gal}(\bar{F}/F)$. We will describe the quaternion extensions N'/F containing K by quadratic characters of Ω_F. The group H_{4n} is given by generators and relations as

$$H_{4n} = \left\langle \sigma, \tau : \sigma^n = \tau^2, \tau^4 = 1, \tau\sigma\tau^{-1} = \sigma^{-1} \right\rangle;$$

it has a faithful irreducible representation T sending

$$\sigma \longrightarrow \begin{pmatrix} \zeta & 0 \\ 0 & \zeta^{-1} \end{pmatrix}$$

$$\tau \longrightarrow \begin{pmatrix} 0 & 1 \\ -1 & 0 \end{pmatrix},$$

ζ some primitive $2n$ - th root of unity. The character ψ of $T_\psi = T$ is called a quaternion character. If there is some extension N/F with Galois group H_{4n}, we may view T_ψ as a representation of Ω_F with fixed field N and we write $N=F_\psi = \bar{F}^{\ker T_\psi}$.

For any homomorphism $\theta : \Omega_F \longrightarrow \pm 1$, $\psi\theta$ is again a quaternion character of Ω_F. Conversely, the next proposition holds.

(2.1) <u>Proposition</u>. [F1] Given a quaternion extension N/F of degree $4n$ with $F \subset K \subset N$, $(N : K) = 2$. Any extension N' containing K with $\mathrm{Gal}(N'/F) = H_{4n}$ has the form $N' = F_{\psi\theta}$, for some $\theta : \Omega_F \longrightarrow \pm 1$.

The Artin root number of a character ψ is defined by the functional equation of the enlarged Artin L-function

$$\Lambda(s,\psi) = W(\psi)\Lambda(1-s,\bar{\psi}).$$

(In the remainder of this article we shall assume all characters are tamely ramified.) It is not unreasonable to expect a relation between $W(\psi\theta)/W(\psi)$ and $(\mathrm{int}\, F_{\psi\theta})(\mathrm{int}\, F_\psi)^{-1} \in C\ell(\mathcal{O}\Delta)$, $\mathcal{O} = \mathrm{int}\, E$, $\Delta = \langle \sigma \rangle$. In fact the character χ of Δ defined by $\chi(\sigma^i) = \zeta^i$ induces ψ and by Frobenius reciprocity $\chi\phi$ induces $\psi\theta$, where ϕ is the restriction of θ to Ω_E. Thus $F_\psi = E_\chi$, $F_{\psi\theta} = E_{\chi\phi}$. Since global root numbers are invariant under induction,

$$W(\psi\theta)/W(\psi) = W(\chi\phi)/W(\chi).$$

Let $E(\alpha)$ be the field E with the values of the character α adjoined. The classgroup $C\ell(\mathcal{O}\Delta)$ of locally free $\mathcal{O}\Delta$-modules is a certain quotient of the group $I(\mathcal{O},\Delta)$ (see p. 429 of [F2]); $I(\mathcal{O},\Delta)$ consists of Ω_E-linear functions g defined on the ring R_Δ of complex characters of Δ such that (i) $g(\alpha)$ is a fractional ideal of the field $E(\alpha)$, $\alpha \in R_\Delta$, and (ii) the numerator and denominator of $g(\alpha)$ are prime to the order of Δ. The class $(\text{int } E_\chi) \in C\ell(\mathcal{O}\Delta)$ is described by resolvents (Theorem 1a, p. 390 of [F2]). For an actual character α of $\Delta \cong \text{Gal}(N/E)$, define the Lagrange resolvent $(b|\alpha)$ by

$$(b|\alpha) = \sum b^\delta \alpha(\delta)^{-1}, \; b \in N,$$

summation over $\delta \in \text{Gal}(N/E)$. By linearity $(b|\alpha)$ is defined for all $\alpha \in R_\Delta$ (take b generating an E-normal basis of N). Define the resolvent module $(\mathcal{O}:\alpha)$ as the int $E(\alpha)$-module generated by the $(b|\alpha)$, $b \in \text{int } N$; actually $(\mathcal{O}:\alpha)$ depends only on α.

(2.2) Proposition. Given tamely ramified extensions E_χ and $E_{\chi\phi}$ of E with Galois groups isomorphic to Δ, the cyclic group of order $2n$. Fix a faithful character χ_1 of Δ and choose an algebraic integer c in the composite field $E_\chi E_{\chi\phi}$ which generates a normal integral basis at primes dividing $2n$. The ratio

$$(\text{int } E_{\chi\phi})(\text{int } E_\chi)^{-1} \in C\ell(\mathcal{O}\Delta)$$

is represented by the function in $I(\mathcal{O},\Delta)$ defined on powers of χ_1 as follows

$$g(\chi_1^i) = \frac{(\mathcal{O}:(\chi\phi)^i)}{(c|(\chi\phi)^i)} \div \frac{(\mathcal{O}:\chi^i)}{(c|\chi^i)} . \tag{2.3}$$

Of course if i is even, $g(\chi_1^i) = (1)$.

In order to compute the ratio of module resolvents we shall assume the conductors $f(\chi)$ and $f(\phi)$ are relatively prime. Then one has [U]

$$(\mathcal{O}:\chi\phi) = (\mathcal{O}:\chi)(\mathcal{O}:\phi). \tag{2.4}$$

Write $E_\phi = E(\sqrt{m})$, $m \in E^*$. Then the resolvent module $(\mathcal{O}:\phi)$ is contained in the one-dimensional E-module $E\sqrt{m}$ with basis $\sqrt{m}$ and

$$(\mathcal{O}:\phi) = (f(\phi)m^{-1})^{\frac{1}{2}} \sqrt{m}, \tag{2.5}$$

where $(f(\phi)m^{-1})^{\frac{1}{2}}$ is an ideal of E.

It is easy to show $(c|\chi)(c|\chi\phi)^{-1} \in E\sqrt{m}$ and

$$(c|\chi)(c|\chi\phi)^{-1} \equiv 1 \bmod 2.$$

(2.6) Corollary to (2.2). Assume the conductors $f(\chi)$ and $f(\phi)$ are relatively prime and normalize $m \equiv 1 \bmod 4$. Then for i odd

$$g(\chi_1^i) = (f(\phi)m^{-1})^{1/2}\, a_i,\ a_i \in E,\ a_i \equiv 1 \bmod 2,$$

where $g(\chi_1^i)$ is an $E(\chi_1^i)$-ideal.

§3. Ratios of modules and root numbers.

From (2.6) we see that if the a_i factor is neglected, then $(\text{int } E_{\chi\phi})(\text{int } E_\chi)^{-1}$ is described simply by the extensions of the E-ideal $(f(\phi)m^{-1})^{1/2}$ to the fields $E(\chi_1^i)$. Moreover, when $m \in F^*$, the E-ideal $(f(\phi)m^{-1})^{1/2}$ is invariant under $\mathrm{Gal}(E/F)$.

Let k be a number field and $I(k)$ the subgroup of the group of invertible ideals of k which are prime to 2. Let $\mathrm{Amb}(E)$ be the subgroup of $I(E)$ of $\mathrm{Gal}(E/F)$ - invariant ideals and let $\mathrm{Amb}'(E)$ be the image of $\mathrm{Amb}(E)$ in $C\ell(\mathcal{O})$. We define a homomorphism $\mathrm{Amb}(E) \longrightarrow C\ell(\mathcal{O}\Delta)$ as follows. Fix a faithful character χ_1 of Δ; for an ideal $A \in I(E(\chi_1))$ define the function $g_A \in I(\mathcal{O},\Delta)$ on actual characters α of Δ by

$$g_A(\alpha) = \begin{cases} A^\omega & \text{if } \alpha = \chi_1^{\ \omega},\ \omega \in \Omega_E \\ (1) & \text{if } \alpha \neq \chi_1^{\ \omega} \end{cases}$$

and extend g_A to R_Δ by linearity. The class of g_A is an element of $C\ell(\mathcal{O}\Delta)$; let

$$j : \mathrm{Amb}(E) \longrightarrow C\ell(\mathcal{O}\Delta) \tag{3.1}$$

be the restriction of the above map to the image of $\mathrm{Amb}(E)$ in $I(E(\chi_1))$.

For a number field k let $(a,b)_v$ denote the Hilbert symbol at the place v of k and $(\frac{b}{A})$ the 2 power residue symbol for $A \in I(k)$, $b \in k^*$. Define the homomorphism $s : \mathrm{Amb}(E) \longrightarrow \pm 1$ by $s(A) = (\frac{-1}{A})_E$.

(3.2) Proposition. Let E/F be a tamely ramified quadratic extension. Assume (1) the relative discriminant $d(E/F)$ factors as d_1d_2, d_i the discriminant of a quadratic extension $F(\sqrt{e_i}) = E_i$ of F and (2) for all places v of F

$$(-1,e_1)_v(-1,e_2)_v(e_1,e_2)_v = 1.$$

Then there is a commutative square

$$\begin{array}{ccc} \mathrm{Amb}(E) & \longrightarrow & C\ell(\mathcal{O}) \\ s\downarrow & & \downarrow u \\ \pm 1 & \cong & \mathrm{Gal}(E_1E_2/E), \end{array}$$

where u is the Artin map of class field theory.

(3.3) Proposition. Let χ_1 be a faithful character of the cyclic 2-group Δ. Assume the extension of ideals map $\mathrm{Amb}'(E) \longrightarrow C\ell(\mathrm{int}\ E(\chi_1))$ is injective. Then $s : \mathrm{Amb}(E) \longrightarrow \pm 1$ defines a homomorphism

$$t : j(\mathrm{Amb}(E)) \longrightarrow \pm 1$$

by $t(j(A)) = s(A)$, $A \in \mathrm{Amb}(E)$.

Proof. Form the commutative diagram below. The top horizontal map is the composite $C\ell(\mathcal{O}\Delta) \longrightarrow C\ell(\text{maximal order}) \longrightarrow C\ell(\mathrm{int}\ E(\chi_1))$.

$$\begin{array}{ccc} C\ell(\mathcal{O}\Delta) & \longrightarrow & C\ell(\mathrm{int}\ E(\chi_1)) \\ j\uparrow & & \uparrow \\ \mathrm{Amb}(E) & \longrightarrow & C\ell(\mathcal{O}) \\ s\downarrow & & \downarrow u \\ \pm 1 & \cong & \mathrm{Gal}(E_1E_2/E). \end{array}$$

It suffices to prove that $\ker j \subset \ker s$. This is immediate from the diagram and the assumed injectivity. Q.E.D.

Write $E_\phi = E(\sqrt{m})$, $m \in F^*$, ϕ = restriction of θ to Ω_E. Factor $m\ \mathrm{int}\ F = IJ^2$ into F-ideals, where I is square free. Next write $I = I_1I_2$, I_1 = product of primes of F ramified in both E and in $F(\sqrt{m})$, so $I_2 = I\,I_1^{-1}$ is the product of primes ramified in $F(\sqrt{m})$ only. Then as E-ideals

$$(f(\phi)m^{-1})^{\frac{1}{2}} = J(I_1\mathcal{O})^{-\frac{1}{2}}.$$

It follows that

$$s((f(\phi)m^{-1})^{\frac{1}{2}}) = \left(\frac{-1}{I_1}\right)_F .$$

(3.4) Lemma. Take $F = \mathbb{Q}$. The set of E-ideals $(d(E(\sqrt{m})/E)m^{-1})^{\frac{1}{2}}$ for all m in $\mathbb{Q}^*$ prime to 2 equals $\mathrm{Amb}(E)$.

Proof. Let m be a subproduct of the ramified primes of E/F times ± 1 so that $m \equiv 1 \bmod 4$. Then $E(\sqrt{m})/E$ is unramified at all finite primes and $f(\phi) = d(E(\sqrt{m})/E) = (1)$. We can get any divisor of the different of E/F in this manner. The desired result follows easily.

Remark. The Steinitz class of the extension $E(\sqrt{m})/E$ is the ideal class of $(f(\phi)m^{-1})^{\frac{1}{2}}$ in $C\ell(o)$.

(3.5) Proposition. [F1] Assume the conductors $f(\chi)$ and $f(\phi)$ are relatively prime. Then

$$W(\psi\theta)/W(\psi) = (\frac{-1}{I_1})_F .$$

(3.6) Theorem. Let E/F be a tamely ramified quadratic extension. Let F_ψ and $F_{\psi\theta}$ be tamely ramified quaternion 2-extensions containing E with Galois group over E isomorphic to the cyclic 2-group Δ. Suppose
(a) the conductors $f(\chi)$ and $f(\phi)$ are relatively prime and
(b) the prime 2 is unramified in E/Q. Then the ratio in $C\ell(o\Delta)$ of rings of integers is

$$(\text{int } F_{\psi\theta})(\text{int } F_\psi)^{-1} = j((f(\phi)m^{-1})^{\frac{1}{2}}) .$$

Assume in addition (c) hypotheses of (3.2) and (d) the map $\text{Amb}'(E) \longrightarrow C\ell(\text{int } E(\chi_1))$ is injective, χ_1 faithful on Δ. Then

$$t[(\text{int } F_{\psi\theta})(\text{int } F_\psi)^{-1}] = W(\psi\theta)/W(\psi) .$$

Sketch of proof. We remark only that 2 unramified in E implies the conductor of the simple component $E(\chi_1)$ of $E\Delta$ is $2 \text{ int } E(\chi_1)$. Thus the a_i factors described in (2.6) may be neglected.

Corollary. Take $F = Q$. The set of ratios $(\text{int } F_{\psi\theta})(\text{int } F_\psi)^{-1}$ for fixed ψ and varying θ is a subgroup of $C\ell(o\Delta)$ isomorphic to the image of $\text{Amb}(E)$ in $C\ell(\text{int } E)$. In particular, it is an elementary abelian 2-group.

Proof. Use the first assertion of the theorem and (3.4).

Example. The hypotheses of the theorem are satisfied for $F = Q$, $E = Q(\sqrt{pq})$, p and q distinct odd primes $\equiv 1 \mod 4$ such that $(\frac{p}{q}) = 1$, and Δ of order 4. Then $C\ell(\text{int } E) \longrightarrow C\ell(\text{int } E(\sqrt{-1}))$ is injective by (3.10) of [U].

Remark. Let L/k be the basic cyclotomic $\mathbb{Z}_2$-extension of a totally real number field k. Let k_i be the intermediate field with $(k_i : k) = 2^i$, $i \geq 0$, and A_i the 2-Sylow subgroup of $C\ell(\text{int } k_i)$. R. Greenberg (Prop. 2 of [G]) has shown that if the order of A_i is bounded as $i \longrightarrow \infty$, then every ideal class of A_0 becomes principal in k_i for some i. As an example, take $k = Q(\sqrt{pq})$ described above. The statement that the order of A_i is bounded is equivalent to the Iwasawa invariants $\lambda(L/k) = \mu(L/k) = 0$. There are no examples known of totally real fields k with nonzero Iwasawa invariants. Thus it is likely that $(f(\phi)m^{-1})^{\frac{1}{2}}$ becomes principal in $E(\chi_1)$ for χ_1 of sufficiently large order; perhaps even χ_1 of order 2^{r+1}, $r \geq 2$, would suffice. If this extended ideal is principal, then we would have $(\text{int } F_{\psi\theta}) = (\text{int } F_{\psi})$ in $C\ell(\mathfrak{o}\Delta)$ even if the corresponding root numbers are different.

On the other hand, when int N is viewed as a $\mathbb{Z}H_{4n}$- module, Fröhlich showed [F3] $(\text{int } N) = 1$ in $C\ell(\mathbb{Z}H_{4n})$, if $n = 2^r$, $r \geq 2$. We conclude with a conjecture.

Conjecture. The class $(\text{int } F_{\psi\theta}) = (\text{int } F_{\psi})$ in $C\ell(\mathfrak{o}\Delta)$ whenever Δ is cyclic of order 2^{r+1}, $r \geq 2$, provided E is totally real.

References

[F1] A. Fröhlich, Artin-root numbers for quaternion characters, Inst. Naz. di Alta Mat., Symposia Mathematica XV(1975), 353-363.

[F2] A. Fröhlich, Arithmetic and Galois module structure for tame extensions, J. Reine Angew. Math. 286/287(1976), 380-440.

[F3] A. Fröhlich, Galois module structure and root numbers for quaternion extensions of degree 2^n, J. Number Theory, to appear.

[G] R. Greenberg, On the Iwasawa invariants of totally real number fields, Amer. J. of Math., 98(1976), 263-284.

[T] M. J. Taylor, On Fröhlich's conjecture for rings of integers of tame extensions, to appear.

[U] S. Ullom, Galois module structure for intermediate extensions, J. London Math. Soc., to appear 1980, vol. 22.

University of Illinois
Urbana, Illinois 61801

An extension of Miyata's Theorem on the transfer map from the classgroup of a finite dihedral group to that of its cyclic maximal subgroup

S. M. J. Wilson

§0. Introduction

The original idea of this work was to find a short proof of the theorem [M] of Miyata to the effect that the map described above is injective. (I had been told that Miyata's proof was rather long although, now I have had a chance to see it, it does not seem to be that long). However, the proof that I found and will here describe generalizes Miyata's result as follows.

Theorem If C is a finite abelian group with cyclic 2-component and if $\Gamma = \langle\tau\rangle$, of order 2, acts on C by inversion then, putting $D = C \rtimes \Gamma$, the restriction (transfer) homomorphism

$$Cl(\mathbb{Z}D) \longrightarrow Cl(\mathbb{Z}C)$$

is injective.

One of the reasons for wanting a result of this sort was to prove normal integral basis theorems for algebraic number fields (c.f. [M]) using Taylor's extension of Hilbert's Theorem [T1]. However, I gather that Taylor's work on the basic conjecture (c.f. [T2]) in this area has finally superceded these algebraic methods.

The result is, none the less, of independant interest and is, of course, available for the investigation of other invariants which take their values in classgroups of group rings.

§1. The idèlic formula for the classgroup

Let A be a semisimple $\mathbb{Q}$-algebra. We shall denote by $\nu = \nu_A$ the reduced norm on A (and on its completions). We have the following formula (first given by Fröhlich [F]) for the locally free classgroup of a $\mathbb{Z}$-order Λ in A:-

$$Cl(\Lambda) \cong \frac{\nu J(A)}{\nu\, U(\Lambda).\nu A^*} \qquad (1)$$

This formula is natural with respect to change of order provided $\nu J(A)$ is viewed as a product of local K_1's (c.f. [W]).

§2. Twisted group rings and the transfer

Let S be a $\mathbb{Z}$-order in a semisimple commutative $\mathbb{Q}$-algebra, L. Let Γ be a finite group with an action $\Gamma \longrightarrow Aut(S) \subseteq Aut(L)$. We form the twisted group ring $S\Gamma$ (no cocycle).

The transfer norm $N = N_{S,\Gamma}$ on $K_1(S\Gamma)$ or $L\Gamma^*$ is defined by the diagram

$$\begin{array}{ccccc} L\Gamma^* & \longrightarrow & K_1(L\Gamma) & \xrightarrow{\text{res}} & K_1(L)^{\Gamma} \\ & K_1(S\Gamma) \nearrow & & \searrow N & \swarrow \det \\ & & & L^{*\Gamma} & \end{array}$$

When $|\Gamma| = 2$, $\Gamma = \langle\tau\rangle$ we find $N(a+b\tau) = a\bar{a} - b\bar{b}$. In particular, $N(1+(1+\tau)a) = 1+a+\bar{a}$. (I shall refer to this observation as McGurn's trick).

We also have a reduced norm on $L\Gamma$ and N factors through ν:-

$$\begin{array}{ccc} K_1(L\Gamma) & \xrightarrow{\nu} & \text{Centre}(L\ )^* \\ & N \searrow & \swarrow \bar{N} \\ & L^{*\Gamma} & \end{array}$$

In particular if $|\Gamma| = 2$ and $L = L_1 \oplus L_2$, where Γ acts faithfully on each invariant component of L_1 and trivially on L_2 then

ν is the identity on $L_2\Gamma$ and N on $L_1\Gamma$

$\bar{N}$ is N on $L_2\Gamma$ and the identity on $L_1{}^{\Gamma}$.

We say $S\Gamma$ is tame if there is an element u in S such that $\sum_{\gamma\in\Gamma} u^{\gamma} = 1$. In this case $e_u = \sum_{\gamma\in\Gamma} u\gamma$ is an idempotent and $S \cong_{S\Gamma} e_u S\Gamma$. So we have

<u>Theorem 1</u> If $S\Gamma$ is tame then $N_{S,\Gamma}$ is split by $S^{*\Gamma} \longrightarrow K_1(S\Gamma)$, $s \longmapsto [S,s]$ (or $S^{*\Gamma} \longrightarrow S\Gamma^*$, $s \mapsto e_u s + 1 - e_u$). Thus $N(S\Gamma^*) = NK_1(S\Gamma) = S^{*\Gamma}$.

We now examine the restriction ('transfer') map $Cl(S\Gamma) \longrightarrow Cl(S)$. We find that the corresponding map for formula (1) is

$$\frac{\nu(J(L\Gamma))}{\nu U(S\Gamma).\,L\Gamma^*} \xrightarrow{\bar{N}} \frac{J(L^*)}{U(S).L^*}$$

<u>Theorem 2</u> This transfer map is injective if

(i) $\ker [\bar{N}:\nu J(L\Gamma) \longrightarrow L^*] \subseteq \nu(U(S\Gamma)).\nu L\Gamma^*$,

(ii) $N(S_2\Gamma^*).S^{*\Gamma} = S_2^{*\Gamma}$ and

(iii) $H^1(\Gamma,S^*) \longrightarrow H^1(\Gamma,U(S))$ is injective.

<u>Proof</u>: With (i) we get $Cl(S\Gamma) \cong \dfrac{NJ(L\Gamma)}{NU(S\Gamma).NL\Gamma^*}$ and, since $S_p\Gamma$ is tame for $p \neq 2$ and $L\Gamma$ is tame, $NU(S\Gamma) = \prod_{p\neq 2} S_p{}^{*\Gamma} \times NS_2\Gamma^*$, $NL\Gamma^* = L^{*\Gamma}$, $NJ(L\Gamma) = J(L)^{\Gamma}$. So the transfer is injective if $(U(S).L^*)^{\Gamma} = NU(S\Gamma).L^{*\Gamma}$ that is if $(U(S).L^*)^{\Gamma} = U(S)^{\Gamma}.L^{*\Gamma}$ (*)

and $U(S)^{\Gamma}.L^{*\Gamma} = (\prod_{p\neq 2} S_p^{*\Gamma} \times NS_2\Gamma^*).L^{*\Gamma}$ (**)

A little thought will show that (ii) implies ** and, as we have an exact sequence $1 \to S^* \to U(S) \times L^* \to U(S).L^* \to 1$, cohomology theory show that (iii) implies (*).

§3. The Work

We take $S = \mathbb{Z}C$, $\Gamma = \Gamma$. Put $C = C_1 \times C_o$ where $|C_1| = n$ is odd and $|C_o| = 2r$, a power of 2. Put $C_o = \langle\sigma\rangle$.and $D_o = C_o \rtimes \Gamma$.

Lemma 1 (i) $N(\mathbb{Z}D^*) = \{1 + a + \bar{a}\ ,\ a \in \mathbb{Z}C\} \cap \mathbb{Z}C^*$

(ii) $N(\mathbb{Z}D^*).\langle\sigma^r\rangle = \mathbb{Z}C^{*\Gamma}$.

(iii) $N(\mathbb{Z}_2 D^*) = \{1 + a + \bar{a}\ , a \in \mathbb{Z}_2 C\}$

(iv) $N(\mathbb{Z}_2 D^*).\langle\sigma^r\rangle = \mathbb{Z}_2 C^{*\Gamma}$.

(v) $NU(\mathbb{Z}D) \cap N\mathbb{Q}D^* = N\mathbb{Z}D^*$.

(vi) Part (ii) of Theorem 2 holds.

Proof: Let T be a set of representatives of the orbits of Γ acting on $C \backslash \{1, \sigma^r\}$. If $u \in \mathbb{Z}C^{*\Gamma}$ then we may put $u = a_1 + \sum_{\gamma \in T} a_\gamma\ (\gamma + \gamma^{-1}) + a_0\,\sigma^r\ ,\ a_i \in \mathbb{Z}$, and, as $\mathrm{aug}(u) = \pm 1$, $a_1 \not\equiv a_0 \bmod 2$.

If $2 | a_0$ then $u = 1 + b + \bar{b}$ $b = \sum_{\gamma \in T} a_\gamma \gamma + \frac{a_1 - 1}{2} + \frac{a_0}{2}\sigma^r$ and u is a norm from $\mathbb{Z}D^*$ by McGurn's trick. If $2 \nmid a_0$ then $\sigma^r u$ is a norm and, as σ^r is not, u is not either.

(i) and (ii) are now clear, (iii) and (iv) are similar and (v) and (vi) are immediate.

Lemma 2 Condition (i) of Theorem 2 holds.

Proof: Put $\Lambda = \mathbb{Z}G/(\Sigma C_1)$ and $A = \mathbb{Q}C_1/(\Sigma C_1)$. We have a cartesian square

$$\begin{array}{ccc} \mathbb{Z}C_1 & \longrightarrow & \mathbb{Z} \\ \downarrow & & \downarrow \\ \Lambda & \longrightarrow & \mathbb{Z}/n \end{array}$$

giving an inclusion $\mathbb{Z}C_1 \subsetneq \Lambda + \mathbb{Z}$ and $\mathbb{Q}C = A \oplus \mathbb{Q}$. We really only need the following equations obtained by forming (twisted) grouprings:- $\mathbb{Q}C = AC_o \oplus \mathbb{Q}C_o$, $\mathbb{Q}D = AD_o \oplus \mathbb{Q}D_o$.

With $\mathbb{Q}D$ split in this way, let $(a,b) \in \nu J(\mathbb{Q}D)$ be in $\ker \bar{N}$. As Γ acts faithfully on the components of A, $a = 1$. Clearly, the class given by b in $Cl(\mathbb{Z}D_o)$ goes to 1 in $Cl(\mathbb{Z}C_o)$. But, by the results of [F,K,W] the transfer map, in this case, reduces to a product of maps of the form $Cl(\mathbb{Z}[\eta + \eta^{-1}]) \longrightarrow Cl(\mathbb{Z}[\eta])$, $\eta = \sqrt[2r]{1}$ which are injective. So the class of b in $Cl(\mathbb{Z}D_o)$ is trivial whence $b = \nu(c).\nu(d)$ for some $c \in U(\mathbb{Z}D_o)$ and $d \in \mathbb{Q}D_o^*$. As $\bar{N}(b) = 1$ $N(c) = N(d)^{-1} \in NU(\mathbb{Z}D_o) \cap N(\mathbb{Q}D_o^*) = N(\mathbb{Z}D_o^*)$ by Lemma 1. Altering c and d by an element of $\mathbb{Z}D_o^*$, therefore, we can assume that $N(c) = N(d) = 1$. Pulling c back to $\mathbb{Z}D$ we find $(\nu_{\mathbb{Q}D}(c) =$ (projection of $N(c)$, $\nu_{\mathbb{Q}D_o}(c)$) $= (1, \nu(c))$ whence $(1,b) = \nu_{\mathbb{Q}D}(c).\nu(1,d) \in \nu U(\mathbb{Z}D)\ .\ \nu\mathbb{Q}D^*$ as required.

We shall be through, once we have:-

Lemma 4 $H^1(\Gamma,\mathbb{Z}C^*) \longrightarrow H^1(\Gamma,\mathbb{Z}_2{C_0}^*)$ is injective. ($\mathbb{Z}_2{C_0}^*$ is a factor of $U(\mathbb{Z}C)$).

Proof: An element of norm 1 in $\mathbb{Z}C^*$ has absolute value 1 everywhere and so it has finite order. Hence it lies in $\pm C$ by Higman's Theorem. Reducing this modulo units of the form $u/\bar{u}$ gives $H^1(\Gamma,\mathbb{Z}C^*) = \{(\pm 1),(\pm\sigma)\}$ - at the worst V_4. A rapid calculation shows that (-1) and $(\pm\sigma)$ live on in $H^1(\Gamma,\mathbb{Z}_2{C_0}^*)$.

References

[F] A. Fröhlich, Locally free modules over arithmetic orders J. reine angew. Math. 274/75 (1975), 112-138.

[FKW] A. Fröhlich, M.E. Keating and S. M. J. Wilson, The Classgroups of quaternion and dihedral 2-groups, Mathematika 21 (1974), 64-71.

[M] T. Miyata, Tohoku Math. Journ. 32 (1980), 49-62.

[T1] M. J. Taylor, Galois module structure of relative abelian extensions, J. reine angew. Math., 303/304 (1978) 97-101.

[T2] M. J. Taylor, On Fröhlich's conjecture for rings of integers of tame extensions, to appear.

[W] S. M. J. Wilson, Reduced norms in the K-Theory of orders, J. Algebra, (46), 1(1977), 1-11.

S. M. J. WILSON
Department of Mathematics
University of Durham
Science Laboratories
South Road
Durham
DH1 3LE

HECKE ACTIONS ON PICARD GROUPS AND CLASS GROUPS

Leonard Scott*
University of Virginia
Charlottesville, Virginia 22903

This article describes joint work with Klaus Roggenkamp, and is largely an abbreviated version of our paper [21] which will appear elsewhere. Essentially I have reproduced those portions of that paper corresponding to my talk at Oberwolfach, together with some remarks on class groups written after the conference.

Our story begins with a paper [9] of Robert Perlis, which gives an intriguing application of the representation theory of finite permutation groups to number theory: Let K be a finite Galois extension of $\mathbb{Q}$ with Galois group G; let H,H' be subgroups of G and F,F' their corresponding fixed fields. Perlis is able to associate with every $\mathbb{Z}G$-homomorphism $\phi:\mathbb{Z}G/H \to \mathbb{Z}G/H'$ (of the permutation modules obtained from the action of G on the coset spaces G/H,G/H') a corresponding homomorphism from the class group of F' to that of F. This procedure is sufficiently powerful to prove, for example, that if ϕ becomes an isomorphism upon localization at a prime p, then the class groups of F and F' have isomorphic Sylow p-subgroups.

Perlis' method was fairly involved, and we felt there might be a more transparent explanation for his construction, perhaps related to the formality that $\mathrm{Hom}_{\mathbb{Z}G}(\mathbb{Z}G/H,\mathbb{Z}G/H)$ acts as a ring on $M^H \cong \mathrm{Hom}_{\mathbb{Z}G}(\mathbb{Z}G/H,M)$ for any $\mathbb{Z}G$-module M. We have indeed found such an explanation, and the more formal proofs which are involved allow Perlis' method to be extended to arbitrary commutative rings (3.4) and in most cases to schemes (3.5), (3.6). Moreover, we make a very simple observation (2.3) which considerably broadens the usefulness of Perlis' method even in the number field case. Unknown to Perlis, Nehrkorn [18] and Walter [19] had used similar but somewhat cruder methods to prove reduction theorems for the computation of class groups. We show how to obtain their results from the Perlis method and go beyond them. For example, we show that any Sylow p-subgroup of the class group of an abelian extension K/$\mathbb{Q}$ is computable from the class groups of subextensions with Galois group a direct product of a cyclic group with a group of order a power

*The author would like to thank NSF and the Thyssen Stiftung Foundation for partial support during the writing of this paper.

of p. Nehrkorn had proved the same result for the case $p \nmid [K:\mathbb{Q}]$.

Our generalization beyond commutative rings to schemes is quite essential to our point of view, for it is only in the geometric analogues of the class group that the arguments become truly transparent. The first two sections of our paper [21], largely reproduced here, very quickly make it conceptually clear what is going on. It is still necessary to pass from "relative" to "absolute" Picard groups, which requires a number of technicalities from commutative algebra. We have omitted these proofs here, and have just stated in §3 some of the main results.

Class groups of orders are not treated in [21], though a possible development of the theory is indicated. We have included these remarks in §4.

We are indebted to A. Fröhlich, E. C. Paige, and D. Carter for supplying valuable references to the literature, and to D. Costa for an improvement in (3.5).

§1. The Hecke category

Let G be any abstract group. If H is a subgroup of G we let $\mathbb{Z}G/H$ denote the permutation module obtained from the action of G on the cosets gH with $g \in G$. Equivalently, $\mathbb{Z}G/H \cong \mathbb{Z}G \otimes_{\mathbb{Z}H} \mathbb{Z}$. Let $\mathcal{H}_G$ denote the category whose objects are the permutation modules $\mathbb{Z}G/H$ with H subgroup of G, and whose morphisms are $\mathbb{Z}G$-module homomorphisms. In the spirit of Yoshida [14], we call $\mathcal{H}_G$ the Hecke category associated with G (though Yoshida reserves this name for a special subcategory when G is infinite). The following property of $\mathcal{H}_G$ is basic.

(1.1) Proposition [14] Let M be a $\mathbb{Z}G$-module. Then there is a contravariant additive function

$$\Phi_M: \mathcal{H}_G \to \text{abelian groups}$$

with $\Phi_M(\mathbb{Z}G/H) = M^H$, the fixed points of H in M, for each subgroup H of G.

Proof We have $M^H \cong \mathrm{Hom}_H(\mathbb{Z},M) \cong \mathrm{Hom}_G(ZG \otimes_{\mathbb{Z}H} \mathbb{Z},M)$, and the result follows.

An important additional comment is that $(\mathbb{Z}G/H,M) \to M^H$ defines a bifunctor, an additive functor of two variables. Thus the assignment of $\mathbb{Z}G/H$ to $(\ )^H$ (and of M to Φ_M) is a functor to functors.

This gives in particular the following result on sheaves.

(1.2) Lemma Let M be a sheaf of $\mathbb{Z}G$-modules on a topological space S. For H a subgroup of G let M^H denote the sheaf with $M^H(U) = M(U)^H$ for U open in S. Then there is a contravariant additive functor

$$\Phi_M: H_G \to \text{abelian sheaves}$$

with $\Phi_M(\mathbb{Z}G/H) = M^H$.

The same result holds for presheaves, mutatis mutandis. In the sequel we will relax notation and just write Φ for Φ_M, Φ_M and similarly defined functors.

§2. Some relative Picard groups

Let $S = (S, \mathcal{O}_S)$ be a ringed space and let A be a commutative $\mathcal{O}_S$-algebra. That is, S is a topological space equipped with a sheaf $\mathcal{O}_S$ of commutative rings, and A is a sheaf of commutative rings over S equipped with a map $\mathcal{O}_S \to A$. The notion of an A-module M is the obvious one. Let $M|_U$ denote the sheaf induced by M on an open set U, and define S-Pic A to be the collection of isomorphism classes $[M]$ of A-modules M with $M|_U \cong A|_U$ as $A|_U$-modules for all U is some open cover of S. These classes $[M]$ form an abelian group under tensor product, and S-Pic $\mathcal{O}_S$ is the usual group Pic S. The group S-Pic A is similar to Fröhlich's "locally free" Picard group for orders [15,§5], in that localization is considered only on the base. Let A^* denote the sheaf of units of A. Then the standard argument [7,III, Exercise 4.5] which shows Pic S $\cong H^1(S, \mathcal{O}_S^*)$ shows as well that S-Pic $A \cong H^1(S, A^*)$.

(2.1) Proposition Let $S = (S, \mathcal{O}_S)$ be a ringed space, A a commutative $\mathcal{O}_S$-algebra, and G a group acting as $\mathcal{O}_S$-algebra automorphisms of A. Then there is a contravariant additive functor

$$\Phi: H_G \to \text{abelian groups}$$

with $\Phi(\mathbb{Z}G/H) = \text{S-Pic } A^H$ for each subgroup H of G.

Proof Since $(A^H)^* = (A^*)^H$, the result follows by composing $H^1(S,-)$ with the functor of (1.2) arising from the action of G on A^*.

For the Perlis situation we take S = Spec $\mathbb{Z}$ and A obtained from the ring of integers in a finite Galois extension $K/\mathbb{Q}$ by localization at the open sets of S. The group G is the Galois group $K/\mathbb{Q}$. It is not hard to see that S-Pic A^H identifies with the class group K^H; in any event this follows from section 3 of [21]. This gives the following corollary, which is implicit in Perlis [9] and summarizes his method. (For a very quick proof for abelian extensions, let J be the idèle group of K and U the subgroup of elements which are units at finite primes; then just note that the class group of K^H is the cokernel of $(U \times K^*)^H \to J^H$ and apply (1.1).)

(2.2) Corollary Let $K/\mathbb{Q}$ be a finite Galois extension with Galois group G. Then there is a contravariant additive functor

$$\Phi: \mathcal{H}_G \to \text{abelian groups}$$

with $\Phi(\mathbb{Z}G/H$ = class group of K^H for each subgroup H of G.

Observe that if $\mathcal{C}$ is any category with $\text{Hom}_{\mathcal{C}}(M,N)$ an abelian group for each M,N in $\mathcal{C}$, and if F is any additive functor from $\mathcal{C}$ to the category of abelian groups, there is a formal way to "tensor" both $\mathcal{C}$ and F with any commutative ring k: Define $k \otimes \mathcal{C}$ to be the category whose objects correspond bijectively to those in $\mathcal{C}$, and are formally labeled $k \otimes M$ with M ranging over $\mathcal{C}$. Define morphisms in $k \otimes \mathcal{C}$ by

$$\text{Hom}_{k\otimes\mathcal{C}}(k \otimes M,\ k\otimes N) \doteq k \otimes_{\mathbb{Z}} \text{Hom}_{\mathcal{C}}(M,N).$$

Thus $k \otimes \mathcal{C}$ is a category with $\text{Hom}_{k\otimes\mathcal{C}}(k\otimes M,\ k\otimes N)$ a k-module for each pair of objects $k \otimes M$, $k \otimes N$ in $k \otimes \mathcal{C}$. We have a k-linear functor

$$k \otimes F: k \otimes \mathcal{C} \to k\text{-modules}$$

with $(k\otimes F)(k\otimes M) = k \otimes_{\mathbb{Z}} F(M)$, and with $(k\otimes F)(\alpha \otimes_{\mathbb{Z}} f) = \alpha \otimes_{\mathbb{Z}} F(f)$ for any $\alpha \in k$ and $f \in \text{Hom}_{\mathcal{C}}(M,N)$.

Thus we always get a functor to abelian groups from the formally defined category $k \otimes \mathcal{C}$, and the issue now is identification of the latter in favorable circumstances. <u>When $\mathcal{C}$ is a category $\mathcal{H}_G$ as in §1, then $k \otimes \mathcal{C}$ identifies naturally with the category of kG-modules of the form $k \otimes_{\mathbb{Z}} M$ for M in $\mathcal{C}$</u> [21,4.2]. The reader can see this is true for G finite and k flat over $\mathbb{Z}$ (additively torsion free) from well-known properties of finitely presented modules; in the general case it follows from the existence of canonical bases for intertwining spaces of permutation modules.

The Perlis result regarding Sylow p-subgroups mentioned in the introduction is now a formal consequence of (2.2): Let p be a prime and let $\mathbb{Z}_p$ denote the complete ring of p-adic integers. From the functor Φ in (2.2) we can construct a functor $\mathbb{Z}_p \otimes \Phi$ as above on $\mathbb{Z}_p \otimes H_G$. Thus $\mathbb{Z}_pG/H \approx \mathbb{Z}_pG/H'$ implies $\mathbb{Z}_p \otimes_{\mathbb{Z}} \Phi(\mathbb{Z}G/H) \approx \mathbb{Z}_p \otimes_{\mathbb{Z}} \Phi(\mathbb{Z}G/H')$. Of course $\mathbb{Z}_p \otimes_{\mathbb{Z}} A$ is isomorphic to the Sylow p-subgroup of A when A is a finite abelian group; hence the class groups of K^H and $K^{H'}$ have isomorphic Sylow p-subgroups.

(2.3) <u>Remark</u> The utility of this method is enhanced considerably if the category H_G is replaced by the category $\hat{H}_G$ of all finite direct sums of objects in H_G. Any additive functor from H_G to an additive category A extends automatically to a functor on $\hat{H}_G$, and there are many more interesting p-local isomorphisms between objects in $\hat{H}_G$ than H_G.

For example, suppose the group G contains an abelian subgroup H of order q^2 and exponent q for some prime $q \neq p$. Let $H_1, \ldots, H_{q+1}$ denote the subgroups of H of order q. Then we have an isomorphism of $\mathbb{C}H$-modules

$$\mathbb{C}H \oplus \mathbb{C}^{(q)} \approx \mathbb{C}H/H_1 \oplus \ldots \oplus \mathbb{C}H/H_{q+1}$$

which is easily checked by decomposing both sides into linear factors. The notation $\mathbb{C}^{(q)}$ means a direct sum of q copies of the trivial module $\mathbb{C}$. Since $p \nmid |H|$, standard results in representation theory allow us to replace $\mathbb{C}$ in the displayed isomorphism with $\mathbb{Z}_p$. Now, tensoring over $\mathbb{Z}_pH$ both sides with $\mathbb{Z}_pG$, we obtain the following isomorphism in $\mathbb{Z}_pG \otimes \hat{H}_G$

$$\mathbb{Z}_pG/1 \oplus (\mathbb{Z}_pG/H)^{(q)} \approx \mathbb{Z}_pG/H_1 \oplus \ldots \oplus \mathbb{Z}_pG/H_{q+1}.$$

Consequently if $G = \mathrm{Gal}(K/\mathbb{Q})$ we have

$$(\mathrm{CL}\ K)_p \oplus (\mathrm{Cl}\ K^H)_p{}^{(q)} \approx (\mathrm{Cl}\ K^{H_1})_p \oplus \ldots \oplus (\mathrm{Cl}\ K^{H_{q+1}})_p$$

where the notation $(\mathrm{Cl}\ K)_p$ refers to the Sylow p-subgroup of the class group of K. If G is abelian, we can always find such a subgroup H unless G is the direct product of a p-group with a cyclic group.

(2.4) <u>Corollary</u> Let $K/\mathbb{Q}$ be a finite abelian extension with Galois group G, and let p be a fixed prime. Then the Sylow p-subgroup of the class group of K may be computed in terms of G and Sylow p-subgroups of class groups for subfields of K with Galois group the direct product of a p-group with a cyclic group.

As mentioned in the introduction, this result is due to Nehrkorn [18] in case $p \nmid |G|$. Our next corollary was first stated by Nehrkorn [18] and proved by Walter [19], but in both instances again with the assumption $p \nmid |G|$. (In which case $\mathbb{Z}_p$ may be replaced by $\mathbb{C}$ in the hypothesis.)

(2.5) <u>Corollary</u> Let $K/\mathbb{Q}$ be a finite Galois extension with Galois group G, and let p be a fixed prime. Let $H_1, \ldots, H_n$ be subgroups of G. Then any isomorphism

$$\bigoplus_{i=1}^{n} (\mathbb{Z}_p G/H_i)^{(a_i)} \approx \bigoplus_{i=1}^{n} (\mathbb{Z}_p G/H_i)^{(b_i)}$$

with nonnegative integers a_i, b_i gives an isomorphism

$$\bigoplus_{i=1}^{n} (\mathrm{Cl}\, K^{H_i})_p^{(a_i)} \approx \bigoplus_{i=1}^{n} (\mathrm{Cl}\, K^{H_i})_p^{(b_i)} .$$

This is just an explicit statement of the main content of (2.3), which we have already applied in the proof of (2.4). Here's another application based on the same principle and the induction theory of permutation modules developed by Conlon and Dress.

(2.6) <u>Corollary</u> Let $K/\mathbb{Q}$ be a finite Galois extension with Galois group G, and let p be a fixed prime. Then the Sylow p-subgroups of class groups of arbitrary subfields K^H with H a subgroup of G are computable from knowledge of G and the structure of these Sylow p-subgroups for all cases in which H is a cyclic extension of a p-subgroup.

This follows from (2.5) and the fact [4, Prop. 9.2] that, for some integer $n > 0$ and $\mathbb{Z}_p G$-modules M,N which are finite direct sums of modules of the form $\mathbb{Z}_p G/H$ with H a cyclic extension of a p-group, we have an isomorphism

$$\mathbb{Z}_p^{(n)} \oplus M \approx N$$

of $\mathbb{Z}_p G$-modules. (Tensoring with an arbitrary $\mathbb{Z}_p G/H$ gives the relation required in the hypothesis of (2.5). Alternately, one can just quote [4] for each $\mathbb{Z}_p G/H$.)

It is not always necessary to apply (2.2), (2.3) for isomorphisms. A split surjection, for example, can yield interesting information. Illustrations are given in (4.7.3) and (4.7.4) of [21].

Finally we mention that all the corollaries above have analogues in the theory of Picard groups of commutative rings because of (3.4) below, and similar remarks apply for schemes in most cases. For infinite Picard groups the reader may want to consider coefficients other than $\mathbb{Z}_p$; there is some further discussion of this in [21, 4.2].

§3. Absolute Picard groups

Results in this section have been given their original labels and numbering from §3 of [21]. We omit all proofs, but everything is ultimately derived from (2.1) by methods of commutative algebra.

If A is a commutative ring, we write Pic A for Pic(Spec A); this is the traditional group of isomorphism classes of invertible A-modules, rank 1 projective A-modules, etc. [1].

(3.4) Theorem Let G be a group of automorphisms of a commutative ring A. Then there is a contravariant additive functor

$$\Phi: \mathcal{H}_G \to \text{abelian groups}$$

with $\Phi(\mathbb{Z}G/H) = \text{Pic } A^H$ for each subgroup H of G.

The next two results are obtained in [21] as consequences of the proof given there for (3.4). For terminology regarding schemes we refer the reader to Hartshorne's book on algebraic geometry [7].

(3.5) Corollary Get $f:X \to S$ be an affine morphism of schemes, and G a group of automorphisms of X/S. Assume X is locally notherian or reduced, or that G is finite. Then there is a contravariant additive functor

$$\Phi: \mathcal{H}_G \to \text{abelian groups}$$

with $\Phi(\mathbb{Z}G/H = \text{Pic}(\text{Spec}(f_*O_X)^H)$.

If X is a scheme over a commutative ring k, and A is a commutative k-algebra, we abbreviate the pullback Spec A $\times_{\text{Spec } k}$ X by $A \otimes_k X$.

(3.6) Corollary Let k be an integral domain, A a domain integral over k, and G the group of all automorphisms of A fixing each element of k. Let X be any scheme flat over k. Then there is a contravariant additive functor

$$\Phi: \mathcal{H}_G \to \text{abelian groups}$$

with $\Phi(\mathbb{Z}G/H) = \text{Pic } A^H \otimes_k X$ for each subgroup H of G.

We view the hypotheses of (3.6) as describing a common "Galois theory" situation for schemes. The existence of Φ, implies among other things, that there are "norm", "restriction", and "conjugation" operators on the various Pic $A^H \otimes_k X$, with the same formal properties familiar from algebraic number theory; see [21,4.3] for details. Our methods also have application to the "Galois theory of commutative rings" in the sense of Samuel [10] and Chase, Harrison, Rosenberg [3]. This is discussed in [21,4.7].

§4. Concluding remarks

The paragraphs below are (4.1) and (4.8) of [21].

Noncommutative theory If $\mathcal{A}$ is a sheaf of noncommutative rings over a ringed space $S = (S,\mathcal{O}_S)$, then one can still define S-Pic $\mathcal{A}$ using bimodules locally isomorphic to $\mathcal{A}$ on S. The result of this is that S-Pic $\mathcal{A}$ $\cong$ S-Pic $\mathcal{Z}$ where $\mathcal{Z}$ assigns to an open U in S the center $Z(\mathcal{A}(U))$ of $\mathcal{A}(U)$. The identification is accomplished through $H^1(S,\mathcal{Z}^*)$. Thus we are led back to sheaves of commutative rings. There are a number of other ways to define a Picard group for bimodules, and we refer the interested reader to [15], [16], [17].

More interesting perhaps is to consider one-sided $\mathcal{A}$-modules locally isomorphic to $\mathcal{A}$ on S. We hope to make a detailed study of this question in a later paper, but for now we can indicate what we think is going on. First, there is still an identification of these isomorphism classes of one-sided modules with the elements of $H^1(S,\mathcal{A}^*)$, though the latter is just a set and the theory of section 1 does not apply. To get started, one has to approximate $H^1(S,\mathcal{A}^*)$ with an abelian group, and the natural choice is the image of $H^1(S,\mathcal{A}^*)$ in $H^1(S,K_1(\mathcal{A}))$. We expect this generalizes the locally free class group Cl Λ, cf. [17], for orders in semisimple algebras. There are still difficulties to overcome, largely due to the fact that K_1 does not in general commute with the fixed point functor of a group action. Nevertheless we expect this does occur often enough so that, if G is a Galois group acting on Λ through the coefficient domain, then there is a functor $\mathbb{Z}G/H \mapsto \text{TCl } \Lambda^H$ on $\mathcal{H}_G$ for a large factor TCl of the class group.

Other Variations on Pic A. It seems likely that the methods of this paper extend without difficulty to Cartier class groups and possibly other kinds of "fractional ideals". We have not, however, considered Azumaya algebras or the Brauer group of a scheme, and leave this project for an interested reader familiar with [6]. It might be possible in some cases to do something with K_0, using roughly the method discussed above for one-sided modules. This should become clearer after the issues already raised for class groups have been more thoroughly investigated.

The bibliography below is that of [21], with the addition of a reference to that paper itself.

Bibliography

1. H. Bass, Algebraic K-Theory, W. A. Benjamin, New York, 1968.

2. N. Bourbaki, Algèbre Commutative, Chpt. I and II, Hermann, Paris, 1961.

3. S. Chase, D. Harrison, A. Rosenberg, Galois theory and Galois cohomology of commutative rings, Memoirs A.M.S. 52 (1965), 15-33.

4. A. Dress, Contributions to the theory of integral representations, Algebraic K-theory, II, edited by H. Bass, Springer Lecture Notes 342 (1973), 183-240.

5. J. Green, Axiomatic representation theory, J. Pure and Applied Algebra 1 (1971), 41-77.

6. A. Grothendieck, Le groupe de Brauer I, II, III, pp. 46-188 in Dix exposés sur la cohomologie des schémas, edited by Grothendieck and Kuiper, North Holland, Amsterdam, 1968.

7. R. Hartshorne, Algebraic Geometry, GTM 52, Springer-Verlag, New York, 1977.

8. M. Kang, Picard groups of some rings of invariants, J. of Alg. 58 (1979), 455-461.

9. R. Perlis, On the class numbers of arithmetically equivalent fields, J. of Number Theory 10, Number 4 (1978), 488-509.

10. P. Samuel, Lectures on unique factorization domains, Tata Inst. Lecture Notes 30 (1964).

11. T. Suwa, The Lyndon-Hochschild-Serre spectral sequences for sheaves with operators, P.A.M.S. 77, Number 1 (1979), 32-34.

12. E. Weiss, Algebraic number theory, McGraw-Hill, New York, 1963.

13. H. Weyl, Algebraic theory of numbers, Princeton University Press, Princeton, 1940.

14. T. Yoshida, On G-Functors (II): Hecke operators, preprint (Dept. Math. Hokkaido University, Sappora 060, Japan).

15. A. Fröhlich, The Picard group of noncommutative rings, in particular of orders, Trans. A.M.S. 180 (1973), 1-45.

16. A. Fröhlich, I. Reiner and S. Ulom, Picard groups and class groups of orders, Proc. L.M.S. (3) 29 (1974), 405-434.

17. I. Reiner, Class groups and Picard groups of group rings and orders, CBMS regional conference lecture notes 26, A.M.S., 1976.

18. H. Nehrkorn, Über absolute Idealklassengruppen und Einheiten in algebraischen Zahlkorpern, Abh. Math. Sem. Hamburg 1933, 318-334.

19. C. Walter, Brauer's class number relation, Acta Arith. 35 (1979), 33-40.

20. H. Cartan and S. Eilenberg, Homological algebra, Princeton Univ. Press, 1956.

21. K. Roggenkamp and L. Scott, Hecke actions on Picard groups, to appear.

USES OF UNITS IN WHITEHEAD GROUPS

Bruce A. Magurn

1. The conjecture

If R is an associative ring with unit, the maps $GL_n(R) \to GL_{n+r}(R)$ defined by sending A to $A \oplus I_r$ form a directed system whose direct limit is known as $GL(R)$. The composite $GL_n(R) \to GL(R) \to K_1(R) = (GL(R))^{\text{abel.}}$ is known as the stabilization map, which we denote by "stab".

For a finite group G, consider the relationship in $K_1(ZG)$ between the cosets in $GL(ZG)$ represented by 1×1 matrices and the kernel $SK_1(ZG)$ of the reduced norm. I conjecture that there is a direct product decomposition: $K_1(ZG) = SK_1(ZG) \times \text{stab}(ZG^*)$. The strongest evidence for this conjecture comes from the case in which G is abelian, where there is an exact sequence:

$$1 \to SK_1(ZG) \to K_1(ZG) \xrightarrow{\det} ZG^* \to 1$$

in which the determinant is split by stabilization. The conjecture has also been verified in the dihedral case as shown below, and the same technique applies to some more general metacyclic groups (see [6]).

Half of the conjecture is a version of G. Higman's Theorem [5, Theorem 3, p. 237], for since the torsion part of $K_1(ZG)$ consists of $SK_1(ZG) \times \text{stab}(\pm G)$, according to [13, Proposition 6.5, p. 611], to say that $SK_1(ZG) \cap \text{stab}(ZG^*) = 1$ is to say that units of ZG which are torsion modulo elementary matrices are trivial units modulo elementary matrices.

In this talk we consider the other half: that $K_1(ZG)/SK_1(ZG)$ is represented by units. This would be a matter of convenience for topological computations. For this purpose it would be helpful to be able to choose representing units of a simple form.

2. Simple homotopy types

For one topological application, let X be a finite CW-complex with fundamental group $\pi_1 X = G$, let $E(X)$ be the group of homotopy classes of homotopy self-equivalences of X, and let the Whitehead group $Wh(G)$ be the quotient group $K_1(ZG)/stab(\pm G)$. The Whitehead torsion of a homotopy equivalence $Y \to X$ is defined to take its values in $Wh(\pi_1 X)$; see [2, §22, p. 72]. According to W. Cockroft and R. Moss [2, §24.4, p. 80] Whitehead torsion $\tau: E(X) \to Wh(G)$ is surjective exactly when the homotopy type and simple homotopy type of X coincide. The connection to stability comes from M. N. Dyer:

Theorem [4, Theorem 2] If stab: $GL_n(ZG) \to K_1(ZG)$ is surjective, then for every $r \geq 2$, $\tau: E(X \vee nS^r) \to Wh(G)$ is surjective.

Dyer's argument is that attaching a wedge of n r-spheres alters the cellular chain complex of the universal cover of X by adding $(ZG)^n$ to the group in dimension n ; then each element M of $GL_n(ZG)$ acts on $(ZG)^n$ to define a chain self-equivalence which is geometrically realizable by an element of $E(X)$ with torsion represented by $M^{(-1)^r}$. □

When G is finite, the hypothesis of this theorem holds for $n \geq 2$ (see [12], or [1, Chapter V, §4]). When $SK_1(ZG)$ vanishes, the conjecture is that it holds for $n = 1$.

3. Dimension of CW-pairs

For a second application, for τ_0 in $Wh(G)$, M. M. Cohen defines (in [3]):

$$\dim(\tau_0) = \min \{ \dim(Y - X) : \tau(Y,X) = \tau_0 \},$$

where (Y,X) ranges over finite CW-pairs for which X is a strong deformation retract of Y, and $\tau(Y,X)$ is the Whitehead torsion of the pair (Y,X) (see [2, §19, p. 62]). Cohen points out that $\dim(\tau_0) = 0$, 2 or 3 and that 0 occurs only for $\tau_0 = 1$. As yet no elements τ_0 have been found with dimension 2.

The most successful approach to computing $\dim(\tau_0)$ has been that of O. S. Rothaus (in [10]). When G is a finite group, the criterion used by Rothaus is as follows. Let g be the composite map:

$$K_1(ZG) \to \bigoplus_{\rho \in R} K_1(\mathbb{R}) \to \bigoplus_{\rho \in R} \mathbb{R}^* \to \bigoplus_{\rho \in R} \{\pm 1\}$$

where R is a full set of inequivalent irreducible real matrix representations of G, the first map applies ρ to the entries of each representative matrix, the second is the determinant, and the third takes the sign of each nonzero real number. Although trivial units are not killed by g, there is an induced map:

$$\overline{g} : Wh(G) \to \bigoplus_{\rho \in R} \{\pm 1\}/g(\mathrm{stab}(\pm G)) .$$

Rothaus proved that the group $Wh^+(G) = \mathrm{kernel}(\overline{g})$ of "positive elements" contains the subset $Wh^*(G)$ of elements with dimension ≤ 2. (It is known that $Wh^*(G)$ is a commutative semigroup with identity, but it is unknown whether or not it is closed under inverses.) So $\dim(\tau_0) = 3$ if τ_0 survives in the quotient $Ro(G) = Wh(G)/Wh^+(G) \cong \mathrm{image}(\overline{g})$, which is an elementary abelian 2-group. Inspecting the definition of the reduced norm ν, we find that $SK_1(ZG)$ is killed even by the first two parts of g. This has two immediate consequences: Since $SK_1(ZG)$ maps isomorphically to the torsion part of $Wh(G)$, $\mathrm{rank}_{F_2} Ro(G) \leq \mathrm{rank}_Z Wh(G)$. And the conjecture implies the map $ZG^* \to Ro(G)$ is surjective.

4. Induction results

According to T. Y. Lam [7, Chapter 4], $K_1(ZG)$ and $Wh(G)$ are modules over the Frobenius functor $G_0(ZG)$; so they are generated by induction from hyperelementary subgroups of G . The induction maps $K_1(ZH) \to K_1(ZG)$ and $Wh(H) \to Wh(G)$, obtained from an inclusion of groups $H \to G$, send $SK_1(ZH)$, $stab(ZH^*)$ and $Wh^+(H)$ into the corresponding groups for G . So both $K_1(ZG)/stab(ZG^*)\cdot SK_1(ZG)$ and $Ro(G)$ are generated by hyperelementary subgroups of G . In addition, both are trivial for abelian groups G (for $Ro(G)$ see [10, Theorem 9, p. 608]). So if $A(G)$ is the Artin exponent of G , the Artin induction theorem implies that $A(G)^2$ is an exponent of $K_1(ZG)/stab(ZG^*)\cdot SK_1(ZG)$, and that $Ro(G)$ is generated by 2-hyperelementary subgroups of G . We cannot do much better, since the dihedral group D_5 of order 10 is a defect group for both $K_1(ZG)$ and $Ro(G)$ (see [10, Theorem 17, p. 610] and [8, Example 8]).

5. Units in K_1 of dihedral groups

The simplest untested case is therefore the dihedral group D_m of order $2m$, with presentation $(a , b : a^m , b^2 , b^{-1}ab = a^{-1})$. Regard the groups $A_m = (a : a^m)$ and $B = (b : b^2)$ as subgroups of D_m. Since D_m is nearly abelian the reduced norm on $M_n(QD_m)$ is nearly the determinant. In particular, left regular representation of QD_m as a right QA-module with basis $\{1 , b\}$ defines the transfer representation $trf: QD_m \to M_2(QA_m)$. Applying trf to entries and taking the determinant, we obtain the vertical maps in the following commutative diagram:

$$\begin{array}{ccccc}
M_n(QD_m) & \to & \bigoplus_{\substack{d|m \\ d>2}} M_n(Q(\zeta_d)\circ B) & \oplus & M_n(QD_{2 \text{ or } 1}) \\
\downarrow{\scriptstyle \det\,\mathrm{trf}} & & \downarrow{\scriptstyle \nu} & \oplus & \downarrow{\scriptstyle \mathrm{norm}\ \nu} \\
(QA)^B & \to & \bigoplus Q(\zeta_d)^B & \oplus & QA_{2 \text{ or } 1}
\end{array}$$

where ν is the reduced norm, which is the determinant on commutative components,

$(-)^B$ denotes the elements fixed under the conjugation action of B, and where the horizontal lines are defined by substituting ζ_d for a to get $Q(\zeta_d)\circ B$, and a (of $A_{2 \text{ or } 1}$) for a (of A_m) to get $D_{2 \text{ or } 1} = D_m$-abelianized. The diagram commutes because the map $\det \circ \text{trf}: M_n(Q(\zeta_d)\circ B) \to Q(\zeta_d)^B$ is the same as ν by a computation of C. T. C. Wall [13, Lemma 3.3, p. 601], and the same map from $M_n(QD_{2 \text{ or } 1})$ to $QA_{2 \text{ or } 1}$ is the determinant ($= \nu$) followed by norm $= \det \circ \text{trf}$ from $QD_{2 \text{ or } 1}$ to $QA_{2 \text{ or } 1}$.

Restricting to integral coefficients, $\det \circ \text{trf} =$ norm (on abelian components) maps $M_n(ZD_m)$ into $(ZA_m)^B$. As with an ordinary determinant, the inverse image of a unit consists of invertible matrices. An easy computation shows that $\det \circ \text{trf}$ takes $GL_2(ZD_m)$ into $\{1 + x + \overline{x} : x \in ZA_m\}^*$ where the bar refers to the involution defined by inverting elements of A_m. But $\det \circ \text{trf}\,(1 + x - bx) = 1 + x + \overline{x}$, and $1 + x - bx$ is a unit if $1 + x + \overline{x}$ is. Finally it turns out that $SK_1(ZD_m) = 1$, and the kernel of $\det \circ \text{trf}$ on $K_1(ZD_m)$ is $\text{stab}(\pm A_m)$. So altogether this proves:

<u>Theorem</u> If $U = (1 + (1 - b)ZA_m)^*$ then the maps $U\cdot(\pm A_m) \to K_1(ZD_m)$ and $U \to Wh(D_m)$ defined by stabilization are surjective. □

(For more details, see [6, Theorem 10].)

6. Consequences for simple homotopy types

The first consequence, immediate from Dyer's theorem, is that if X is a finite connected CW-complex with fundamental group $\pi_1(X) = D_m$, then the simple homotopy type and homotopy type of $X \vee S^r$ coincide for any $r \geq 2$. Using the form of the units in U, we have obtained a more delicate result: If X and Y are 2-dimensional finite connected CW-complexes with fundamental group D_m (m odd), then X and Y have the same simple homotopy type if and only if they have the same homotopy type, and that occurs exactly when they have the same Euler characteristic (see [6]).

7. Consequences for $\dim(\tau_0)$

It is immediate from the theorem that stabilization $U \to Ro(D_m)$ is surjective. From the particular form of units in U, it follows (see [8]) that the kernel of $U \to Ro(D_m)$ equals the kernel of $h: U \to \bigoplus_{2 \dim \rho} \{\pm 1\}$ defined to be $g \circ \mathrm{stab} \mid U$ followed by dropping off the components for 1-dimensional representations ρ. Thus there is an injective homomorphism λ completing the commutative diagram:

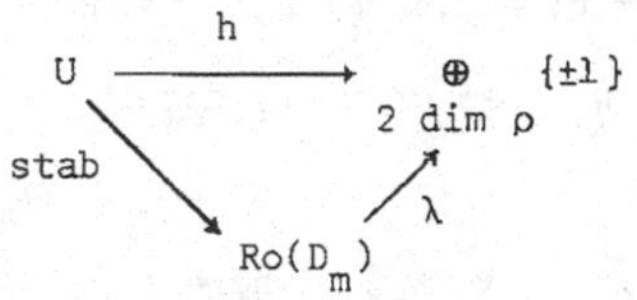

So the image across the top is isomorphic to $Ro(D_m)$.

Studying the 2-dimensional representations ρ, we find that $h(1 + (1 - b)p(a))$ has n^{th} coordinate $\mathrm{sgn}(1 + p(\zeta^n) + p(\zeta^{-n}))$ for integers n with $0 < |n| < m/2$, where $\zeta = \zeta_m$. If $(s,m) = 1$, then $a \to a^s$, $b \to b$ is an automorphism of ZD_m, and $h(1 + (1 - b)p(a^s))$ has n^{th} coordinate $\mathrm{sgn}(1 + p(\zeta^{sn}) + p(\zeta^{-sn}))$. Thus if $1 + (1 - b)p(a) \in ZD_m^*$ and we apply the function $\mathrm{sgn}(1 + p(\zeta^x) + p(\zeta^{-x}))$ to a multiplication table of $(Z/mZ)^*$, we obtain a matrix in which the number of distinct rows is a lower bound for $|Ro(D_m)|$.

For example, when m is not divisible by 2 or 3, then $-1 + a + a^{-1} = \det \circ \mathrm{trf}(1 + (1 - b)(a - 1))$ is a unit in ZA^B. So we may take $p(x) = x - 1$. Then $\mathrm{sgn}(1 + p(\zeta^x) + p(\zeta^{-x})) = \mathrm{sgn}(-1 + \zeta^x + \zeta^{-x})$, which is determined by the position of ζ^x on the unit circle, as indicated in the diagram:

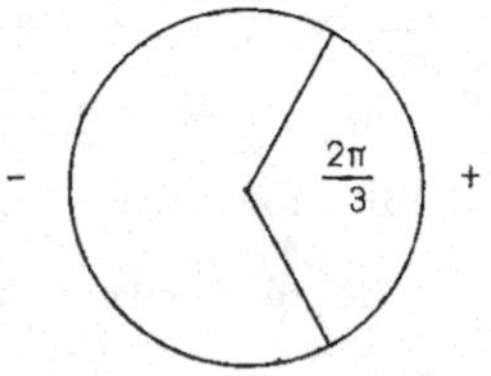

Such a search for elements of U provides enough to show that $|Ro(D_m)| \geq \varphi(m)/4$. When m is a prime > 3, this is the computation used by Rothaus (in [10]) to find the first elements of dimension 3 in a Whitehead group.

8. Postscript

The dihedral group is too nearly abelian to be a very serious test of the conjecture. Perhaps a better challenge can be mounted using generalized quaternion groups $Q_m = (a , b : a^m = b^2 , b^4 , b^{-1}ab = a^{-1})$, especially since R. Oliver has recently shown (in [9]) that $SK_1ZQ_m = 1$. The failure of K_0 stability at Q_8 as shown by R. G. Swan (in [11]) suggests that Q_8 might provide a counter-example to the proposed stability of units in K_1ZQ_8 . However there appears to be no formal connection, since cancellation of free summands is equivalent to $GL_2(ZG)$ acting transitively on unimodular elements of $ZG \times ZG$; but surjective stability of units modulo $SK_1(ZG)$ is equivalent to $SL_2(ZG)$ acting transitively on rows of matrices in $GL_2(ZG)$.

REFERENCES

1 H. Bass, Algebraic K-Theory, Benjamin, New York, 1968.

2 M. M. Cohen, A Course in Simple Homotopy Theory, Springer-Verlag GTM 10, 1973.

3 __________, Whitehead torsion, group extensions, and Zeeman's conjecture in high dimensions, Topology 16, 1977, 79-88.

4 M. N. Dyer, Simple homotopy types for (G,m) complexes, to appear.

5 G. Higman, The units of group rings, Proc. London Math. Soc. 2 , 46 , 1940, 231-248.

6 S. Jajodia and B. A. Magurn, Surjective stability of units and simple homotopy type, Jour. Pure and Appl. Algebra 18, 1980, 45-58.

7 T. Y. Lam, Induction theorems for Grothendieck groups and Whitehead groups of finite groups, Ann. sci. Ecole norm. sup., IV. Ser. 1, 1968, 91-148.

8 B. A. Magurn, Nonpositive elements in Whitehead groups, to appear in J. Pure and Appl. Algebra.

9 R. Oliver, SK_1 for finite group rings: II, to appear in Inventiones math.

10 O. S. Rothaus, On the non-triviality of some group extensions given by generators and relations, Ann. of Math. 106, 1977, 599-612.

11 R. G. Swan, Projective modules over group rings and maximal orders, Ann. of Math. 76, No. 1, 1962, 55-61.

12 L. N. Vaserstein, On the stabilization of the general linear group over a ring, Math. USSR, Sbornik 8, 1969, 383-400.

13 C. T. C. Wall, Norms of units in group rings, Proc. London Math. Soc. 3, 29, 1974, 593-632.

University of Oklahoma
Norman, Oklahoma 73019

Integral Representations in the theory of finite CW-complexes

C.B. Thomas

In this survey we propose to sketch some of the connections between the theory of finitely generated projective $\mathbb{Z}\Gamma$-modules and their automorphisms on the one hand, and the theory of finitely dominated CW-complexes on the other. The main algebraic object of study is the Swan subgroup $T(\mathbb{Z}\Gamma)$ of the projective class group, the precise structure of which (when known) yields information about CW-complexes with fundamental group Γ. Conversely, if Γ has periodic cohomology, by using classical representation theory (over $\mathbb{R}$ or $\mathbb{C}$) to construct elliptic manifolds (with constant positive curvature), we obtain partial information about $T(\mathbb{Z}\Gamma)$. Perhaps the most interesting question is the extent to which the theory over $\mathbb{Z}$ is approximated by the theory over $\mathbb{R}$. In topological terms this is reflected in the relation between the set of homotopy classes of elliptic manifolds, and the set of homotopy classes of CW-complexes covered by homotopy spheres.

Detailed proofs for most of the results mentioned here can be found in the references. The exceptions are the gloss on the work of G.Mislin and K.Varadarajan [10], Theorem 5 & Complement, and the result on the geometric realisability of $\mathbb{Z}D_{2p}$-complexes in the last section, Theorem 9. Also through lack of space we make no mention of Hermitian forms and surgery theory.

1. Summary of results from algebra.

Let Γ be a finite group of order n and A some coefficient ring, usually $\mathbb{Z}$ or $\mathbb{Q}$. Let $\mathfrak{M}_\Gamma$ be some maximal order in the rational group algebra $\mathbb{Q}\Gamma$,which contains the integral group ring $\mathbb{Z}\Gamma$. As usual we write

$K_0(A\Gamma)$ = Grothendieck group associated to the semigroup of finitely generated $A\Gamma$ -projective modules.

$\tilde{K}_0(A\Gamma) = K_0(A\Gamma)$ (modulo free modules), the projective class group. The modules M and N map to the same class in $\tilde{K}_0$ if and only if there exist finitely generated free modules F_1 and F_2 such that $N + F_1 \cong M + F_2$.

$K_1(A\Gamma) = (\lim_r GL(r, A\Gamma))_{ab}$, the abelianised general linear group.

We shall be interested in the following subgroups:

$D(\mathbb{Z}\Gamma) = \mathrm{Ker}(\tilde{K}_0(\mathbb{Z}\Gamma) \twoheadrightarrow \tilde{K}_0(\mathfrak{M}_\Gamma))$,

$T(\mathbb{Z}\Gamma) = \{(r,\Sigma) : r \in \mathbb{Z},\ (r,n) = 1,\ \Sigma = \gamma_1 + \gamma_2 + \cdots \gamma_n\}$.

LEMMA 1. (a) <u>If</u> $r \equiv r' \pmod n$ <u>then</u> $(r,\Sigma) \cong (r',\Sigma)$,

(b) $(r,\Sigma) + (r',\Sigma) \cong \mathbb{Z}\Gamma + (rr',\Sigma)$, and

(c) <u>there is an isomorphism</u> φ , <u>such that the composition</u>

$$\mathbb{Z} \rightarrowtail \mathbb{Z} + (r,\Sigma) \underset{\varphi}{\cong} \mathbb{Z} + \mathbb{Z}\Gamma \twoheadrightarrow \mathbb{Z}$$

<u>is multiplication by</u> r, see [14] .

Note that (b) defines the group structure on the subset $\{(r,\Sigma)\}$, and that $T(\mathbb{Z}\Gamma) \subseteq D(\mathbb{Z}\Gamma) \subseteq \tilde{K}_0(\mathbb{Z}\Gamma)$. It is convenient in (c) to say that φ is a map of degree r.

Turning to K_1, we set $SK_1(\mathbb{Z}\Gamma)$ equal to the kernel of the natural map $K_1(\mathbb{Z}\Gamma) \to K_1(\mathbb{Q}\Gamma)$, and $K_1'(\mathbb{Z}\Gamma)$ equal to the image.

LEMMA 2. <u>There is an exact sequence</u>

$$0 \to K_1'(\mathbb{Z}\Gamma) \longrightarrow \{\pm 1\} + K_1'(\mathbb{Z}\Gamma/\langle\Sigma\rangle) \longrightarrow (\mathbb{Z}/n)^{\times} \xrightarrow[\partial]{} \tilde{K}_0(\mathbb{Z}\Gamma)$$, see [16].

Note that the subgroup $T(\mathbb{Z}\Gamma)$ coincides with the image of the map ∂.

The functors K_0 and K_1 both satisfy Frobenius reciprocity with respect to the integral representation ring, defined by finitely generated $\mathbb{Z}\Gamma$-modules, which are torsion free as abelian groups. Recall that we have the Artin induction theorem:

If χ is a rational character of Γ, then there exist integers $\{a_1 \ldots a_q\}$ such that for some $m \in Z^+$, $m\chi = \sum_{k=1}^{q} a_k \mu_k$, where μ_k is induced up from the trivial character of π_k, as π_k runs through a representative family of cyclic subgroups of Γ, one from each conjugacy class. (For a proof, see, for example [13].)

This theorem motivates the following definition (due to T.Y. Lam, [6]):

The Artin Exponent $A(\Gamma)$ is the least value of m in the equation above, such that a solution $\{a_k\}$ in integers exists.

For an arbitrary finite group Γ the calculation of $A(\Gamma)$ proceeds in two stages: first reduce to semidirect products of the form $(Z/r) \tilde{\times}$ (p-group) with $(p,r) = 1$, and then to p-groups. A typical general result is the following:

If the Sylow subgroup Γ_p is cyclic of order p^t, then $A(\Gamma)_p = p^s$, where $s = \min(\Delta$ = p-free cyclic subgroup of Γ & $x \in \Gamma_p \cap N(\Delta)$

$$\Rightarrow x^{p^s} \in C(\Delta)).$$

Examples. (i) Let p and q be dictinct primes, such that q divides $(p-1)$, and let $D^*_{pq^2}$ be the central extension of $\mathbb{Z}/q$ by the non-abelian group of order D_{pq} of order pq. Then $A(D_{pq}) = A(D^*_{pq^2}) = q$. In order to fix the notation we shall use the presentation

$$D^*_{pq^2} = \{A, B : A^p = B^{q^2} = 1, \quad BAB^{-1} = A^s, \quad s^q \equiv 1 (\mathrm{mod}\ p)\}. \qquad (*)$$

D_{pq} then has generators $A, \bar{B}$, where $\bar{B}$ is the image of B under the projection map. Note that if $q = 2$, D^*_{4p} is a quaternion group.

(ii) Write $Q(8p; q, 1)$ for the non-trivial extension of $\mathbb{Z}/q$ by a quaternion group of order 8p, where p and q are distinct odd primes. Then $A(Q(8p; q, 1))$ equals 4.

LEMMA 3. <u>If</u> $x \in \tilde{K}_0(\mathbb{Z}\Gamma)$ <u>restricts to</u> 0 <u>for each elementary subgroup of</u> Γ, <u>and</u> φ <u>is the Euler function, then</u> $(\varphi(n), A(\Gamma))x = 0$.

Proof. This depends on Frobenius reciprocity and is essentially contained in [13].

2. Some calculations for T and for SK_1.

(i) If Γ is cyclic, then (r, Σ) is free. In the exact sequence of Lemma 2 one shows that $\mathrm{Im}(\partial) = 0$, because $\mathbb{Z}(\zeta)^\times = K_1'(Z\Gamma/\langle\Sigma\rangle)$ maps onto $(\mathbb{Z}/n)^\times$.

(ii) Let p and q be distinct primes, $q \mid (p-1)$. Then $T(\mathbb{Z}D_{pq})$ is cyclic of order q (q = odd), and $T(D_{2p})$ is trivial. Furthermore the natural projection map induces a monomorphism $T(ZD^*_{pq^2}) \rightarrowtail T(\mathbb{Z}D_{pq})$.

Proof. The first part is contained in [4], modified by Wall in [16], and depends on a Mayer-Vietoris sequence associated to the square

$$\begin{array}{ccc} \mathbb{Z}D_{pq} & \longrightarrow & \mathbb{Z}(\zeta)C_q^{\sim} = \Lambda \\ \downarrow & & \downarrow \\ \mathbb{Z}C_q & \longrightarrow & \mathbb{F}_p C_q \end{array},$$

where C_q is cyclic of order q, $\zeta = \exp(2\pi i/p)$ and $(^{\sim})$ denotes twisting in the group ring. The second part is due to Martin Taylor (unpublished). However in the case $q = 2$, it is easy to give a proof using the facts (a) that (r,Σ) has order at most two (see section four below) and (b) that the natural homomorphism $\tilde{K}_0(\mathbb{Z}D^*_{4p}) \longrightarrow \tilde{K}_0(\mathbb{Z}(\zeta,j)) + \tilde{K}_0(ZD_{2p})$ is a monomorphism on 2-torsion, see also [3] .

(iii) $SK_1(ZD_{pq}) = SK_1(ZD^*_{pq^2}) = 0$.

Proof. This is a special case of [11, Theorem 2] . But as Oliver himself points out, for a metacyclic group with periodic cohomology the triviality of SK_1 follows from Mayer-Vietoris arguments of the type used in (ii) together with

LEMMA 4. <u>If $\mathfrak{A} \subseteq \mathbb{Z}\Gamma$ is an hereditary order in $\mathbb{Q}\Gamma$, then</u>

$$SK_1(\mathbb{Z}\Gamma) = \mathrm{Ker}\left[K_1(\mathbb{Z}\Gamma) \longrightarrow K_1(\mathfrak{A})\right].$$

Proof. [op.cit.]

3. Finiteness conditions for CW-complexes.

Recall that (roughly speaking) a CW-complex is a topological space K obtained by attaching cells $\bar{e}^n = \left\{ \underline{x} \in \mathbb{R}^n : \|\underline{x}\| \leq 1 \right\}$ to K^{n-1}

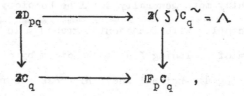

by means of maps $\varphi : S^{n-1} \to K^{n-1}$, and assigning the weak topology to the whole. Slightly more generally let X be homotopy equivalent to a connected CW-complex with fundamental group Γ. In [15] Wall studied the problem of replacing X up to homotopy by a simpler CW-complex, reflecting dimensional and finiteness restrictions on X. Since we assume throughout that Γ is finite, we shall state the main result under a simplified version of Wall's hypotheses. Thus:

since Γ is finite, the ring $\mathbb{Z}\Gamma$ is Noetherian [5]. Write $\tilde{X}$ for the universal covering space of X, on which Γ acts as a group of covering transformations.

Hypothesis NF(n) : $H_i(\tilde{X})$ is a finitely generated $\mathbb{Z}\Gamma$-module for all $2 \leq i \leq n$.

Hypothesis D(n) : $H_i(\tilde{X}) = 0$ for $i > n$ and $H^{n+1}(X; \mathcal{B}) = 0$ for all coefficient bundles $\mathcal{B}$. (Again assume $n \geq 2$ to avoid non-abelian complications with $\mathcal{B}$.)

THEOREM 4. If $n \geq 3$ and X satisfies NF(n) & D(n), then X is dominated by a finite CW-complex. X is homotopy equivalent to a finite CW-complex K^n if and only if a certain obstruction $\sigma(X) \in \tilde{K}_0(\mathbb{Z}\Gamma)$ vanishes. The class $\sigma(X)$ depends only on the homotopy type of X.

The idea of the proof is to construct K inductively, dimension by dimension; at each stage r ($r \leq n$) the hypothesis NF(n) implies the existence of an (r-1)-connected map $\varphi : K^{r-1} \to X$ such that the relative homotopy group $\pi_r \varphi$ is a finitely generated $\mathbb{Z}\Gamma$-module. Let $\{\alpha_1 \ldots \alpha_k\}$ be a family of $\mathbb{Z}\Gamma$-generators. Now an element of $\pi_r \varphi$ is a homotopy class of commutative diagrams

$$\begin{array}{ccc} S^{r-1} & \hookrightarrow & \bar{e}^{r} \\ \beta \downarrow & & \downarrow \gamma \\ K^{r-1} & \longrightarrow & X , \end{array}$$

and a finite family of pairs (β_j, γ_j), $1 \leq j \leq k$, can be used to both add r-cells to K^{r-1} and extend the map φ. Under hypothesis D(n) this construction stops at stage n with $\pi_n \varphi$ a finitely generated projective $\mathbb{Z}\Gamma$-module, and the obstruction $\sigma(X)$ measures the departure from freeness of the class of $\pi_n \varphi$ in $K_0(\mathbb{Z}\Gamma)$. "Domination" in this context means that we can always construct a homotopy equivalence $K^n \longrightarrow X$ at the price of allowing countably many n-cells in the complex K. (If Q is a projective inverse to $P = \pi_n \varphi$, then $P + (Q + P) + (Q + P) + \ldots$ is free on countably many generators.)

It is easy to see that in general every element in $\tilde{K}_0(\mathbb{Z}\Gamma)$ occurs as the finiteness obstruction for some complex, so the problem is to impose further topological restrictions on X in the hope of reducing the range of values taken by $\sigma(X)$. The general philosophy is, that for some suitable category of spaces, the finiteness obstructtion σ should lie in the subgroup $D(\mathbb{Z}\Gamma)$ and that varying the homotopy type should vary σ by an element of $T(\mathbb{Z}\Gamma)$. Let I be the kernel of the augmentation map of $\mathbb{Z}\Gamma$.

Definition. The $\mathbb{Z}\Gamma$-module M is <u>nilpotent</u> if $I^k M = 0$ for some $k > 0$.

THEOREM 5. If $n \geq 3$, X satisfies NF(n) & D(n) and $H_i \tilde{X}$ is nilpotent for all $i \leq n$, then $(\varphi|\Gamma|, A(\Gamma))\sigma(X) \in D(\mathbb{Z}\Gamma)$.
(As in Lemma 3 $\varphi(\ )$ denotes Euler's function.)

Proof. This is a slight sharpening of the main result in [10] .

The result is stated there for nilpotent fundamental groups, and is proved by decomposing Γ as $\prod_p \Gamma_p$, and then considering the decomposition of $\mathfrak{M}_{\Gamma_p}$ corresponding to the decomposition of $Q\Gamma_p$ into simple factors, for each prime separately. However, since an elementary subgroup is nilpotent, Lemma 3 enables us to extend the theorem to arbitrary finite groups, at least up to the multiple shown. For certain solvable finite groups one can do somewhat better.

COMPLEMENT. <u>If the fundamental group Γ of the homologically nilpotent space X in Theorem 5 is isomorphic to D_{2p}, then</u> $\sigma(X) \in D(\mathbf{Z}\Gamma)$.

Proof. Use the decomposition of $\mathbf{Z}$, $\mathfrak{M}_\Gamma$ or $Q\Gamma$ given by the pullback diagram on page 5. If $\alpha = \exp(2\pi i/p)$, then $\mathfrak{M}_\Gamma$ splits as $2Z + M_2(Z(\alpha+\bar\alpha))$. The third summand contains the matrix representation of the twisted group ring Λ . Specialising the exact sequence in [4] we have that $\sigma(X)$ is detected by its image in $\tilde{K}_0(\Lambda)$. By Morita equivalence we can work over $\mathbf{Z}(\alpha+\bar\alpha)$, which topologically amounts to lifting $\sigma(X)$ to the regular covering complex with fundamental group $\mathbf{Z}^A/p$. (Note that the homology, etc., is invariant under the action of $\bar{\mathbf{B}}$.) The result now follows as in [9, Lemma 3.4] .

Clearly we could obtain further results of this kind. In particular one has the result of J. Milgram [8] , which is very similar in spirit. If Γ is isomorphic to the solvable group $Q(8p;q,1)$, then at least for certain simple spaces X ($k = 1$ in the nilpotency hypothesis), $\sigma(X)$ rather than $4\sigma(X)$ belongs to $D(\mathbf{Z}\Gamma)$.

4. Periodic projective resolutions.

In this section we illustrate the problem of varying the homotopy type of a finite CW-complex. Let Γ be a group with periodic cohomology, that is, the odd order Sylow subgroups Γ_p are cyclic and Γ_2 is either cyclic or generalised quaternion. It is not too hard to see that the Artin exponent of Γ either equals the cohomological period (in the case of $Q(8p;q,1)$ for example) or half the cohomological period (D_{pq} or D^*_{pq}). If $m = 2A(\Gamma)$, the cohomology of the discrete group Γ can be calculated from a finitely generated, free, periodic resolution of the form

$$0 \to \mathbb{Z} \to C_{m-1} \to \cdots \quad C_1 \to C_0 \to \mathbb{Z} \to 0,$$ see [14] & [7].

Since $\operatorname{Ext}^m_{\mathbb{Z}\Gamma}(\mathbb{Z},\mathbb{Z}) \cong H^m(\Gamma,\mathbb{Z}) = \mathbb{Z}g_0/|\Gamma|$, the extension is determined by a generator $g = rg_0$, where g_0 is the class of some fixed reference complex. Assuming the existence of a topological realisation X of this resolution, the generator g coincides with the first k-invariant of a Postnikov decomposition for X, and along with Γ determines its homotopy type.

PROPOSITION 6. <u>The homotopy type rg_0 is represented by a finite complex if and only if $(r,\Sigma) \sim 0$ in $\tilde{K}_0(\mathbb{Z}\Gamma)$.</u>

Proof. Use assertion (c) in Lemma 1 to modify the reference complex with k-invariant g_0:

$$\begin{array}{ccccc} 0 \longrightarrow & \mathbb{Z} & \longrightarrow C_{m-1} & \longrightarrow C_{m-2} & \cdots \\ & \oplus & \oplus & & \\ & (r,\Sigma) & \mathbb{Z}\Gamma & & \end{array}$$

Since (r,Σ) is $\mathbb{Z}\Gamma$-weakly injective, an averaging argument shows that it is a $\mathbb{Z}\Gamma$-direct summand of $C_{m-1} + \mathbb{Z}\Gamma$. If ψ is a retraction, the sequence below is still exact :

$$\begin{array}{ccccccc} 0 \longrightarrow Z \longrightarrow & C_{m-1} & \longrightarrow & C_{m-2} & \longrightarrow \\ & + & \searrow^{\psi} & + & \\ & \mathbb{Z}\Gamma & \longrightarrow & (r,\Sigma) & . \end{array}$$

Assuming that the original complex was geometrically realisable (with $m \geqslant 4$), so is the modified complex. The sketch proof of Theorem 4 shows that only finitely many cells are needed, provided $(r,\Sigma) + F_1 \cong F_2$, with F_1 and F_2 finitely generated free modules. Thus, up to sign, the class of (r,Σ) equals the finiteness obstruction for the modified complex, which has k-invariant rg_0, since the map φ in Lemma 1(c) has degree r.

The significance of this result for the study of $T(\mathbb{Z}\Gamma)$ lies in the homotopy invariance of $\sigma(X)$. If we can finitely realise the generator rg_0 in some topological way, Proposition 6 will imply that the corresponding module (r,Σ) is stably free. For example, in this way we can recover half of Martin Taylor's result on $D^*_{pq^2}$, quoted in section two above. For simplicity assume that $q \geqslant 3$.

PROPOSITION 7. <u>If</u> r <u>is such that</u> $r \not\equiv u^q \pmod p$ <u>and</u> $r \equiv v^q \pmod{q^2}$, <u>then</u> (r,Σ) <u>maps to zero in</u> $\tilde{K}_0(ZD^*_{pq^2})$.

Proof. Using the presentation (*) on page four above, define a representation $\pi_{\cdot,\cdot}$ by transferring a faithful 1-dimensional representation of the normal subgroup generated by AB^q up to $D^*_{pq^2}$.

It is easy to check that in matrix form the possibilities are

$$\pi_{u,v}(A) = \begin{pmatrix} \alpha^{u} & & & 0 \\ & \alpha^{us} & & \\ & & \ddots & \\ 0 & & & \alpha^{us^{q-1}} \end{pmatrix}, \quad \pi_{u,v}(B) = \begin{pmatrix} 0 & 1 & \dots & 0 \\ 0 & 0 & 1 & 0 \\ \cdot & \cdot & \cdot & \cdot \\ 0 & \dots & 0 & 1 \\ \beta^{v} & \dots & 0 & 0 \end{pmatrix}$$

where α and β are primitive pth. and qth. roots of unity, $(u,p) = 1$ and $(v,q) = 1$. Inspection of characteristic values shows that $\pi_{u,v}$ defines a free linear action of the group on S^{2q-1}. The chain complex determined by an equivariant cellular decomposition of the sphere gives a periodic free resolution for the cohomology of $D^{*}_{pq^{\lambda}}$, whose k-invariant may be identified with $c_q(\pi_{u,v})$, the top dimensional Chern class of the representation. If we set $g_o = c_q(\pi_{1,1})$, then an easy calculation with characteristic values and the Cartan formula for the Chern classes of bundle sums shows that $c_q(\pi_{u,v})$ equals rg_o with r satisfying the congruences of the hypothesis. Taylor's more delicate algebraic arguments actually prove that over $ZD^{*}_{pq^{\lambda}}$

$$(r,\Sigma) \sim 0 \text{ if and only if } r \equiv u^{q} (\text{mod } p).$$

In other words the reduction modulo q is irrelevent. Topologically this stronger result is most interesting, since combined with calculations of the surgery obstruction groups, it shows the existence of free topological actions of $D^{*}_{pq^{\lambda}}$ on S^{2q-1}, homotopically distinct from the linear actions exhibited above, see [18] .

The argument of the proposition above can be generalised to supply evidence for the following conjecture:

Let Γ be a metacyclic group of odd order mn with periodic cohomology, and presentation (*) of the form above. (Replace p by m, q by n, subject to the restrictions $s^n \equiv 1(m)$, $(n(s-1),m) = 1$, and write d for the order of s in $(Z/m)^{\times}$.) Then $(r,\Sigma)\sim 0$ if and only if r is a dth. power modulo m.

Evidence. First the condition can be seen to be necessary from the exact sequence of Lemma 2. Topologically this says that the algebraic invariants of a CW-complex, realising a modified periodic free resolution must be invariant under B when lifted to the covering complex with fundamental group Z^A/m. Next note that the reduction of r modulo n is irrelevent in the sense, that if $r' \in (Z/n)^{\times}$ is arbitrary, there is <u>some</u> $r \in (Z/mn)^{\times}$ reducing to r' with $(r,\Sigma)\sim 0$ - use Lemma 2 again. Turning to possible values of the reduction modulo m, suppose that Γ admits fixed point free representations of type π (thatis, every prime dividing d also divides n/d). Then, as above, any pair (u^d,v^d) corresponds to a stably free module over $Z\Gamma$. At the other extreme, if $n/d = 1$, and m is prime, then the conjecture is true [4]. A complete proof now amounts to showing that any pair (u^d,r') is finitely realisable. That this is not obvious can be seen from the example of the group $Z/3 \times D_8^*$. Here $(\pm 7,\Sigma) \not\sim 0$ in the projective class group, even though the module is free over every proper subgroup. The proof of this fact again depends on Lemma 2, and is due to Swan [14].

5. Low dimensional topological applications.

I. 3-dimensional Poincaré complexes

Let Γ be either a finite subgroup of SU(2) or a generalised binary tetrahedral group. Recall that T^*_v $(v \geq 1)$ is an extension of D^*_8 by $Z/3^v$, where the generator X acts as the cycle $A \mapsto B \mapsto AB \mapsto A$.

THEOREM 8. If $\Gamma \cong C_n$, $D^*_{2^l}$ or T^*_v $(v \geq 2)$, any finite 3-dimensional CW-complex, which satisfies Poincaré duality and has fundamental group isomorphic to Γ, is homotopy equivalent to an elliptic manifold.

Proof. This is similar to Proposition 7. We compare the list of finitely realisable generators $rg_o \in H^4(\Gamma, \mathbb{Z})$ with that of second Chern classes of fixed point free representations of Γ in SO(4). The result for T^*_v depends on the example of Swan mentioned in the "evidence" above. For more details see [17].

For the remaining groups Γ in our list the situation is rather interesting:

(i) $\Gamma \cong D^*_{2^t 3^u}$, T^*_1, O^*_1 (binary octahedral group). For these groups there exist homotopically exotic complexes in the sense that a free topological action of Γ on S^3 must be conjugate to a free linear action [12].

(ii) $\Gamma \cong D^*_{4t}$(t divisible by some prime ≥ 5), I^* (binary icosahedral group). Very little is known for certain, although in the case of I^* of order 120 it may be possible to use the methods of [12] to prove that a free action on S^3 is simple homotopy equivalent to a free linear action.

II. <u>Geometric realisability in dimension two.</u>

Although very useful in practice Theorem 4 above does not provide a canonically simple model for the homotopy type of X. Ideally we would like to represent $\tilde{X}$ by a complex such that the number of cells in each dimension is determined, via some simple formula, by a "minimal" family of generators for $H_i(\tilde{X},\mathbb{Z})$, $i = 0,1,2$ .. If $\Gamma = \{1\}$, this can be done [15, Prop. 4.1] ;more generally the situation is complicated by the fact that a specific algebraic model for the cohomology of $\tilde{X}$ up through dimension 2 may not admit geometric realisation, see [2] for an example with the non-finite group $\mathbb{Z}/5 * \mathbb{Z}$. If $\Gamma = \{1\}$ or C_r this problem can be solved, see [1] , and our last theorem extends their result to the dihedral group D_{2p}.

In order to fix the notation let C_* be a 2-dimensional algebraic complex, and let $\{a_i : 1 \leq i \leq m\}$, $\{b_j : 1 \leq j \leq n\}$ and c be free $\mathbb{Z}\Gamma$-generators for C_2, C_1 and C_0. We shall say that C_* is <u>stably realisable</u> as the chains of a 2-dimensional CW-complex, if the complex below, where $\alpha : F \longrightarrow F$ is an isomorphism of the finitely generated free module F is so realisable,

$$F + C_2 \xrightarrow[\alpha + d_2]{} F + C_1 \xrightarrow[0 + d_1]{} C_0 \longrightarrow \mathbb{Z} \longrightarrow 0 \ .$$

Note that this a weakening of the original definition in [1] ,where α is taken to be the identity.

Hypothesis W. (i) $H_1(C_*) = 0$, (ii) $H_0(C_*) \cong \mathbb{Z}$ and (iii) there is a set of generators $\{g_1 \ldots g_n\}$ for Γ such that $d_2(b_j) = (g_j - 1)c$.

THEOREM 9. <u>Let</u> $\Gamma \cong D_{2p}$ <u>and</u> C_* <u>be a 2-dimensional algebraic complex satisfying</u> (W). <u>Then</u> C_* <u>is stably realisable.</u>

Proof. We first use the limited family of Tietze transformations allowed in [1] to pass from the given set of generators of Γ to the set $\{1, \ldots 1, B, A\}$, where $\Gamma = \{A, B : B^2 = 1,\ BAB = A^{p-1}\}$ is a minimal presentation of the group. Since the only elements to have order p are powers A^r with $1 \leq r \leq p-1$, by possibly first applying the automorphism $(A \mapsto A^r,\ B \mapsto B)$, we may assume that some $g_j = A$. Move this generator into position n (transformation T_1). If B also occurs in the given list of generators, move B into position (n-1). Otherwise each generator is of the form A^sB or A^t with $s \neq 0$, and at least one g_k must be of the first type. Replace $A^sB = BA^{-s}$ with B, by postmultiplying by the generator A, in position n, s times (transformation T_3). Now move B into position (n-1), and kill the remaining generators in positions 1, ... (n-2) by repeated use of (T_3). The set of generators is now $\{1, 1, \ldots, B, A\}$, and we may assume that $d_1(b_j) = 0$ $(1 \leq j \leq n-2)$, $d_1(b_{n-1}) = (B-1)c$ and $d_1(b_n) = (A-1)c$.

If we apply the free differential calculus to the minimal presentation of Γ, we obtain the following algebraic complex, which is well known to be geometrically realisable:

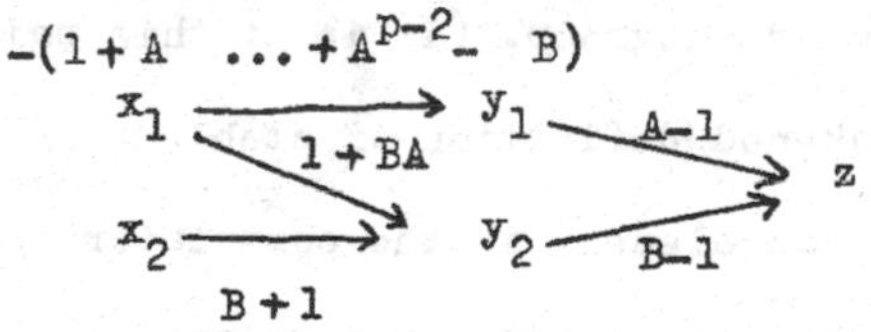

equals $C_*(\tilde{Y}_2)$.

And our aim is to show that the original complex, after application of the Tietze transformations above is stably equivalent to this reference complex, modified by trivial direct summands in dimensions one and two.

$$\mathrm{Ker}(d_1) = \langle b_1, \dots b_{n-2}\rangle + \langle -(1+A \dots A^{p-2} - B)b_n + (1+BA)b_{n-1}, \; (B+1)b_{n-1}\rangle .$$

Choose an independent set of elements $\{a'_1, \dots a'_{n-2}\}$ in C_2 such that $d_2(a'_j) = b_j$ $(1 \leq j \leq n-2)$, so splitting the short exact sequence

$$0 \to K \to C_2 \to \langle b_1, \dots b_{n-2}\rangle \to 0 .$$

The module K is thus stably free, hence free since the Eichler condition holds for the ring $\mathbb{Z}\Gamma$. We can now find a free basis $\{a'_{n-1}, a'_n, \dots a'_m\}$ of K such that $d_2(a'_{n-1}) = (B+1)b_{n-1}$, $d_2(a'_n) = -(1 \;\; A \dots A^{p-2} - B)b_n + (1+BA)b_{n-1}$, and $d_2(a'_i) = 0$ for $i > n$.

It remains to show that the unimodular matrix describing the basis change $[\underline{a}' | \underline{a}]$ can be decomposed as a product of elementary matrices in $GL_m(\mathbb{Z}\Gamma)$. Suppose that the change of basis determines an element u^* in $K_1(\mathbb{Z}\Gamma)$. Since $SK_1(\mathbb{Z}\Gamma) = 0$, there is a unit u in $\mathbb{Z}\Gamma$, such that the matrix $\mathrm{diag}(u^{-1}, 1, \dots 1) \in GL_{m+1}(\mathbb{Z}\Gamma)$ also maps to u^*. Adding a dummy copy of $\mathbb{Z}\Gamma$ to each of C_1 and C_2 and twisting d_2 by this diagonal matrix implies that we may assume the basis change to be in the commutator subgroup. (It is at this point that we need to work with the weakened definition of stably realisable above.) Since $m+1 \geq 3$, an element in the commutator subgroup is expressible as a product of elementary matrices. Geometrically we have proved, that starting from $\widetilde{Y}_2 \cup$ (2-spheres) $\cup$

(2-discs), we can change the cellular structure in such a way as to obtain a complex whose homology coincides with that of C_* (modified by $\alpha : F \longrightarrow F$).

This result can clearly be further generalised. The first part of the argument will apply to more general metacyclic groups, cancellation of free summands over $\mathbb{Z}\Gamma$ can be avoided by adding more cells, and by [11] we know that SK_1 vanishes for metacyclic groups. But even when $1 \neq u^* \in SK_1(\mathbb{Z}\Gamma)$ the stable result can probably be salvaged by representing u^* by a 2×2 matrix rather than by a unit.

REFERENCES

1. W.H. Cockcroft, R.M.S. Moss: On the 2-dimensional realisability of chain complexes, J.London Math.Soc.(2) 11 (1975)257-262.
2. M.J. Dunwoody, Relation modules, Bull.London Math.Soc.4(1972)151-5.
3. A. Frohlich, Module invariants & root numbers for quaternion fields of degree $4\ell^r$, Proc.Camb.Phil.Soc.76(1974)393-9.
4. S. Galovich, I. Reiner, S. Ullom, Class groups for integral representations of metacyclic groups, Mathematika 19, (1972) 105-111.
5. P. Hall, Finiteness conditions for solvable groups, Proc. London Math.Soc. (3) 4 (1954) 419-436.
6. T.Y. Lam, Induction theorems for Grothendieck groups & Whitehead groups of finite groups, Ann.Sci.E.N.S.(4)1(1968)99-148.

7. I. Madsen, C. Thomas, C.T.C. Wall, Topological spherical space form problem III: dimensional bounds& smoothing, to appear.

8. J. Milgram, The Swan finiteness obstruction for periodic groups, preprint, Stanford University, 1980.

9. G. Mislin, Wall's obstruction for nilpotent spaces, Topology 14 (1975) 311-317.

10. G. Mislin, K. Varadarajan, The finiteness obstructions for nilpotent spaces lie in D($\mathbb{Z}\Gamma$), Inv.Math. 53(1979)185-191.

11. R. Oliver, SK_1 for finite group rings I, preprint, Aarhus University, 1979.

12. J. Rubinstein, Free actions of some finite groups on S^3, Math.Annalen 240(1979)165-175.

13. R.G. Swan, Induced representations & projective modules, Ann. of Math. 71(1960)552-578.

14. R.G. Swan, Periodic resolutions for finite groups, ibid, 72 (1960) 267-291.

15. C.T.C. Wall, Finiteness conditions for CW-complexes, ibid, 81(1965) 56-69.

16. C.T.C. Wall, Periodic projective resolutions, Proc. London Math. Soc. (3) 39 (1979) 509-533.

17. C.B. Thomas, Homotopy classification of free actions by finite groups on S^3, Proc. London Math. Soc.(3)40(1980)384-397.

18. C.B. Thomas, Classification of free actions by some metacyclic groups on S^{2n-1}, to appear in Ann. Sci. de l'E.N.S.

LENSTRA'S CALCULATION OF $G_o(R\pi)$, AND

APPLICATIONS TO MORSE-SMALE DIFFEOMORPHISMS

by Hyman Bass

Contents

1. Morse-Smale diffeomorphisms and the group SSF

A diffeomorphism $f: M \to M$ of a compact smooth manifold is said to be Morse-Smale if it is structurally stable and has only finitely many non-wandering points, which are fixed. (A point $x \in M$ wanders if, for some neighborhood U of x, $U \cap f^n(U) = \emptyset$ for all $n \neq 0$.) Shub and Sullivan [SS] showed then that all eigenvalues of $f_*: H_*(M,\mathbb{Q}) \to H_*(M,\mathbb{Q})$ are roots of unity.

Shub and Franks investigated the converse proposition, and constructed an obstruction which lives in a group we shall denote SSF. It is defined as follows.

Let $\mathfrak{S}$ denote the (abelian) category of pairs (H,u) where H is a finitely generated $\mathbb{Z}$-module, $u \in \mathrm{Aut}(H)$, and all eigenvalues of $\mathbb{Q} \otimes u$ are roots of unity; the morphisms are evident. Call (H,u) a permutation module if H has a $\mathbb{Z}$-basis permuted by u. Then

$$SSF = K_0(\mathfrak{S})/P \quad ,$$

the quotient of the Grothendieck group of $\mathfrak{S}$ by the subgroup P generated by classes of permutation modules. Write $[H,u]$ for the class of (H,u) in SSF.

(1.1) Theorem (Shub-Franks [SF]): *Let $f: M \to M$ be a diffeomorphism of a compact smooth manifold such that all eigenvalues of f_* on $H_*(M,\mathbb{Q})$ are roots of unity. Consider the "Lefschetz invariant"*

$$L(f) = \sum_{i \geq 0} (-1)^i [H_i(M,\mathbb{Z}), f_i] \in SSF \quad .$$

a) *If f is homotopic to a Morse-Smale diffeomorphism then* $L(f) = 0$.

b) *Suppose that M is simply connected and of dimension ≥ 5. If $L(f) = 0$ then f is isotopic to a Morse-Smale Diffeomorphism.*

With this theorem in hand, Shub posed to me the problem of calculating SSF. It was not even clear whether or not $SSF \neq 0$.

I applied some more or less standard K-theoretic techniques to the problem and came up with a presentation of SSF in terms of ideal class groups of cyclotomic fields [B3]. I learnt later that essentially equivalent calculations, in a different guise, had been made in the Sao Paulo lecture notes of I. Reiner [R1].

Mike Stein and J. Franks meanwhile noticed that a paper of Grayson [G1] could be applied to the problem. Grayson persued this suggestion and obtained among other things an isomorphism

$$SSF \simeq SK_1(R) \quad ,$$

where R is a rather remarkable principal ideal domain [G2]. From algebraic K-theory this yields a "Mennicke symbol" presentation of SSF, quite different in appearance from the above cyclotomic presentation. However we failed, using either presentation, to decide whether or not $SSF \neq 0$.

I transmitted the problem and circumstances to Keith Dennis at Cornell. During the summer of 1979, H. Lenstra visited Cornell and learned of the problem. He quickly found an ingenious idea for manipulating the cyclotomic presentation to show, in particular, that $SSF \neq 0$. Over the next few months he refind his method to obtain, finally, the following beautiful answer.

<u>(1.2) Theorem (Lenstra [L])</u>:

$$SSF \simeq \bigoplus_{n\geq 1} \operatorname{Pic}(\mathbb{Z}[\zeta_n, \tfrac{1}{n}]) \quad .$$

From this and known results in class field theory Lenstra was able, almost completely, to determine the abelian group structure of SSF. For this it suffices to describe the p-primary part $(SSF)_p$ for each prime p.

<u>(1.3) Theorem (Lenstra [L])</u>:

1. <u>If</u> p <u>is odd then</u> $(SSF)_p \simeq \bigoplus_{h\geq 1} (\mathbb{Z}/p^h\mathbb{Z})^{(\aleph_0)}$.
2. <u>There is a sequence of integers</u> $1 \leq h_1 < h_2 < h_3 < \dots$ such that $(SSF)_2 \simeq \bigoplus_{i\geq 1} (\mathbb{Z}/2^{h_i}\mathbb{Z})^{(\aleph_0)}$.

Conjecturally, $h_i = i$ in part 2. Related calculations of cyclotomic ideal class groups can be found in Brumer [Br].

<u>(1.4) Problem</u>: What is the Galois module structure of SSF ? More generally, for each $n \in \hat{\mathbb{Z}}$, we have an endomorphism of SSF defined by $[H,u] \mapsto [H,u^n]$ whenever u has finite order. Determine the structure of SSF as a module over the multiplicative monoid of $\hat{\mathbb{Z}}$.

The methods used to calculate SSF yield more, as we now indicate. For any right noetherian ring A let

$\mathfrak{M}_A$ = the (abelian) category of finitely generated right A-modules

and

$$G(A) = K_0(\mathfrak{M}_A) \quad .$$

The calculation of SSF is first (routinely) reduced to the calculation of $G(\mathbb{Z}[\pi])$, where π is a finite cyclic group (see section 2). In fact Lenstra's main result calculates $G(R[\pi])$ where R is an arbitrary right noetherian ring and π is any finite abelian group; see section 5, Theorem (5.1) for the precise statement. Theorem (1.2) above is deduced in section 5 as an easy corollary of Theorem (5.1).

Since my preprint [B3] is almost completely superceded by Lenstra's work it seemed more useful mathematically to publish instead this survey of work on the above problem, incorporating the few features of my paper whose possible interest survives, mainly sections 4 and 7 below. I am grateful to H.Lenstra and to D.Grayson for permission to present some of their work here.

2. $SSF \simeq \varinjlim_{n} \tilde{G}(\mathbb{Z}[T]/(T^n-1))$.

Let T be an indeterminate. For each integer $n \geq 1$ we have, in $\mathbb{Z}[T]$, the factorization

(1) $$T^n - 1 = \prod_{d|n} \varphi_d(T)$$

into cyclotomic polynomials. By Mobius inversion we have further

(2) $$\varphi_n(T) = \prod_{d|n} (T^d - 1)^{\mu(n/d)} .$$

Thus $(T^n-1)_{n\geq 1}$ and $(\varphi_n(T))_{n\geq 1}$ are two $\mathbb{Z}$-bases for the same free abelian group in $\mathbb{Q}(T)^*$.

Let S denote the multiplicative set in $\mathbb{Z}[T]$ generated by the cyclotomic polynomials $\varphi_n(T)$ $(n\geq 1)$. Let $\mathfrak{M}^S_{\mathbb{Z}[T]}$ denote the full subcategory of $\mathfrak{M}_{\mathbb{Z}[T]}$ formed by S-torsion modules M (i.e. $sM = 0$ for some $s \in S$). Since each $s \in S$ is a monic polynomial, the condition $sM = 0$ implies that M is a finitely generated $\mathbb{Z}$-module. If $T_M \in \mathrm{End}_{\mathbb{Z}}(M)$ denotes multiplication by T or M , then the condition $s(T_M) = 0$ implies further that all eigenvalues of $\mathbb{Q} \otimes T_M$ are roots of unity. Thus (M, T_M) belongs to the category $\mathfrak{S}$ used above to define SSF.

Conversely, let (M,u) be an object of $\mathfrak{S}$. We can then give M the $\mathbb{Z}[T]$-module structure such that $T_M = u$. Since the eigenvalues of $\mathbb{Q} \otimes u$ are roots of unity, its characteristic polynomial s belongs to S . By the Cayley-Hamilton Theorem $s(\mathbb{Q} \otimes u) = 0$; equivalently, $sM \subset M_o$, where M_o is the torsion submodule of M . Since M_o is finite, the automorphism u restricted to M_o has finite order, say q . Putting $s' = (T^q-1)s \in S$ we then have $s'M = 0$, so that $M \in \mathfrak{M}^S_{\mathbb{Z}[T]}$.

In summary, we have identified the two categories $\mathfrak{S}$ and $\mathfrak{M}^S_{\mathbb{Z}[T]}$. Let $\mathfrak{S}_o$ denote the full subcategory of $\mathfrak{S}$ formed by pairs (M,u) such that u has finite order. These correspond to $\mathbb{Z}[T]$-modules M

annihilated by T^n-1 for some n. In other words

(3) $$\mathfrak{S}_0 = \bigcup_{n\geq 1} \mathfrak{M}_{\mathbb{Z}[T]/(T^n-1)}$$

and this union is filtering if we order the integers n by divisibility.

We first note that

(4) $$K_0(\mathfrak{S}_0) \to K_0(\mathfrak{S}) \quad ,$$

induced by the inclusion $\mathfrak{S}_0 \subset \mathfrak{S}$, is an isomorphism. Indeed one easily sees that each object of $\mathfrak{S}$ has a finite filtration (or composition series) with successive quotients in $\mathfrak{S}_0$. Since these categories are abelian, it follows from the "Devissage Theorem" in algebraic K-theory (see [Bl], p.402, or [Q], Th.4) that (4) is an isomorphism.

Combining this with (3) and the fact that K_0 commutes with filtered inductive limits, we conclude that

(5) $$K_0(\mathfrak{S}) \simeq \varinjlim_{n} G(\mathbb{Z}[T]/(T^n - 1)) \quad ,$$

where the isomorphism (5) induced by the inclusions $\mathfrak{M}_{\mathbb{Z}[T]/(T^n-1)} \subset \mathfrak{S}$ described above.

To use (5) to decribe SSF we must interprete permutation modules as $\mathbb{Z}[T]$-modules. Every permutation module is a direct sum of modules $(\mathbb{Z}^n, u)$ where u cyclically permutes the basis of $\mathbb{Z}^n$. The latter, as $\mathbb{Z}[T]$-module, is evidently isomorphic to $\mathbb{Z}[T]/(T^n-1)$.

Thus the subgroup P_n of $G(\mathbb{Z}[T]/(T^n-1))$ generated by classes of permutation modules in $\mathfrak{M}_{\mathbb{Z}[T]/(T^n-1)}$ is generated by the elements

(6) $$[\mathbb{Z}[T]/(T^d - 1)] \qquad (d|n) \quad .$$

Define $\tilde{G}(\mathbb{Z}[T]/(T^n-1))$ by the exact localization sequence

(7) $$0 \to \tilde{G}(\mathbb{Z}[T]/(T^n-1)) \to G(\mathbb{Z}[T]/(T^n-1)) \xrightarrow{\chi} G(\mathbb{Q}[T]/(T^n-1)) \to 0.$$

Since $\mathbb{Q}[T]/(T^n-1) \simeq \prod_{d|n} \mathbb{Q}[T]/(\varphi_d(T))$, we can identify $G(\mathbb{Q}[T]/(T^n-1))$ with the free abelian subgroup of $\mathbb{Q}[T]^*$ having $\mathbb{Z}$-basis $(\varphi_d(T))_{d|n}$. With this identification, the map χ in (7) can be described as

(8) $$\chi : [M] \to \text{char.poly.}(\mathbb{Q} \otimes T_M) \quad .$$

In particular $\chi[\mathbb{Z}[T]/(T^d-1)] = T^d-1$. It follows therefore from (6) and (2) above that χ maps the group P_n of permutation modules isomorphically onto $G(\mathbb{Q}[T]/(T^n-1))$, so that

(9) $$G(\mathbb{Z}[T]/(T^n-1)) = P_n \oplus \widetilde{G}(\mathbb{Z}[T]/(T^n-1)) \quad .$$

Combining (5) and (9) we conclude that

(10) $$\begin{aligned} SSF &\simeq \varinjlim_n G(\mathbb{Z}[T]/(T^n-1))/P_n \\ &\simeq \varinjlim_n \widetilde{G}(\mathbb{Z}[T]/(T^n-1)) \quad . \end{aligned}$$

We shall use this expression for SSF to prove Theorem (1.2) (at the end of section 5).

3. $SSF \simeq SK_1(\mathbb{Z}[T]_{S_+})$ (Grayson's Theorem)

Let S_+ denote the multiplicative set in $\mathbb{Z}[T]$ generated by all $\varphi_n(T)$, $n \geq 0$, where we agree to put $\varphi_o(T) = T$.

(3.1) Theorem (Grayson [G2])

a) The ring $R = \mathbb{Z}[T]_{S_+}$ is a principal ideal domain.

b) There is a natural isomorphism $SSF \simeq SK_1(R)$.

The proof of b) uses some slightly sophisticated techniques from algebraic K-theory. Assertion a) is rather startling at first sight, but it has a beautifully elementary proof as follows. Since R is a unique factorization domain, being a localization of one, we need only show that it has Krull dimension ≤ 1. For this we need only show that every maximal ideal M of $\mathbb{Z}[T]$ meets S_+; for then it will follow that $\dim R < \dim \mathbb{Z}[T] = 2$. But it is well known, by a variant of the Nullstellensatz, that $F = \mathbb{Z}[T]/M$ is a finite field. Thus the image of T in F is either 0, so $\varphi_o \in M$, or a root of unity, so some $T^n - 1 \in M$.

We thus obtain a simple and concrete example of a principal ideal domain which has a large SK_1 (in view of Theorem (1.2)) and which is therefore very far from being euclidean.

From Grayson's Theorem and the localization of the embedding of R in its field of fractions $\mathbb{Q}(T)$ one obtains an exact sequence

$$(1) \qquad K_2(\mathbb{Q}(T)) \xrightarrow{\lambda = (\lambda_p)} \bigoplus_p k(p)^* \xrightarrow{\mu} \underset{\displaystyle SSF}{\overset{\displaystyle SK_1(R)}{\|}} \to 0 \quad .$$

Here p runs over the prime elements, modulo units, of R, and $k(p) = R/pR$ is the residue field at p. If $u \in R$ represents $\bar{u} \in k(p)^*$ then $\mu(\bar{u})$ is the class in $SK_1(R)$ of any matrix $\begin{pmatrix} p & u \\ * & * \end{pmatrix} \in SL_2(R)$. If $f, g \in \mathbb{Q}(T)^*$ then the symbol $\{f,g\} \in K_2(\mathbb{Q}(T))$ maps under λ_p to

$(f,g)_p = \bar{u} \in k(p)^*$, where $u = (-1)^{rs} f^s/g^r \in R_{(p)}$, with $r = v_p(f)$ and $s = v_p(g)$, v_p being the p-adic valuation on $\mathbb{Q}(T)$. The exactness of (1) expresses the universality of μ among homomorphisms satisfying the product formula, or "reciprocity law",

$$(2) \qquad \prod_p \mu(f,g)_p = 1$$

for all $f,g \in \mathbb{Q}(T)^*$. Thus each homomorphism φ: SSF $\to$ C , C an abelian group, yields a reciprocity law on R with values in C . In this light Lenstra's calculation (Theorem (1.2)) furnishes non trivial and independent reciprocity laws on R with values in $\mathrm{Pic}(\mathbb{Z}[\zeta_n, \frac{1}{n}])$ for all $n \geq 1$.

The prime elements of R are of two types, arithmetic and geometric. The rational primes $p \in \mathbb{Z}$ are the arithmetic ones. For these $k(p) = \mathbb{F}_p(T)$. The geometric primes are of the form $p = p(T)$ where $p(T)$ is a primitive irreducible polynomial, which we may normalize modulo units to make its leading coefficient positive, and such that $p(T)$ is not a cyclotomic polynomial $\varphi_n(T)$ for any $n \geq 0$. In this case $k(p) = \mathbb{Q}[T]/(p(T))$. The units of R are the direct product of $\{\pm 1\}$ with the free abelian group with basis $(\varphi_n(T))_{n \geq 0}$.

From the fact that $\mathbb{Z}$ and $\mathbb{Q}[T]$ are both euclidean, and hence have trivial SK_1's, one can deduce that a non trivial reciprocity law on R can not be supported exclusively on the arithmetic or on the geometric primes; it must effectively mingle the two. More precisely, the restrictions of μ above to $\bigoplus_{p \text{ arithmetic}} k(p)^*$ and to $\bigoplus_{p \text{ geometric}} k(p)^*$ are both surjective. Thus the reciprocity laws produced above from Lenstra's calculation are quite unlike any of the classical examples produced in number theory and in algebraic geometry.

4. A Mayer-Vietoris sequence for G(A).

We present here a useful technical tool (Theorem (4.2)) to be used in calculating $G(R[\pi])$ in section 5. Only the statement of (4.2) is needed in the sequel.

Consider a right noetherian ring A and two sided ideals $J_1,\ldots,J_n$ in A. We have the intersection

(1) $$J = J_1 \cap \ldots \cap J_n$$

and the "product"

(2) $$J' = \sum_{\sigma} J_{\sigma(1)} \ldots J_{\sigma(n)}$$

where σ varies over permutations of $(1, \ldots, n)$. When the J_i are generated by central elements of A, e.g. when A is commutative, we have $J' = J_1 \ldots J_n$, Clearly

(3) $$J^n \subset J' \subset J \; .$$

From "Devissage" ([Q], Th.4; see also [B1], p.402) we conclude:

(4.1) <u>Proposition</u>: If $J' \subset J'' \subset J$ <u>then the natural homomorphism</u> $G((A/J'') \to G(A/J)$ <u>is an isomorphism</u>.

We define a sequence

(4) $$\bigoplus_{1\le i<j\le n} G(A/J_i + J_j) \xrightarrow{\Delta} \bigoplus_{1\le i\le n} G(A/J_i) \xrightarrow{\sigma} G(A/J) \to 0$$

as follows: σ on $G(A/J_i)$ is induced by the natural projection $A/J \to A/J_i$. If $\delta_{i,j}: G(A/J_i + J_j) \to G(A/J_i)$ is the map induced by the natural projection $A/J_i \to A/J_i+J_j$ then, for $x \in G(A/J_i+J_j)$ we define $\Delta(x) = \delta_{i,j}(x) - \delta_{j,i}(x) \in G(A/J_i) \oplus G(A/J_j)$ whenever $1\le i<j\le n$. Evidently, $\sigma \cdot \Delta = 0$ since the diagrams

$$\begin{array}{ccc} A/J & \to & A/J_i \\ \downarrow & & \downarrow \\ A/J_j & \to & A/J_i+J_j \end{array}$$

are commutative.

(4.2) Theorem: The sequence (4) above is exact. If J_o is another two sided ideal of A the following diagram is commutative

$$
(5)\qquad
\begin{array}{ccccccc}
\bigoplus_{1\leq i<j\leq n} G(A/J_i+J_j) & \xrightarrow{\Delta} & \bigoplus_{1\leq i\leq n} G(A/J_i) & \xrightarrow{\sigma} & G(A/J) & \rightarrow & 0 \\
\downarrow \alpha_2 & & \downarrow \alpha_1 & & \downarrow \alpha_o & & \\
\bigoplus_{o\leq i<j\leq n} G(A/J_i+J_j) & \xrightarrow{\Delta} & \bigoplus_{o\leq i\leq n} G(A/J_i) & \xrightarrow{\sigma} & G(A/J_o\cap J) & \rightarrow & 0 \quad .
\end{array}
$$

Here α_1 and α_2 are the obvious inclusions and α_o is induced by the natural projection $A/(J_o \cap J) \rightarrow A/J$.

The commutativity of the left hand square in (5) is immediate, and that of the right hand square follows from the commutative diagrams of natural projections

$$
\begin{array}{ccc}
A/J_o \cap J & \longrightarrow & A/J \\
& \searrow & \downarrow \\
& & A/J_i
\end{array} \quad .
$$

We shall prove the exactness of (4) by induction on n . The case $n = 1$ is trivial. The case $n = 2$ is covered by the next lemma.

(4.3) Lemma: For any two sided ideals J_o and J of A the sequence

$$
(6)\qquad G(A/J_o+J) \xrightarrow{\Delta} G(A/J_o) \oplus G(A/J) \xrightarrow{\sigma} G(A/J_o\cap J) \rightarrow 0
$$

is exact.

We shall construct an inverse τ to the induced map

$$
\frac{G(A/J_o) \oplus G(A/J)}{\Delta\, G(A/J_o + J)} \xrightarrow{\sigma'} G(A/J_o \cap J) \quad .
$$

For any $M \in \mathfrak{M}_{A/J_o\cap J}$ we have the exact sequence

$$
(7)\qquad 0 \rightarrow JM \rightarrow M \rightarrow M/JM \rightarrow 0
$$

with $J_o\cdot(JM) = 0$ and $J\cdot(M/JM) = 0$, whence $JM \in \mathfrak{M}_{A/J_o}$ and $M/JM \in \mathfrak{M}_{A/J}$. Define

$$
\tau(M) = ([JM],[M/JM])_\Delta \quad ,
$$

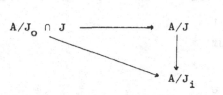

the class modulo $\mathrm{Im}(\Delta)$ of $([JM],[M/JM]) \in G(A/J_o) \oplus G(A/J)$. We claim that τ is additive over exact sequences

$$0 \to M' \to M \to M'' \to 0$$

in $\mathfrak{M}_{A/J_o \cap J}$. For this we consider the following diagram with exact rows and column.

$$\begin{array}{ccccccccc}
 & & 0 & & & & & & \\
 & & \uparrow & & & & & & \\
0 & \to & \dfrac{M' \cap JM}{JM'} & \to & M'/JM' & \to & M/JM & \to & M''/JM'' \to 0 \\
 & & \uparrow & & & & & & \\
0 & \to & M' \cap JM & \to & JM & \to & JM'' & \to & 0 \\
 & & \uparrow & & & & & & \\
 & & JM' & & & & & & \\
 & & \uparrow & & & & & & \\
 & & 0 & & & & & &
\end{array}$$

The top row is in $\mathfrak{M}_{A/J}$ while the bottom row and column are in $\mathfrak{M}_{A/J_o}$. Therefore in $G(A/J_o) \oplus G(A/J)$ we have

$$([JM],[M/JM]) - ([JM'],[M'/JM']) - ([JM''],[M''/JM''])$$
$$= \left(\left[\frac{M' \cap JM}{JM'}\right], -\left[\frac{M' \cap JM}{JM'}\right]\right) \in \mathrm{Im}(\Delta) \quad .$$

Thus τ is indeed additive, and hence defines a homorphism, also denoted τ, from $G(A/J_o \cap J)$ to $\mathrm{Coker}(\Delta)$. We have

$$\sigma'\tau[M] = \sigma'([JM],[M/JM])_\Delta$$
$$= [JM] + [M/JM] = M \quad .$$

Suppose that $M \in \mathfrak{M}_{A/J}$. Then

$$\tau\sigma'(0,[M])_\Delta = \tau[M] = ([JM],[M/JM])_\Delta$$
$$= (0,[M])_\Delta \quad .$$

Suppose that $M \in \mathfrak{M}_{A/J_o}$. Then

$$\tau\sigma'([M],0)_\Delta = \tau[M] = ([JM],[M/JM])_\Delta$$
$$= ([JM] + [M/JM],0)_\Delta$$

(note that $M/JM \in \mathfrak{M}_{A/J_o+J}$)

$$= ([M],0)_\Delta \quad .$$

Thus τ is indeed inverse to σ' . This proves Lemma (4.3).

Now to prove Theorem (4.2) by induction, we assume the exactness true for a given $n \geq 1$, and we propose to prove it for $n+1$ ideals $J_o, J_1, \ldots, J_n$. For this purpose we introduce $\bar{A} = A/J_o$, and denote by $\bar{X}$ the image in $\bar{A}$ of any set X in A . The functoriality of the sequence (4) yields a commutative diagram

$$\begin{array}{lccccccc}
 & \bigoplus\limits_{1\le i<j\le n} G(\bar{A}/\bar{J}_i+\bar{J}_j) & \xrightarrow{\Delta_{\bar{A}}} & \bigoplus\limits_{1\le i\le n} G(\bar{A}/\bar{J}_i) & \xrightarrow{\sigma_{\bar{A}}} & G(\bar{A}/\bar{J}_1 \cap \ldots \cap \bar{J}_n) & \to & 0 \\
(8) & \downarrow & & \downarrow \beta & & \downarrow & & \\
 & \bigoplus\limits_{1\le i<j\le n} G/A/J_i+J_j) & \xrightarrow{\Delta_A} & \bigoplus\limits_{1\le i\le n} G(A/J_i) & \xrightarrow{\sigma_A} & G(A/J) & \to & 0
\end{array} .$$

Its rows are exact by our inductive hypothesis. We have $\bar{J} \subset \bar{J}_1 \cap \ldots \cap \bar{J}_n$ and

$$\begin{aligned}
(\bar{J}_1 \cap \ldots \cap \bar{J}_n)^n &\subset \sum_\pi \bar{J}_{\pi(1)} \cdots \bar{J}_{\pi(n)} \\
&= \sum_\pi \overline{J_{\pi(1)} \cdots J_{\pi(n)}} \subset \bar{J} \quad ,
\end{aligned}$$

where π varies over permutations of $\{1,\ldots,n\}$. It follows therefore from Proposition (4.1) that $G(\bar{A}/\bar{J}_1 \cap \ldots \cap \bar{J}_n) \to G(\bar{A}/\bar{J})$ is an isomorphism, so we may replace $\bigoplus\limits_{1\le i\le n} G(\bar{A}/\bar{J}_i) \xrightarrow{\sigma_{\bar{A}}} G(\bar{A}/\bar{J}_1 \cap \ldots \cap \bar{J}_n)$ in (8) by $\bigoplus\limits_{1\le i\le n} G(\bar{A}/\bar{J}_i) \xrightarrow{\sigma'} G(\bar{A}/\bar{J})$ without upsetting commutativity or exactness.

Now consider the diagram

$$\begin{array}{ccccc}
 & & & & G(\bar{A}/\bar{J}) \\
 & & & & \| \\
 & & \bigoplus\limits_{1\le i\le n} G(\bar{A}/\bar{J}_i) & \xrightarrow{\sigma'} & G(A/J_o+J) \to 0 \\
 & & \downarrow \Delta' & & \downarrow \Delta \\
\bigoplus\limits_{1\le i<j\le n} G(A/J_i+J_j) & \xrightarrow{(0,\Delta_A)} & G(\bar{A}) \oplus \bigoplus\limits_{1\le i\le n} G(A/J_i) & \xrightarrow{1_{G(\bar{A})} \oplus \sigma_A} & G(\bar{A}) \oplus G(A/J) \to 0 \\
 & & & & \downarrow \sigma \\
 & & & & G(A/J_o \cap J) \\
 & & & & \downarrow \\
 & & & & 0
\end{array} ,$$

where the right hand column is the exact sequence of Lemma (4.3), and where $\Delta'(x) = (\gamma(x), -\beta(x)$ for $x \in G(\bar{A}/\bar{J}_i)$ with β as in (8) and

$$\gamma : G(\bar{A}/\bar{J}_i) = G(A/J_o+J_i) \to G(\bar{A}) = G(A/J_o)$$

induced by the natural projection $A/J_o \to A/J_o+J_i$. It is then easily checked that the diagram is commutative. The exactness of its horizontal sequences is immediate from the exactness of the rows of (8). It now follows from Lemma (4.4) below that we have an exact sequence

$$\begin{array}{c}
\bigoplus_{1\le i\le n} G(\bar{A}/\bar{J}_i) \oplus \bigoplus_{1\le i<j\le n} G(A/J_i+J_i) \\
\Big\downarrow \begin{pmatrix}\Delta' \\ (0,\Delta_A)\end{pmatrix} \\
G(\bar{A}) \oplus \bigoplus_{1\le i\le n} G(A/J_i) \\
\Big\downarrow \sigma\circ(1_{G(\bar{A})}\oplus\sigma_A) \\
G(A/J_o \cap J) \\
\big\downarrow \\
0
\end{array}$$

Rewriting $\bar{A}$ as A/J_o and $\bar{A}/\bar{J}_i$ as $A/J_o + J_i$ the above is seen to be the exact sequence we sought to establish.

(4.4) <u>Lemma</u>: <u>Suppose that the diagram of abelian group homomorphisms</u>

$$\begin{array}{ccccccc}
 & & X_1 & \xrightarrow{\sigma_X} & X_o & \longrightarrow & 0 \\
 & & \downarrow{\scriptstyle\Delta'} & & \downarrow{\scriptstyle\Delta} & & \\
Y_2 & \xrightarrow{\Delta_Y} & Y_1 & \xrightarrow{\sigma_Y} & Y_o & \longrightarrow & 0 \\
 & & & & \downarrow{\scriptstyle\sigma} & & \\
 & & & & Z_o & & \\
 & & & & \downarrow & & \\
 & & & & 0 & &
\end{array}$$

is commutative with exact rows and column. Then the sequence

$$X_1 \oplus Y_2 \xrightarrow{\binom{\Delta'}{\Delta_Y}} Y_1 \xrightarrow{\sigma\circ\sigma_Y} Z_o \longrightarrow 0$$

is exact.

We have

$$\sigma\sigma_Y \circ \binom{\Delta'}{\Delta_Y} = \binom{\sigma\sigma_Y\Delta'}{\sigma\sigma_Y\Delta_Y} = \binom{\sigma\Delta\sigma_X}{\sigma 0} = \binom{0\sigma_X}{\sigma 0} = 0 \quad .$$

Moreover $\sigma\sigma_Y$ is an epimorphism since σ and σ_Y are. Suppose finally that $y_1 \in \mathrm{Ker}(\sigma\sigma_Y)$. Then $\sigma_Y y_1 = \Delta x_o$ for some $x_o \in X_o$, and $x_o = \sigma_X x_1$ for some $x_1 \in X_1$. Let $y_1' = \Delta' x_1$. Then $\sigma_Y y_1' = \sigma_Y \Delta' x_1 = \Delta\sigma_X x_1 = \Delta x_o = \sigma_Y y_1$, so $\sigma_Y(y_1 - y_1') = 0$. Thus $y_1 - y_1' = \Delta_Y y_2$ for some $y_2 \in Y_2$, and then $y_1 = y_1' + (y_1 - y_1') = \Delta' x_1 + \Delta_Y y_2 \in \mathrm{Im}\binom{\Delta'}{\Delta_Y}$. This proves the lemma.

(4.5) Remark: For commutative rings A the Mayer-Vietoris sequence of Lemma (4.3) can be generalized to schemes, and deduced naturally from the localization sequence. For these observations I am indebted to Dan Grayson.

Let X be a noetherian separated scheme and let Z be a closed subscheme with open complement U. Write $\mathfrak{M}_X$ for the category of coherent sheaves on X, and $\mathfrak{M}_Z^X$ the subcategory of those with support in Z. We have an exact sequence

$$(9) \qquad 0 \to \mathfrak{M}_Z^X \to \mathfrak{M}_X \to \mathfrak{M}_U \to 0$$

of abelian categories. It follows by devissage ([Q],Th.4) that the inclusion $\mathfrak{M}_Z \subset \mathfrak{M}_Z^X$ induces isomorphisms $K_i(\mathfrak{M}_Z) \to K_i(\mathfrak{M}_Z^X)$. Writing $G_i(X) = K_i(\mathfrak{M}_X)$, etc., the localization exact sequence of (9) ([Q], Th.5) then takes the form

$$\ldots \to G_i(Z) \to G_i(X) \to G_i(U) \to G_{i-1}(Z) \to \ldots \quad .$$

Now let Z_o be another closed subscheme and assume that $X = Z_o \cup Z$.

Then $U = X - Z = Z_o - (Z_o \cap Z)$. The commutative diagram

$$\begin{array}{ccc} Z & \subset & X \\ \cup & & \cup \\ Z_o \cap Z & \subset & Z_o \end{array}$$

then yields a map of localization sequences

$$\begin{array}{ccccccccc} \ldots \to & G_i(Z) & \to & G_i(X) & \to & G_i(U) & \to & G_{i-1}(Z) & \to \ldots \\ (10) & \uparrow & & \uparrow & & \| & & \uparrow & \\ \ldots \to & G_i(Z_o \cap Z) & \to & G_i(Z_o) & \to & G_i(U) & \to & G_{i-1}(Z_o \cap Z) & \to \ldots \end{array}$$

Then by a standard procedure one obtains from (10) an exact Mayer-Vietoris sequence

$$\ldots \to G_i(Z_o \cap Z) \to G_i(Z_o) \oplus G_i(Z) \to G_i(Z_o \cup Z) \to G_{i-1}(Z_o \cap Z) \to \ldots$$

which terminates in the exact sequence

$$(11) \quad \ldots \to G_o(Z_o \cap Z) \to G_o(Z_o) \oplus G_o(Z) \to G_o(Z_o \cup Z) \to 0$$

of (4.3).

Now just as in the inductive proof of Theorem (4.2) one can use (11) to deduce, for closed subschemes $Z_1, \ldots, Z_n$ of X an exact sequence

$$(12) \qquad \bigoplus_{1 \le i < j \le n} G_o(Z_i \cap Z_j) \to \bigoplus_{1 \le i \le n} G_o(Z_i) \to G_o(Z_1 \cup \ldots \cup Z_n) \to 0 \ .$$

5. $G(R[\pi]) \simeq \bigoplus_{\rho} G(R\langle\rho\rangle)$ (Lenstra's Theorem)

Let π be a finite abelian group. By $C(\pi)$ we denote the set of cyclic quotient groups of π .

Let R be a right noetherian ring. Then the group ring $R[\pi]$ is also right noetherian. Let $\rho \in C(\pi)$. Say ρ has order n and generator t . Then we put

$$(1)\qquad R(\rho) = \frac{R[\rho]}{\varphi_n(t)R[\rho]} = R \otimes_{\mathbb{Z}} \frac{\mathbb{Z}[T]}{(T^n-1,\varphi_n(T))} = R \otimes_{\mathbb{Z}} \mathbb{Z}[\zeta_n] .$$

Note that the ideal $\varphi_n(t)R[\rho]$ does not depend on the choice of generator t , since this is so when $R = \mathbb{Z}$. We further put

$$(2)\qquad R\langle\rho\rangle = R(\rho)[\tfrac{1}{n}] \simeq R \otimes_{\mathbb{Z}} \mathbb{Z}[\zeta_n, \tfrac{1}{n}] .$$

Let

$$(3)\qquad J_\rho^R = \operatorname{Ker}(R[\pi] \to R(\rho)) .$$

We have the homomorphism

$$(4)_R\qquad R[\pi] \to \prod_{\rho \in C(\pi)} R(\rho)$$

with kernel $\bigcap_\rho J_\rho^R$. It is well known and easy to see, that $(4)_{\mathbb{Z}}$ is injective, whence $\bigcap_\rho J_\rho^{\mathbb{Z}} = 0 = \prod_\rho J_\rho^{\mathbb{Z}}$. Since, evidently, $J_\rho^R = J_\rho^{\mathbb{Z}} \cdot R[\pi]$, it follows that $J_\rho^R J_{\rho'}^R$, for $\rho, \rho' \in C(\pi)$, and $\prod_\rho J_\rho^R = 0$. Further, if $N = \operatorname{Card} C(\pi)$, then $(\bigcap_\rho J_\rho^R)^N \subset \prod_\rho J_\rho^R$. Thus $\bigcap_\rho J_\rho^R$ is nilpotent, so it follows by devissage that, writing now J_ρ for J_ρ^R ,

$$G\left(\frac{R[\pi]}{\bigcap_\rho J_\rho}\right) \xrightarrow{\ \simeq\ } G(R[\pi])$$

is an isomorphism. Thus the "Mayer-Vietoris sequence" of Theorem (4.2) yields the exact sequence

$$(5)\qquad \bigoplus_{\rho \neq \rho'} G\left(\frac{R[\pi]}{J_\rho + J_{\rho'}}\right) \xrightarrow{\ \Delta\ } \bigoplus_\rho G(R(\rho)) \xrightarrow{\ \sigma\ } G(R[\pi]) \to 0 .$$

We shall use this to prove Lenstra's Main Theorem:

(5.1) Theorem (Lenstra [L]): For any finite abelian group π there is an isomorphism

$$\sigma_\pi : \bigoplus_{\rho \in C(\pi)} G(R\langle\rho\rangle) \to G(R[\pi])$$

which is functorial with respect to R.
If $\tilde{\pi} \to \pi$ is a surjective homomorphism of finite abelian groups, so that $C(\pi)$ is canonically embedded in $C(\tilde{\pi})$, then the diagram

$$\begin{array}{ccc} \bigoplus_{\rho \in C(\pi)} G(R\langle\rho\rangle) & \xrightarrow{\sigma_\pi} & G(R[\pi]) \\ \downarrow & & \downarrow \\ \bigoplus_{\rho \in C(\tilde{\pi})} G(R\langle\rho\rangle) & \xrightarrow[\sigma_{\tilde{\pi}}]{} & G(R[\tilde{\pi}]) \end{array}$$

commutes, where the left vertical arrow is the obvious inclusion of a direct summand. Hence $G(R[\pi]) \to G(R[\tilde{\pi}])$ is a split monomorphism.

Theorem (5.1) will be proved in the following sections. We first draw some consequences.

If A is a ring which is torsion free of finite rank as a $\mathbb{Z}$-module put

$$\tilde{G}(A) = \operatorname{Ker}(G(A) \twoheadrightarrow G(\mathbb{Q} \otimes_{\mathbb{Z}} A)).$$

In fact $G(A) \simeq G(\mathbb{Q} \otimes_{\mathbb{Z}} A) \oplus \tilde{G}(A)$. If A is commutative and regular, and hence, in this case, a product of rings of S-integers in number fields, then there is a natural isomorphism

$$\operatorname{Pic}(A) \simeq \tilde{G}(A)$$

sending the class of an invertible A-module P to $[P]-[A] \in \tilde{G}(A)$.

Suppose that R is the ring of S-integers in a number field. Then for any $n \geq 1$, $R \otimes_{\mathbb{Z}} \mathbb{Z}[\zeta_n, \frac{1}{n}]$ is regular. This comes down to showing that if we embed R in $\mathbb{C}$ and set $\zeta_n = e^{2\pi i/n}$ then $R[\zeta_n, \frac{1}{n}] \subset \mathbb{C}$ is integrally closed in its field of fractions. This follows since

n always belongs to the conductor of $R[\zeta_n]$.

Combining the above remarks with (2) and Theorem (5.1) applied to $R \to \mathbb{Q} \otimes_{\mathbb{Z}} R$, we obtain:

(5.2) <u>Corollary:</u> <u>Let</u> R <u>be a ring of S-integers in a number field.</u> <u>Let</u> π <u>be a finite abelian group. There is an isomorphism</u>

$$\sigma_\pi : \bigoplus_{\rho \in C(\pi)} \mathrm{Pic}(R\langle\rho\rangle) \to \widetilde{G}(R[\pi]) .$$

<u>If</u> $\widetilde{\pi} \to \pi$ <u>is a surjective homomorphism of finite abelian groups then the diagram</u>

$$\begin{array}{ccc} \bigoplus_{\rho \in C(\pi)} \mathrm{Pic}(R\langle\rho\rangle) & \to & \widetilde{G}(R[\pi]) \\ \downarrow & & \downarrow \\ \bigoplus_{\rho \in C(\widetilde{\pi})} \mathrm{Pic}(R\langle\rho\rangle) & \to & \widetilde{G}(R[\widetilde{\pi}]) \end{array}$$

<u>commutes. Hence</u> $\widetilde{G}(R[\pi]) \to \widetilde{G}(R[\pi])$ <u>is a split monomorphism.</u>

Let t_n denote the image of T in $\mathbb{Z}[T]/(T^n-1) = \mathbb{Z}[\pi_n]$, where π_n is the cyclic group generated by t_n . From Corollary (5.2) with $R = \mathbb{Z}$ we obtain:

(5.3) <u>Corollary:</u> <u>There is an isomorphism</u>

$$\bigoplus_{d|n} \mathrm{Pic}(\mathbb{Z}[\zeta_d, \tfrac{1}{d}]) \to \widetilde{G}(\mathbb{Z}[T]/(T^n-1))$$

<u>which, on passage to the inductive limit with respect to</u> n <u>(ordered by divisibility), induces an isomorphism</u>

$$\bigoplus_{d\geq 1} \mathrm{Pic}(\mathbb{Z}[\zeta_d, \tfrac{1}{d}]) \to \varinjlim_n \widetilde{G}(\mathbb{Z}[T]/(T^n-1)) .$$

Combining this with (10) of section 2 we obtain the isomorphism

$$\bigoplus_{n\geq 1} \mathrm{Pic}(\mathbb{Z}[\zeta_n, \tfrac{1}{n}]) \to \mathrm{SSF}$$

of Theorem (1.2), which is thus proved modulo Theorem (5.1) .

<u>Remark:</u> For any multiplicative set R of monic polynomials in $\mathbb{Z}[T]$ let $\mathfrak{M}^R_{\mathbb{Z}[T]}$ denote the category of finitely generated R-torsion $\mathbb{Z}[T]$-

modules. These can be identified with pairs (M,u) where $M \in \mathfrak{M}_{\mathbb{Z}}$ and $u \in \mathrm{End}_{\mathbb{Z}}(M)$ is such that $r(u) = 0$ for some $r \in R$. Similarly define $\mathfrak{M}^R_{\mathbb{Q}[T]}$, and put

$$\widetilde{G}(R) = \mathrm{Ker}(K_o(\mathfrak{M}^R_{\mathbb{Z}[T]} \to K_o(\mathfrak{M}^R_{\mathbb{Q}[T]})) .$$

If $R \subset R'$ we have $\mathfrak{M}^R_{\mathbb{Z}[T]} \subset \mathfrak{M}^{R'}_{\mathbb{Z}[T]}$, and so a homomorphism $G(R) \to \widetilde{G}(R')$.

For each $n \geq 1$ let S_n denote the multiplicative set generated by all $\varphi_d(T)$ $(d|n)$. From Corollary (5.3) and the discussion in section 2 one sees that the homomorphisms

$$\widetilde{G}(S_n) \to \widetilde{G}(S_{nm})$$

are <u>split monomorphisms</u>, with inductive limit $G(\widetilde{S}) \simeq G(\widetilde{S}_+) \simeq SSF$, where $S = \bigcup_n S_n$, as in section 2.

In contrast with this growth of $\widetilde{G}(S_n)$ as n increases, suppose we enlarge S_+ to larger and larger multiplicative sets R. Then the maps $\widetilde{G}(S_+) \to \widetilde{G}(R)$ collapse the groups, eventually to $\widetilde{G}(R) = 0$ when R consists of <u>all</u> monic polynomials. Indeed in this case $\mathfrak{M}^R_{\mathbb{Z}[T]}$ can be identified with the category of pairs (M,u) with $M \in \mathfrak{M}_{\mathbb{Z}}$ and $u \in \mathrm{End}_{\mathbb{Z}}(M)$ <u>arbitrary</u>. We have the homomorphism

$$K_o(\mathfrak{M}^R_{\mathbb{Z}[T]}) \xrightarrow{\chi} \mathbb{Q}(T)^*$$

defined by $\chi[M,u] = \mathrm{char.poly.}(\mathbb{Q} \otimes u)$, whose kernel is just $\widetilde{G}(R)$. But Almkvist [A] (cf. also [G1] has shown that χ is injective.

6. Proof of Lenstra's Theorem (5.1)

Let π , $C(\pi)$, and R be as in section 5. For each $\rho \in C(\pi)$ we have the exact sequence

$$O \to J_\rho \to R[\pi] \to R(\rho) \to O$$

and the exact sequence ((5) of section 5)

$$(1) \qquad \bigoplus_{\rho \neq \rho} G\left(\frac{R[\pi]}{J_\rho + J_{\rho'}}\right) \xrightarrow{\Delta} \bigoplus_{\rho} G(R(\rho)) \xrightarrow{\sigma} G(R[\pi]) \to O$$

where, for $M \in \mathfrak{M}_{R[\pi]/J_\rho + J_{\rho'}}$, $\Delta[M]_{\rho,\rho'} = [M]_\rho - [M]_{\rho'} \in G(R(\rho)) \oplus G(R(\rho'))$, and the indices tell which group an element lives in.

(6.1) Lemma: Let $\rho,\rho' \in C(\pi)$, $\rho \neq \rho'$ and suppose that $J_\rho + J_{\rho'} \neq R[\pi]$. There is a unique prime p such that $\rho/\rho_p = \rho'/\rho'_p$ (as quotients of π), where ρ_p,ρ'_p denote Sylow p-subgroups. Moreover $p \in J_\rho + J_{\rho'}$ and there is a commutative diagram

$$\begin{array}{ccc} & R[\pi] & \\ \swarrow & & \searrow \\ R(\rho)/pR(\rho) & \xrightarrow[h]{} & R(\rho/\rho_p)/pR(\rho/\rho_p) \end{array}$$

with Ker(h) nilpotent.

This lemma will be proved in section 7 below. We shall use it now to conclude the proof of Lenstra's Theorem (5.1).

Let ρ,ρ',p be as in Lemma (6.1), and let $N = \mathrm{Ker}(h)$. Let $M \in \mathfrak{M}_{R[\pi]/J_\rho + J_{\rho'}}$. The finite filtration $M \supset NM \supset N^2M \supset \dots$ has successive quotients annihilated by $J_{\rho/\rho_p} = J_{\rho'/\rho'_p}$. Thus $G\left(\frac{R[\pi]}{J_\rho + J_{\rho'}}\right)$ is generated by classes of modules M killed by J_{ρ/ρ_p}. With Δ as in (1) above, we have, for such an M ,

$$\begin{aligned} \Delta[M]_{\rho,\rho'} &= [M]_\rho - [M]_{\rho'} \\ &= ([M]_\rho - [M]_{\rho/\rho_p}) - ([M]_{\rho'} - [M]_{\rho'/\rho'_p}). \end{aligned}$$

This shows that:

Im(Δ) <u>is generated by elements</u>

(2) $$\Delta_{\rho,p}(M) = [M]_{\rho} - [M]_{\rho/\rho_p}$$

<u>where</u> $\rho \in C(\pi)$, p <u>is a prime divisor of</u> $|\rho|$, and $M \in \mathfrak{M}_{R[\pi]/J_\rho + J_{\rho/\rho_p}}$.

If $\rho \in C(\pi)$ we shall identify $C(\rho)$ with a subset of $C(\pi)$, via the natural projection $\pi \to \rho$. For $\rho, \rho' \in C(\pi)$ we write

(3) $$\rho | \rho' \quad \text{if} \quad C(\rho) \subset C(\rho') ,$$

(i.e., if ρ is a quotient of ρ'), and

(4) $$\rho \| \rho' \quad \text{if} \quad \rho|\rho' \text{ and } |\rho| \,\|\, |\rho'| .$$

(For integers $n, m \geq 1$ the relation $n \| m$ signifies that $n | m$ and $\gcd(n, m/n) = 1$.)

Suppose that $\rho \| \rho'$. Then the natural projection $\rho' \to \rho$ admits a unique section $f: \rho \to \rho'$. The corresponding homomorphism $f: R[\rho] \to R[\rho']$ induces, by passage to quotients, a homomorphism $f: R(\rho) \to R(\rho')$. To see this one need only check it for $R = \mathbb{Z}$, and then tensor with R in general. In case $R = \mathbb{Z}$ the construction amounts to a commutative diagram of the following form, where t_r denotes the image of T in $\mathbb{Z}[T]/(T^r-1)$ for any $r \geq 1$, and ζ_r its image mod $\varphi_r(T)$.

$$\begin{array}{ccc} \mathbb{Z}[t_n] & \xrightarrow{f} & \mathbb{Z}[t_{nm}] \\ \downarrow & & \downarrow \\ \mathbb{Z}[\zeta_n] & \xrightarrow[\bar{f}]{} & \mathbb{Z}[\zeta_{nm}] \end{array} .$$

Here n and m are relatively prime integers, $f(t_n) = t_{nm}^e$, and $\bar{f}(\zeta_n) = \zeta_{nm}^e$, where $e \equiv 1 \bmod n$ and $e \equiv 0 \bmod m$.

The homomorphism $\bar{f}: R(\rho) \to R(\rho')$ defines, by restriction, a homomorphism

(5) $$N_\rho^{\rho'}: \; G(R(\rho')) \to G(R(\rho)) .$$

These homomorphsims are clearly functorial with respect to the natural projections $\rho' \to \rho$ $(\rho, \rho' \in C(\pi),\ \rho \| \rho')$.

Let $\rho \in C(\pi)$ and let p be a prime divisor of $|\rho|$.
Let $M \in \mathfrak{M}_{R[\pi]/J_\rho + J_{\rho/\rho_p}}$, so that $\Delta_{\rho,p}(M) = [M]_\rho - [M]_{\rho/\rho_p}$ is one of the generators (2) of $\mathrm{Im}(\Delta)$. Consider the commutative diagram of projections

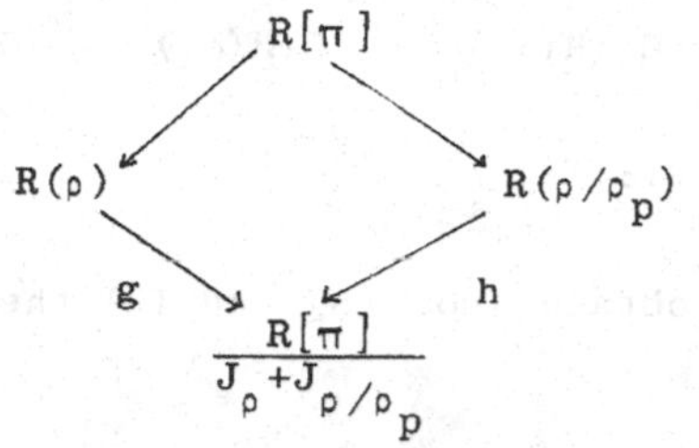

Since the homomorphism $\bar{f}\colon R(\rho/\rho_p) \to R(\rho)$ constructed above (note that $\rho/\rho_p \| \rho$) comes from a section $\rho/\rho_p \to \rho$ of the projection $\rho \twoheadrightarrow \rho/\rho_p$, and since ρ_p maps to 1 in $R[\pi]/J_\rho + J_{\rho/\rho_p}$, it follows that $g \circ \bar{f} = h$, and consequently $N^\rho_{\rho/\rho_p}[M]_\rho = [M]_{\rho/\rho_p}$. Thus in (2) we have

(6) $$\Delta_{\rho,p}(M) = [M]_\rho - N^\rho_{\rho/\rho_p}[M]_\rho \quad .$$

We come now to Lenstra's main idea. Define $N \in \mathrm{End}(\bigoplus_\rho G(R(\rho)))$ as follows: For $x \in G(R(\rho))$

(7) $$N(x) = \sum_{\rho_1 \| \rho} N^\rho_{\rho_1}(x) \quad .$$

First note that N <u>is an automorphism</u>. Indeed, we can totally order $C(\pi)$ so that $\rho | \rho' \Rightarrow \rho \le \rho'$. Then (7) shows that, for $x \in G(R(\rho))$, $N(x) = x + \sum_{\rho_1 \| \rho,\, \rho_1 \neq \rho} N^\rho_{\rho_1}(x) \equiv x \bmod \bigoplus_{\rho_1 < \rho} G(R(\rho_1))$. Thus the matrix of N relative to the ordered decomposition $\bigoplus_\rho G(R(\rho))$ is unipotent upper triangular, hence invertible.

It follows now from the exact sequence (1) that

(8) $$G(R[\pi]) \simeq (\bigoplus_\rho G(R(\rho)))/N(\mathrm{Im}(\Delta)) \quad .$$

For each $\rho \in C(\pi)$ let $G'(R(\rho))$ denote the subgroup of $G(R(\rho))$ generated by classes of modules $M \in \mathfrak{M}_{R(\rho)}$ annihilated by some power of $|\rho|$. The localization sequence of

$$R(\rho) \to R\langle\rho\rangle = R(\rho)[\tfrac{1}{|\rho|}]$$

yields an exact sequence

$$(9) \qquad 0 \to G'(R(\rho)) \to G(R(\rho)) \to G(R\langle\rho\rangle) \to 0 .$$

<u>Claim:</u> $N(\mathrm{Im}(\Delta)) = \bigoplus_{\rho} G'(R(\rho))$.

Once this is shown we obtain from (8) and (9) the desired isomorphism

$$\sigma_{\pi} : \bigoplus_{\rho} G(R\langle\rho\rangle) \xrightarrow{\ \simeq\ } G(R[\pi]) .$$

<u>Proof that</u> $N(\mathrm{Im}(\Delta)) \subset \bigoplus_{\rho} G'(R(\rho))$. Let ρ, p , and $M \in \mathfrak{M}_{R[\pi]/J_{\rho}+J_{\rho/\rho_p}}$ be as in (2) and put $x = [M]_{\rho} \in G(R(\rho))$. By (6),

$$\Delta_{\rho,p}(M) = x - N^{\rho}_{\rho/\rho_p}(x) ,$$

and by (2), such elements generate $\mathrm{Im}(\Delta)$.

$$N(x) = \sum_{\rho_1 \| \rho} N^{\rho}_{\rho_1}(x)$$

$$= \sum_{\rho_1\|\rho,\ p \mid |\rho_1|} N^{\rho}_{\rho_1}(x) + \sum_{\rho_1\|\rho,\ p\nmid|\rho_1|} N^{\rho}_{\rho_1}(x) .$$

If $\rho_1 \| \rho$ and $p \nmid |\rho_1|$ then $\rho_1 \| \rho/\rho_p$, and $N^{\rho}_{\rho_1}(x) = N^{\rho/\rho_p}_{\rho_1} N^{\rho}_{\rho/\rho_p}(x)$,

so $$N(\Delta_{\rho,p}(M)) = \sum_{\rho_1\|\rho,\ p \mid |\rho_1|} N^{\rho}_{\rho_1}[M]_{\rho} .$$

Since $pM = 0$ we have $N^{\rho}_{\rho_1}[M]_{\rho} \in G'(R(\rho_1))$ when $p \mid |\rho_1|$.

<u>Proof that</u> $\bigoplus_{\rho} G'(R(\rho)) \subset N(\mathrm{Im}(\Delta))$. Clearly $G'(R(\rho))$ is generated by classes of modules $M \in \mathfrak{M}_{R(\rho)}$ annihilated by some prime p dividing $|\rho|$. Since, by Lemma (6.3), the projection $R(\rho)/pR(\rho) \to R(\rho/\rho_p)/pR(\rho/\rho_p)$ has nilpotent kernel we can make a devissage and further restrict M

to be annihilated by J_{ρ/ρ_p} . Then with $x = [M]_\rho$ the calculation above shows that

$$N\Delta_{\rho,p}(M) = \sum_{\rho_1 \| \rho,\, p \| |\rho_1|} N^\rho_{\rho_1}[M]_\rho$$

$$= [M]_\rho + \sum_{\substack{\rho_1 \| \rho,\, p \| |\rho_1| \\ \rho_1 \neq \rho}} N^\rho_{\rho_1}[M]_\rho \quad .$$

If $\rho_1 \| \rho$, $\rho_1 \neq \rho$, and $p \| |\rho_1|$ then $N^\rho_{\rho_1}[M]_\rho \in G'(R(\rho_1))$ which, by induction on the number of prime divisors of $|\rho|$, is contained in $N(\mathrm{Im}(\Delta))$. The calculation above then shows that $[M]_\rho \in N(\mathrm{Im}(\Delta))$, as was to be shown.

This proves the first part of Theorem (5.1).

To complete the proof of Theorem (5.1) let us abbreviate the exact sequence (1) as

$$\Sigma\Sigma(\pi) \xrightarrow{\Delta(\pi)} \Sigma(\pi) \xrightarrow{\sigma(\pi)} G(R[\pi]) \to 0 \quad .$$

Suppose that $h : \tilde{\pi} \to \pi$ is a surjective homomorphism of finite abelian groups. Then it defines an embedding of $C(\pi)$ into $C(\tilde{\pi})$ so that the following diagram commutes

$$\begin{array}{ccccccc} \Sigma\Sigma(\pi) & \xrightarrow{\Delta(\pi)} & \Sigma(\pi) & \xrightarrow{\sigma(\pi)} & G(R[\pi]) & \to & 0 \\ \downarrow \Sigma\Sigma(h) & & \downarrow \Sigma(h) & & \downarrow G(h) & & \\ \Sigma\Sigma(\tilde{\pi}) & \xrightarrow[\Delta(\tilde{\pi})]{} & \Sigma(\tilde{\pi}) & \xrightarrow[\sigma(\tilde{\pi})]{} & G(R[\tilde{\pi}]) & \to & 0 \end{array} \tag{10}$$

where $\Sigma(h)$ and $\Sigma\Sigma(h)$ are the obvious inclusions of direct summands. Next note that the diagram

$$\begin{array}{ccc} \Sigma(\pi) & \xrightarrow{N(\pi)} & \Sigma(\pi) \\ \Sigma(h)\downarrow & & \downarrow\Sigma(h) \\ \Sigma(\tilde{\pi}) & \xrightarrow[N(\tilde{\pi})]{} & \Sigma(\tilde{\pi}) \end{array}$$

commutes, where $N(\pi)$ and $N(\tilde{\pi})$ are the automorphisms defined as in (7) above. This is immediate from (7) and the obvious fact that if $\rho \in C(\pi)$, $\rho_1 \in C(\tilde{\pi})$, and $\rho_1 \| \rho$, then $\rho_1 \in C(\pi)$. It follows now that (10) remains commutative with exact rows if we replace $\Delta(\pi)$ by $N(\pi)\circ\Delta(\pi)$ and $\Delta(\tilde{\pi})$ by $N(\tilde{\pi})\circ\Delta(\tilde{\pi})$. The commutativity of the resulting diagram

$$\begin{array}{ccc} \mathrm{Coker}(N(\pi)\circ\Delta(\pi)) & \xrightarrow{\sigma_\pi} & G(R[\pi]) \\ \downarrow & & \downarrow G(h) \\ \mathrm{Coker}(N(\tilde{\pi})\circ\Delta(\tilde{\pi})) & \xrightarrow[\sigma_{\tilde{\pi}}]{} & G(R[\tilde{\pi}]) \end{array}$$

is just the last assertion of Theorem(5.1), whose proof is now complete.

7. The ideals $\mathbb{Z}[T]\varphi_n(T) + \mathbb{Z}[T]\varphi_m(T)$.

We prove here some useful facts about cyclotomic polynomials (cf. [B2],[D],[R2] and [K-O]).

(7.1) Proposition: Let m be an integer > 1, and let $N: \mathbb{Q}(\zeta_m) \to \mathbb{Q}$ denote the norm.

1) If m is composite then $1 - \zeta_m$ is a unit of $\mathbb{Z}[\zeta_m]$ and $N(1-\zeta_m) = \varphi_m(1) = 1$.
2) If m is a power of a prime p then $N(1-\zeta_m) = \varphi_m(1) = p$ and $\mathbb{Z}[\zeta_m](1-\zeta_m)^{\varphi(m)} = \mathbb{Z}[\zeta_m]p$.

This is essentally proved in [B2]; we sketch the proof for convenience. Let Φ_m denote the set of primitive m^{th} roots of unity; there are $\varphi(m)$ of them.

$$\varphi_m(T) = \prod_{\zeta\in\Phi_m} (T-\zeta)$$

so

$$\varphi_m(1) = \prod_{\zeta\in\Phi_m} (1-\zeta) = N(1-\zeta_m) \quad .$$

If p is a prime then $\varphi_p(T) = 1+T+\ldots+T^{p-1}$ and $\varphi_{p^r}(T) = \varphi_p(T^{p^{r-1}})$ for $r \geq 1$. Hence $\varphi_{p^r}(1) = p$.

If $\zeta,\zeta' \in \Phi_m$ they generate the same cyclic group, so the congruences $\zeta \equiv 1$ and $\zeta' \equiv 1$ are equivalent conditions, i.e. $\mathbb{Z}[\zeta_m](1-\zeta) = \mathbb{Z}[\zeta_m](1-\zeta')$, and $1-\zeta'/1-\zeta$ is a unit of $\mathbb{Z}[\zeta_m]$. When $m = p^r$ as above we have $\mathbb{Z}[\zeta_m]p = \mathbb{Z}[\zeta_m](\prod_{\zeta\in\Phi_m}(1-\zeta)) = \mathbb{Z}[\zeta_m](1-\zeta_m)^{\varphi(m)}$.

This proves (2). To prove (1) it suffices to show that $\varphi_m(1) = 1$ when m is composite. We do this by induction on m. We have

$$1 + T + \ldots + T^{m-1} = \frac{T^m-1}{T-1}$$

$$= \prod_{d|m, d\neq 1} \varphi_d(T)$$

so

$$m = \prod_{d|m,d\neq 1} \varphi_d(1) = \varphi_m(1) \cdot \prod_{d|m,d\neq 1,m} \varphi_d(1) \quad .$$

If $m = p_1^{r_1} \dots p_s^{r_s}$ is the prime factorization of m then the factors $\varphi_{p_i^t}(1) = p\,(1 \le t \le r_i)$ contribute $p_i^{r_i}$ in the latter expression. For composite $d < m$, $\varphi_d(1) = 1$ by induction. Hence we have $\prod_{d|m, d\neq 1, m} \varphi_d(1) = m$, whence $\varphi_m(1) = 1$, as claimed.

For any prime p and integer $n \ge 1$ we write

$$n = n_{p'} n_p \tag{1}$$

where n_p is a power of p and $n_{p'}$ is prime to p.

(7.2) <u>Proposition</u>: <u>Let</u> p <u>be a prime and</u> n <u>an integer</u> ≥ 1. <u>Then</u>

$$\varphi_n(T) \equiv \varphi_{n_{p'}}(T)^{\varphi(n_p)} \bmod p .$$

We shall work in $\mathbb{F}_p[T]$. Let $q = n_{p'}$ and $p^s = n_p$ $(s \ge 0)$. Then

$$\begin{aligned} T^n - 1 &= \prod_{d|q} \prod_{0 \le i \le s} \varphi_{dp^i}(T) \\ &\parallel \\ (T^q - 1)^{p^s} &= \prod_{d|q} \varphi_d(T)^{p^s} . \end{aligned} \tag{2}$$

Since $T^q - 1$ has distinct roots over $\mathbb{F}_p$ the various $\varphi_d(T)\,(d|q)$ are relatively prime over $\mathbb{F}_p$. If $\zeta \in \Phi_{dp^i}$ (the primitive (dp^i)th roots of unity) then $\zeta^{p^i} \in \Phi_d$. It follows that $\varphi_{dp^i}(T)$ divides $\varphi_d(T^{p^i})$ in $\mathbb{Z}[T]$. In $\mathbb{F}_p[T]$ we have $\varphi_d(T^{p^i}) = \varphi_d(T)^{p^i}$. Hence the irreducible factors of φ_{dp^i} are among those of φ_d in $\mathbb{F}_p[T]$. From formula (2) and the remarks above we deduce that for each divisor d of q,

$$\varphi_d(T)^{p^s} = \prod_{0 \le i \le s} \varphi_{dp^i}(T) \quad \text{in } \mathbb{F}_p[T] . \tag{3}$$

Our claim that

$$\varphi_{dp^s}(T) = \varphi_d(T)^{\varphi(p^s)}$$

will be proved by induction on $s \geq 0$. It is clear for $s = 0$. For $s > 0$ we deduce from (3) and induction that

$$\varphi_d(T)^{p^s} = \varphi_{dp^s}(T) \cdot \prod_{0 \leq i \leq s} \varphi_d(T)^{\varphi(p^i)}$$
$$= \varphi_{dp^s}(T) \cdot \varphi_d(T)^e ,$$

where $e = \Sigma_{0 \leq i \leq s}\, \varphi(p^i) = p^{s-1}$. Thus $\varphi_{dp^s} = \varphi_d^{p^s - p^{s-1}} = \varphi_d^{\varphi(p^s)}$, as claimed.

The proof of Theorem (5.1) in section 6 relied on Lemma (6.1). We now give the:

Proof of Lemma (6.1): Let $\rho = \pi/\sigma$ and $\rho' = \pi/\sigma'$ be two distinct cyclic quotient groups of π such that $J\rho + J\rho' \neq R[\pi]$. In order to prove the assertions of Lemma (6.1) it clearly suffices to do so when $R = \mathbb{Z}$, the general case being then deducible by base change $\mathbb{Z} \to R$.

If ρ has order n we can identify $\mathbb{Z}(\rho)$ with $\mathbb{Z}[\zeta]$ where ζ is a primitive n^{th} root of unity, image of a generator of ρ in $\mathbb{Z}[\rho]$. Let $\chi : \pi \to \mathbb{Z}(\rho)^*$ be the projection homomorphism. Either $\sigma \not\subset \sigma'$ or $\sigma' \not\subset \sigma$; say there is an $s \in \sigma'$, $s \notin \sigma$. Then $\chi(s) \neq 1$ is a root of unity in $\mathbb{Z}[\zeta]$. In the quotient $\mathbb{Z}[\pi]/J_\rho + J_{\rho'}$ of $\mathbb{Z}[\zeta]$, $\chi(s)$ maps to 1. Since, by assumption, $J_\rho + J_{\rho'} \neq \mathbb{Z}[\pi]$, $1-\chi(s)$ is not a unit. Therefore by Proposition (7.1), $\chi(s)$ has order a power of the (unique) prime p that annihilates $\mathbb{Z}[\pi]/J_\rho + J_{\rho'}$. It follows that $\sigma\sigma'/\sigma$ is a p-group. Similarly $\sigma\sigma'/\sigma'$ is a p-group, and so $\sigma\sigma'/\sigma \cap \sigma'$ is a p-group. Consequently $\rho = \pi/\sigma$ and $\rho' = \pi/\sigma'$ coincide modulo their Sylow p-subgroups, ρ_p and ρ'_p respectively.

Write $\mathbb{Z}(\rho)$ in the form $\mathbb{Z}[T]/(\varphi_n(T))$, and $\mathbb{Z}(\rho/\rho_p)$ correspondingly as $\mathbb{Z}[T]/(\varphi_{n_{p'}}(T))$. Reducing mod p we obtain

$$\mathbb{Z}(\rho/\rho_p)/p\mathbb{Z}(\rho/\rho_p) = \mathbb{F}_p[T]/(\varphi_{n_{p'}}(T))$$

and

$$\mathbb{Z}(\rho)/p\,\mathbb{Z}(\rho) = \mathbb{F}_p[T]/(\varphi_n(T))$$
$$= \mathbb{F}_p[T]/(\varphi_{n_{p'}}(T)^{\varphi(n_p)})$$

in view of Proposition (7.2). Thus there is a natural projection $h: \mathbb{Z}(\rho)/p\mathbb{Z}(\rho) \to \mathbb{Z}(\rho/\rho_p)/p\mathbb{Z}(\rho/\rho_p)$ with nilpotent kernel. This completes the proof.

We record here, more precisely, the information given by the above argument.

(7.3) <u>Theorem</u>: <u>For integers</u> $1 \leq n < m$, <u>put</u>

$$q_{n,m} = \mathbb{Z}[T]\varphi_n(T) + \mathbb{Z}[T]\varphi_m(T) .$$

<u>We have</u> $q_{n,m} = \mathbb{Z}[T]$ <u>unless, for some prime</u> p <u>and</u> $r \geq 1$ <u>we have</u> $m = np^r$. <u>In this case we have</u>

$$q_{n,np^r} = \mathbb{Z}[T]p + \mathbb{Z}[T](\varphi_{n_{p'}}(T))^{\varphi(n_p)}$$

<u>(notation as in (1) above)</u>. <u>(This depends on</u> n <u>and</u> p <u>but not</u> r.)
<u>Hence</u>

$$\mathbb{Z}[T]/q_{n,np^r} \simeq F_p[T]/F_p[T](\varphi_{n_{p'}}(T))^{\varphi(n_p)} .$$

<u>The radical of</u> q_{n,np^r} <u>is</u>

$$\sqrt{q_{n,np^r}} = \mathbb{Z}[T]p + \mathbb{Z}[T]\varphi_{n_{p'}}(T)$$

<u>so</u>

$$\mathbb{Z}[T]/\sqrt{q_{n,np^r}} \simeq \mathbb{Z}[\zeta_{n_{p'}}]/p\mathbb{Z}[\zeta_{n_{p'}}] .$$

Assume that $q_{n,m} \neq \mathbb{Z}[T]$. Let $d = \gcd(m,n)$ and write $n = dn'$ and $m = dm'$. We have $\mathbb{Z}[T]/q_{n,m} = \mathbb{Z}[\zeta_m]/\mathbb{Z}[\zeta_m]\varphi_n(\zeta_m)$. The image t of T in this ring satisfies $t^m = 1 = t^n$ whence $t^d = 1$ and so $\mathbb{Z}[\zeta_n]\varphi_n(\zeta_m)$ contains $1 - \zeta_m^d = 1 - \zeta_{m'}$. Our assumption implies that $1 - \zeta_{m'}$ is not a unit. It follows therefore from Proposition (7.1) that m' is a power, p^r, of some prime p, and $\mathbb{Z}[T]/q_{n,m}$ then has characteristic p. By symmetry we conclude that n' must be a power of the same prime p, whence $m = np^r$. Let $n_{p'} = q$ and $n_p = p^s$. Modulo p we have, from Proposition (7.2) $\varphi_n \equiv \varphi_q^{\varphi(p^s)}$ and $\varphi_m \equiv \varphi_q^{\varphi(p^{s+r})}$.

Therefore we can write

$$\begin{aligned} q_{n,m} &= \mathbb{Z}[T]p + \mathbb{Z}[T]\varphi_q^{\varphi(p^s)} + \mathbb{Z}[T]\varphi_q^{\varphi(p^{s+r})} \\ &= \mathbb{Z}[T]p + \mathbb{Z}[T]\varphi_q^{\varphi(p^s)} \end{aligned}$$

and $\mathbb{Z}[T]/q_{n,m} = \mathbb{F}_p[T]/\mathbb{F}_p[T]\varphi_q^{\varphi(p^s)}$. Since φ_q has distinct roots over F_p the ring $\mathbb{F}_p[T]/\mathbb{F}_p[T]\varphi_q = \mathbb{Z}[\zeta_q]/p\mathbb{Z}[\zeta_q]$ has zero nil radical, whence $\sqrt{q_{n,m}}$ is generated by p and φ_q. This proves Theorem (7.37).

(7.4) <u>Corollary</u>: <u>Let</u> $1 \leq n < m$ <u>be integers. If</u> m <u>is not a prime power multiple of</u> n <u>then</u> $\varphi_m(\zeta_n)$ <u>is a unit of</u> $\mathbb{Z}[\zeta_n]$, <u>and</u> $\varphi_n(\zeta_m)$ <u>is a unit of</u> $\mathbb{Z}[\zeta_m]$. <u>Suppose, on the other hand, that</u> $m = np^r$ <u>for some prime</u> p <u>and</u> $r \geq 1$. <u>Then</u>

$$\mathbb{Z}[\zeta_n]\varphi_m(\zeta_n) = (\prod_{\mathfrak{p}|p} \mathfrak{p})^{\varphi(n_p)} = \mathbb{Z}[\zeta_n]p$$

<u>and</u>

$$\mathbb{Z}[\zeta_m]\varphi_n(\zeta_m) = (\prod_{\mathfrak{p}|p} \mathfrak{p})^{\varphi(n_p)} .$$

We have

$$\mathbb{Z}[\zeta_n]/\mathbb{Z}[\zeta_n]\varphi_m(\zeta_n) = \mathbb{Z}[\zeta_m]/\mathbb{Z}[\zeta_m]\varphi_n(\zeta_m) = \mathbb{Z}[T]/q_{n,m}$$

which, by Theorem (7.3) is zero unless $m = np^r$ as above, in which case it is isomorphic to $F_p[T]/F_p[T]\varphi_{n_p'}(T)^{\varphi(n_p)} = F_p[T]/F_p[T]\varphi_n(T)$ (by Proposition (7.2)) $= \mathbb{Z}[\zeta_n]/p\mathbb{Z}[\zeta_n]$. It follows that the prime divisors of $\varphi_m(\zeta_n)$ in $\mathbb{Z}[\zeta_n]$ are the primes above p, and that they all occur with exponent $\varphi(n_p)$. One can similarly deduce the prime factorization of $\mathbb{Z}[\zeta_m]\varphi_n(\zeta_m)$.

For more refined results on the degrees over $\mathbb{Q}$ of the units $\varphi_m(\zeta_n)/p$, as above, see [K-O].

References:

[A] Almkvist, G.: "The Grothendieck ring of endomorphisms", J.Algebra 28 (1974), 375-388.

[B1] Bass, H.: "Algebraic K-theory", W.A.Benjamin, New York (1968).

[B2] ------ : "The Dirichlet unit theorem, induced characters and Whitehead groups of finite groups", Topology 4 (1966), 391-410.

[B3] ------ : "The Grothendieck group of the category of abelian group automorphisms of finite order", Preprint, Columbia University (1979).

[Br] Brumer, A.: "The class group of all cyclotomic integers",

[CR] Curtis, C. - I.Reiner: "Representation theory of finite groups and associative algebras", Interscience, New York (1962).

[D] Diederichsen, F.-E.: "Über die Ausreduktion ganzzahliger Gruppendarstellungen bei arithmetischer Äquivalenz", Abh.Math.Sem.Univ.Hamburg 13 (1940), 357-412.

[FS] Franks, J. - M.Shub: "The existence of Morse-Smale diffeomorphisms", Topology (to appear).

[G1] Grayson, D.:"The K-theory of endomorphisms", J.Algebra 48 (1977), 439-446.

[G2] ------ : "SK_1 of an interesting principal ideal domain", preprint, Columbia University.

[K-O] Kurshan, R.P. - A.M.Odlyzko: "Values of cyclotomic polynomials at roots of unity", preprint, Bell Laboratories, Muray Hill, New Jersey (1980).

[L] Lenstra, H.: "Grothendieck groups of abelian group rings", J.Pure and Applied Algebra (to appear)

[M] Milnor, J.: "Introduction to algebraic K-theory, I", Annals of Math.Studies, Princeton Univ.Press (1971).

[Q] Quillen, D.: "Higher algebraic K-theory, I" Proc.Battelle Conf. on Alg.K-theory, Springer Lecture Notes 341 (1973), 85-147.

[R1] Reiner, I.: "Topics in integral representation theory", Springer LN 744 (1979), 1-143.

[R2] ------ : "On Diederichsen's formula for extensions of lattices", preprint, Univ.of Illinois, Urbana, Illinois.

[SS] Shub, M. - D.Sullivan: "Homology theory and dynamical systems", Topology 14 (1975), 109-132.

ALGEBRAIC GEOMETRY OF TERNARY QUADRATIC FORMS AND ORDERS IN QUATERNION ALGEBRAS

Juliusz Brzezinski

1. INTRODUCTION. Let f be a ternary quadratic form with integral coefficientes. The form f defines two objects: a Z-scheme $M(f)$ and a Z-order $O(f)$, where Z denotes the integers. If, for example, $f = X^2+Y^2+Z^2$, then $M(f) = \mathrm{Proj}(Z[X,Y,Z]/(f))$, and $O(f) = Z + Zi + Zj + Zk$ is the ring of the Lipschitz integral quaternions. Our aim is to discuss some relations between these two kinds of objects and show how to use them in order to prove the following result:

MAIN THEOREM. Let E be a regular extension of genus 0 of the field of fractions F of a perfect Dedekind ring A (that is, the field A/p is perfect for each prime ideal $p \neq (0)$ of A). If M and M' are two relatively minimal models of E over A (see definitions below), then there is a sequence of relatively minimal models $M_0 = M, M_1, \ldots, M_n = M'$ such that M_{i+1} is an elementary transform of M_i for $0 \leq i < n$.

This result corresponds to a similar one in the algebraic geometry of surfaces over fields, which gives exactly the same picture of relations between the relatively minimal models of an extension E/F, where F is an algebraically closed field, E has transcendence degree 2 over F, and there are many relatively minimal models of E/F (this is the case if and only if E is the field of rational functions on a ruled surface over F, that is, $E = F(C \times P^1)$, where C is a curve and P^1 is a projective line over F). For this classical result see [8, Chap. V, Theorem 1] or [7, Theorem 2.1]. The case of an extension E/F of genus 0 is just the case when there are many relatively minimal arithmetical surfaces, which are models of the extensions E/A. If the genus of E over F is at least one, there is exactly one relatively minimal model of E/A (see [9, p. 155]).

In order to simplify notations and some formal constructions, we shall often assume that $A = Z$. Complete proofs of the results discussed here will be given in [3] and [4].

2. MODELS. $M(f)$ is a Spec(Z)-scheme but it will be more convenient to look at $M(f)$ as a model of a suitable extension E/Z in the following sense:

DEFINITION. Let E be a field containing a (Dedekind) ring A. An affine model of E/A is the set M of all localizations $R_{\underline{p}}$, $\underline{p} \in \mathrm{Spec}(R)$, where the ring R contains A, is contained in E and has E as its field of fractions. A model of E/A is a union M of a finite number of affine models satisfying the following condition: if $A \subset V \subset E$, where V is a valuation ring with field of fractions E, then there is at most one local ring $\mathcal{O} \in M$ such that V dominates $\mathcal{O}$ (notation: $V > \mathcal{O}$). A model M of E/A is complete if for each valuation ring V (as above), there is exactly one $\mathcal{O} \in M$ such that $V > \mathcal{O}$. We define the dimension of M as the maximum of the Krull dimensions $\dim \mathcal{O}$, where $\mathcal{O} \in M$.

If $f \in Z[X_0,X_1,X_2]$ is an arbitrary irreducible ternary quadratic form of rank 3, then $M(f)$ is defined in the following way. We consider the ring $Z[x_0,x_1,x_2] = Z[X_0,X_1,X_2]/(f)$, its field of fractions E^*, and inside of it the rings $R_0 = Z[x,y]$, $R_1 = Z[1/x,y/x]$ and $R_2 = Z[1/y,x/y]$, where $x = x_1/x_0$, $y = x_2/x_0$. Let $E(\subset E^*)$ be the (common) field of fractions of the rings R_i. $M(f)$ is the model of E/Z which is the union of the affine models defined by the rings R_i. $M(f)$ is a complete model of E/Z, and since its dimension is equal to two, while the ground ring $A = Z$ is a Dedekind ring, we say that $M(f)$ is an arithmetical surface.

Since the form f is equivalent over Q to a diagonal one, we have $E = Q(x,y)$, where $ax^2+by^2 = 1$, $ab \neq 0$. This is a general form of finitely generated regular extensions E/Q of genus 0. Now we can reformulate our problem: we have an extension E/Z, where $E = Q(x,y)$, $ax^2+by^2 = 1$, $ab \neq 0$, and we want to investigate the models of E/Z. We can always construct a model $M(f)$ of E/Z by taking e.g. $f = a_0X_0^2+a_1X_1^2+a_2X_2^2$, where $a_1/a_0 = -a$, $a_2/a_0 = -b$, $a_i \in Z$. Such a model is in fact singular - a model M is regular if all its local rings are regular. But we have:

THEOREM 1. ([2],[3]) Each regular extension E/Z of genus 0 has a regular model $M(f)$, where f is an integral ternary quadratic form. Moreover, $M(f)$ is regular if and only if the discriminant $\delta(f)$ of f is square-free.*)

If M and M' are two models of E/A, then we say that M dominates M' (notation: $M > M'$) if each local ring of M dominates a local ring of M'. This gives a partial ordering in the set of models of E/A. A relatively minimal model of E/A is a complete regular model which is minimal in the set of complete regular models with respect to this partial ordering.

THEOREM 2. ([3]) Relatively minimal models of a regular extension E/Z of genus 0 exist and each relatively minimal model of E/Z is equal to a model $M(f)$ for an integral ternary quadratic form f. Moreover, $M(f)$ is relatively minimal if and only if $\delta(f) = \delta_{E/Z}$, where $\delta_{E/Z} = \Pi\, p$, $(a,b)_p = -1$ ($(a,b)_p$ denotes the Hilbert symbol and $E = Q(x,y), ax^2+by^2 = 1$).

The existence of relatively minimal models of E/Z was proved in [9] (for arbitrary genus and arbitrary perfect Dedekind ring A instead of Z), while the problem of investigation of such models when the extension has genus 0 was left open (see [9, pp. 157-8]).

Now a few words about orders.

3. ORDERS. Instead of quadratic forms f over a (Dedekind) ring A with field of fractions F it will be more convenient to consider quadratic spaces (V,q) over F and A-lattices on V. Here $q(x) =: f(x)$, where $x = (x_i)$, $x = \Sigma x_i e_i$, e_i is a basis of V. The quadratic space (V,q) defines its (even)

*) If $f = \Sigma a_{ij}X_iX_j$, $0 \le i \le j \le 2$, then $\delta(f) = (1/2)|\det[b_{ij}]|$, where $b_{ij} = a_{ij}$ for $i \neq j$ and $b_{ii} = 2a_{ii}$.

Clifford algebra $C_0(V,q) = (\bigotimes T^{2r}(V))/(x \otimes x - q(x))$, where $T^i(V)$ denotes the i-th tensor power of V. We shall also write $C_0(f)$ instead of $C_0(V,q)$. We will assume that $ch(F) \neq 2$ even if this assumption is not necessary (see [3]). We denote by $[x_1,\ldots,x_{2r}]$ the image of $x_1 \otimes \cdots \otimes x_{2r}$ in $C_0(V,q)$. $C_0(V,q)$ is a generalized quaternion algebra over F and if L is an A-lattice on V, then it defines the corresponding A-order $O(L)$ in $C_0(V,q)$: $O(L)$ is generated as an A-modul by 1 and the products $a[x_1,\ldots,x_{2r}]$, where $x_i \in L$ and $an(L)^r \subseteq A$. Here $n(L)$ is an A-ideal in F generated by $q(x)$, $x \in L$. The definition of $O(L)$ was given by Eichler ([5, p. 96]). The first important question is, which orders in $C_0(V,q)$ correspond to lattices L on V. The answer is given by

THEOREM 3. ([4]). _Let_ Λ _be an A-order in_ $C_0(V,q)$. _Then_ $\Lambda = O(L)$ _for an A-lattice_ L _on_ V _if and only if_ Λ _is a Gorenstein order_.

(Recall that Λ is a Gorenstein order if Λ is Λ-injective in the category of (left) Λ-lattices.)

If f is a quadratic form corresponding to a basis of a free A-lattice L, we shall write $O(f)$ instead of $O(L)$. We have

LEMMA. ([4]) $\delta(O(f)) = (\delta(f))$, ($\delta(\Lambda)$ denotes the discriminant of the order Λ).

Now we shall discuss some relations between the objects $M(f)$ and $O(f)$.

4. _MODELS AND ORDERS_. Let $M(f)$ be a model of E/Z, and $O(f)$ the corresponding Z-order in $C_0(f)$.

THEOREM 4. ([3]) (a) $M(f)$ _is a regular model if and only if the order_ $O(f)$ _is hereditary_.

(b) $M(f)$ _is a realtively minimal model if and only if the order_ $O(f)$ _is maximal_.

PROOF. $M(f)$ is regular (relatively minimal) if and only if $\delta(f)$ is square-free ($\delta(f) = \delta_{E/Z}$), so by the Lemma, $M(f)$ is regular (relatively minimal) if and only if $\delta(O(f))$ is square-free ($(\delta(O(f)) = (\delta_{E/Z}))$. It is well known that an order Λ in a quaternion algebra is hereditary if and only if its discriminant is square-free, while it is maximal if and only if its discriminant is equal to the ground ideal of the algebra, which is exactly $(\delta_{E/Z})$ (see [3, Corollary 4.3]).

Let us consider two relatively minimal models M and M' of E/Z. Since we want to investigate some relations between them and, in particular, to show how to prove the Main Theorem, we have to recall some notions concerning elementary transformations of models. If M is a model of E/Z, then the fiber $M_{(p)}$, for a prime ideal $(p) \neq (0)$ is the set of local rings $\mathcal{O} \in M$ such that $\mathcal{O} > Z_{(p)}$. If $M = M(f)$, then to each local ring $\mathcal{O} \in M_{(p)}$ corresponds a uniquely determined homogeneous prime ideal $\neq (x_0,x_1,x_2)$ of $Z/(p)[x_0,x_1,x_2]$, where $f(x_0,x_1,x_2) = 0$, and conversely. If f modulo p has rank 3, the conic $f(x_0,x_1,x_2) = 0$ over $Z/(p) =: F_p$ is isomorphic to a projective line, so we say that the fiber $M_{(p)}$ is a projective line over F_p. If f modulo p is a product of two linear factors, $M_{(p)}$ is a union of two projective lines over F_p.

DEFINITION. By a <u>blowing-up</u> of M at an F_p-rational point of $M_{(p)}$ (that is, a point $\mathcal{O}_0 \in M_{(p)}$ such that $\mathcal{O}_0/\text{max.id.}(\mathcal{O}_0) \cong F_p$), we mean a model $\overline{M}$ such that a) $\overline{M} > M$, b) $\overline{M}_{(p)} = M_{(p)} \cup \ell$, where ℓ is a projective line over F_p, c) if $\mathcal{O} \in \ell$, then $\mathcal{O} > \mathcal{O}_0$, and $\overline{M} \setminus \ell = M \setminus \{\mathcal{O}_0\}$ (Hence the model $\overline{M}$ has a projective line which is contracted to the point $\mathcal{O}_0$ of M, while all the remaining points of the two models are equal.) By an <u>elementary transform</u> of M we mean a model M' such that there is a model $\overline{M}$, which is a blowing-up of both M and M' in such a way that $\overline{M}_{(p)} = M_{(p)} \cup \ell = M'_{(p)} \cup \ell'$, ℓ and ℓ' are projective lines, $\ell \neq \ell'$, ℓ is contracted to an F_p-rational point $\mathcal{O}_0$ of M, while ℓ' is contracted to an F_p-rational point $\mathcal{O}'_0$ of M'.

The following result connects models and orders:

THEOREM 5. ([3]). *Let* $M = M(f)$ *be a relatively minimal model of* E/Z *and* $\Gamma = O(f)$ *a corresponding maximal order. If* Λ *is a hereditary order* $\Lambda \subsetneq \Gamma$, *then* $\Lambda = O(\overline{f})$, *where* $\overline{f}$ *is an integral ternary quadratic form, and* $\overline{M} =: M(\overline{f})$ *is a blowing-up of* M *at an* F_p*-rational point* $\vartheta_0 \in M_{(p)}$, *where* $p \nmid \delta_{E/Z}$.

If $\Lambda = \Gamma \cap \Gamma'$, *where* Γ' *is the second maximal order containing the hereditary order* Λ, *then* $\Gamma' = O(f')$ *for an integral ternary quadratic form* f' *and* $M(f') =: M'$ *is a relatively minimal model of* E/Z *which is an elementary transform of* M *(at* ϑ_0*).*

Now it is easy to give a proof of the Main Theorem. If $M = M(f)$, $M' = M(f')$ are two relatively minimal models of E/Z and $\Gamma = O(f)$, $\Gamma' = O(f')$ are corresponding maximal orders, we construct a chain of maximal orders $\Gamma_0 = \Gamma$, $\Gamma_1, \ldots, \Gamma_n = \Gamma'$ such that the intersections $\Gamma_i \cap \Gamma_{i+1} = \Lambda_{i+1}$ are hereditary for $i = 0, \ldots, n-1$:

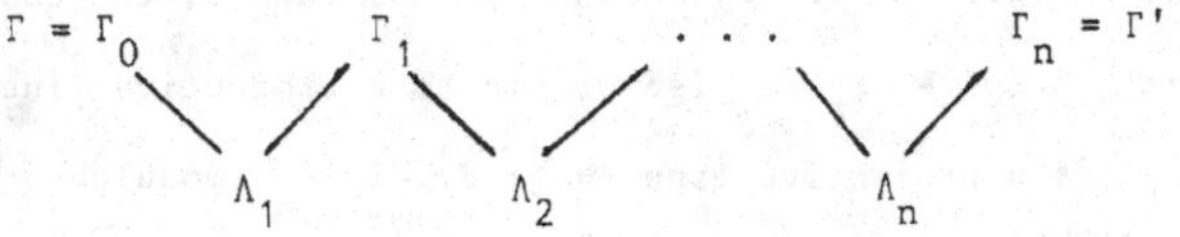

Such chains of orders were considered by Harada [6], who called them "paths of maximal orders". The construction of a path is quite easy: we take the left distance ideal I from Γ' to Γ, that is, all x in the quaternion algebra such that $x\Gamma' \subset \Gamma$, then we factorize I into a product of integral maximal ideals $I = P_1 \ldots P_n$, and define $\Gamma_i = O_r(P_i)$ the right order of P_i for $i = 1, \ldots, n$. It is easy to check that Λ_{i+1} is hereditary, so Theorem 5 can be applied.

The next result shows that our proof is not accidental.

THEOREM 6. ([3]). *If* $M = M(f)$ *is a relatively minimal model of* E/Z *and* $\Gamma = O(f)$ *is a corresponding maximal order, then there is a one-to-one correspondence between the* F_p*-rational points of the fiber* $M_{(p)}$ *and the integral left ideals of* Γ *with norm* (p) *such that the elementary transforms of* M *at two* F_p*-rational points* $\vartheta_1, \vartheta_2 \in M_{(p)}$ *are equal if and only if the right orders of the* left ideals of Γ corresponding to these points are isomorphic.

Hence each elementary transform of $M = M(f)$ can be described in the following way: we take an F_p-rational point $\mathcal{O}_0 \in M_{(p)}$, the corresponding integral left ideal P of $\Gamma = O(f)$ with norm (p), then the right order $O_r(P) = \Gamma'$. The corresponding model $M(f')$ with $O(f') = \Gamma'$ is an elementary transform of M at $\mathcal{O}_0$.

Our considerations were limited by some regularity assumptions: quadratic forms f should have square-free discriminant, models $M(f)$ should be regular, and orders $O(f)$ hereditary. It is possible to extend these considerations to arbitrary integral ternary quadratic forms, corresponding models and Gorenstein orders in quaternion algebras. This will be discussed in the second part of [3].

REFERENCES

[1] A. Białynicki-Birula, Remarks on relatively minimal models of fields of genus 0, I, II, III, Bull. Acad. Polon. Sci. Sér. Sci. Math. Astronom. Phys. 15(1967), 301-307; 16(1968), 81-85; 17(1969), 419-423.

[2] J. Brzezinski, Models for some fields of genus 0 determined by forms, Bull. Acad. Polon. Sci. Sér. Sci. Math. Astronom. Phys. 17(1969), 473-475.

[3] J. Brzezinski, Arithmetical quadratic surfaces of genus 0, I, Math. Scand. (to appear).

[4] J. Brzezinski, A characterization of Gorenstein orders in quaternion algebras (to appear).

[5] M. Eichler, Quadratische Formen und Ortogonale Gruppen, Springer-Verlag, Berlin-Heidelberg-New York, 1974.

[6] M. Harada, Multiplicative ideal theory in hereditary orders, J. Math. Osaka City Univ. 14(1963), 83-106.

[7] R. Hartshorne, Curves with high self-intersection on algebraic surfaces, Publ. Math. I.H.E.S. 36(1969), 111-126.

[8] I.R. Šafarevič, Algebraic surfaces, Proc. Steklov Inst. Math. 75(1965).

[9] I.R. Šafarevič, Lectures on minimal models and birational transformations of two dimensional schemes, Tata Inst. of Fund. Research, Bombay, 1966.

Preprojective Lattices over Classical Orders

M. Auslander

and

S. O. Smalø

Introduction

We assume throughout these lectures that R is a complete discrete valuation ring with field of quotients K. By an R-order, we mean an R-algebra which is a finitely generated free R-module such that $K \otimes_R \Lambda$ is semisimple. A Λ-lattice is a finitely generated Λ-module which is a free R-module. It is well known that a Λ-module M is a lattice if and only if it is isomorphic to a submodule of a finitely generated projective Λ-module. The category of all finitely generated Λ-modules is denoted by mod Λ and the full subcategory of mod Λ whose objects are the lattices is denoted by $L(\Lambda)$. More generally, by a subcategory $\underline{\underline{C}}$ of mod Λ we always mean a full subcategory of mod Λ having the property that if $C_1 \coprod C_2 \simeq C$ with C in $\underline{\underline{C}}$ then C_1 and C_2 are in $\underline{\underline{C}}$, where $C_1 \coprod C_2$ means the (direct) sum of C_1 and C_2.

One of the main results in these lectures is that there is a unique sequence of subcategories $\underline{\underline{P}}_0, \underline{\underline{P}}_1, \ldots, \underline{\underline{P}}_\infty$ of ind $L(\Lambda)$, the subcategory of $L(\Lambda)$ consisting of the indecomposable lattices, having the following properties:

1) ind $L(\Lambda) = \bigcup_{i \geq 0} \underline{\underline{P}}_i$

2) $\underline{\underline{P}}_i \cap \underline{\underline{P}}_j = \emptyset$ if $i \neq j$

3) Each $\underline{\underline{P}}_i$ with $i < \infty$ has the following properties:

a) $\underline{\underline{P}}_i$ has only a finite number of nonisomorphic objects;

b) For each X in $\bigcup_{j \geq i} \underline{\underline{P}}_j$, there is a surjection $\coprod_{k \in K} U_k \to X$, where $\{U_k\}_{k \in K}$ is a finite family of objects in $\underline{\underline{P}}_i$ (some of the U_k may be isomorphic);

c) If $\underline{\underline{P}}'_i$ is a proper subcategory of $\underline{\underline{P}}_i$, then $\underline{\underline{P}}'_i$ does not satisfy b), i.e., there is an X in $\bigcup_{j \geq i} \underline{\underline{P}}_j$ such that there is no surjection $\coprod_{k \in K} U_k \to X$ with $\{U_k\}_{k \in K}$ a finite family of lattices in $\underline{\underline{P}}'_i$.

This uniquely determined sequence of subcategories of $L(\Lambda)$ is called the <u>preprojective partition</u> of $L(\Lambda)$. Also a lattice L is said to be a <u>preprojective lattice</u> if it is isomorphic to a sum of lattices in $\bigcup_{i < \infty} \underline{\underline{P}}_i$.

Clearly, the preprojective partition of $L(\Lambda)$ is the exact analogue of the preprojective partitions of the categories of finitely generated modules over artin algebras introduced in [4] as are the rest of the main results in these lectures. In particular, if $\underline{\underline{P}}_0, \underline{\underline{P}}_1, \ldots, \underline{\underline{P}}_\infty$ is the preprojective partition of $L(\Lambda)$, then

I) $\underline{\underline{P}}_i = \emptyset$ for some $i < \infty$ if and only if ind $L(\Lambda)$ has only a finite number of nonisomorphic objects;

II) Every lattice is preprojective, i.e., $\underline{\underline{P}}_\infty = \emptyset$, if and only if ind $L(\Lambda)$ has only a finite number of nonisomorphic objects;

III) If $\underline{\underline{P}}_\infty \neq \emptyset$, then given any X in $\underline{\underline{P}}_\infty$ there is a surjection $\coprod_{k\in K} U_k \to X$ which is not a splittable surjection with $\{U_k\}_{k\in K}$ a finite family of objects in $\underline{\underline{P}}_\infty$. From this it will follow that given any subcategory $\underline{\underline{C}}$ of $\underline{\underline{P}}_\infty$ with only a finite number of nonisomorphic objects, there is an X in $\underline{\underline{P}}_\infty$ such that there is no surjection $\coprod_{k\in K} U_k \to X$ with $\{U_k\}_{k\in K}$ a finite family of objects in $\underline{\underline{C}}$.

Not only are these results modeled on those for artin algebras but so are their proofs, subject in some cases to suitable modifications. In those cases where the necessary modifications are nonexistent or obvious, we will simply refer to the work on artin algebras. No effort has been made to give analogues for lattices of all the results obtained for artin algebras in [4]. Only the most basic results have been dealt with; the rest are left to the interested reader. In [1] and [2] higher dimensional orders and lattices were introduced. The fact that preprojective partitions also exist for these types of lattices will be shown in another publication [3]. The proof given here is essentially the same but is included for the convenience of the reader.

Finally, we would like to thank K. Roggenkamp for the opportunity to present this material at this conference.

1. Covers

Let mod Λ be the category of finitely generated modules over the R-order Λ. If $\underline{\underline{D}}$ is a subcategory of mod Λ, we denote by ind $\underline{\underline{D}}$ the subcategory of mod Λ consisting of the indecomposable objects in $\underline{\underline{D}}$ (remember $\underline{\underline{D}}$ is closed under summands) and by add $\underline{\underline{D}}$ the subcategory of mod Λ consisting of modules isomorphic to finite sums of objects in ind $\underline{\underline{D}}$. We denote ind mod Λ by ind Λ. Clearly add ind Λ = mod Λ. The definitions, results and proofs given in this section are essentially the same as those given in [4] for artin algebras.

Suppose $\underline{\underline{D}}$ is a subcategory of ind Λ. By a cover of $\underline{\underline{D}}$ we mean a subcategory $\underline{\underline{C}}$ of $\underline{\underline{D}}$ such that for each D in $\underline{\underline{D}}$ there is a surjection $C \to D$ with C in add $\underline{\underline{C}}$. A cover $\underline{\underline{C}}$ of $\underline{\underline{D}}$ is said to be a <u>minimal cover</u> of $\underline{\underline{D}}$ if no proper subcategory of $\underline{\underline{C}}$ is a cover of $\underline{\underline{D}}$. We will be mainly concerned with minimal covers, one of the reasons being that if they exist they are unique. To explain why this is the case, it is necessary to introduce the notion of a splitting projective.

An object D in $\underline{\underline{D}}$ is said to be a splitting projective in $\underline{\underline{D}}$ if every surjection D' → D with D' in add D is a splittable surjection. We denote by $\underline{\underline{P}}_0(\underline{\underline{D}})$ the subcategory of $\underline{\underline{D}}$ consisting of the splitting projectives in $\underline{\underline{D}}$.

As in [4, section 2] one can show the following.

Proposition 1.1. Let $\underline{\underline{C}}$ be a cover for $\underline{\underline{D}}$. Then

a) $\underline{\underline{C}} \supset \underline{\underline{P}}_0(\underline{\underline{D}})$

b) $\underline{\underline{C}}$ is a minimal cover of $\underline{\underline{D}}$ if and only if $\underline{\underline{C}} = \underline{\underline{P}}_0(\underline{\underline{D}})$.

As an immediate consequence of Proposition 1.1, we have the following.

Corollary 1.2. $\underline{\underline{D}}$ has a minimal cover if and only if $\underline{\underline{P}}_0(\underline{\underline{D}})$ is a cover for $\underline{\underline{D}}$. Moreover, if $\underline{\underline{P}}_0(\underline{\underline{D}})$ is a cover for $\underline{\underline{D}}$, then $\underline{\underline{P}}_0(\underline{\underline{D}})$ is the unique minimal cover of $\underline{\underline{D}}$.

Consquently a nonempty subcategory $\underline{\underline{D}}$ ind Λ with $\underline{\underline{P}}_0(\underline{\underline{D}}) = \emptyset$ has no minimal cover. For example, if $\Lambda = R$ and $\underline{\underline{D}}$ consists of every module isomorphic to some $R/\underline{m}^n$ where n is a positive integer and $\underline{m}$ is the unique maximal ideal of R, then $\underline{\underline{D}}$ has no splitting projective objects and hence has no minimal cover.

Finally, a subcategory $\underline{\underline{D}}$ of ind Λ is said to be finite if it has only a finite number of nonisomorphic objects. A cover $\underline{\underline{C}}$ of $\underline{\underline{D}}$ is said to be a finite cover of $\underline{\underline{D}}$ if $\underline{\underline{C}}$ is finite. Clearly, if $\underline{\underline{C}}$ is a finite cover of $\underline{\underline{D}}$, then there is a subcategory $\underline{\underline{C}}'$ of $\underline{\underline{C}}$ which is the finite minimal cover of $\underline{\underline{D}}$. Therefore, in order to show that $\underline{\underline{P}}_0(\underline{\underline{D}})$ is the unique finite minimal cover of $\underline{\underline{D}}$, it suffices to show that $\underline{\underline{D}}$ has a finite cover. This is the procedure we will generally use in proving that subcategories of ind Λ have finite minimal covers.

2. Existence of preprojective partitions

This section is devoted to showing that for an R-order Λ the subcategory $L(\Lambda)$ of mod Λ consisting of all Λ-lattices has a preprojective partition in the sense described in the introduction. While the proofs need some modifications, the definitions and results given in this section are essentially the same as those given in [4, section 3]. To simplify notation, we denote ind $L(\Lambda)$ by L in the rest of this section.

In the language of covers developed in the previous section, the description of the preprojective partition of L given in the introduction can be formulated as follows.

A preprojective partition of L is a sequence of subcategories $\underline{\underline{P}}_0$, $\underline{\underline{P}}_1$, ..., $\underline{\underline{P}}_\infty$ of having the following properties:

a) $L = \bigcup_{i \geq 0} \underline{\underline{P}}_i$

b) $\underline{\underline{P}}_i \cap \underline{\underline{P}}_j = \emptyset$ for $i \neq j$

c) $\underline{\underline{P}}_i$ is the finite minimal cover for $\bigcup_{j \geq i} \underline{\underline{P}}_j$ for all $i < \infty$.

It is not hard to see that such a sequence of subcategories of L can also be described by the following properties:

i) For each $i < \infty$, the subcategory $L - \bigcup_{j<i} \underline{\underline{P}}_j$ has a finite cover and $\underline{\underline{P}}_i = \underline{\underline{P}}_0(L - \bigcup_{j<i} \underline{\underline{P}}_j)$, the unique finite minimal cover of $L - \bigcup_{j<i} \underline{\underline{P}}_j$.

ii) $\underline{\underline{P}}_\infty = L - \bigcup_{i<\infty} \underline{\underline{P}}_i$.

From this description it is obvious that such a sequence of subcategories is unique if it exists since minimal covers are unique. We now turn our attention to showing that such a sequence of subcategories does exist.

For $i < \infty$, define the sequence of subcategories $\underline{\underline{P}}_0$, $\underline{\underline{P}}_1$, ... of L by induction as follows: Let $\underline{\underline{P}}_0 = \underline{\underline{P}}_0(L)$ and $\underline{\underline{P}}_i = \underline{\underline{P}}_0(L - \bigcup_{j<i} \underline{\underline{P}}_j)$ for $i > 0$. Using our second description of the preprojective partition of L, it is easily seen that the sequence of subcategories $\underline{\underline{P}}_0$, $\underline{\underline{P}}_1$, ..., $\underline{\underline{P}}_\infty$, where $\underline{\underline{P}}_\infty = L - \bigcup_{i<\infty} \underline{\underline{P}}_i$, is the preprojective partition of L provided that each of the subcategories $\underline{\underline{P}}_i$ is finite for all finite i. That this is indeed the case is an immediate consequence of the following general result whose proof will be given shortly.

Proposition 2.1. If $\underline{\underline{A}}$ is a finite subcategory of L, then $L - \underline{\underline{A}}$ has a finite cover. Hence $\underline{\underline{P}}_0(L - \underline{\underline{A}})$ is the finite minimal cover of $L - \underline{\underline{A}}$.

This gives the following.

Theorem 2.2. Let Λ be an order over the complete discrete valuation ring R. Then L, the category of indecomposable Λ-lattices has a preprojective partition.

The rest of this section is devoted to proving Proposition 2.1. The proof is modeled on that given in section 3 of [4] for the existence of preprojective partitions in the case of artin algebras.

Let $\underline{\underline{C}}$ be a subcategory of mod Λ. Since mod Λ is a Krull-Schmidt category (i.e., every module is a sum of indecomposables each of which has a local endomorphism ring) the radical of $\underline{\underline{C}}$, denoted by $\underline{r}(\underline{\underline{C}})$, can be described as consisting of all morphism $f: A \to B$ in $\underline{\underline{C}}$ having the property that for each indecomposable Z in $\underline{\underline{C}}$ every composition $Z \to A \xrightarrow{f} B \to Z$ is in rad End Z. For each $n \geq 0$ we define $\underline{r}^n(\underline{\underline{C}})$ to consist of all morphisms in $\underline{\underline{C}}$ which are finite sums of morphisms each of which is a composition of at least n morphisms in $\underline{r}(\underline{\underline{C}})$. The basic properties of the $\underline{r}^n(\underline{\underline{C}})$ given in section 3 of [4] are equally valid in our situation and will be used freely.

Now let $\underline{\underline{C}} = L(\Lambda)$. Then by [1] or [2] (see also [6] and [7]) we know that for each indecomposable B there is a left almost split morphism $g: B \to C$ in L, i.e., the morphism $g: B \to C$ in L is not a splittable injection and any morphism $f: B \to Z$ in $L(\Lambda)$ which is not a splittable injection can be factored through $g: B \to C$. Then the proof of Lemma 3.14 of [4] can be copied verbatim to give

<u>Lemma</u> 2.3. Suppose $\underline{\underline{D}}$ is a subcategory of $L(\Lambda)$. For each C in $L(\Lambda)_{\underline{\underline{D}}}$, where $L(\Lambda)_{\underline{\underline{D}}}$ consists of all lattices having no indecomposable summands in $\underline{\underline{D}}$, and integer $n \geq 0$, there is a morphism $g: C \to D \coprod Z$ with D in add $\underline{\underline{D}}$ and Z in $L(\Lambda)_{\underline{\underline{D}}}$ satisfying the following conditions: a) Im $(g, X) = (C, X)$ for all X in add $\underline{\underline{D}}$ and b) the composition $C \to D \coprod Z \xrightarrow{p} Z$, where p is the projection morphism, is in $\underline{r}^n(C, Z)$.

We shall also have need of the following.

<u>Lemma</u> 2.4. Let $\underline{\underline{A}}$ be a finite subcategory of ind $L(\Lambda)$. Then there is an integer $n \geq 0$ such that if $A_1 \xrightarrow{f_1} A_2 \to \cdot \xrightarrow{f_s} A_{s+1}$ is a chain of s nonisomorphisms with $s \geq n$, then $f_s \ldots f_1$ is in $\underline{m}(A_1, A_{s+1})$ where $\underline{m}$ is the unique maximal ideal of R.

<u>Proof</u>: Let $A = \coprod_{i=1}^{t} A_i$ with the A_i a complete set of nonisomorphic modules in $\underline{\underline{A}}$. Then $\mathrm{End}_\Lambda(A)$ is an R-algebra which is a finitely generated R-module. Hence there is an integer $n > 0$ such that $(\mathrm{rad}\,\mathrm{End}_\Lambda(A))^n \subset \underline{m}\,\mathrm{End}_\Lambda(A)$. This n is our desired n.

Combining the previous two lemmas we obtain

<u>Proposition</u> 2.5. Let $\underline{\underline{A}}$ be a finite subcategory of ind $L(\Lambda)$ and let $\underline{\underline{D}} = \mathrm{ind}\, L(\Lambda) - \underline{\underline{A}}$. For each Y in $\underline{\underline{A}}$ there is a morphism $g: Y \to Y_{\underline{\underline{A}}} \coprod Y_{\underline{\underline{D}}}$ with $Y_{\underline{\underline{A}}}$ in add $\underline{\underline{A}}$ and $Y_{\underline{\underline{D}}}$ in add $\underline{\underline{D}}$ satisfying:

a) The composition $Y \xrightarrow{g} Y_{\underline{\underline{A}}} \coprod Y_{\underline{\underline{D}}} \xrightarrow{p} Y_{\underline{\underline{A}}}$ is in $m\,(Y, Y_{\underline{\underline{A}}})$, where p is the projection morphism.

b) If X is in $\underline{\underline{D}}$ and $t: Y \to X$ is an arbitrary morphism, then there is an $h: Y_{\underline{\underline{A}}} \coprod Y_{\underline{\underline{D}}} \to X$ such that

$$\begin{array}{ccc} Y & \xrightarrow{g} & Y_{\underline{\underline{A}}} \coprod Y_{\underline{\underline{D}}} \\ t \downarrow & \swarrow h & \\ X & & \end{array}$$

commutes, which implies

$$t(Y) \subset h\, f_{\underline{\underline{A}}}(Y) + h\, f_{\underline{\underline{D}}}(Y) \subset \underline{m}(X) + h(Y_{\underline{\underline{D}}}).$$

Proof: Since $\underline{\underline{A}}$ is finite, it follows from Lemma 2.4 that there is an integer k such that $\underline{r}^k(A_i, A_j) \subset \underline{m}(A_i, A_j)$ for all A_i and A_j in $\underline{\underline{A}}$. Then letting $n = k$ in Lemma 2.3, we obtain our desired result.

We now show how Proposition 2.1 follows from Proposition 2.5.

Let $\underline{\underline{A}}$ be a finite subcategory of ind $L(\Lambda)$ and let $\underline{\underline{D}} = \text{ind } L(\Lambda) - \underline{\underline{A}}$. We want to show that $\underline{\underline{D}}$ has a finite cover. Let $P_1, \ldots, P_s$ be a complete set of nonisomorphic indecomposable projective modules in $\underline{\underline{A}}$ and let $P_{s+1}, \ldots, P_t$ be a complete set of nonisomorphic indecomposable projective modules in $\underline{\underline{D}}$. For each $1 \leq i \leq s$ let $g_i: P_i \to P_{i_{\underline{\underline{A}}}} \coprod P_{i_{\underline{\underline{D}}}}$ be the morphisms described in Proposition 2.5.

We claim that $\underline{\underline{C}}$ consisting of the modules isomorphic to those in $\bigcup_{i=1}^{s} \text{ind}(P_{i_{\underline{\underline{D}}}}) \cup \{P_{s+1}, \ldots, P_t\}$ is a finite cover for $\underline{\underline{D}}$. That it is finite is obvious. Now let X be in $\underline{\underline{D}}$. Then there are morphisms $t_{ij}: P_{ij} \to X$ such that the induced morphism $\coprod t_{ij}: \coprod P_{ij} \to X$ is surjective. If $i \leq s$, then by Proposition 2.5 there are morphisms $h_{ij}: P_{i_{\underline{\underline{A}}}} \coprod P_{i_{\underline{\underline{D}}}} \to X$ such that $t_{ij}(P_i) \subset \underline{m}X + h_{ij}(P_{i_{\underline{\underline{D}}}})$. Hence $X = \sum_{i=1}^{t} t_{ij}(P_i) \subset \underline{m}(X) + \sum_{i=1}^{s} h_{ij}(P_{i_{\underline{\underline{D}}}}) + \sum_{i=s}^{t} t_{ij}(P_i)$. Since X is a finitely generated R-module and $\underline{m} = \text{rad } R$, it follows that $X = \sum_{i=1}^{s} h_{ij}(P_{i_{\underline{\underline{D}}}}) + \sum_{i=s}^{t} t_{ij}(P_i)$. Hence $\underline{\underline{C}}$ is a finite cover for $\underline{\underline{D}}$. This finishes the proof of Proposition 2.1.

3. A criterion for finite lattice type

We recall that an R-order Λ is said to be of finite lattice type if there is only a finite number of nonisomorphic indecomposable lattices. This section is devoted to giving criteria for when Λ is of finite lattice type. These criteria will be used later on to show that Λ is of finite lattice type if and only if every indecomposable lattice is preprojective. Throughout this section $L(\Lambda)$ denotes the category of Λ-lattices and ind $L(\Lambda)$ denotes the category of indecomposable Λ-lattices.

We recall that an indecomposable lattice is said to be preprojective if it is in $\bigcup_{i<\infty} \underline{\underline{P}}_i$ where $\underline{\underline{P}}_0, \underline{\underline{P}}_1, \ldots, \underline{\underline{P}}_\infty$ is the preprojective partition of $L(\Lambda)$. We now give a description of when an indecomposable lattice is preprojective which does not depend on knowing the preprojective partition. The following definition is convenient for stating this result.

Let $\underline{\underline{C}}$ be a subcategory of mod Λ and let $F:\underline{\underline{C}} \to \text{Ab}$ be a contravariant or covariant function from $\underline{\underline{C}}$ to Abelian groups. By Supp F, the <u>support of</u> F, we mean the subcategory of ind $\underline{\underline{C}}$ consisting of those objects X in ind $\underline{\underline{C}}$ such that $F(X) \neq 0$. We say that F has <u>finite support</u> if Supp F is a finite category.

<u>Proposition</u> 3.1. The following are equivalent for an indecomposable lattice A.

a) A is preprojective

b) There is a nonzero morphism $f:A \to S$ with S a simple Λ-module such that Supp Im (, f) $< \infty$.

c) There is a nonzero surjection $f:A \to A''$ with $L(A'') < \infty$ (where $L(A'')$ denotes the length of A") and Supp Im (, f) $< \infty$.

<u>Proof:</u> The proof of the equivalence of a) and b) is the same as that for the corresponding result for artin algebras (see [4, Theorem 5.1]). The equivalence of b) and c) is obvious.

These descriptions of preprojective lattices suggest the following. For a lattice A let $S(A)$ to be the set of all sublattices A' of A such that $L(A/A') < \infty$ and Supp Im (, $k_{A/A'}$) $< \infty$ where $k_{A/A'}:A \to A/A'$ is the canonical surjection. Clearly $S(A)$ is closed under finite intersections and if $A' \subset A'' \subset A$ and A' is in $S(A)$, then A" is also in $S(A)$. Define $A_0 = \bigcap_{A' \text{ in } S(A)} A'$.

As an easy consequence of these definitions and Proposition 3.1, we have the following.

Lemma 3.2. a) If A is an indecomposable lattice, then $A_0 \neq A$ if and only if A is preprojective.

b) If $f: A \to B$ is a morphism of lattices, then $f(A_0) \subset B_0$.

From Lemma 3.2 it follows that $A \mapsto A_0$ defines a functor $L(\Lambda) \to L(\Lambda)$.

Before giving the main result of this section it is convenient to have the following.

Lemma 3.3. The following are equivalent for a lattice A.

a) $A_0 = (0)$

b) $S(A)$ consists of all the sublattices A' of A such that $L(A/A') < \infty$.

Proof: a) implies b). We first show that we can construct a chain

$$A_1 \supset A_2 \supset \dots \supset A_n \supset \dots$$

of sublattices of A in $S(A)$ such that $\bigcap_{i=1}^{\infty} A_i = (0)$. For each $n \geq 0$ consider the family F_n of all sublattices A' of A in $S(A)$ containing $\underline{m}^n A$. Since $A/\underline{m}^n A$ is of finite length, there is an $A_n \in F_n$ such that $A_n/\underline{m}^n A$ is minimal in the family of all $A'/\underline{m}^n A$ with A' in F_n. From the fact that $S(A)$ is closed under finite intersections, it follows that $A_n \subset A'$ for all $A' \in F_n$. Because $\underline{m}^n A \supset \underline{m}^{n+1} A$ for n, it follows that

$$A_1 \supset A_2 \supset \dots \supset A_n \supset \dots .$$

We now show that $(0) = A_0 = \bigcap_{A' \in S(A)} A'$ implies that $\bigcap_{i=1}^{\infty} A_i = 0$. For if x is a non-zero element of A, there is an A' in $S(A)$ such that $x \notin A'$. But A' contains $\underline{m}^n A$ for some n since $L(A/A') < \infty$. Hence $A' \supset A_n$ and so x is not in A_n. Hence $\bigcap_{i=1}^{\infty} A_i = (0)$.

Now the fact that A is complete and separated in the $\underline{m}$-adic topology implies that given any i there is an n such that $A_n \subset \underline{m}^i A$ (see [5, Chap. III, section 2, Proposition 8]). Hence if A' is a sublattice of A such that $L(A/A') < \infty$, then $A' \supset \underline{m}^i A$ for some i and so $A' \supset A_n$ for some n. Therefore A' is in $S(A)$ since A_n is in $S(A)$. Thus we have shown that a) implies b).

b) implies a). Since $L(A/\underline{m}^i A) < \infty$ for all i we have that $\underline{m}^i A$ is in $S(A)$ for all i. Hence $\bigcap \underline{m}^i A \supset A_0$. But $\bigcap_{i=0}^{\infty} \underline{m}^i A = (0)$, so $A_0 = (0)$ our desired result. This finishes the proof of the lemma.

Remark: The proof that a) implies b) actually proves the following more general result. Let R be a complete noetherian local ring and A a finitely generated

R-module. Suppose $\mathcal{S}(A)$ is a family of submodules A' of A with $L(A/A') < \infty$, which is closed under finite intersections and such that $\bigcap_{A' \in \mathcal{S}(A)} A' = 0$. Then $\underline{m}^i A$ contains an A_i in $\mathcal{S}(A)$ for all i. The above proof was suggested by M. Artin. A slightly different proof was also suggested by J. Becker.

We now use these observations to prove the following characterizations of when Λ is of finite lattice type.

<u>Proposition</u> 3.4. The following statements are equivalent for an R-order Λ.

a) Λ is of finite lattice type.

b) Supp $(\ , S) | \mathcal{L}(\Lambda) < \infty$ for all simple Λ-modules S, where $(\ , S) | \mathcal{L}(\Lambda)$ is the functor $\mathcal{L}(\Lambda) \to \mathrm{Ab}$ given by $A \mapsto (A, S)$ for all A in $\mathcal{L}(\Lambda)$.

c) Supp $(\ , M) | \mathcal{L}(\Lambda) < \infty$ for all Λ-modules M of finite length.

d) $A_0 = (0)$ for all lattices A.

<u>Proof</u>: That a) implies b) and b) implies c) is obvious. That c) implies d) follows from Lemma 3.3.

d) implies a). It follows from [3, Proposition].4] that for each simple Λ-module S, there is a surjective morphism $f_S : A_S \to S$ with A_S in $\mathcal{L}(\Lambda)$ such that $(X, f_S) : (X, A_S) \to (X, S)$ is surjective for all X in $\mathcal{L}(\Lambda)$. Since $(A_S)_0 = (0)$ it follows from Lemma 3.3 that Ker f_S is in $\mathcal{S}(A_S)$. Thus Supp Im $(\ , f_S) | \mathcal{L}(\Lambda) < \infty$. Since there are only a finite number of nonisomorphic simple Λ-modules and for each indecomposable lattice X there is a simple S such that Im $(X, f_S) \neq 0$, it follows that there is only a finite number of nonisomorphic indecomposable lattices.

As an immediate consequence of this result we have the following criterion for an order to be of finite lattice type.

<u>Corollary</u> 3.5. An order Λ is of finite lattice type if and only if $\Lambda_0 = (0)$.

<u>Proof</u>: Suppose $\Lambda_0 = 0$. Since every lattice C is a sublattice of a free Λ-module F, it follows that $C_0 \subset F_0 = (0)$ for every lattice C. Hence by Proposition 3.4, it follows that Λ is of finite lattice type.

The proof, in the other direction, is also a trivial consequence of Proposition 3.4.

<u>Remark</u>: The results and proofs given in this section are also valid for lattices over higher dimensional orders in the sense of [1] and [2].

4. Properties of $\underline{\underline{P}}_\infty$

Our aim in this and the next section is to prove the following. Suppose Λ is an order not of finite lattice type. Then a) $\underline{\underline{P}}_\infty \neq \emptyset$ (i.e., not every lattice is preprojective) and b) $\underline{\underline{P}}_\infty$ does not have splitting projectives and therefore does not have a finite cover.

Since Λ is not of finite lattice type we know by Corollary 3.5 that $\Lambda_0 \neq (0)$. The fact that Λ_0 is sent into itself under all endomorphisms of Λ, which are just multiplications on the right by the elements of Λ, shows that Λ_0 is a two-sided ideal in Λ. This section is devoted to proving a) and b) under the assumption that Λ/Λ_0 is of finite length, or what is the same thing $K \otimes_R \Lambda_0 = K \otimes_R \Lambda$, where K is the field of quotients of R. In section 5 we use this result to establish a) and b) even when Λ/Λ_0 is not of finite length. It is worth noting that if $K \otimes_R \Lambda$ is simple, then the hypothesis $K \otimes_R \Lambda_0 = K \otimes_R \Lambda$ is satisfied since $K \otimes_R \Lambda_0$ is a nonzero two-sided ideal in $K \otimes_R \Lambda$.

The following observation gives the main reason why the hypothesis that Λ/Λ_0 is of finite length is helpful.

Lemma 4.1. Let Λ be an order. Then $L(\Lambda/\Lambda_0) < \infty$ if and only if $L(C/C_0) < \infty$ for all lattices C. Moreover if $L(\Lambda/\Lambda_0) < \infty$, then C_0 is in $\mathcal{S}(C)$ for all lattices C.

Proof: Suppose $L(\Lambda/\Lambda_0) < \infty$. If C is a lattice, then there is a surjection $f: n\Lambda \to C$ which induces a surjection $n\Lambda/(n\Lambda)_0 \to C/C_0$ (recall that $f(n\Lambda_0) \subset C_0$). Hence $L(C/C_0) < \infty$ since $L(\Lambda/\Lambda_0) < \infty$ implies $L(n\Lambda/(n\Lambda)_0) < \infty$. The rest of the proof is obvious.

For the rest of this section we assume that the order Λ satisfies $L(\Lambda/\Lambda_0) < \infty$, or what is the same thing, $L(C/C_0) < \infty$ for all Λ-lattices C. Of course, this hypothesis implies that Λ is of infinite lattice type (see Corollary 3.5).

The fact that $\underline{\underline{P}}_\infty \neq \emptyset$ is an obvious consequence of the following characterization of the sublattice A_0 of a preprojective lattice A.

Proposition 4.2. Suppose $L(\Lambda/\Lambda_0)$ is finite. Let A be a preprojective lattice. Then A_0 is in add $\underline{\underline{P}}_\infty$ and contains all sublattices A' of A in add $\underline{\underline{P}}_\infty$. Hence A_0 is the unique maximal sublattice of A contained in add $\underline{\underline{P}}_\infty$.

Proof: Suppose A' is an indecomposable sublattice of A. If $A' \not\subset A_0$, then we have the commutative exact diagram

$$\begin{array}{ccc}
0 \to A' & \xrightarrow{\text{inc}} & A \\
\downarrow k & & \downarrow k_0 \\
0 \to A'/A' \cap A_0 & \longrightarrow & A/A_0 . \\
\downarrow & & \downarrow \\
0 & & 0
\end{array}$$

Since $A' \not\subset A_0$, the monomorphism $0 \to A'/A' \cap A_0 \to A/A_0$ shows that $A'/A' \cap A_0$ is a nonzero module of finite length. We also have that Supp Im (, k) $< \infty$ since Im (, k) $\subset$ Im (, k_0) and Supp Im (, k_0) $< \infty$ because A is preprojective. Therefore A' is preprojective by Proposition 3.1. Consequently if A' is not preprojective, then $A' \subset A_0$. We now complete the proof of the proposition by showing that A_0 is in add $\underline{\underline{P}}_\infty$, or what is the same thing $(A_0)_0 = A_0$.

Let $B = (A_0)_0$. Then we have the exact commutative diagram

$$\begin{array}{ccccccc}
0 \to A_0 & \longrightarrow & A & \xrightarrow{k} & A/A_0 & \longrightarrow & 0 \\
\downarrow g & & \downarrow f & & \| & & \\
0 \to A_0/B & \xrightarrow{j} & A/B & \xrightarrow{k} & A/A_0 & \longrightarrow & 0 \\
\downarrow & & \downarrow & & & & \\
0 & & 0 & & & &
\end{array}$$

Since $L(A_0/B) < \infty$ and $L(A/A_0) < \infty$, it follows that $L(A/B) < \infty$. Also it is easily checked that Supp (Im (, f)) is contained in Supp (Im (, kf)) $\cup$ Supp Im (, g). Hence Supp Im (, f) $< \infty$ since Supp Im (, g) $< \infty$ and Supp Im (, kf) $\subset$ Supp Im (, k) $< \infty$. Thus B is in $\mathcal{S}(A)$, which implies that $B \supset A_0$. Hence $A_0 \supset B \supset A_0$, so $(A_0)_0 = B = A_0$. This finishes the proof of the proposition.

As an easy consequence of this result we have the following.

<u>Proposition</u> 4.3. If $L(\Lambda/\Lambda_0)$ is finite, then there are nonpreprojective lattices.

<u>Proof</u>: Since $L(\Lambda/\Lambda_0)$ is finite, Λ_0 is a nonzero lattice which by Proposition 4.2 is in add $\underline{\underline{P}}_\infty$. Hence $\underline{\underline{P}}_\infty \neq \emptyset$.

We now turn our attention to showing that if Λ is not of finite lattice type, then $\underline{\underline{P}}_\infty$ has no splitting projectives in the case Λ/Λ_0 is of finite length. The proof of this fact is based on the following.

<u>Proposition</u> 4.4. Suppose $L(\Lambda/\Lambda_0)$ is finite. Let A be an indecomposable lattice and $f: A \to A''$ a nonzero surjective morphism with $L(A'')$ finite. Then the

following are equivalent:

a) Ker f $\supset A_0$;

b) Supp Im (, f) is finite;

c) Supp Im (, f) consists only of preprojective lattices.

Moreover if any of these conditions holds, A is preprojective.

<u>Proof</u>: a) implies b). By Lemma 4.1 we have that A_0 is in $\mathcal{S}(A)$. Since Ker f $\supset A_0$ we have that Ker f is in $\mathcal{S}(A)$ which implies that Supp Im (, f) is finite.

b) implies c). Suppose X is in Supp Im (, f). Let $g:X \to A$ be a morphism such tht the composition $t = fg:X \to A''$ is not zero. Then we have the exact commutative diagram

$$\begin{array}{ccc} & X & \xrightarrow{t} & A \\ & \downarrow & & \downarrow \\ 0 \to & \mathrm{Im}\, t & \longrightarrow & A'' \\ & \downarrow & & \downarrow \\ & 0 & & 0 \end{array}$$

from which it follows that $0 < L(\mathrm{Im}\, t) < \infty$ and Supp Im (, t) $< \infty$ since Im (, t) is isomorphic to a subfunctor of Im (, f). Hence X is preprojective.

c) implies a). Since Supp Im (, f) consists solely of preprojective lattices, A is preprojective. Suppose Ker f $\not\supset A_0$. Then there is an indecomposable summand B of A_0 not contained in Ker f. Thus the composition $B \xrightarrow{\text{inc}} A \xrightarrow{f} A''$ is not zero. This means that B is in Supp Im (, f) and is therefore preprojective. But this contradicts the fact that A_0 is in $\underline{\underline{P}}_\infty$ (Proposition 4.2). Hence Ker f $\supset A_0$.

As a consequence of this proposition, we obtain the following result from which the fact that $\underline{\underline{P}}_\infty$ has no splitting projectives follows easily.

<u>Proposition</u> 4.5. Suppose $L(\Lambda/\Lambda_0)$ is finite. If C is in $\underline{\underline{P}}_\infty$ and $f:C \to S$ is not zero with S a simple module, then there is a D in $\underline{\underline{P}}_\infty$ and a nonisomorphism $g:D \to C$ such that the composition $D \xrightarrow{g} C \xrightarrow{f} S$ is not zero.

<u>Proof</u>: Since (, C) $\to$ Im (, f) $\to$ 0 is exact, we have that Im (, f) is a finitely generated functor. Since C is indecomposable, we have that (, C)/$\underline{r}$(, C) = F is a simple functor with Supp F = {C} where $\underline{r}$(, C) is the radical of (, C). From this it follows that Im (, f)/$\underline{r}$ Im (, f) = F where $\underline{r}$ Im (, f) is the radical of Im (, f). More specifically we have the following

exact commutative diagram

$$\begin{array}{ccccccc} 0 \to \underline{r}(\ ,C) & \to & (\ ,C) & \to & F & \to & 0 \\ \downarrow & & \downarrow & & \| & & \\ 0 \to \underline{r}\,\mathrm{Im}(\ ,f) & \to & \mathrm{Im}(\ ,f) & \to & F. & & \\ \downarrow & & \downarrow & & & & \\ 0 & & 0 & & & & \end{array}$$

Clearly our desired result is nothing more than the assertion that Supp Im (, f) contains some nonpreprojective lattices, which is what we now show.

Let $\coprod_{i=1}^{u} D_i \xrightarrow{g} C$ be a minimal right almost split morphism in $L(\Lambda)$ with the D_i indecomposable lattices. Since C is not preprojective, C is not projective and so $\coprod D_i \xrightarrow{g} C$ is surjective. Consequently the composition $\coprod D_i \xrightarrow{g} C \longrightarrow S$ is not zero. Moreover the fact that Im ((, $\coprod D_i$) → (, C)) = $\underline{r}$(, C) implies that Im ((, $\coprod D_i$) → (, S)) = $\underline{r}$ Im (, f). Let i = 1, ..., s be all the i's such that the composition $D_i \xrightarrow{g_i} C \xrightarrow{f} S$ is not zero where $g_i = g|D_i$. Then each $fg_i{:}D_i \to S$ is a nonzero morphism with Supp Im (, fg_i) contained in Supp $\underline{r}$ Im (, f). Therefore if Supp $\underline{r}$ Im (, f) consists only of preprojective lattices, then each Supp Im (, fg_i) consists only of preprojective lattices. Since L(S) is finite we have by Proposition 4.4 that each Supp Im (, fg_i) is finite. Since (, $\coprod_{i=1}^{s} D_i$) ⟶ $\underline{r}$ Im (, f) → 0 is exact, it follows that $\coprod_{i=1}^{s}$ Im (, fg_i) → $\underline{r}$ Im (, f) → 0 is exact. Hence Supp $\underline{r}$ Im (, f) is finite and so Supp Im (, f) = Supp $\underline{r}$ Im (, f) ∪ {C} is finite. This means by Proposition 4.4 that C is preprojective, which contradicts the hypothesis that C is in $\underline{\underline{P}}_\infty$. Therefore there is a nonpreprojective D in Supp $\underline{r}$ Im (, f). Hence there is a nonisomorphism D → C with D in $\underline{\underline{P}}_\infty$ such that the composition $D \to C \xrightarrow{f} S$ is not zero, which is our desired result.

As an immediate consequence of this result we have the following.

<u>Proposition</u> 4.6. If $L(\Lambda/\Lambda_0) < \infty$, then $\underline{\underline{P}}_\infty$ has no splitting projective lattices. In particular, $\underline{\underline{P}}_\infty$ has no finite cover.

<u>Proof</u>: Suppose C is in $\underline{\underline{P}}_\infty$. It is an easy consequence of Proposition 4.5 that there is a nonsplittable surjection D → C with D in add $\underline{\underline{P}}_\infty$. Therefore C is not a splitting projective in $\underline{\underline{P}}_\infty$.

5. A reduction

This section is devoted to showing that Propositions 4.3 and 4.6 are valid for an arbitrary order, not just those orders Λ such that Λ/Λ_0 is of finite length. The method of proof is to show that the general case can be reduced to the special case handled in the previous section. We assume in this section that Λ is an arbitrary order over a complete discrete valuation ring R with field of quotients K such that $\Gamma = K \otimes_R \Lambda$ is a semisimple algebra.

We began by recalling without proofs some basic facts about lattices.

Lemma 5.1. The following are equivalent for two lattices C and D:

a) $\mathrm{Hom}_\Lambda(C, D) \neq (0)$;

b) $\mathrm{Hom}_\Gamma(K \otimes_R C, K \otimes_R D) \neq (0)$;

c) $\mathrm{Hom}_\Gamma(K \otimes_R D, K \otimes_R C) \neq 0$;

d) $\mathrm{Hom}_\Lambda(D, C) \neq 0$.

Lemma 5.2. The following are equivalent for two lattices C and D:

a) $K \otimes_R C \underset{\Gamma}{\simeq} K \otimes_R D$;

b) There is an exact sequence $0 \to C \to D \to U \to 0$ with $L(U)$ finite;

c) There is an exact sequence $0 \to D \to C \to V \to 0$ with $L(V)$ finite.

As an immediate consequence of Lemma 5.2 we have the following.

Lemma 5.3. Suppose C and D are lattices such that $K \otimes_R C \underset{\Gamma}{\simeq} K \otimes_R D$. Then $C_0 = (0)$ if and only if $D_0 = (0)$.

Proof: Suppose $D_0 = (0)$. Since $K \otimes_R C \simeq K \otimes_R D$, we know there is a monomorphism $C \to D$ which induces a monomorphism $C_0 \to D_0$. Therefore $C_0 = (0)$.

We say that a lattice C is simple if $K \otimes_R C$ is a simple Γ-module. The following observations are well known and easily seen.

Lemma 5.4. a) A nonzero lattice C is simple if and only if $L(C/C')$ is finite for each nonzero sublattice C' of C.

b) If S is a simple Γ-module, then there is a simple lattice C such that $S \simeq K \otimes_R C$.

These remarks show that the simple Γ-modules are of two mutually exclusive types. Suppose S is a simple Γ-module. Then either $S \simeq K \otimes_R C$ with C a simple lattice such that $C_0 = (0)$ or $S \simeq K \otimes_R C$ with C a lattice such that $L(C/C_0)$ is finite. Let $\mathcal{U}$ be the set of simple Γ-modules of the form $K \otimes_R C$ where C is a simple lattice with $C_0 = (0)$ and let $\mathcal{V}$ be the set of simple Γ-modules of the form $K \otimes_R C$ where C is a simple lattice with $L(C/C_0)$ finite. We denote by $L_{\mathcal{U}}$ the full subcategory of L consisting of those lattices C such that the simple factors of $K \otimes_R C$ are in $\mathcal{U}$. We denote by $L_{\mathcal{V}}$ the full subcategory of L consisting of those lattices C such that the simple factors of $K \otimes_R C$ are in $\mathcal{V}$. Clearly $L_{\mathcal{U}}$ and $L_{\mathcal{V}}$ are closed under sublattices, factor lattices and extensions. Moreover, if C is in $L_{\mathcal{U}}$ and D in $L_{\mathcal{V}}$, then $\mathrm{Hom}_\Lambda(C, D) = (0) = \mathrm{Hom}_\Lambda(D, C)$.

We now give some other descriptions of the objects in $L_{\mathcal{U}}$ and $L_{\mathcal{V}}$.

Suppose C is a lattice. Then $K \otimes_R C = \coprod_{j=1}^{u} S_j$ with the S_j simple Γ-modules. For each projection $K \otimes_R C \to S_j$ we denote the simple lattice which is the image of the composition $C \to K \otimes_R C \to S_j$ by C_j. Then the surjections $C \to C_j$ induce an exact commutative diagram

$$\begin{array}{ccc} & 0 & & 0 \\ & \downarrow & & \downarrow \\ 0 \to & C & \longrightarrow & \coprod_{j=1}^{n} C_j \\ & \downarrow & & \downarrow \\ 0 \to & K \otimes_R C & \longrightarrow & \coprod_{j=1}^{n} K \otimes_R C_j \to 0 \end{array}$$

from which it follows that $\mathrm{Coker}\,(C \to \coprod_{j=1}^{n} C_j)$ is of finite length. Therefore we also have that there is a monomorphism $0 \to \coprod C_j \to C$ whose cokernel is also of finite length (see Lemma 5.2).

Now the monomorphisms $0 \to \coprod C_j \to C$ and $0 \to C \to \coprod C_j$ induce the monomorphisms $0 \to \coprod (C_j)_0 \to C_0$ and $0 \to C_0 \to \coprod (C_j)_0$. Hence $\coprod (C_j)_0$ and C_0 have the same rank. Therefore since $\coprod C_j$ and C also have the same rank, we have the exact commutative diagram

$$\begin{array}{ccc} & 0 & & 0 \\ & \downarrow & & \downarrow \\ 0 \to & \coprod_j K \otimes_R (C_j)_0 & \longrightarrow & K \otimes_R C_0 \to 0 \\ & \downarrow & & \downarrow \\ 0 \to & \coprod K \otimes_R C_j & \longrightarrow & K \otimes_R C \to 0. \end{array}$$

From this it follows that $L(C/C_0)$ is finite if and only if $L(C_j/(C_j)_0)$ is finite for all j. But $L(C_j/(C_j)_0)$ is finite if and only if S_j is in $\mathcal{V}$. Hence we have shown the following.

Proposition 5.5. A lattice C is in $\mathcal{L}_V$ if and only if L (C/C_0) is finite.

We now give some other descriptions of the objects in $\mathcal{L}_U$.

Proposition 5.6. The following are equivalent for a nonzero lattice C.

a) C is in $\mathcal{L}_U$;

b) $C_0 = (0)$;

c) Every sublattice of C is preprojective;

d) If A is an indecomposable lattice such that $\mathrm{Hom}_\Lambda(A, C) \neq 0$, then A is preprojective.

Proof: a) implies b). As in the proof of Proposition 5.5, we have that there is a monomorphism $0 \to C \to \coprod C_j$ with the C_j simple lattices such that $K \otimes_R C \to \coprod K \otimes_R C_j$ is an isomorphism. Since C is in $\mathcal{L}_U$ we have that $(C_j)_0 = (0)$ for all j. But the monomorphism $C \to \coprod C_j$ induces a monomorphism $C_0 \to \coprod (C_j)_0$. Hence $C_0 = (0)$.

b) implies c). Trivial.

c) implies a). As in the proof of Proposition 5.5 we have that there is a monomorphism $0 \to \coprod C_j \to C$ with the C_j simple lattices such that the induced morphism $\coprod K \otimes_R C_j \to K \otimes_R C$ is an isomorphism. Since each C_j is a submodule of C we have that $(C_j)_0 = (0)$. Hence $K \otimes_R C_j$ is in U for all j which shows that C is in $\mathcal{L}_U$.

c) implies d). Suppose $f:A \to C$ is not zero with A indecomposable. Then Im f is preprojective since it is a nonzero sublattice of C. Let B be an indecomposable factor of Im f. Then there is a surjection $A \to B$ which shows that A is preprojective.

d) implies c). Trivial.

Suppose now that C is a lattice. Then $K \otimes_R C$ can be written as a sum $U \coprod V$ of Γ-submodules where every simple summand of U is in U and where every simple summand of V is in V. Let $C_U = U \cap C$ and $C_V = C \cap V$. Then it is easily seen that C_U is the unique maximal sublattice of C in $\mathcal{L}_U$ and C_V is the unique maximal sublattice of C in $\mathcal{L}_V$. Moreover, since $\mathcal{L}_U$ and $\mathcal{L}_V$ are closed under factor lattices, it follows that if $f:C \to D$ is a morphism of lattices, then $f(C_U) \subset D_U$ and $f(C_V) \subset D_V$. Finally C/C_U is a lattice in $\mathcal{L}_V$ since C/C_U is contained in V and C/C_V is a lattice in $\mathcal{L}_U$ since C/C_V is contained in U.

We now use these observations to show that there are only a finite number of nonisomorphic indecomposable lattices not in L_V. We begin with the following special case.

Lemma 5.7. There is an integer n such that $\bigcup_{i=1}^{n} \underline{\underline{P}}_i$ contains the indecomposable lattices in L_U.

Proof: We know by [3, Proposition 1.4] that for each simple Λ-module S, there is a surjection $f_S : A_S \to S$ with A_S a lattice such that if $g : X \to S$ is a morphism with X a lattice, then there is an $h : X \to A_S$ such that $f_S h = g$. Now if X is in L_U, then $\operatorname{Im} h \subset (A_S)_U$. Hence the induced morphisms $(f_S)_U : (A_S)_U \to S$ have the property that if X is in L_U, then given $g : X \to S$, there is an $h : X \to (A_S)_U$ such that $g = (f_S)_U h$. Since $((A_S)_U)_0 = (0)$ by Proposition 5.6, it follows that Supp Im (, $(f_S)_U$) is finite. Since there are only a finite number of nonisomorphic simple Λ-modules S, it follows that there are only a finite number of indecomposable X in L_U. This combined with the fact that every lattice in L_U is preprojective (Proposition 5.6) gives our desired result.

We now apply this result to prove the following.

Proposition 5.8. There is an integer n such that $\bigcup_{i=0}^{n} \underline{\underline{P}}_i$ contains all the indecomposable lattices not in L_V.

Proof: By Lemma 5.7 we know there is an integer n such that $\bigcup_{i=0}^{n} \underline{\underline{P}}_i$ contains the indecomposable lattices in L_U. Now suppose C is an indecomposable lattice not in L_V. Then C/C_V is a nonzero lattice in L_U. So there is a surjective morphism $C \to B$ where B is an indecomposable summand of C/C_V. Hence C is in $\bigcup_{i=0}^{n} \underline{\underline{P}}_i$ since B is in $\bigcup_{i=0}^{n} \underline{\underline{P}}_i$. Thus $\bigcup_{i=0}^{n} \underline{\underline{P}}_i$ contains the indecomposable lattices not in L_V.

Since C_U is functorial in C, we have that Λ_U is a two-sided ideal in Λ. It is also easy to see that Λ/Λ_U is an R-order and that Λ/Λ_U is in L_V. In addition we have the following.

Proposition 5.9. A Λ-lattice C is in L_V if and only if $\Lambda_U \cdot C = 0$. Hence $L_V = L(\Lambda/\Lambda_U)$.

Proof: Suppose C is in L_V. Since Λ_U is in L_U we have that $\operatorname{Hm}_\Lambda(\Lambda_U, C) = 0$. Thus for each c in C, the morphism $\Lambda_U \to C$ given by $\lambda \longmapsto \lambda c$ is zero, which means that $\Lambda_U \cdot C = 0$. On the other hand if $\Lambda_U \cdot C = 0$, then C is a Λ/Λ_U-lattice and so there is a surjection $m(\Lambda/\Lambda_U) \to C$ for some integer m. Since Λ/Λ_U is in L_V and L_V is closed under factor lattices, it follows that C is in L_V.

As a consequence of this result we have the following which enables us to show that the main results of the previous section are valid for arbitrary orders.

Proposition 5.10. Let Λ be an order and let Σ be the order Λ/Λ_U. Then Σ has the following properties:

a) There are only a finite number of indecomposable (up to isomorphism) Λ-lattices which are not Σ-lattices and all of these are preprojective Λ-lattices.

b) If C is a Σ-lattice, then $S_\Sigma(C) = S_\Lambda(C)$ and hence the sublattices C_0 of C is the same over Λ and Σ.

c) Viewing Σ as a Σ-module we have that Σ/Σ_0 is of finite length.

d) Suppose C is a Σ-lattice. Then

i) C is a preprojective Σ-lattice if and only if it is a preprojective Λ-lattice.

ii) $\underline{\underline{P}}_\infty(\Sigma) = \underline{\underline{P}}_\infty(\Lambda)$.

Proof: a) This is a restatement of Proposition 5.9.

b) Suppose C' is a Σ-sublattice of C in $S_\Sigma(C)$. Then L(C/C') is finite and Supp ((, C) $\to$ (, C/C')) is finite in ind $L(\Sigma)$. However, from the fact that ind $L(\Sigma)$ and ind $L(\Lambda)$ differ only by a finite number of objects (up to isomorphism), it follows that Supp ((, C) $\to$ (, C/C')) is finite in ind $L(\Lambda)$. Therefore C' is in $S_\Lambda(C)$. Hence $S_\Sigma(C) \subset S_\Lambda(C)$. The fact that $S_\Lambda(C) \supset S_\Sigma(C)$ is obvious.

c) Viewing Σ as a Λ-lattice, we know that Σ is in L_V and so Σ/Σ_0 is of finite length. Therefore by b), we have that Σ/Σ_0 is of finite length when Σ is viewed as a Σ-lattice.

d) This is an easy consequence of b).

We now apply this result to obtain the main theorem of this section, the generalization of Propositions 4.3 and 4.6 to arbitrary orders Λ.

Theorem 5.11. Let Λ be an arbitrary order.

a) Λ is of finite lattice type if and only if $\underline{\underline{P}}_\infty = \emptyset$.

b) If Λ is of infinite lattice type, then $\underline{\underline{P}}_\infty$ has no splitting projectives and hence has no finite cover.

Proof: a) Let $\Sigma = \Lambda/\Lambda_U$. Since ind $L(\Lambda)$ and ind $L(\Sigma)$ differ by only a finite number of objects (up to isomorphism), we have that Λ is of finite lattice type if and only if Σ is of finite lattice type. Suppose Λ and hence Σ is of infinite lattice type. Since Σ/Σ_0 is of finite length (Proposition 5.10), we have by Proposition 4.3, that $\underline{\underline{P}}_\infty(\Sigma) \neq \emptyset$. But (again by Proposition 5.10) we have that $\underline{\underline{P}}_\infty(\Lambda) = P_\infty(\Sigma)$. Hence $\underline{\underline{P}}_\infty(\Lambda) \neq \emptyset$ if Λ is of infinite lattice type. The rest of a) is trivial.

b) By Proposition 5.10, we know that $\underline{\underline{P}}_\infty(\Lambda) = \underline{\underline{P}}_\infty(\Sigma)$. By Proposition 4.6 we know that if $\underline{\underline{P}}_\infty(\Sigma) \neq \emptyset$, then $\underline{\underline{P}}_\infty(\Sigma)$ has no splitting projectives since Σ/Σ_0 is of finite length. Hence if $\underline{\underline{P}}_\infty(\Lambda) \neq \emptyset$, then $\underline{\underline{P}}_\infty(\Lambda)$ has no splitting projectives.

Remark: It would be interesting to know if each R-order Λ is isomorphic to a product $\Lambda_i \times \Lambda_2$ of R-orders where Λ_1 is of finite lattice type and $L(\Lambda_2/(\Lambda_2)_0)$ is finite.

M. Auslander
Brandeis University
Waltham, Massachusetts

S. O. Smalø
Trondheim University
Trondheim, Norway

References

[1] M. Auslander, "Existence theorems for almost split sequences," Ring Theory II: Proceedings of the Second Oklahoma Conference, Marcel Dekker, New York and Base., 1977.

[2] M. Auslander, "Functors and morphisms determined by objects," Representation Theory of Algebras: Proceedings of the Philadelphia Conference, Marcel Dekker, New York and Basel, 1978.

[3] M. Auslander and S. O. Smalø, "Lattices over orders: Finitely presented functors and preprojective partitions,"

[4] M. Auslander and S. O. Smalø, "Preprojective modules over Artin algebras," J. of Algebra, in press.

[5] N. Bourbaki, Éléments de matématique; Algébra Commutative, Hermann, Paris, 1961.

[6] K. W. Roggenkamp, "The construction of almost split sequences for integral group rings and orders," Communications in Algebra 5 (1977), 1363-1373.

[7] K. W. Roggenkamp and J. Schmidt, "Almost split sequences for integral group rings and orders," Communications in Algebra 4 (1976), 893-917.

POSET REPRESENTATIONS

L.A. Nazarova

Let $\mathfrak{N} = \{\alpha_1, \dots, \alpha_n\}$ be a finite partially ordered set (poset). A representation of the poset $\mathfrak{N}$ is given over the field k if to each element $\alpha_i \in \mathfrak{N}$ there is assigned a subspace V_i of the linear finite-dimensional vector space V such that if $\alpha_i < \alpha_j$, then $V_i \subseteq V_j$.

Representations of the poset $\mathfrak{N}$ form an additive category. So, equivalent and indecomposable representations are naturally definded. Representations of posets were introduced by Roiter and myself [1] but the given definition was suggested by Prof. Gabriel [13]. The dimension of a representation $\mathfrak{S} = (V, V_1, \dots, V_n)$ is the n+1 - dimensional integral vector $(v_0, v_1, \dots, v_n)$, where v_0 is the dimension of V, v_i is the dimension of the factor-space $V_i / \sum_{\alpha_j < \alpha_i} V_j$.

A poset $\mathfrak{N}$ has finite (infinite) type if it has a finite number (infinitely many) indecomposable representations.

The general definition of a poset representation was given in connection with the second Brauer-Thrall conjecture. The first important step to the proof of this conjecture was the proof of the same statement for posets. In [1] we proved

Theorem: If a poset $\mathfrak{N}$ has infinite type (over an infinite field) then there are infinitely many dimensions for each of which there are infinitely many indecomposables.

To prove Theorem 1 we've constructed the algorithm of differentiation of posets.

Let α_1 be a maximal element of $\mathfrak{N}$ and α_1 does not belong to a subset of the form $\mathfrak{N}_4 = oooo$. $\mathfrak{N}'_{\alpha_1}$ is the new poset obtained from $\mathfrak{N}$ by the elimination of α_1 and formal addition of the least upper bound $\alpha_i \cup \alpha_j$ of every incomparable triple of points $(\alpha_1, \alpha_i, \alpha_j)$. Let $T = (V, V_{\alpha_1}, \ldots, V_{\alpha_n})$ be a representation of $\mathfrak{N}$. We assign to T the representation $T' = (\bar{V}, \bar{V}_{\beta_1}, \ldots, \bar{V}_{\beta_m})$ of the poset $\mathfrak{N}'_{\alpha_1}$ in the following way:

1.) $\bar{V} = V_{\alpha_1}$

2.) $\bar{V}_{\beta_i} = V_{\beta_i} \cap V_{\alpha_1}$, if $\beta_i \in \mathfrak{N}$;

3.) $\bar{V}_{\beta_j} = V_{\alpha_1} \cap (V_{\alpha_i} + V_{\alpha_j})$, if $\beta_j = \alpha_i \cup \alpha_j$

We proved [1]:

Lemma 1: This correspondence is a one-to one correspondence between all indecomposables of $\mathfrak{N}'_{\alpha_1}$ and all indecomposables of $\mathfrak{N}$ with the exception of a finite number of idecomposables for which $V_{\alpha_1} = \{0\}$.

Proposition 1: A poset $\mathfrak{N}$ vanishes after several differentiations or there exists a subset $\mathfrak{S} \subseteq \mathfrak{N}$ for which $\mathfrak{S}^{(n)} \supseteq \mathfrak{N}_4$, where $^{(n)}$ is an n-th derivative poset.

Theorem 1 follows from Lemma 1, Proposition 1 and the classification of the representations of $\mathfrak{N}_4$, which was done independently by Gelfand, Ponomarev [14] and myself [15] .

Kleiner [2] proved

Theorem 2: A poset $\mathfrak{N}$ has finite type iff it does not contain one of the following subsets:

1.) 0 0 0 0 ; 2.) 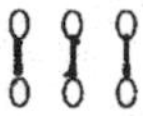 ; 3.) 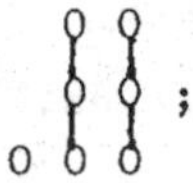 ; 4.) 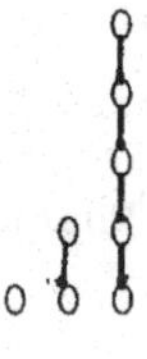 ;

5.) 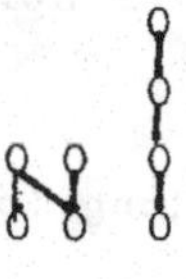

We'll call the posets 1.)-5.) critical. In fact Kleiner proved the following two statements.

<u>Proposition 2:</u> If $\mathfrak{N}$ contains no critical subsets, neither does $\mathfrak{N}'_a$.

<u>Lemma 2:</u> The critical posets have infinite type.

The Theorem 2 immediately follows from the Lemmas 1, 2 and Propositions 1, 2. Really let $\mathfrak{N}$ contain no critical subsets. Then $\mathfrak{N}'_a$ does not contain them either (Prop. 2) and according to Proposition 1 after several differentiations $\mathfrak{N}$ vanishes. But there exists the one-to-one correspondence between indecomposables of $\mathfrak{N}'$ and almost all (with the exception of a finite number) indecomposables of $\mathfrak{N}$. So $\mathfrak{N}$ has finite type.

A representation of $\mathfrak{N}$ is said to be exact if its vector-dimension d does not contain zero-coordinates. A poset having at least one exact indecomposable is called exact. Kleiner has described all exact posets of finite type. There is a finite number of exact posets of finite type [13].

We may try to use the algorithm of differentiation in the finite case. For example:

$\hat{\mathfrak{N}} =$ 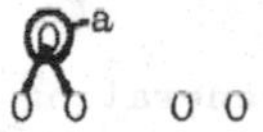 $\rightarrow \hat{\mathfrak{N}}'_a =$ 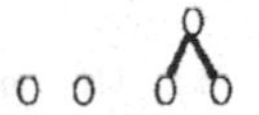 ; $\mathfrak{N}'_a \simeq \mathfrak{N}$

Using the fact that $\mathfrak{N}' \simeq \mathfrak{N}$ it is easy to see that after several differentiations every representation of $\hat{\mathfrak{N}}$ of dimension d turns into a representation of $\mathfrak{N}_4$. This may happen, for instance, after the first step, if $V_a = \{0\}$. So we can reduce the classification of representations of $\hat{\mathfrak{N}}$ to classifying the representations of $\mathfrak{N}_4$.

The situation when $\mathfrak{N}^{(m)} \simeq \mathfrak{N}^{(s)}$ is typical for posets of infinite type, related to quivers of tame type. I had used this fact to prove that the quivers of tame type correspond to the extended Dynkin diagrams [3]. But in this case, it was useful to consider the dual differentiation by a minimal point or the differentiation by a pair of points introduced by Zawadskii [15].

The properties of representations of posets are in fact independent of the field k but we'll consider now only infinite fields.

As with algebras and quivers, posets of infinite type are divided into two classes, tame and wild. To define tame type we have to consider poset representations over the ring $k[x]$ of polynomials in one variable. Corresponding definitions can be given if we consider representations by finite-dimensional $k[x]$-modules instead of by vector spaces.

We'll say that a representation $\mathfrak{S} = (V, V_1, \dots, V_n)$ over the field k is generated by the representation $\bar{\mathfrak{S}} = (\bar{V}, \bar{V}_1, \dots, \bar{V}_n)$ of $\mathfrak{N}$ over the ring $k[x]$ if there exists a finite-dimensional module B such that $V = \bar{V} \otimes_{k[x]} B$, $V_i = \bar{V}_i \otimes_{k[x]} B$

$V_i = \operatorname{Im}(\bar{V}_i \otimes_{k[x]} B \xrightarrow{f_i \otimes 1} V \otimes_{k[x]} B)$, where $f_i : V_i \to V$ is a natural inclusion.

A finite set $\mathfrak{D} = \{T_1, \dots, T_m\}$ of representations of $\mathfrak{N}$ over $k[x]$ will be called a generating set of order m if almost all indecomposables of dimension d can be generated by representations of $\mathfrak{D}$.

A poset $\mathfrak{N}$ has tame type over k if there exists a generating set in every dimension d.

$\mathfrak{N}$ has wild type if the category of its representations contains as a full subcategory the category of representations of the quiver

 .

It is clear that the first important problem in the theory of representations of posets of infinite type was to find the necessary and sufficient conditions for posets to have tame type. Fortunately, besides the class of posets arising in connection with representations of tame quivers it was known that the posets ,

G: 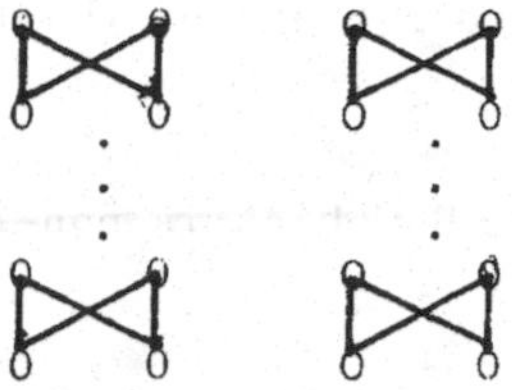

has tame type. It follows from the result of Roiter's and myself [6] on the Gelfand's problem about the classification of the representations of the quiver

(1) 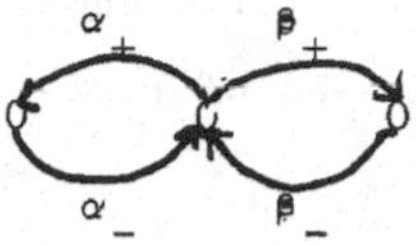

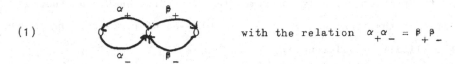

with the relation $\alpha_+\alpha_- = \beta_+\beta_-$

We can consider $\mathfrak{R}_4$ the special case of the poset G . If $\mathfrak{R}$ is an arbitrary poset we'll denote by $\widetilde{G}$ the poset with the same number of points as G and more strong relation of order, i.e. $\widetilde{G}$ is obtained from G by introducing non-trivial ordering for some formerly non-comparable elements. Of course, the posets $\widetilde{G}$ also have tame type.

Essentially using this result and differentiations I've proved [5] :

<u>Theorem 2':</u> $\mathfrak{R}$ has tame type if $\mathfrak{R}$ does not contain subsets of the form:

1.) 0 0 0 0 ; 2.) 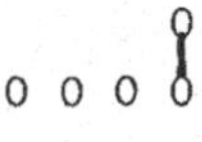 ; 3.) 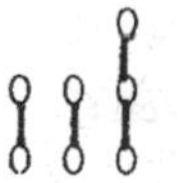 ; 4.) 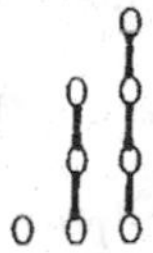 ;

5.)  ; 6.) 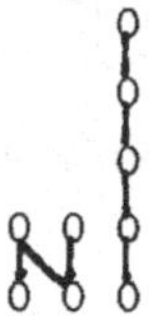

and wild type otherwise.

Similar to the finite case, the Theorem 2' follows very easily from the following statements.

Proposition 2': If $\mathfrak{N}$ contains no subsets of the form 1.)-6.) (Th.2), neither does $\mathfrak{N}'$.

Lemma 1': Every exact tame poset $\mathfrak{N}$ which is non-isomorphic to $\widetilde{G}$. differentiable.

Lemma 2': The posets 1.)-6.) have wild type.

After proving Theorem 2' as in the finite case, it was natural to describe representations of tame posets. But unlike the finite case, where the problem was only to enumerate the finite number of indecomposables, in the tame case we have some classes of posets having essential qualitative differences.

The posets of tame type are naturally divided into two classes:

1. The posets of finite growth for which orders of generating sets can be bounded for all dimensions.
2. The posets of infinite growth for which orders of generating sets increase infinitely with increasing dimension.

The posets of finite growth are divided into two classes also:

1. Posets of domestic type for which we can choose the same generating set for all dimensions.
2. The rest - non-domestic.

Zawadskiy and I have proved:

Theorem 3: A poset $\mathfrak{N}$ of tame type is a poset of finite growth iff it does not contain subsets of the form G_6 =
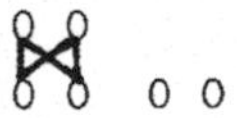

Moreover, we can choose no more than one generating set for every dimension.

To formulate the criterion of domestic type, I have to consider the Tits form of posets. If $\mathfrak{N} = \{\alpha_1, \ldots, \alpha_n\}$, Tits form of $\mathfrak{N}$ is the quadratic form in n+1 variables:

$$f = \sum_{i=o}^{n} x_i^{\,2} + \sum_{\alpha_i \leq \alpha_j} x_i x_j - x_o \sum_{i=o}^{n} x_i$$

The Tits form was introduced by Tits and Gabriel for quivers and then by Drozd for posets in a similar way. Drozd proved that:

1. $\mathfrak{N}$ has finite type iff its Tits form is weakly positive (positive on non-negative vectors).
2. If $\mathfrak{N}$ has finite type and d is a vector-dimension for which $f(d) = 1$, there exists only one indecomposable representation of dimension d [3].

The Theorem 2' can be formulated also in the following way:

A poset $\mathfrak{N}$ has tame type iff its Tits form is weakly non-negative (non-negative on non-negative vectors) [9].

For tame quivers, the dimensions of indecomposables are the roots of the equations $f(x)=1$, $f(x)=0$. If $f(d)=1$ there exists only one indecomposable of dimension d, if $f(d)=0$, there are infinitely many indecomposables of dimension d.

However, in [11] is given an example of a poset $\mathfrak{N}_8$, for which $f(d)=n$, given any natural number n, and d is the dimension-vector of an indecomposable : $\mathfrak{N}_8$:

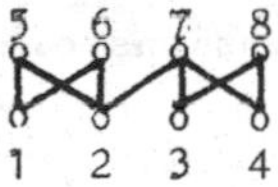

$$f(x_o, \dots, x_8) = \sum_{i=o}^{8} x_i^{\,2} + x_1x_5 + x_1x_6 + x_2x_5 + x_2x_6 + x_2x_7 + x_3x_7 + x_3x_8 + x_4x_7 + x_4x_8 - x_o \sum_{i=o}^{8} x_i$$

$d = (6n, n, n, 2n, 2n, 2n, 2n, n, n)$ $\qquad f(d) = n^2$

But it seems that studing connections between the Tits form and the indecomposables is interesting. In this direction Roiter and I have proved a result for special case of representations of DGC, more general than poset representations: if $f(d)=1$, there exists an indecomposable representation of dimension d; if $f(d)=0$ there are infinitely many indecomposables of dimension d. This result is extensively used in the theory of representations of posets, but we could not prove it inside of this theory. From this statement and results of [11], we get

Theorem 4: Let $\mathfrak{N}$ be a poset of finite growth. There are infinitely many indecomposables in dimension d iff $f(d)=0$.

Let F be any quadratic form over the field $\mathbb{Q}$ of rational numbers. Ann F is the set of solutions of the equation $F(x)=0$; ker F is the space of solutions of the simultaneous equations: $\{ \frac{\partial F}{\partial x_i} = 0$, $i = 0, 1, \ldots, n \}$.

If F is non-negative, then $\ker F = \mathrm{Ann}\, F$, but if F is only weakly non-negative this formula is unfortunately not true. In this case Ann F is not a vector space, but of course **ker** F is a vector space and we have $\ker F \leq \mathrm{Ann}\, F$.

Let's define $\mathrm{Ann}^+F = \mathrm{Ann}\, F \cap V^+$, where V^+ is the set of non-negative vectors. We shall say that some subset T of the space of rational vectors is a half space if for every $t_1, t_2 \in T$ and every pair of rational numbers $\alpha \geq 0$, $\beta \geq 0$ $\alpha t_1 + \beta t_2 \in T$. The dimension of the half space is the dimension of the space generated by T. Let $r(T)$ be the maximum of the dimensions of all the half spaces belonging to T.

Theorem 5: Let $\mathfrak{N}$ be a poset of finite growth, f its Tits form.

1. $r(\mathrm{Ann}^+f) \leq 2$.
2. $\mathfrak{N}$ is domestic if and only if $r(\mathrm{Ann}^+f) = 1$.

The first statement of Theorem 5 follows from the direct description of the poset of finite growth with infinite number of exact indecomposable representations.

To prove the criterion of domestic type, at first we have to prove

Proposition 3: Assume that $\mathfrak{N}$ is of finite growth, there exists a vector-dimension d for which $f(d)=0$, where f is Tit's form of $\mathfrak{N}$,

d does not contain zero-coordinates. Then the Tits form of $\mathfrak{N}$ is not negative.

Proof: Since $2f = \sum_i x_i \frac{\partial f}{\partial x_i} \Big|_d$ and $x_i \neq 0$ for every i, $d \in \ker f$. Since $d > 0$ there exists a neighborhood $\epsilon = (d-\alpha , d+\alpha)$ such that $x \in \epsilon$, if $x > 0$. Since f is weakly non-negative, f is non-negative in ϵ, i.e. $f(d+x) \geq 0$, $x \in \epsilon$. But $d \in \ker f$ and we have $f(d+x) = f(x)$, $x \in \epsilon$. So $f(x) \geq 0$, $x \in \epsilon$. Hence f is non-negative in a neighborhood of 0 and f is non-negative.

Let $r(\mathrm{Ann}^+ f) \geq 1$. Then there exists an infinite set of vectors of the form $\Lambda_i = g_1 + ug_2$, for which $f(\Lambda_i) = 0$. It follows now from Theorem 4 that $\mathfrak{N}$ is non-domestic.

Let $r(\mathrm{Ann}^+ f) = 1$. It suffices to prove our Theorem for exact subsets of $\mathfrak{N}$, for which there are infinitely many indecomposables. But for them $\ker f = \mathrm{Ann}\, f$. Since $\dim(\ker f) = 1$, using Proposition 3, we can prove that the only subsets of $\mathfrak{N}$, having infinitely many exact indecomposables in a fixed dimension are critical. Since the critical posets are domestic [3] so is $\mathfrak{N}$.

Zawadskiy has obtained the complete list of exact posets of finite growth, Bondarenko of infinite growth.

It follows from Lemma 1' that every representation of a tame poset of dimension d turns into a representation of a poset of the form $\widetilde{G}$ after several differentiations.

In fact all the representations of finite growth of dimension d turn in such a way into representation of $\mathfrak{N}_4$. The reason is that representations of posets of finite growth are in some sense similar to representations of $\mathfrak{N}_4$. Representations of infinite growth may be reduced to representations of other $\widetilde{G}$'s and so the properties of representations of infinite growth are similar to those of the representations of the quiver (1). Comparing the lists of Bondarenko and

Zawadskii we obtain the following statement: there exists only a finite number of finite exact posets of finite growth and infinitely many exact posets of infinite growth. We can prove this statement only after all calculations are done. To prove this Theorem directly seems to me an interesting problem.

Bibliography

[1] Nazarova, L.A. - Roiter, A.V.: "Representations of partially ordered sets"; Zap. Nauen. Sem. Leningrad, Otdel.math. Inst. Steklov, 28 (1972) , 5-31 , Soviet Math. 3 (1975) 585-606 , MR 49=4877 .

[2] Kleiner, M.M.: "On the explit representation of partially ordered sets of finite type"; Zap. Nauen. Sem. Leningrad, Otdel. math. Inst. Steklov, 28 (1972) , 32-42 , Soviet Math. 3 (1975) , 607-615 .

[3] Nazarova, L.A.: "Representations of quivers of infinite type"; Izv. Acad. Nauk SSSR Ser. Mat. 37 (1973) , 752-791 , Math. USSR Izv. 7 (1973) , 749-792 , MR 49=2785.

[4] Ringel, C.M.: "Tame algebras, Representation theory I "; Proceedings, Ottawa, Carleton University, 1979 Springer, Lecture Notes 831

[5] Nazarova, L.A.: "Partially ordered sets of infinite type"; Izv. Akad. Nauk. SSSR Math. Tom 39 (1975) No. 5, 911-938.

[6] Nazarova, L.A. - Roiter, A.V.: "A certain problem of I.M. Gelfand"; Funkcional. Anal. Appl. 7 (1973) 299-312 (1974)

[7] Gelfand, I.M.: "The cohomology of infinite dimensional Lie algebras, some questions of integral geometry"; Proc. Internat. Congr. Math. (Nice 1970) , vol. 1 Chanthier-Villars, Paris, 1971 , pp. 95-111 .

[8] Drozd, U.A.: "Coxeter transformations and representation of partially ordered sets"; Funkcional. Anal. i Prilozen. 8 (1974), 34-42 .

[9] Nazarova, L.A. - Zavadskiy, A.G.: "Partially ordered sets of tame type, Matrix problems"; Kiev, (1977) , 122-143 .

[10] Otrasevskaja, V.V.: "On criteria for a partially ordered set to be a one-parameter set"; Proc. All-Union Algebra Collog.(Gomel 1975)

[11] Bondarenko, V.M. - Zavadskiy, A.G. - Nazarova, L.A.: to appear

[12] Kleiner, M.M.: "On faithful representations of partially ordered sets of finite type"; Zap. Nauen. Sem. Leningrad, Otdel. math. Inst. Steklov, (LOM), 28 (1972) , 42-59 , Soviet Math. 3 (1975) , 616-628 .

[13] Gabriel, P.: "Représentations indécomposables des ensembles ordonés"; Séminaire Dubreil (Algèbre) , 26e annee , 1972/73 , No 13.

[14] Gelfand, I.M. - Ponomarev, V.I.: "About representations of 0 0 0 0 ".

[15] Nazarova, L.A.: "Representations of 0 0 0 0 "; Isv. Acad. of USSR 1967 .

[16] Zawadskiy, A.G.: "Differentiation with respect to pairs of points 'Matrix problems' "; Kiev 1977, 115-121.

GROTHENDIECK GROUPS AND ALMOST SPLIT SEQUENCES

M. C. R. Butler

1. INTRODUCTION

This paper's main objectives are the theorem stated at the end of the introduction and the corollary 7.2. The theorem provides a description involving only almost split sequences of the Grothendieck groups of certain types of module and lattice categories of finite representation type, that is, categories with only finitely many indecomposable objects. The corollary shows, for an artin algebra of finite representation type, that the combinatorics of its Auslander-Reiten quiver already suffice to determine the multiplicity of each simple module as a composition factor in each indecomposable module. These two results, and the ad-hoc proof presented at the Oberwolfach Conference 'Orders and their applications' in June 1980, were first obtained by Sheila Brenner, David Hughes and the author. The exposition here, though containing essentially the same proof, places the theorem in the much broader context of higher algebraic K-theory in the sense of Quillen [8], this being applied in Sections 3, 4 and 5 to categories of finitely presented functors and to some subcategories of functors of finite composition length. The key reference for finitely presented functors (there called coherent functors) is Auslander's paper [1]. For their relationship to almost split sequences in module and lattice categories, see [2], [3], [4], [5], [6] and [9].

First, it is convenient to fix some notation and terminology. The term <u>category</u> is restricted in this paper to mean a small, full, additive subcategory, closed under extensions, of an abelian category. For such a subcategory $\underline{C}$ of an abelian category $\underline{M}$, the restriction to $\underline{C} \times \underline{C}$ of $\mathrm{Ext}^1_{\underline{M}}$ will be denoted by $\mathrm{Ext}_{\underline{C}}$, and <u>a class</u>, Θ, <u>of extensions in</u> $\underline{C}$ is defined to be the union over all pairs (C, C') of objects in $\underline{C}$ of subsets $\Theta(C, C')$ of $\mathrm{Ext}_{\underline{C}}(C, C')$ containing 0. By abuse of language, short exact sequences which represent elements of Θ will be referred to as elements of Θ.

A category $\underline{C}$ and class Θ of extensions in $\underline{C}$ determine, in the usual way, a Grothendieck group $K(\underline{C}, \Theta)$: it is the abelian group with generators $[C], \ldots$ in (1, 1)-correspondence with the isomorphism classes of objects in $\underline{C}$, and with relations $[C'] - [C] + [C''] = 0$, one for each short exact sequence $0 \to C' \to C \to C'' \to 0$ in Θ. For the special case $\Theta = \mathrm{Ext}_{\underline{C}}$, the notation $K(\underline{C}, \mathrm{Ext}_{\underline{C}})$ will be shortened to $K(\underline{C})$. If $\Theta \subset \Theta'$ is an inclusion of classes, there is an evident epimorphism $K(\underline{C}, \Theta) \to K(C, \Theta')$.

Throughout the paper, Λ denotes a left noetherian ring, mod Λ the category of finitely generated left Λ-modules, and $\underline{A}$ a subcategory of mod Λ in the sense above, but with the further properties that $\underline{A}$ is closed under kernels and contains the full subcategory, pmod Λ, of projectives in mod Λ.

The main result may now be stated.

THEOREM Suppose either (1) that Λ is an artin algebra of finite representation type and $\underline{A}$ = mod Λ, or (2) that Λ is an R-order of finite lattice type and $\underline{A}$ the category of Λ-lattices, where R is a complete discrete valuation ring, with field of quotients L such that $L \otimes_R \Lambda$ is semisimple. If Θ is any class of extensions in $\underline{A}$ which contains all the split and almost split short exact sequences, then the canonical map

$$K(\underline{A}, \Theta) \to K(\underline{A})$$

is an isomorphism.

Acknowledgement In a letter following the 3rd International Conference on the Representations of Algebras (ICRA III) in Puebla, Mexico, in August 1980, Maurice Auslander outlined a proof that, for Λ an artin algebra and $\underline{A}$ = mod Λ, the conclusion of the theorem holds only if Λ has finite representation type. His letter also contained some helpful technical suggestions now incorporated in the paper, and these are acknowledged at the end of §4 .

2. EXACT CATEGORIES AND K-GROUPS

In [8], Quillen constructs K-groups $K_n(\underline{C}, \Theta)$, $n = 0, 1, 2, \ldots$, for each exact category $(\underline{C}, \Theta)$, that is, for each pair $(\underline{C}, \Theta)$ in which $\underline{C}$ is a category and Θ a class of extensions in $\underline{C}$ satisfying the following conditions for being admissible:

2.1 (a) Θ is an abelian group valued subfunctor of $\mathrm{Ext}_{\underline{C}}$; (b) for each short exact sequence $0 \to C' \to C \to C'' \to 0$ in Θ and each object X in $\underline{C}$, the two sequences

$$\Theta(X, C') \to \Theta(X, C) \to \Theta(X, C'') \quad \text{and} \quad \Theta(C'', X) \to \Theta(C, X) \to \Theta(C', X)$$

are exact.

The K_n's are abelian group valued covariant functors with respect to exact functors $(\underline{C}, \Theta) \to (\underline{C}', \Theta')$, that is, functors $\underline{C} \to \underline{C}'$ which map exact sequences in Θ to exact sequences in Θ'. The only aspect of the construction of the K_n's given in [8] relevant to this paper is as follows:

2.2 For each exact category $(\underline{C}, \Theta)$, there exists a natural equivalence of $K_0(\underline{C}, \Theta)$ with the Grothendieck group $K(\underline{C}, \Theta)$.

Obviously the class $\mathrm{Ext}_{\underline{C}}$ is admissible in the sense of 2.1, and the notation $K_n(\underline{C}, \mathrm{Ext}_{\underline{C}})$ will be abbreviated to $K_n(\underline{C})$.

A first application of Quillen's fundamental theorems of K-theory is to the inclusion functor of the category $\underline{A}$ into $\mathrm{mod}\ \Lambda$, with Λ and $\underline{A}$ as in §1. Since $\underline{A}$ is closed under kernels and contains $\mathrm{pmod}\ \Lambda$, the method of reduction by resolution [8, §4] shows that the induced maps of K-groups,

2.3 $\alpha_n : K_n(\underline{A}) \to K_n(\mathrm{mod}\ \Lambda)$,

are isomorphisms for all n .

3. FINITELY PRESENTED FUNCTORS

Most of the exact categories discussed in this paper depend, in one way or another, on categorical properties of the finitely presented functors on the base category $\underline{A}$, and the properties used in this and the next section are all contained in Auslander's paper [1]: the proofs given there in the case of an abelian base category are valid under the weaker assumptions of §1 on $\underline{A}$. Write $[\underline{A}^{op}, \text{Mod}\ \mathbb{Z}]$ for the abelian category of all (contravariant, additive, abelian group valued) functors on $\underline{A}$. Since $\underline{A}$ is full in mod Λ, this functor category has the representable functors,

$$(-, A) : X \mapsto \text{Hom}_{\Lambda}(X, A) \quad \text{for } X \in \underline{A},$$

one for each A in $\underline{A}$, as a family of projective generators. The functor F is said to be <u>finitely presented</u> (<u>coherent</u> in [1]) if it is the cokernel of a map $(-, A_1) \to (-, A_2)$ between representable functors. Since $\underline{A}$ is closed under kernels, F is finitely presented if and only if there exists a left exact sequence

3.1.(a) $$A_* \ : \ 0 \mapsto A_0 \to A_2 \to A_3 \to 0$$

and a projective resolution of F,

3.1.(b) $$0 \to (-, A_0) \to (-, A_1) \to (-, A_2) \to F \to 0 ;$$

this relationship will be described by saying that A_* <u>supports</u> F, and F <u>is supported by</u> A_*.

The next result is proved in [1].

3.2 <u>The full subcategory</u>, $\underline{F}$, <u>of all finitely presented functors in</u> $[\underline{A}^{op}, \text{Mod}\ \mathbb{Z}]$ <u>is an abelian category with enough projectives, and has global dimension</u> 0 <u>or</u> 2 .

As a K-theory consequence, one may apply the method of reduction by resolution to the inclusion functor,

$$p\underline{F} \to \underline{F},$$

into $\underline{F}$ of the full subcategory $p\underline{F}$ of projectives in $\underline{F}$ and conclude that <u>the induced maps</u>

3.3 $$\mu_n \ : \ K_n(p\underline{F}) \to K_n(\underline{F})$$

<u>are isomorphisms for all</u> n .

Since idempotents split in $\underline{A}$, the projectives in $\underline{F}$ are just the representable functors. Let 0 denote the admissible class of split exact sequences in $\underline{A}$. Then $A \to (\underline{A}, 0)$ is an equivalence of exact categories,

$$(\underline{A}, 0) \xrightarrow{\sim} p\underline{F}$$

<u>and the induced maps</u>

3.4 $$\rho_n : K_n(\underline{A}, 0) \to K_n(p\underline{F})$$

are isomorphisms for all n **.**

Lastly, the identity functor on $\underline{A}$ **gives an exact functor of exact categories,**

$$(\underline{A}, 0) \to \underline{A} ,$$

inducing K-group homomorphisms

3.5 $$\iota_n : K_n(\underline{A}, 0) \to K_n(\underline{A})$$

for each n **. By 2.2,** ι_0 **is surjective.**

4. A LOCALISATION EXACT SEQUENCE

For each functor F in $[\underline{A}^{op}, \text{Mod}\,\mathbb{Z}]$, the right action of Λ on itself induces a left Λ-module structure on the abelian group $F(\Lambda)$. For F finitely presented, $F(\Lambda)$ is a finitely generated Λ-module (since $\underline{A}$ consists of finitely generated Λ-modules), and one obtains a functor

4.1 $$\pi : \underline{F} \to \text{mod}\,\Lambda; \quad F \to F(\Lambda) ,$$

which may be viewed as an exact functor between exact categories. Let $\underline{E}$ denote its kernel, that is, the full subcategory of functors F in $\underline{F}$ satisfying the following, evidently equivalent, conditions:

4.2 (a) $\pi(F) = 0$, (b) $F(\Lambda) = 0$, (c) if A_* supports F, then A_* is a short exact sequence in $\underline{A}$.

As the kernel of an exact functor between abelian categories, $\underline{E}$ is automatically a Serre subcategory of $\underline{F}$ (dense subcategory in [1] and [7]) in the sense that it is full in $\underline{F}$, and closed in $\underline{F}$ under subobjects, quotient objects and extensions. The following result is the basis for proving existence of a long exact sequence of K-groups.

4.3 The functor $\pi : \underline{F} \to \text{mod}\,\Lambda$ is the universal quotient functor of abelian categories associated with the Serre subcategory $\underline{E}$ of $\underline{F}$.

Proof The proof is routine, and will be merely outlined here. The closure properties of $\underline{A}$ ensure that every object (morphism) in $\text{mod}\,\Lambda$ is equivalent to an object (morphism) in the image of π, and that π is full. The universality of π may then be verified directly: given any exact functor, σ, from $\underline{F}$ to an abelian category such that $\sigma(F) = 0$ for F in $\underline{E}$, one constructs another functor τ, which is unique up to equivalence, such that $\sigma = \tau \circ \pi$.

General descriptions may be found in [7] and [10] of the universal quotient abelian category of a Serre subcategory, and more specifically for functor categories in [1]. Note that the Serre subcategory $\underline{E}$ of $\underline{F}$ is actually a localising subcategory in Gabriel's sense, since π has the functor $M \mapsto \text{Hom}_\Lambda(-, M)|_{\underline{A}}$ as a right adjoint.

In view of 4.3, Quillen's localisation theorem [8, §5] applies to the sequence of exact functors,

$$\underline{E} \xrightarrow{\text{incl}} \underline{F} \xrightarrow{\pi} \text{mod}\,\Lambda,$$

to give a long exact sequence of K-groups,

$$\ldots \to K_{n+1}(\text{mod}\,\Lambda) \xrightarrow{\partial_{n+1}} K_n(\underline{E}) \xrightarrow{\nu_n} K_n(\underline{F}) \xrightarrow{\pi_n} K_n(\text{mod}\,\Lambda) \to \ldots$$

$$\ldots \to K_1(\text{mod}\,\Lambda) \xrightarrow{\partial_1} K_0(\underline{E}) \xrightarrow{\nu_0} K_0(\underline{F}) \xrightarrow{\pi_0} K_0(\text{mod}\,\Lambda) \to 0 ,$$

where ν_n is induced by the inclusion $\underline{E} \to \underline{F}$ and π_n by π. This may be re-written, using 2.3, 3.4, 3.5 and 3.6, as

4.4 The localisation exact sequence: there is a long exact sequence

$$\cdots \to K_{n+1}(\underline{A}) \overset{\delta_{n+1}}{\to} K_n(\underline{E}) \overset{\gamma_n}{\to} K_n(\underline{A}, 0) \overset{\iota_n}{\to} K_n(\underline{A}) \to \cdots$$

$$\cdots \to K_1(\underline{A}) \overset{\delta_1}{\to} K(\underline{E}) \overset{\gamma}{\to} K(\underline{A}, 0) \overset{\iota}{\to} K(\underline{A}) \to 0$$

of K-groups, where $\delta_n = \partial_n \alpha_n$, $\gamma_n = \rho_n^{-1} \mu_n^{-1} \nu_n$, $\iota_n = \alpha_n^{-1} \pi_n \mu_n \rho_n$, and the K_0's have been identified with the Grothendieck groups of the corresponding exact categories.

4.5 In 4.4, the map ι of Grothendieck groups is induced by the identity function on $\underline{A}$, and the map γ is given on the generators of $K(\underline{E})$ by the formula

$$\gamma([F]) = [A_0] - [A_1] + [A_2],$$

for the functor F in $\underline{E}$ with the short exact sequence A_* as support.

Note The proof of the theorem in the introduction given at the meeting ICRA III depended on a direct construction of the Grothendieck group portion of 4.4 using the formulae for ι, γ given in 4.5. The role of the 'Cartan' homomorphism 3.3 and of the map $F \to F(\Lambda)$ was pointed out by Maurice Auslander in his letter referred to in the introduction.

5. FUNCTORS OF FINITE LENGTH

Let $\underline{F}_0$ denote the full subcategory of $\underline{F}$ consisting of the functors of finite (composition) length. This obviously is a Serre subcategory of $\underline{F}$, and so also is

5.1 $$\underline{E}_0 = \underline{E} \cap \underline{F}_0 .$$

Let $\underline{F} \to \underline{F}/\underline{E}_0$ be the universal quotient functor of abelian categories associated with the Serre subcategory $\underline{E}_0$ of $\underline{F}$. Since $\pi : \underline{F} \to \operatorname{mod} \Lambda$ is exact and vanishes on $\underline{E}_0$, there is an exact functor $\underline{F}/\underline{E}_0 \to \operatorname{mod} \Lambda$ and a commutative diagram of exact categories

$$\begin{array}{ccccc} \underline{E}_0 & \to & \underline{F} & \to & \underline{F}/\underline{E}_0 \\ \downarrow & & \| & & \downarrow \\ \underline{E} & \to & \underline{F} & \xrightarrow{\pi} & \operatorname{mod} \Lambda \end{array} ,$$

in which the three left hand arrows are inclusion functors. Quillen's localisation theorem then shows that there is a morphism between the long exact sequences of K-groups of the top and bottom rows, and 2.3, 3.4, 3.5 and 3.6 may be used again to replace the K-groups of $\underline{F}$ and $\operatorname{mod} \Lambda$ by those of $(\underline{A}, 0)$ and $\underline{A}$. The result, so far as Grothendieck groups are concerned, is that *there exists a commutative diagram*

5.2 $$\begin{array}{ccccccccc} \dots \to K_1(\underline{F}/\underline{E}) & \to & K(\underline{E}_0) & \xrightarrow{\gamma'} & K(\underline{A}, 0) & \xrightarrow{\iota'} & K(\underline{F}/\underline{E}_0) & \to & 0 \\ \downarrow & & \downarrow & & \| & & \downarrow & & \\ \dots \to K_1(A) & \to & K(\underline{E}) & \xrightarrow{\gamma} & K(\underline{A}, 0) & \xrightarrow{\iota} & K(\underline{A}) & \to & 0 \end{array}$$

with exact rows, with ι *and* γ *given by 4.5 and* γ' *being given by the obvious analogue of the formula for* γ.

In fact the proof of the theorem in §1 will depend on a precise description of the inclusion $\operatorname{Im} \gamma' \subset \operatorname{Im} \gamma$, and it is convenient to introduce two new classes Φ and Σ of extensions in $\underline{A}$ on which to base this description. By 4.2, the functors in $\underline{E}$ are precisely the functors supported by short exact sequences in $\underline{A}$, so one may use the definition 5.1 of $\underline{E}_0$ to determine a class Φ of extensions in $\underline{A}$ as follows:

5.3 *Definition* *The short exact sequence* A_* *belongs to* Φ *if and only if the functor it supports has finite length in* $\underline{F}$ *(or, equivalently, in* $\underline{E}$*)*.

Now $K(\underline{E}_0)$ is generated by the images of functors of finite length, so is already generated by the images of the *simple* functors in $\underline{E}$ (if any such exist! If not, $K(\underline{E}_0) = 0$). This leads to the next definition.

5.4 *Definition* *Let* Σ *denote the class of short exact sequences in* $\underline{A}$ *which support functors of length* 0 *or* 1 .

The required descriptions of Im γ and Im γ' now follow from 4.5, 5.2, 5.3, 5.4 and the fact that Im γ' is generated by the images under γ' of the generators described above of $K(\underline{E}_0)$.

5.5 (a) Im γ <u>is the subgroup of</u> $K(\underline{A}, 0)$ <u>generated by the elements</u> $[A_0] - [A_1] + [A_2]$, <u>one for each short exact sequence</u> A_* <u>in</u> $\underline{A}$.

5.5 (b) Im γ' <u>is the subgroup of</u> $K(\underline{A}, 0)$ <u>generated by the elements</u> $[A_0] - [A_1] + [A_2]$, <u>one for each short exact sequence</u> A_* <u>in</u> Φ, <u>or more economically, one for each short exact sequence</u> A_* <u>in</u> Σ.

It then follows from 5.5(b), 5.2, and the definitions of the Grothendieck groups involved that

5.6 $$K(\underline{F/E}_0) \cong K(\underline{A}, \Phi) \cong K(\underline{A}, \Sigma).$$

<u>Remarks</u> 1. Since Σ is not closed under the direct sum of extensions, it is certainly not admissible in the sense of §2. However, it is a routine exercise to verify that Φ does satisfy the criterion 2.1 of admissibility. It would be of interest to know whether the first isomorphism of 5.6 can be generalised to obtain some relation between the higher K-groups of the exact categories $\underline{F/E}_0$ and $(\underline{A}, \Phi)$.

2. The definition of Φ is in terms of contravariant functors on $\underline{A}$, and one may define another class, Ψ, of extensions in $\underline{A}$ in terms of the covariant functors of finite length on $\underline{A}$. The following argument shows that $\Phi = \Psi$ when Λ is an artin algebra and $\underline{A} = \text{mod}\,\Lambda$. Let A_* be a short exact sequence in mod Λ, and $F = \text{Coker}\,[(-, A_1) \to (-, A_2)]$ the contravariant functor and $G = \text{Coker}\,[(A_1, -) \to (A_0, -)]$ the covariant functor supported by A_*. One may verify, for each non-projective indecomposable module X, that

$$F(X) = 0 \text{ if and only if } G(\text{Dtr}\,X) = 0.$$

Therefore Supp F (the set of isomorphism classes of indecomposable modules X with $F(X) \neq 0$) and Supp G have the same cardinality. Since for artin algebras a finitely presented functor H has finite length if and only if Supp H is a finite set [4, Proposition 3.4], it follows that either both or neither of F and G have finite length, and that $A_* \in \Phi$ if and only if $A_* \in \Psi$.

6. PROOF OF THE THEOREM

For the rest of this paper, it will be assumed either (1)' that Λ is an artin algebra and $\underline{A} = \text{mod}\,\Lambda$, or (2)' that Λ is an R-order and $\underline{A}$ is the category of Λ-lattices, where R is a complete discrete valuation ring, with field of quotients L such that $L \otimes_R \Lambda$ is a semisimple algebra.

In both cases, the category $\underline{F}$ of finitely presented functors on $\underline{A}$ has projective covers, and the almost split sequences in $\underline{A}$ may be characterised as the exact sequences which support minimal projective resolutions in $\underline{F}$ of simple functors ([3], [4], [5], [9]). Hence

6.1 A short exact sequence in $\underline{A}$ belongs to Σ if and only if it is direct sum of a split exact sequence and at most one almost split sequence.

The following reulst is also well-known.

6.2 If $\underline{A}$ has only finitely many isomorphism types of indecomposables, then $\Phi = \text{Ext}^1_{\underline{A}}$.

Proof The statement asserts that each finitely presented functor which has some short exact sequence as support is, in fact, of finite lenth, so in case (1)' is a special case of one implication in Auslander's theorem [3] that an artin algebra has finite representation type if and only if each finitely presented functor on its module category has finite length. In case (2)', suppose M is a direct sum of one representative of each isomorphism class of indecomposable lattices in $\underline{A}$, and let F be a functor supported by some short exact sequence A_* of lattices. Then F is a subfunctor of $\text{Ext}^1(-, A_0)$, and since A_0 is a summand of a direct sum of copies of M, F will have finite length provided $\text{Ext}^1(-, M)$ has finite length. This is certainly the case, for $\text{Ext}^1(-, M)$ has length bounded by the R-length of $\text{Ext}^1(M, M)$ and it is a standard fact about lattices that the values of Ext^m-groups are finitely generated torsion R-modules for $m \geq 1$.

The theorem stated in the introduction is now easy to prove. Its hypotheses, and 6.2, imply that $K(\underline{A}, \Phi) \to K(A)$ is an isomorphism. By 5.6, $K(\underline{A}, \Sigma) \to K(A)$ is an isomorphism, and then 6.1 shows that $K(\underline{A}, \Theta) \to K(\underline{A})$ is an isomorphism for any class Θ of extensions in $\underline{A}$ which contains all the split and almost split sequences in $\underline{A}$. This concludes the proof.

7. THE AUSLANDER-REITEN QUIVER OF $\underline{A}$

For a category $\underline{A}$ of the type considered in §6 there is currently much interest in the structure of the so-called Auslander-Reiten quiver, $Q(\underline{A})$, of $\underline{A}$. This quiver may be regarded as an oriented graph possessing distinguished cells. Its vertices $a, b, c, \ldots$ are the isomorphism classes of a full set of pairwise non-isomorphic indecomposable objects $A, B, C, \ldots$ of $\underline{A}$, and its arrows and cells depend on the structure of almost split sequences in the following way. Let $0 \to A \to X \to C \to 0$ be such a sequence. Then A and C are indecomposable and each determines the entire sequence uniquely to within isomorphisms of exact sequences [6], so they also determine uniquely pairwise nonisomorphic indecomposable modules $B_1, \ldots, B_p$ and positive integers $n_1, \ldots, n_p$ such that $X \cong \bigoplus_{i=1}^{p} n_i B_i$. Then, corresponding to the isomorphism class of this almost split sequence, $Q(\underline{A})$ has a cell $(a; b_p^{n_1}, \ldots, b_p^{n_p}; c)$, and arrows $a \to b_i$, $b_i \to c$ $(i; 1, \ldots, p)$.

This purely combinatorial object $Q(\underline{A})$ may also be used to define a group $KQ(\underline{A})$, namely, the group generated by the vertices of the quiver, subject to one relation $a - \sum_{i=1}^{p} n_i b_i + c = 0$ for each cell $(a; b_1^{n_1}, \ldots, b_p^{n_p}; c)$. It is a trivial exercise to verify the following relationship between $KQ(\underline{A})$ and $K(\underline{A}, \Sigma)$ using 5.4, 6.1, the uniqueness of almost split sequences, and the validity of the Krull-Schmidt theorem in the categories $\underline{A}$ under discussion:

7.1 The map $a \to [A]$ induces an isomorphism of $KQ(A)$ onto $K(\underline{A}, \Sigma)$.

The next and last statement follows from 7.1 in the case that $\underline{A} = \mathrm{mod}\ \Lambda$, with Λ an artin algebra of finite representation type, so that 5.6 and 6.2 give

$$K(\underline{A}, \Sigma) = K(\mathrm{mod}\ \Lambda).$$

In this case $K(\mathrm{mod}\ \Lambda)$ is a free abelian group with a basis

$$\{[S_1], \ldots, [S_r]\}$$

consisting of the images of the pairwise non-isomorphic simple modules $S_1, \ldots, S_r$, and it is clear that this basis is uniquely distinguished amongst all bases of $K(\mathrm{mod}\ \Lambda)$ by the property that, for each indecomposable module A, $[A]$ is a non-negative integral linear combination of $[S_1], \ldots, [S_r]$. By 7.1, the map $a \to [A]$ is an isomorphism of $KQ(\underline{A})$ onto $K(\underline{A}, \Sigma) = k(\mathrm{mod}\ \Lambda)$, so one obtains the following rather remarkable property of $Q(\mathrm{mod}\ \Lambda)$:

7.2 Corollary If Λ is an artin algebra of finite representation type, then $KQ(\mathrm{mod}\ \Lambda)$ is a free abelian group, and contains a basis $\{s_1, \ldots, s_r\}$ which is uniquely characterised by the following two statements: (i) $s_1, \ldots, s_r$ are vertices of $Q(\mathrm{mod}\ \Lambda)$; (ii) each vertex of $Q(\mathrm{mod}\ \Lambda)$ is a non-negative integral linear combination of $s_1, \ldots, s_r$.

This result makes precise the meaning of the third sentence of the introduction. The purely combinatorial object, Q(mod Λ), determines uniquely which vertices correspond to the simple modules, and also determines the multiplicity of each simple module as a composition factor of each indecomposable module. Similar conclusions may also be drawn for those lattice categories whose Grothendieck groups are free on bases of 'simple' lattices.

REFERENCES

1. Auslander, M. Coherent functors, Proc. Conf. on Categorical Algebra, La Jolla, 189-231. Springer-Verlag (1966).

2. Auslander, M. Representation theory of artin algebras, I, Comm. Algebra 1 (1974), 177-268.

3. Auslander, M. Representation theory of artin algebras, II, Comm. Algebra 2 (1974), 269-310.

4. Auslander, M. Large modules over artin algebras, Algebra, Topology and Categories, 1-17. Academic Press (1976).

5. Auslander, M. Existence theorems for almost split sequences, Proc. Comp. on Ring Theory II, Oklahoma. Marcel Dekker (1977).

6. Auslander, M and Reiten, I. Representation theory of artin algebras III: Almost split sequences. Comm. Algebra, 3 (1975), 239-294.

7. Popescu, N. Abelian categories with applications to rings and modules, L.M.S. Monograph No. 3, Academic Press (1973).

8. Quillen, D. Higher algebraic K-theory: I. Proc. Conf. on Algebraic K-theory, Seattle, vol. 1. Lecture Notes in Mathematics, 341, 77-139. Springer-Verlag (1973).

9. Roggenkamp, K. W. The construction of almost split sequences for integral group rings and orders, Comm. Algebra 5 (1975), 1363-1373.

10. Swan, R. G. Algebraic K-theory, Lecture Notes in Mathematics, 76. Springer-Verlag (1968).

Department of Pure Mathematics,
University of Liverpool,
P.O. Box 147,
Liverpool.
L69 3BX
England.

REPRESENTATION TYPES OF GROUP RINGS
OVER COMPLETE DISCRETE VALUATION RINGS

Ernst Dieterich

1. Introduction.

1.1 The starting-point.

The present paper contributes to the problem of classification of lattices over orders. In this respect we will be content if we know the representation type of the lattice categories which we are interested in.

I will restrict myself to those orders which are given as group rings $\Lambda = RG$, where G is a finite p-group and R a complete discrete valuation ring.

Definition. the category $L = {}_{\Lambda}L$ of (left) Λ-lattices (or the order Λ, respectively) is said to be

(i) of finite representation type if there exists only a finite number of isomorphism classes of indecomposable Λ-lattices,

(ii) of wild representation type if there exists a full subcategory U of L together with a representation equivalence $F : U \to W(F)$, for some field F,

(iii) of tame respresentation type if it is neither of finite nor of wild representation type.

Here, $W(F)$ denotes "the" wild category over F : namely the category of all finitely generated modules over the free associative algebra in two variables over F. Recall that an additive functor is said to be a representation equivalence if it is full, dense and isomorphism-reflecting. Consequently, a representation equivalence $F : A \to B$ induces a bijection between the isomorphism classes of indecomposable objects of A and B. Therefore we deduce from the above definition that given an order Λ of wild representation type, the problem of classifying L contains as a subproblem the matrix pair problem (over F). The general view is that this latter problem is very difficult. Some mathematicians therefore consider the indecomposable objects in a category of wild representation type as being "not classifiable".

On the other hand, it is expected that the objects in a category of tame representation type can be classified. And in practice, in order to prove that a category is of tame representation type, one usually gives a complete classification of its objects.

1.2 Historical survey.

Let $\Lambda = RG$ be given as in 1.1. Our aim is now to determine the representation type of $_{\Lambda}L$.

I begin with a brief survey on the orders RG whose representation type is known from literature: In the following table G ranges horizontally over all finite p-groups and v(p) (v being the (exponential) valuation of R) ranges vertically over all possible values $v(p) \in \mathbb{N} \cup \{0,\infty\}$. Each one of the connected areas encloses only group rings of the indicated representation type. The numbers relate to the literature (see references) where these cases have been investigated.

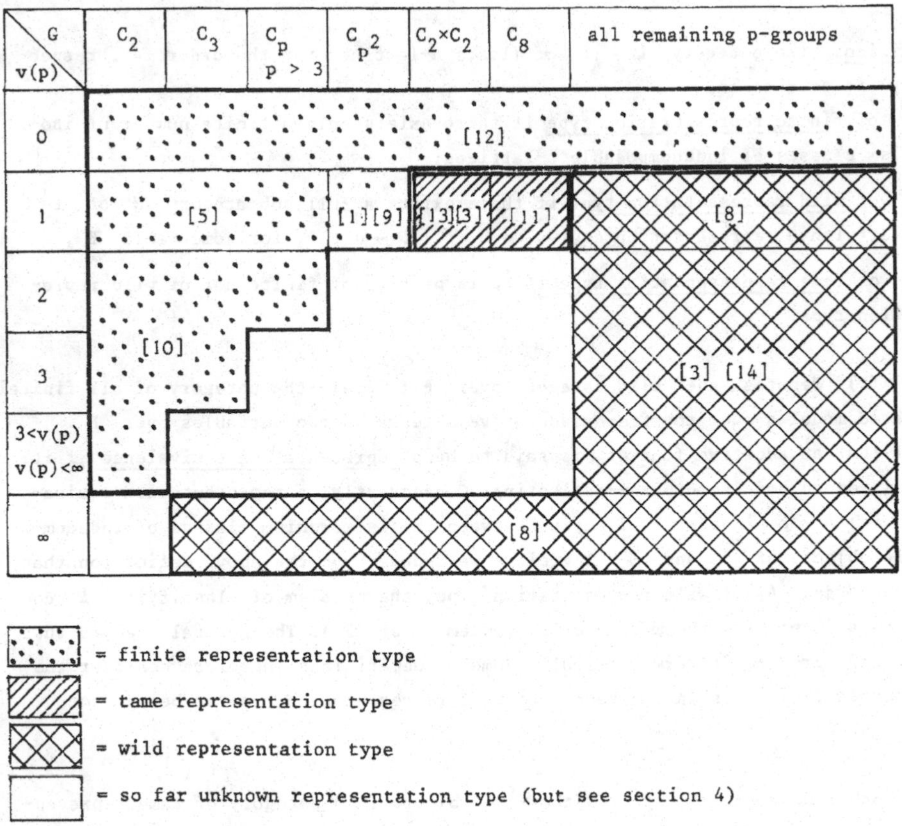
G
v(p)
C_2
C_3
C_p
p > 3
C_{p^2}
$C_2 \times C_2$
C_8
all remaining p-groups
0
1
2
3
3<v(p)
v(p)<∞
∞
[12]
[5]
[1][9]
[13][3]
[11]
[8]
[10]
[3] [14]
[8]
= finite representation type
= tame representation type
= wild representation type
= so far unknown representation type (but see section 4)

Remarks.

(i) The case $\Lambda = R(C_2 \times C_2)$, 2 unramified in R , was first solved by Nazarova [13] . But an easier proof may be found in [3].

(ii) The case $\Lambda = RG$, where $G \neq C_p, C_{p^2}, C_2 \times C_2, C_8$, and p unramified in R , was first solved by Gudivok [8]. But he gives only hints for the proof which - if written down explicitely- turns out to be rather technical and lengthy. Easier proofs for the wild representation type of these group rings - which work in the more general setting, where p may be unramified in R or not - can be deduced from Butler [3] (in the abelian group case), or from Ringel and Roggenkamp [14] (which also covers the non-abelian situation).

(iii) The modular case (for $G \neq C_2$) has been pointed out by Gudivok [8] , yet without proof.

(iv) Due to Jacobinski [10] we know that the white area encloses only group rings of infinite representation type (if $v(p) < \infty$). Some of these cases will be further distinguished according to tame and wild representation type in section 3 of the present paper.

1.3 Leading ideas. Notation.

In the attempt to determine the representation type of any one of our group rings, the leading idea is to reduce this problem to some matrix problem which is easier to handle than the original problem, or the representation type of which is already known. In particular one tries to find representation equivalences between "large" full subcategories of the lattice category and categories of representations of quivers, partially ordered sets, or species. Concepts and techniques in this direction have been developed by Butler [3] and Ringel/Roggenkamp [14]. In the main part of the present paper I will illustrate these ideas in a few variations by applying them to some of the group rings of table 1.2, thus exhibiting their representation type. In this way, we will (in 3.2 and 3.3) get new proofs for some of the group rings which are already known to be of finite representation type ([10]). In 3.1 and 3.4 we will achieve new results (in the sense of remark (iv) above).

Notation: With a complete discrete valuation ring R and a finite p-group G we associate the following data:

v = (exponential) valuation of R

ℓ = $v(p)$

π = parameter of R

k = residue class field of R

K = quotient field of R

$r := r(KG)$ = number of non-isomorphic irreducible KG-modules

$\{e_o,\ldots,e_{r-1}\}$ = the complete set of primitive orthogonal idempotents of KG .

We always assume that e_o is the trivial idempotent: $e_o = \frac{1}{|G|} \sum_{g \in G} g$.

$\Lambda = RG$

$J = \mathrm{rad}\Lambda = \pi\Lambda + \sum_{g \in G} R(g-1)$

Moreover, if Ξ is any R-order, then ${}_{\Xi}L$ denotes the category of (left) Ξ-lattices. If S is any ring, then u(S) denotes the group of units of S . All modules which appear in the sequel are assumed to be finitely generated.

2. Preliminaries.

2.1 Some Lemmata from algebraic number theory.

It turns out that the knowledge of $\Delta = \mathrm{intclos}_R\{K(\alpha)\}$ (where $K(\alpha)$ is the p-th cyclotomic field over K) plays a key rôle in many of the following investigations. Therefore this subsection essentially deals with the calculation of Δ.

Lemma 1. Let $\Lambda = RC_{p^n}$, g a generating element of C_{p^n}, $1 < v(p) < \infty$. Let $X^{p^n}-1 = \varphi_o(X)^{n_o} \cdot \ldots \cdot \varphi_{s-1}(X)^{n_{s-1}}$ be the prime factorization of $X^{p^n}-1$ in $K[X]$, and let $\alpha_i : \Lambda \to \Lambda e_i$, $i = 0,..,r-1$ be the canonical ring surjections. Then:

(i) $n_o = \ldots = n_{s-1} = 1$

(ii) $s = r(KC_{p^n})$

(iii) $\ker\alpha_i = \langle\varphi_i(g)\rangle$.

Proof.

(i) By the Chinese remainder theorem we have

$$KC_{p^n} \cong K[X] / \langle X^{p^n}-1\rangle \cong \bigoplus_{i=0}^{s-1} K[X] / \langle\varphi_i(X)\rangle^{n_i} = : A .$$

Since p is ramified in R, char K = 0, hence KC_{p^n} is semisimple (Maschke's Theorem). Therefore each of the direct summands $B_i := K[X] / \langle\varphi_i(X)\rangle^{n_i}$ is (commutative and) semisimple, hence is isomorphic to a direct sum of fields (Wedderburns Theorem). On the other hand, each of the B_i is a local ring. This implies (i).

(ii) By (i), the $B_o,\ldots,B_{s-1}$ are the simple components of A. This implies (ii).

(iii) By (i), we have a canonical isomorphism $\rho : KC_{p^n} \to \bigoplus_{i=0}^{r-1} K[X] / \langle\varphi_i(X)\rangle = : A$.

The primitive orthogonal idempotents of A are given by $f_j=(\delta_{ij}+ \langle\varphi_i(X)\rangle)_{i=0,..,r-1}$. (Here, δ_{ij} is the Kronecker-symbol).

For $e_o,\ldots,e_{r-1}$ we choose an order such that $\rho(e_j) = f_j$.

Then $\rho(\varphi_j(g)) = (\varphi_j(X) + \langle\varphi_i(X)\rangle)_{i=0,\ldots,r-1} = : a_j \in A$.

$a_jf_j = 0 \Rightarrow \varphi_j(g)\cdot e_j = 0$.

Since $\varphi_j(X) \in R[X]$, by Gauss' Lemma, $\langle\varphi_j(g)\rangle \subset \ker\alpha_j$ follows.

Conversely, let $\lambda(g) \in \ker\alpha_j$. Then we have

$\lambda(g)\cdot e_j = 0 \Rightarrow \rho(\lambda(g))\cdot\rho(e_j) = 0 \Rightarrow (\rho(\lambda(g)))_j = \lambda(X) + \langle\varphi_j(X)\rangle = 0 \Rightarrow \lambda(X) = \varphi_j(X)\cdot\mu(X)$, for some $\mu(X) \in K[X]$. Therefore $\lambda(g) = \varphi_j(g)\cdot\mu(g)$. Hence it remains to show that $\mu(X) \in R[X]$.

Since $\varphi_j(X)$ is a monic polynominal in $R[X]$, we may apply the division algorithm in $R[X]$, and we get: $\lambda(X) = \varphi_j(X)\cdot\beta(X)+\gamma(X)$, for some polynomials $\beta(X)$, $\gamma(X)\in R[X]$, such that $\deg\gamma(X) < \deg\varphi_j(X)$.

Hence $\varphi_j(X)\cdot\mu(X) = \lambda(X) = \varphi_j(X)\cdot\beta(X) + \gamma(X)$

$\Rightarrow \varphi_i(X)(\mu(X)-\beta(X)) = \gamma(X)$.

$\deg\gamma(X) < \deg\varphi_j(X) \Rightarrow \mu(X) = \beta(X) \in R[X]$. q.e.d.

<u>Lemma 2.</u> Let R be a complete discrete valuation ring and let α be a primitive p-th root of unity. Assume that p is ramified in R . With R and α we associate the following data:

$\psi(X) = \text{min.pol.}_K\alpha$, $d = \deg\psi(X)$

$s-1 = \frac{p-1}{d}$ = number of prime factors of $\phi_p(X)$ in $K[X]$

$R[\alpha] \cong R[X] / \langle\psi(X)\rangle$

$\Delta = \text{intclos}_R\{K(\alpha)\}$

π_o = parameter of Δ

w = normed discrete valuation of Δ

$e = w(\pi)$ (ramification index)

$f = (\Delta/\pi_o\Delta:k)$ (residue class degree)

Then:

(i) $s = r(KC_p)$.

(ii) $p = \vartheta(\alpha-1)^{p-1}$ for some $\vartheta \in u(R[\alpha])$.

(iii) $\ell = (r-1)\cdot f\cdot w(\alpha-1)$.

<u>Proof.</u> X^p-1 factorizes into s irreducible polynomials over K . Hence (i) follows from Lemma 1. The proof of (ii) is analogous to the proof of (21.11) in [4]. As to (iii), observe that $d = (K(\alpha):K) = e\cdot f$. Because of (ii) we have $w(p) = (p-1)\, w(\alpha-1) \Rightarrow \ell\cdot e = (s-1)\cdot d\cdot w(\alpha-1) \Rightarrow \ell = (r-1)\cdot f\cdot w(\alpha-1)$. q.e.d.

<u>Lemma 3.</u> Let $\Lambda = RC_p$, with $p > 2$, $v(p) = 2$, $r(KC_p) = 3$. Then, Λe_o, Λe_1, Λe_2 are discrete valuation rings.

Proof. With a primitive p-th root of unity α associate the situation of Lemma 2. Consider the following inclusion diagram:

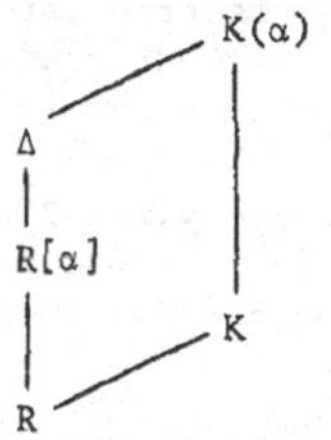

From Lemma 2 we deduce that $2 = 2\cdot f\cdot w(\alpha-1)$. Hence $K(\alpha)$ is an Eisenstein extension of K, and $\alpha-1$ is a parameter of Δ. Clearly $R[\alpha]$ is contained in Δ. But because $e = d = \frac{p-1}{2}$ and $w(\alpha-1) = 1$, the equality $R[\alpha] = \Delta$ can be shown by arguments analogous to the proof of (21.13) (first part) in [4].

From the assumption $r(KC_p) = 3$ and Lemma 1 we know that X^p-1 has the prime factorization $X^p-1 = (X-1)\varphi_1(X)\varphi_2(X)$ in $K[X]$. Applying Lemma 1 again, we deduce that: $\Lambda e_i \cong \Lambda / \ker\alpha_i = \Lambda / \langle \varphi_i(g) \rangle \cong R[X] / \langle\varphi_i(X)\rangle$; $i = 0,1,2$.

Therefore: $\Lambda e_o \cong R[X] / \langle X-1\rangle \cong R$

$\Lambda e_i \cong R[X] / \langle \varphi_i(X)\rangle \cong R[\alpha] = \Delta$; $i = 1,2$

Since R and Δ are discrete valuation rings, this finishes the proof. q.e.d.

Lemma 4. Let $p > 2$ be a prime number and $d := \frac{p-1}{2}$. Let R be a complete discrete valuation ring such that $v(p) = 2$ and $\phi_p(X)$ is irreducible in $K[X]$. With a primitive p-th root of unity α associate $R[\alpha] \subset K(\alpha)$, $J = \pi R[\alpha] + (\alpha-1)R[\alpha] = \text{rad } R[\alpha]$, and the R-order $\Delta := R[\alpha] + \frac{(\alpha-1)^d}{\pi} R[\alpha]$ in $K(\alpha)$.
Then:

(i) $\text{rad } \Delta = (\alpha-1)\Delta$

(ii) $\Delta = \text{intclos}_R\{K(\alpha)\}$

(iii) $f := (\Delta / \text{rad } \Delta : k) = 2$

Proof.
(i) We know from Lemma 2 that $\pi^2 = \vartheta\cdot(\alpha-1)^{p-1}$ for some $\vartheta \in u(R[\alpha])$. Therefore $(\alpha-1)^{p-1}\Delta \subset \pi\Delta \Rightarrow (\alpha-1)\Delta \subset \text{rad}\Delta$. Assume that $(\alpha-1)\Delta \neq \text{rad}\Delta$. Then there exists an element $x = 1 + s\frac{(\alpha-1)^d}{\pi} \in \text{rad}\Delta \setminus (\alpha-1)\Delta$; $s \in u(R)$. $x^m \in \pi\Delta \subset (\alpha-1)\Delta$ for some $m \in \mathbb{N}$, since $x \in \text{rad}\Delta$. On the other hand, we claim that the assertion

(*) : $x^m - 2^{m-1}x \in (\alpha-1)\Delta$ is valid for all $m \in \mathbb{N}$.

We prove this by induction:

(*) is true for $m = 1$.

$x^2 = x + s\frac{(\alpha-1)^d}{\pi} + s^2\vartheta = 2x + y_2$, with $y_2 = (s^2\vartheta - 1)$.

$y_2 \in R[\alpha] \cap \mathrm{rad}\Delta \subset J \subset (\alpha-1)\Delta$. Hence (*) is true for $m = 2$.

Assume that (*) is true for all $m = 1,\dots,i$.

Then $x^{i+1} = x(2^{i-1}x + y_i) = 2^{i-1}(2x + y_2) + y_i x = 2^i x + y_{i+1}$, with

$y_{i+1} = 2^{i-1}y_2 + y_i x \in (\alpha-1)\Delta$. This proves (*) for all $m \in \mathbb{N}$.

$x^m \in (\alpha-1)\Delta$ and (*) $\Rightarrow 2^{m-1}x \in (\alpha-1)\Delta$. By assumption $2 \neq p$. Therefore $2 \in u(R)$.
Hence $x \in (\alpha-1)\Delta$, contradicting the assumption. This proves (i).

(ii) The ring of multipliers of $\mathrm{rad}\Delta$ coincides with Δ , due to (i). This implies (ii).

(iii) Applying Lemma 2 to our situation we deduce that $2 = f\cdot w(\alpha-1) = f$, due to (i).

q.e.d.

2.2 The representation equivalences of Butler and of Ringel/Roggenkamp.

For the convenience of the reader I include some of the definitions and statements from [3] and [14] which will be needed in section 3. As to the quotations from [3], G is assumed to be an abelian p-group.

Definition. ([3], 2.2) Let $\Lambda = RG$ be given. Then $\mathcal{V}$ is the category of the Λ-module representations $V_* = (V \mid V_o,\dots,V_{r-1})$ of the partially ordered set

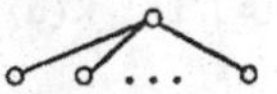
0 1 ... r-1

, which satisfy the following conditions:

(a) $\frac{|G|}{\pi}\cdot V = 0$.

(b) $\sum_{\substack{i=0\\ i\neq j}}^{r-1} V_i = V$, for each $j \in \{0,\dots,r-1\}$.

(c) If α_i denotes the canonical ring surjection $\alpha_i : \Lambda \to \Lambda e_i$, then $(\ker\alpha_i)V_i = 0$ for each $i \in \{0,\dots,r-1\}$.

Theorem 1. ([3], 3.2) There exists a full subcategory $\mathcal{L}_\delta$ of $\mathcal{L}$ which is representation equivalent to $\mathcal{V}$.

<u>Proposition 1.</u> ([3], 3.3(a)) If each of $\Lambda e_o,\ldots,\Lambda e_{r-1}$ is a discrete valuation ring, then L_δ consists of those Λ-lattices which have no direct summand isomorphic to Λ, $\Lambda e_o,\ldots,$ or Λe_{r-1} .

<u>Proposition 2.</u> ([3], § 4) Let t be an indeterminate and $s,m \in N$. Consider the categories K and M which are defined as follows:
K is the category of the $k[t] / \langle t^m \rangle$ - module representations $K_* = (K \mid K_o,\ldots,K_{s-1})$ of the partially ordered set 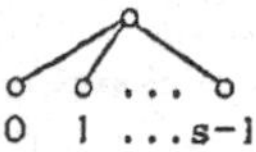 , which satisfy the following conditions:

(i) $tK_o = 0$.

(ii) $tK \subset \bigcap_{i=1}^{s-1} tK_i$.

M is the category of all k-representations of the partially ordered set

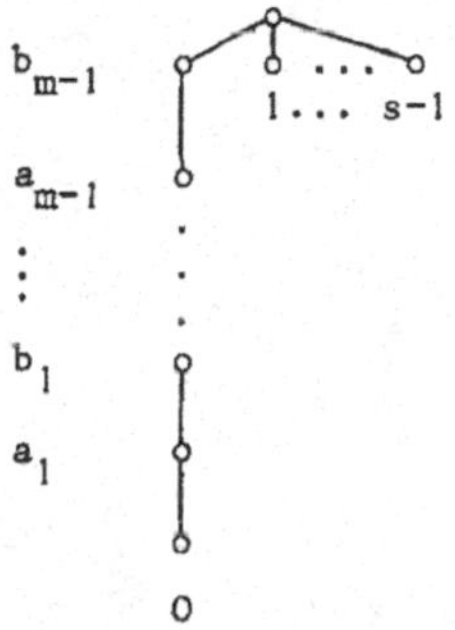

Moreover, consider the functor $F: K \to M$ which is given by reduction modulo t . To be precise: if $K_* \in K$ and K' is a submodule of K , then we write $\overline{K'} := (K' + tK) / tK$. Then F is defined by

$$FK_* = (M \mid M_i, M_{a_j}, M_{b_j})_{\substack{i=0,\ldots,s-1 \\ j=1,\ldots,m-1}} \quad \text{and} \quad M := \overline{K}\ ,\ M_i := \overline{K_i}\ ,\ M_{a_j} := \overline{t^{-j}(0)}\ ,$$
$$M_{b_j} := \overline{t^{-j}(K_o)}\ .$$

Then F is a representation equivalence.

<u>Remark.</u> Butler has not stated Proposition 2 in this form. Yet it is evidently inherent in [3], § 4.

Definition. ([14], 2.1) An R-order Ξ in a semisimple K-algebra A is said to be a Bäckström-order if there exists a hereditary R-order $\Gamma \supsetneq \Xi$ with $\operatorname{rad}\Gamma = \operatorname{rad}\Xi$

To any Bäckström-order Ξ with corresponding hereditary order Γ there can be associated a certain species S . For the definition of S see [14], 2.4. By $\mathcal{L}(S)$ we denote the category of all representations of S without simple direct summands.

Theorem 2. ([14], 1.2 and 2.9) Let Ξ be a Bäckström-order with corresponding hereditary order Γ . Let S be the species of $\Xi \subset \Gamma$. Then there exists a representation equivalence between ${}_{\Xi}L$ and $\mathcal{L}(S)$.

Remark. For the explicit description of this representation equivalence see [14], 1.1 and 2.9.

3. Explicit determination of the representation type in some special cases .

The following subsections are all concerned with cyclic p-groups. Throughout this section, g is assumed to be a generating element of G , and $t := g-1 \in \Lambda$.

3.1 If $p > 2$ and $1 < v(p) < \infty$, then $\Lambda = RC_{p^2}$ is of wild representation type.

Proof. Let $X^{p^2}-1 = (X-1)\phi_p(X)\phi_{p^2}(X)$ be the decomposition of $X^{p^2}-1$ into cyclotomic polynomials. Since p is ramified in R , we do not know whether this is already the prime factorization of $X^{p^2}-1$ over K . Therefore we distinguish two possible cases:

Case A. $\phi_{p^2}(X)$ is reducible in $K[X]$.

Let s be the number of irreducible factors in the prime factorization of $X^{p^2}-1$. From Lemma 1 we know that $s = r$. Thus, for the proof of the wild representation type of Λ , it will be sufficient to show that $s \geq 5$ (see [3], 3.4 (e)) .
If $\phi_p(X)$ is reducible, then clearly $s \geq 5$.
Let $\phi_p(X)$ be irreducible. Let $\phi_{p^2}(X) = \psi_1(X)\cdot \ldots \cdot\psi_m(X)$ be the prime factorization of $\phi_{p^2}(X)$ over K . Because for each $i \in \{1,\ldots,m\}$, the degree of $\psi_i(X)$ is the degree of the splitting field for $X^{p^2}-1$ over K , we conclude that $\deg\psi_1(X) = \ldots = \deg\psi_m(X) = \frac{p(p-1)}{m}$. Choose a zero ϑ of $\phi_{p^2}(X)$, and consider the field extensions $K \subset K(\vartheta^p) \subset K(\vartheta)$, together with the corresponding Galois-groups. Then:

$$\left.\begin{aligned} |\mathrm{Gal}(K(\vartheta)/K)| &= \frac{p(p-1)}{m} \\ |\mathrm{Gal}(K(\vartheta^p)/K)| &= p-1 \end{aligned}\right\} \Rightarrow |\mathrm{Gal}(K(\vartheta)/K(\vartheta^p))| = \frac{p}{m} \in \mathbb{N} .$$

Since $m > 1$ (by assumption), we conclude that $m = p$.
Since $p > 2$ (by assumption), it follows that $s = m+2 \geq 5$. q.e.d.

Case B. $\phi_{p^2}(X)$ is irreducible in $K[X]$.

(Remark that this implies that $\phi_p(X)$ is irreducible, due to $\phi_{p^2}(X) = \phi_p(X^p)$. In particular: $r = 3$.)
We shall exhibit a full subcategory $\mathcal{U}$ of $\mathcal{L}$ which is representation equivalent to $\mathcal{W}(k)$. This will be achieved in several steps which are sketched in the following diagram of categories and functors (the sign "~" shall mean that the indicated functor is a representation equivalence):

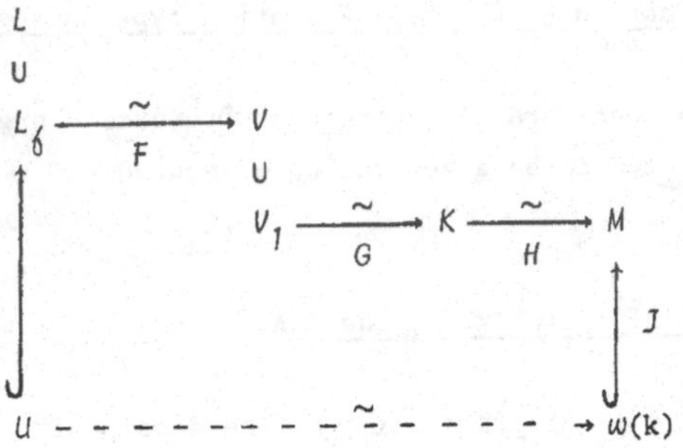

<u>Step 1:</u> We start with an application of Theorem 1 (see 2.2): There exists a full subcategory L_δ of L and a representation equivalence $F : L_\delta \to V$. For the definition of V see 2.2. In our present situation, the conditions imposed on an object $V_* = (V \mid V_o, V_1, V_2)$ of V are the following:

(a) $\pi^{2\ell-1} V = 0$;

(b) $\sum_{\substack{i=0\\ i\neq j}}^{2} V_i = V$; for all $j = 0,1,2$;

(c_o) $tV_o = 0$,

(c_1) $\phi_p(g)V_1 = 0$,

(c_2) $\phi_{p^2}(g)V_2 = 0$.

Here, (a) and (b) is immediate from the definition of V, and (c) follows by means of Lemma 1.

V is a certain category of Λ-module representations of the "3-subspace-diagram". Therefore to each object V_* of V there is attached a π-operation and a t-operation on V. The idea for the next two steps is that these operations can give rise to additional subspaces which then will be realized by modules over a reduced ring structure. For this objective, we have to find a subcategory V_1 of V with such properties that on the one hand some of the awkward conditions (a), (b), (c) vanish, and that on the other hand reduction modulo π (and modulo t, afterwards) in fact can be formulated as a representation equivalence.

<u>Step 2:</u> Let V_1 be the full subcategory of V which consists of the objects V_* of V satisfying the additional conditions:

(d) $\langle \pi^2, \pi t, t^2 \rangle \cdot V = 0$;

(e) $\pi V \subset \bigcap_{i=0}^{2} V_i$;

(f) $\pi V \cap tV = 0$.

Observe that (d) $\Rightarrow$ (a) .

Also, consider the following commutative diagram of canonical homomorphisms:

$$\begin{array}{ccc} \mathbb{F}_p[X] & \overset{\bar{\iota}}{\hookrightarrow} & (R/\pi^2 R)[X] \\ \uparrow\!\!\uparrow \rho & & \uparrow\!\!\uparrow \sigma \\ \mathbb{Z}[X] & \overset{\iota}{\hookrightarrow} & R[X] \end{array}$$

We conclude: $\sigma(\phi_p(X)) = \bar{\iota}\rho(\phi_p(X)) = (X-1)^{p-1} \Rightarrow \phi_p(g) \equiv t^{p-1} \pmod{\pi^2\Lambda}$;

$\sigma(\phi_{p^2}(X)) = \bar{\iota}\rho(\phi_{p^2}(X)) = (X-1)^{p(p-1)} \Rightarrow \phi_{p^2}(g) \equiv t^{p(p-1)} \pmod{\pi^2\Lambda}$.

This shows that (d) $\Rightarrow$ (c_1) and (c_2).

Therefore V_1 is the category of the Λ-module representations of the partially ordered set  which satisfy the conditions (b), (c_o), (d), (e) and (f).

Next, let K be the category of the $k[t]/\langle t^2 \rangle$-module representations $K_* = (K \mid K_o, \ldots, K_3)$ of the partially ordered set  which satisfy the conditions

(i) $\sum_{\substack{i=0 \\ i \neq j}}^{2} K_i = K$, for all $i \neq j$;

(ii) $tK_o = 0$;

(iii) $tK \subset K_3$.

Then there exists a representation equivalence $G : V_1 \to K$, where G is given by reduction modulo π . To be precise: for $V_* \in V_1$ and any submodule U of V , let $\overline{U} := (U + \pi V)/\pi V$. Then G is defined by : $GV_* = (K \mid K_o, K_1, K_2, K_3)$, and $K := \overline{V}$; $K_i := \overline{V_i}$ for all $i = 0,1,2$; $K_3 := \overline{\pi^{-1}(0)}$.

It is clear that G is a functor: take the operation t on K to be $\bar{t}$, the operation on $\overline{V}$ induced by $t = g-1$. (b) and (e) implies (i) ;

(c_o) implies (ii) ;

(d) implies (iii) .

A routine verification shows that G is full, dense and isomorphism-reflecting.

Step 3: Since $tK = t(K_o + K_i) = tK_i \subset K_i$ for $i = 1,2$, we can apply Proposition 2 (see 2.2) to the category K : there exists a representation equivalence $H : K \to M$. Here, M is the category of the k-representations $M_* = (M \mid M_i, M_a, M_b)_{i=0,\dots,3}$ of the partially ordered set

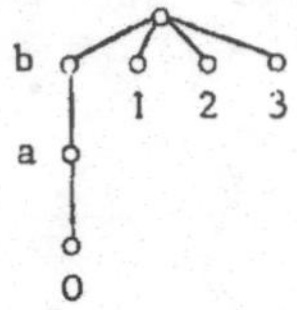

, which satisfy the condition :

$$(*) \qquad \sum_{\substack{i=0\\ i\neq j}}^{2} M_i = M \text{ , for all } j = 0,1,2 \text{ .}$$

Disregarding condition (*) , it is well-known that this last category would be of wild representation type (e. g. see [15]). Indeed it turns out that also M is of wild representation type, as we will see in the next step.

Step 4: A full and faithful embedding functor $J : W(k) \to M$ is given by:
$J(V;x,y) = (M \mid M_i, M_a, M_b)_{i=0,\dots,3}$;

and

$M := V \times V \times U \times V$,
$M_a := M_b := V \times 0 \times U \times V$,
$M_o := \{(v,0,u,u) \mid v \in V , u \in U\}$,
$M_1 := \{(w,v,xv,w+yv) \mid v,w \in V\}$,
$M_2 := \{(v,v,0,w) \mid v,w \in V\}$,
$M_3 := 0 \times V \times U \times 0$,
where $U := \operatorname{im} x$.

Morphisms are mapped by J in the obvious way. It is easy to check that J is a faithful functor. Verification that J is a full functor is routine linear algebra.

Step 5: Combining all the previous steps it is now clear that there exists a full subcategory U of L_δ which is representation equivalent to $W(k)$. Therefore L is of wild representation type.

q.e.d.

<u>3.2 If $p > 2$, $v(p) = 2$, $r(KC_p) = 3$, then $\Lambda = RC_p$ is of finite representation type. The number of non-isomorphic indecomposable Λ-lattices is $2p + 3$.</u>

<u>Proof.</u> We know from Lemma 3 that each of Λe_i $(i = 0,1,2)$ is a discrete valuation ring. Applying Theorem 1 and Proposition 1 (see 2.2) we get: There exists a full subcategory L_δ of L, cofinite in L, and a representation equivalence $F : L_\delta \to V$. For the definition of L_δ and V see 2.2. In our present situation the conditions imposed on an object $V_* = (V \mid V_o, V_1, V_2)$ of V are the following:

(a) $\pi V = 0$;

(b) $\sum_{\substack{i=0 \\ i \neq j}}^{2} V_i = V$, for all $j = 0,1,2$;

(c) $tV_o = 0$,

$t^d V_i = 0$, for $i = 1,2$ and $d := \frac{p-1}{2}$.

(a) and (b) is immediate from the definition of V. As to (c), we know that $X^p - 1$ has the prime factorization $X^p - 1 = (X-1)\varphi_1(X)\varphi_2(X)$ over K (combining the assumption $r = 3$ with Lemma 1). Application of Lemma 1 shows:

(c) $tV_o = 0$,

$\varphi_i(g)V_i = 0$, for $i = 1,2$.

But $\varphi_i(X) \equiv (X-1)^d \pmod{\pi R[X]}$ implies $\varphi_i(g) \equiv t^d \pmod{\pi\Lambda}$. Since $\pi V = 0$, we have $t^d V_i = \varphi_i(g)V_i$; for $i = 1,2$.

Now it is clear that V is (isomorphic to) the category W of the $k[t]/\langle t^d\rangle$-module representations $W_* = (W \mid W_o, W_1, W_2)$ of the partially ordered set 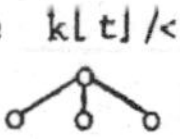 which satisfy the conditions:

(i) $\sum_{\substack{i=0 \\ i \neq j}}^{2} W_i = W$; for all $j = 0,1,2$;

(ii) $tW_o = 0$.

Since $tW = t(W_o + W_i) = tW_i \subset W_i$, for $i = 1,2$, we can apply Proposition 2 (see 2.2). We get: There exists a representation equivalence $G : W \to D_{p+1}$.

Here, D_{p+1} is the category of the k-representations $X_* = (X \mid X_i, X_{a_j}, X_{b_j})_{\substack{i=0,1,2 \\ j=1,..,d-1}}$

of the partially ordered set

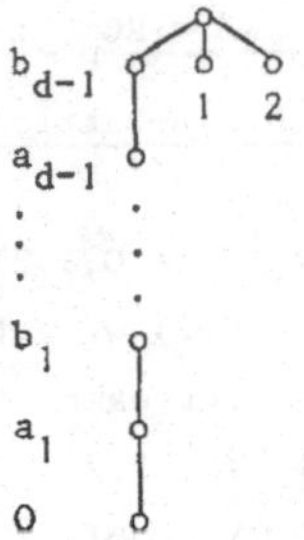

which satisfy the condition $\sum_{\substack{i=0 \\ i \neq j}}^{2} X_i = X$; for all $j = 0,1,2$.

This partially ordered set is of Dynkin type D_{p+1} . Therefore $\mathcal{D}_{p+1}$ is of finite representation type and the indecomposable objects can be determined ([1], [7]). There are $2p-1$ isomorphism classes of indecomposable objects in $\mathcal{D}_{p+1}$. By means of GF there are $2p-1$ isomorphism classes of indecomposable lattices in L_δ . Up to isomorphism, Λ, Λe_i $(i = 0,1,2)$ are the only indecomposable Λ-lattices not being an object of L_δ (see Prop. 1) Thus we obtain a total of $2p+3$ isomorphism classes of indecomposable Λ-lattices.

q.e.d.

<u>3.3 If $v(3) = 2$ and $r(KC_3) = 2$, then $\Lambda = RC_3$ is of finite representation type. The number of non-isomorphic indecomposable Λ-lattices is 7.</u>

In order to prove this we begin with a calculation of a certain class of Bäckström-orders. (For definition of Bäckström-orders see 2.2).

<u>Lemma 5.</u> Let $p > 2$ be a prime number, $v(p) = 2$ and $r(KC_p) = 2$. Let $d := \frac{p-1}{2}$, and $s := \sum_{i=0}^{p-1} g^i \in \Lambda$. Consider

$$\Xi := \Lambda + \frac{s}{\pi}\Lambda + \frac{t^{d+1}}{\pi}\Lambda ,$$

$$\Gamma := e_o\Lambda \oplus e_1(\Lambda + \frac{t^d}{\pi}\Lambda) ,$$

$$I := e_o\pi\Lambda \oplus e_1(J + \frac{t^{d+1}}{\pi}\Lambda)$$. Then:

(i) Ξ and Γ are R-orders in KC_p and I is an ideal of Ξ and of Γ .

(ii) Γ is the unique maximal order in KC_p .

(iii) $I = \text{rad}\,\Gamma = \text{rad}\,\Xi$.

(iv) Ξ is a Bäckström-order in KC_p , with corresponding hereditary order Γ.

Proof.

(i) Using the equalities $e_o = \frac{s}{p}$, $e_1 t = t$, and $e_o\Lambda \oplus e_1\Lambda = e_o\Lambda + \Lambda$, we find that $I = J + \frac{s}{\pi}\Lambda + \frac{t^{d+1}}{\pi}\Lambda$. With this description of I , the proof of (i) is a routine verification.

(ii) Let Ω be the unique maximal order of KC_p . Consider the canonical isomorphism $\rho : Ke_o \oplus KC_p e_1 = KC_p \cong K[X]/\langle X-1\rangle \oplus K[X]/\langle \phi_p(X)\rangle \cong K \oplus K(\alpha) = : A$,

where α is a primitive p-th root of unity. The maximal order of A is given by $\Omega_A = R \oplus \Delta$, where $\Delta = \text{intclos}_R\{K(\alpha)\}$. We know from Lemma 4 that

$\Delta = R[\alpha] + \frac{(\alpha-1)^d}{\pi} R[\alpha]$.

Therefore $\Omega = \rho^{-1}(\Omega_A) = \Lambda e_o \oplus e_1(\Lambda + \frac{t^d}{\pi}\Lambda) = \Gamma$.

(iii) $\rho(I) = \pi R \oplus ((\alpha-1)R[\alpha] + \pi R[\alpha] + \frac{(\alpha-1)^{d+1}}{\pi} R[\alpha]) = \pi R \oplus (\alpha-1)\Delta = \text{rad}\,\Omega_A$

(see Lemma 4). Hence $I = \text{rad}\,\Gamma$.

Using the description of I noted in part (i) of the proof, it is easily seen that $\Xi/I \cong k$. Hence $\text{rad}\,\Xi \subset I$.

Conversely, it is not difficult to verify that $e_1 I^p \subset e_1 \pi\Lambda$.

(One writes $e_1 I^p$ as a binomial sum and one checks the inclusion for each summand.)

Therefore $I^{2p} \subset \pi^2(e_o\Lambda \oplus e_1\Lambda) \subset s\Lambda + \pi^2\Lambda \subset \pi\Xi$. Hence $I \subset \text{rad}\,\Xi$.

(iv) Evidently $\Xi \subsetneq \Gamma$. Thus (iv) follows from (i), (ii) and (iii) . q.e.d.

Remark. It follows from the definition of a Bäckström-order and from the previous Lemma that any Bäckström-order in KC_p must contain $I + R = \Xi$. Therefore Ξ is the unique minimal Bäckström-order in KC_p .

We are now in the position to apply Theorem 2 (see 2.2) to our situation: in the proof of the previous Lemma we have seen that $\Xi/\text{rad}\,\Xi \cong k$; $\Gamma/\text{rad}\,\Gamma \cong k \oplus \bar{\Delta}$, where $\bar{\Delta} := \Delta/\text{rad}\,\Delta$ and $\Delta = \text{intclos}_R\{K(\alpha)\}$ as in part (ii) of the proof. Moreover, we know from Lemma 4 that $(\bar{\Delta}:k) = 2$. Therefore the species S of $\Xi \subset \Gamma$ is

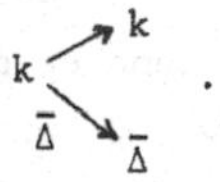

S is of type B_3 : 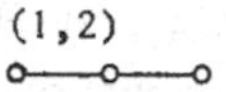

. Therefore S is of finite representation type: There are only 6 isomorphism classes of indecomposable non-simple representations of S (see [6]) . By Theorem 2 then there are only 6 isomorphism classes of indecomposable Ξ-lattices.

In particular, let $p = 3 : t^2 = s - 3g \Rightarrow \Xi = \Lambda + \frac{s}{\pi}\Lambda$.
Recall that a Λ-lattice M has a non-zero free summand if and only if $sM \not\subset \pi M$, that is to say if and only if M is not a Ξ-lattice. Therefore any indecomposable Λ-lattice is either an indecomposable Ξ-lattice or isomorphic to Λ. Thus Λ is of finite representation type: we obtain a total of exactly 7 isomorphism classes of indecomposable Λ-lattices.

q.e.d.

Remark. In case $p > 3$, the knowledge of the Bäckström-order Ξ does not solve the classification problem for the Λ-lattices. However, by different methods the following generalization of 3.3 can be proved: Let $\Lambda = RC_p$, $p > 2$, $v(p) = 2$, $r(KC_p) = 2$. Then Λ is of finite representation type. The number of isomorphism classes of indecomposable Λ-lattices is $2p+1$.

3.4 If $\operatorname{char} R = 2$, then $\Lambda = RC_2$ is of tame representation type. A complete list of all indecomposable Λ-lattices is given by the matrices (1), $\begin{pmatrix} 1 & \pi^n \\ & 1 \end{pmatrix}$; $n \in N_o$.

Proof. Let M be any Λ-lattice. The aim is to find a normal form for the g-operation on M. $\ker t$ is a pure R-submodule of M. Decompose M as R-module into $M = \ker t \oplus N$. Choose an R-basis of M which is adapted to this decomposition. Relative to this R-basis, the matrix of the g-operation has the form

$$(g) = \begin{pmatrix} E & X \\ & E \end{pmatrix}.$$

(Here we have to observe that $t^2 = 0$ due to $\operatorname{char} R = 2$, hence $tN \subset tM \subset \ker t$). It remains to find a normal form for X. Let S and T be unimodular R-matrices.

Then $\begin{pmatrix} S & \\ & T^{-1} \end{pmatrix} \begin{pmatrix} E & X \\ & E \end{pmatrix} \begin{pmatrix} S^{-1} & \\ & T \end{pmatrix} = \begin{pmatrix} E & SXT \\ & E \end{pmatrix}$. Thus we are reduced to the following easy matrix problem: Consider an arbitrary R-Matrix X which may be transformed by such operations on the rows (on the columns respectively) which correspond to unimodular R-matrices S (T respectively).
As solution of this matrix problem we obtain: X decomposes into matrices of the type (π^n), $n \in N_o$, or $\vdash\!\dashv$, I. (Here, $\vdash\!\dashv$ denotes the matrix with zero rows and one column; I denotes the matrix with one row and zero columns.) This shows that each indecomposable Λ-lattice is isomorphic to one of the lattices which are given by the matrices (1), $\begin{pmatrix} 1 & \pi^n \\ & 1 \end{pmatrix}$, $n \in N_o$. Conversely, it is elementary to show that these lattices are indecomposable and pairwise non-isomorphic.

q.e.d.

4. Conclusion.

Table 1.2, brought up to date, presents the following picture (observe that in the ramified situation some of the vertical strips are now subdivided into finer strips which correspond to the numbers r(KG) , noted at the bottom of the refined strips):

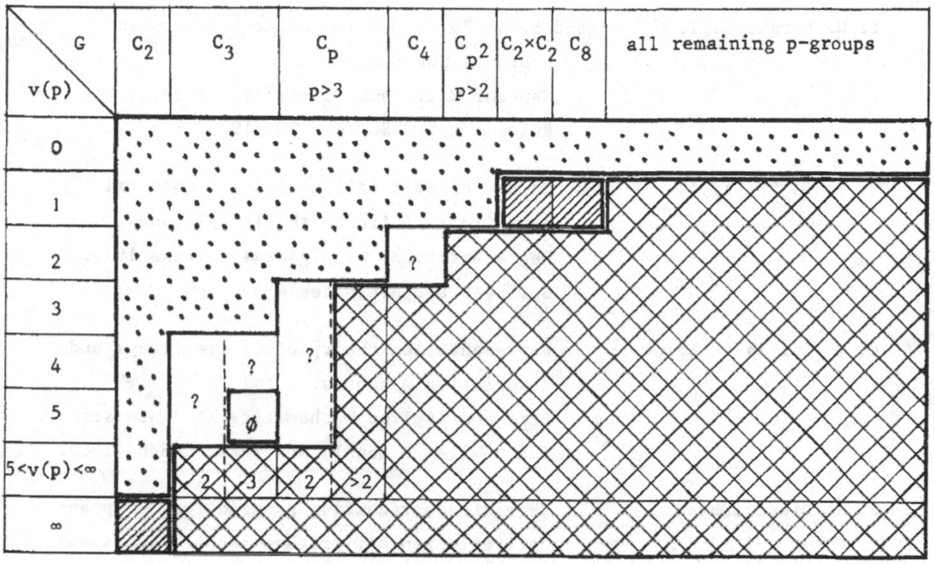
G
v(p)
C_2
C_3
C_p
p>3
C_4
C_{p^2}
p>2
$C_2 \times C_2$
C_8
all remaining p-groups
0
1
2
3
4
5
5<v(p)<∞
∞
?
?
?
?
Ø
2
3
2
>2

The symbol "Ø" means that there exists no group ring at this place. The symbol "?" denotes those group rings the representation type of which is not yet clear. However, it is true that all of them are of infinite representation type (see remark (iv) in 1.2). Moreover it seems that among them there are some group rings of tame representation type.

References

[1] S. D. Berman/P. M. Gudivok : Integral representations of finite groups. Dokl. Akad. Nauk SSSR 145 (1962), 1199-1201. Soviet Math. Dokl. 3 (1962), 1172-1174.

[2] I. N. Bernstein/I. M. Gelfand/B. A. Ponomarev : Coxeter functors and Gabriels Theorem. Uspechi Mat. Nauk 28 (1973), 19-38. Russian Math. Surveys 28 (1973), 17-32.

[3] M. C. R. Butler : On the classification of local integral representations of finite abelian p-groups. Representations of algebras, Ottawa 1974. Springer Lecture Notes 488.

[4] C. W. Curtis/I. Reiner : Representation theory of finite groups and associative algebras. Pure and Applied Mathematics XI. Interscience (John Wiley & Sons), New York, London 1962.

[5] F. E. Diederichsen : Über die Ausreduktion ganzzahliger Gruppendarstellungen bei arithmetischer Äquivalenz. Hamb. Abh. 14 (1938), 357-412.

[6] V. Dlab/C. M. Ringel : Indecomposable representations of graphs and algebras. Memoires Amer. Math. Soc. No. 173 (1976).

[7] P. Gabriel : Unzerlegbare Darstellungen I. Manuscripta Mathematica, 6 (1972), 71-103

[8] P. M. Gudivok : On modular and integral representations of finite groups. Dokl. Adad. Nauk SSSR, Tom 214 (1974), No. 5. Soviet Math. Dokl. Vol. 15 (1974), No. 1.

[9] A. Heller/I. Reiner : Representations of cyclic groups in rings of integers I. Ann. of Math. 76 (1962), 73-92.

[10] H. Jacobinski : Sur les ordres commutatifs avec une nombre fini de résaux indecomposables. Acta Math. 118 (1967), 1-31 .

[11] A. V. Jacovlev : Classification of the 2-adic representations of the cyclic group of order 8.
Zap. Nauc. Sem. Leningrad. Otdel. Mat. Inst. Steklov (LOMI) 28 (1972), 93-129.
Journal of Soviet Math., Vol. 3, No. 5, May 1975.

[12] H. Maschke : Über den arithmetischen Charakter der Coeffizienten der Substitutionen endlicher linearer Substitutionsgruppen.
Math. Ann. 50 (1898), 482-498.

[13] L. A. Nazarova : Representations of tetrads.
Izv. Akad. Nauk SSSR Ser. Mat. 31 (1967), 1361-1378. Math. USSR Izv. 1 (1967), 1305-1321 (1969).

[14] C. M. Ringel/K. W. Roggenkamp : Diagrammatic methods in the representation theory of orders.
Journal of algebra, Vol. 60, No. 1 (1979), 11-42.

[15] A. G. Zavadskij/L. A. Nazarova : Partially ordered sets of tame type.
Matrix problems, Kiev (1977), 122-143.

ON BLOCKS OF FINITE LATTICE TYPE II

Christine Bessenrodt (Essen)

Let G be a finite group, p a prime dividing the order of G and R a finite unramified extension of $\mathbb{Z}_p$, the complete ring of p-adic integers. For the other notations we refer to [3,8].
The title implies the canonical question: which blocks of RG are of finite lattice type? The answer to this is given in the following

Theorem [1]: A block of RG is of finite lattice type iff its defect group is cyclic and of order $\leq p^2$.

Now the second quite natural question is: what is the number of non-isomorphic indecomposable lattices in a block B of finite type? If $\delta(B) = \{1\}$ there is only one indecomposable lattice. For B with $|\delta(B)| = p$ we have the

Theorem [1]: Let B be a block of RG of defect 1 and of inertial degree t. Then B has exactly $3t$ non-isomorphic indecomposable RG-lattices.

This was independently obtained by K.W. Roggenkamp [9].

From now on let R be as above and moreover such that $R/J(R) =: F$ is a splitting field for G and its subgroups (this always exists). Furthermore assume that R contains a primitive $(p-1)^2$-th root of 1. Then:

<u>Theorem:</u> Let B be a block of RG with cyclic defect group of order p^2 and inertial degree t. Then B has exactly (4p+1) non-isomorphic indecomposable RG-lattices and each of these is generated by at most two elements.

For the proof of the theorem it is necessary to take a closer look at the lattices over the cyclic group of order p^2.

Using [4] the indecomposable lattices can be computed (see also [5]). The complete list is given below.

Let $D = \langle d \rangle$ be a cyclic group of order p^2, Y its subgroup of order p. Define for $X \leq D$: $\hat{X} = \sum_{x \in X} x$.

KD has the primitive idempotents $e_1 = \frac{1}{p^2} \hat{D}$, $e_2 = \frac{1}{p} \hat{Y} - \frac{1}{p^2} \hat{D}$, $e_3 = 1 - \frac{1}{p} \hat{Y}$. With this notation we can state the list of the 4p+1 indecomposable RD-lattices:

RDe_1, RDe_2, RDe_3, $RD(e_1+e_2)$, $RD(e_1+e_3)$, $RD(e_2+e_3)$, RD

$RD(e_1+(1-d)^i e_3) + RD(e_2+e_3)$, $0 \leq i \leq p-2$

$RD + (1-d)^i RDe_3$, $1 \leq i \leq p-1$

$RD(e_2+e_3) + (1-d)^i RDe_3$, $1 \leq i \leq p-2$

$RD(u_1+u_2+u_3) + RD(v_1+(1-d)^i u_3)$, $1 \leq i \leq p-2$,

where this is considered as an RD-sublattice of

$RDe_1u_1 \oplus RDe_2u_2 \oplus RDe_3u_3 \oplus RDe_1v_1$.

This list together with a stronger version of [3,49.7] and elementary properties of conjugation can be used to obtain:

<u>Proposition:</u> Let D be a cyclic p-group of order $\leq p^2$. Then

a) All indecomposable RD-lattices are absolutely indecomposable (that is, they remain indecomposable under finite extensions of R).

b) All indecomposable RD-lattices are stable under conjugation, that means: if $D \trianglelefteq G$, then $U^g \underset{RD}{\simeq} U$ for every indecomposable RD-lattice U and $g \in G$.

With this in mind, we state a more general result on blocks with cyclic defect groups, which gives the theorem on blocks of cyclic defect 2 almost immediately.

Theorem: Let B be a block of RG with cyclic defect group D and inertial degree t. Let W be an indecomposable RD-lattice, Y = vx W and U an RY-source of W. Suppose that

(i) U is absolutely indecomposable

(ii) If $t \neq 1$, then: $U^g \underset{RY}{\simeq} U$ for all $g \in N_G(Y)$, and

$U \simeq R \underset{\mathbb{Z}_p}{\otimes} \tilde{U}$ with an indecomposable $\mathbb{Z}_pY$-lattice $\tilde{U}$.

Then W "induces" exactly t non-isomorphic indecomposable RG-lattices in B, that is, W^G has exactly t non-isomorphic indecomposable direct summands in B. (We call these "induced" lattices in B.)

Sketch of proof: Let us first fix some notation. Set $C := C_G(Y)$, $H := N_G(Y)$. Let B_1 be the block of RH corresponding to B. Then there is a unique - up to H-conjugation - block $b \leftrightarrow f$ of RC with defect group D, $B_1 = \sum_g b^g$ where g runs through a set of coset representatives of $T := T_H(b) = \{h \in H \mid f^h = f\}$ in H, and b has only one projective indecomposable RC-lattice

f_1RC [7] .

Now using (i) it can be proved by [6] that U induces exactly one indecomposable RC-lattice in b, namely $U \otimes_{RY} f_1RC =: V$.

Because U is stable under conjugation, $T_H(V) := \{h \in H \mid V^h \underset{RC}{\simeq} V\} = T$.

Set $A := End_{RC}(V)$, $E := End_{RT}(V^T)$. To obtain the decomposition of V^T it suffices by [2,6] to determine the primitive idempotents of $E/_{J(A)E}$ (notice that A is canonically isomorphic to a subring of E). This is a twisted group algebra of $T/_C$ over F ((i)!) and with factor set in GF(p) ((ii)!). Now $T/_C$ is cyclic, $t = |T:C|$ divides p-1 and F contains a primitive $(p-1)^2$-th root of 1, hence by [2]: $E/_{J(A)E} \simeq F^T/_C$. Since this decomposes into t non-isomorphic indecomposables, so does V^T. Tensoring with RH does not change this. Then Green correspondence finishes the proof.

Now we turn our interest to the relation between Heller-orbits over RD and those in B. Let Ω denote Heller's operator.

<u>Theorem:</u> Let B be a block of RG with cyclic defect group D and inertial degree t. Let W be a non-projective indecomposable RD-lattice as above and suppose that in addition to (i) and (ii) we have:

(iii) If $t \neq 1$, then $U/_{pU}$ is a direct sum of indecomposable FY-modules of different dimensions.

Then

a) Ω-orbits of indecomposable non-projective RD-lattices with properties (i),(ii),(iii) and Ω-orbits of their "induced" indecomposable RG-lattices in B correspond via a vertex-preserving map.

b) From W "induced" indecomposable RG-lattices in B have Ω-period t or 2t corresponding to Ω-period 1 or 2 of U.

Moreover, all Ω-orbits in B are finite.

Proof: Suppose the Ω-period of U is 2 (the case of Ω-period 1 is similar). With the notation of the preceding theorem let $V^T = \bigoplus_{i=1}^{t} V_i$, the V_i indecomposable RT-lattices. First observe that if $\bar{U} = U/_{pU}$ decomposes into m indecomposables, then so does each $\bar{V}_i$, say $\bar{V}_1 = \bigoplus_{j=1}^{m} V_{1j}$. Now (iii) implies $\dim_F V_{1i} \neq \dim_F V_{1j}$ for all $i \neq j$, and so $\Omega^{2l} V_1 \simeq V_1$, $l \leq t$, gives $\Omega^{2l} V_{1i} \simeq V_{1i}$ for all i. But the block of FT corresponding to fRT is uniserial with t simple modules [7], so all V_{1i} have Ω-period 2t. Hence $l = t$. Again tensoring with RH and applying Green correspondence complete the proof. - It is easy to show the last assertion.

The next statements can be obtained by using the explicit description of the lattices.

Proposition: Let D be a cyclic p-group of order $\leq p^2$.

a) If U is an indecomposable RD-lattice, then $U/_{pU}$ is a direct sum of at most two indecomposable FD-modules of different dimensions.

b) Every indecomposable non-projective RD-lattice has Ω-period 2.

Together with the last theorem this gives the

Corollary: Let B be a block of RG with a cyclic defect group of order $\leq p^2$ and inertial degree t. Then

a) There is a bijection between the set of Ω-orbits of indecomposable non-projective RD-lattices and the set of Ω-orbits of indecomposable non-projective RG-lattices in B.

b) Every indecomposable non-projective RG-lattice in B has Ω-period 2t.

The proofs will soon be published in detail in an article on "Indecomposable lattices in blocks with cyclic defect groups" in "Communications in Algebra".

The results of this report are contained in the author's doctoral dissertation written under Prof. G. Michler, Essen 1980. Parts of them were obtained later by A. Wiedemann using other techniques.

References

[1] C. Bessenrodt, On blocks of finite lattice type, Archiv der Math. 33 , 334-337 (1980)

[2] S.B. Conlon, Twisted group algebras and their representations, J. Australian Math. Soc. 4, 152-173 (1964)

[3] L. Dornhoff, Group representation theory, Part B, New York 1972

[4] E.L. Green, I. Reiner, Integral representations and diagrams, Mich. Math. J. 25, 53-84 (1978)

[5] A. Heller, I. Reiner, Representations of cyclic groups in rings of integers I, Annals of Math. 76, 73-92 (1962)

[6] R. Knörr, On the vertices of irreducible modules, Annals of Math. 110, 487-499 (1979)

[7] G. Michler, Green correspondence between blocks with cyclic defect groups I, J. Algebra 39, 26-51 (1976)

[8] G. Michler, Green correspondence between blocks with cyclic defect groups II, in "Representations of Algebras", LNM 488, 210-235 (1975)

[9] K.W. Roggenkamp, Representation theory of blocks of defect 1, Carleton Math. Lecture Notes 25, 2001-2024 (1980)

Christine Bessenrodt
Fachbereich 6 - Mathematik -
Universität Essen - GHS
Universitätsstr. 3

4300 Essen 1

THE AUSLANDER-REITEN-GRAPH OF BLOCKS WITH CYCLIC DEFECT TWO

A. Wiedemann
Mathematisches Institut B
der Universität Stuttgart
West-Germany

This paper contains the main results of the second part of my Thesis, written in Stuttgart 1980.

§ 1 Introduction

The concept of the Auslander-Reiten-Graph (cf. [A2, Ri, W1, W2]) gives the following useful criterion for finite lattice type [W1,W2]:

Theorem: Let R be a complete Dedekind domain with quotient field K, and let Λ be an indecomposable R-order in a separable K-algebra. If the Auslander-Reiten-graph $\mathfrak{A}(\Lambda)$ of Λ has a finite connected component $\mathfrak{C}$, then $\mathfrak{A}(\Lambda) = \mathfrak{C}$, and Λ is of finite type.

Let $\hat{Z}_p$ be the p-adic integers, G a finite group and let $\mathfrak{B}$ be a block of the group ring $\hat{Z}_p G$ with a cyclic defect group of order p^2 and t simple modules. In this paper we use the above criterion in order to show that $\mathfrak{B}$ has $(4p+1)t$ indecomposable lattices. (This result was independently obtained also by Chr. Bessenrodt [Be] in the modular splitting case.) Moreover, it will be indicated how to construct explicitely the Auslander-Reiten-graph of $\mathfrak{B}$.

It should be noted that the Auslander-Reiten-graph (in the following abbreviated by "AR-graph") gives information not only about the indecomposable Λ-lattices, but also on the morphisms between them.

If Λ is of finite lattice type, each morphism between Λ-lattices can be written as a sum of compositions of irreducible morphisms. Hence in this case the AR-graph of Λ reflects the whole category of Λ-lattices.

Let R and Λ be as in the above Theorem. From [A1, Ro1] we recall the following: For each indecomposable non-projective Λ-lattice M there exists an almost split sequence in the category of Λ-lattices with cokernel M . The kernel of this sequence will be denoted by AM , the Auslander-translate of M . It can be constructed by the following process:

Let $0 \to \Omega M \to P \to M \to 0$ be a projective cover sequence of M , then AM is the dual (with respect to R) of the cokernel C of the following exact sequence:

$$0 \to \mathrm{Hom}_\Lambda(M,\Lambda) \to \mathrm{Hom}_\Lambda(P,\Lambda) \to C \to 0 \quad ;$$

and $AM \simeq C^* := \mathrm{Hom}_R(C,R)$.

In case that Λ is isomorphic to Λ^* as Λ-Λ-bimodule, AM is isomorphic to ΩM . This holds in particular for $\Lambda = RG$, the group ring of a finite group G with coefficients in R .

Moreover, $\mathrm{Ext}^1_\Lambda(M,AM)$ has simple socles as $\mathrm{End}_\Lambda(M)$- respectively $\mathrm{End}_\Lambda(AM)$-module which actually coincide as sets. The almost split sequence of M is then - up to isomorphism - represented by a nonzero element of this socle.

Hence for $\Lambda = RG$ the almost split sequence $\mathcal{E}$ of M can be constructed from the above projective cover sequence as pullback induced by an endomorphism φ of M which can not be factored through $P \to M$ but has the property that the composed endomorphism $\varphi\rho$ factors through the projective cover of M for each non-isomorphism $\rho \in \mathrm{End}_\Lambda(M)$:

$$\begin{array}{ccccccccc} & 0 & \to & \Omega M & \to & P & \to & M & \to & 0 \\ & & & \| & & \uparrow & & \uparrow\varphi & & \\ \mathfrak{E} : & 0 & \to & AM & \to & E & \to & M & \to & 0 \end{array} .$$

§ 2 The Auslander-Reiten-graph of the cyclic group of order p^2.

Let

$R = \hat{\mathbb{Z}}_p$ = p-adic integers,

$P = C_{p^2}$ = cyclic group of order p^2,

$\Lambda = RP$ = p-adic group ring of P,

δ (resp. η) = p-th (resp. p^2-th) primitive root of unity,

$\alpha_l : R[\delta] \to R[\delta]/(1-\delta)^l R[\delta]$,

$\beta_l : R[\eta] \to R[\eta]/(1-\eta)^l R[\eta]$,

the canonical projections for $l = 1, \ldots, p-1$.

Since $\mathbb{Z}/p\mathbb{Z}\, C_{p^2}$ is uniserial, there is an isomorphism

$$\varphi_l : R[\delta]/(1-\delta)^l R[\delta] \to R[\eta]/(1-\eta)^l R[\eta], \quad 1 \le l \le p-1 .$$

Moreover, for $l = 1, \ldots, p-1$ let

U_l be the kernel of the Λ-morphism

$$(\alpha_l\varphi_l, -\beta_l) : R[\delta] \oplus R[\eta] \to R[\eta]/(1-\eta)^l R[\eta] .$$

Then Λ can be described as follows (for details see [W2]):

<u>Lemma 1:</u> As subring of $R \oplus R[\delta] \oplus R[\eta]$ we have the equality

$$\Lambda = R \cdot (1,1,1) + (1-\delta,\ 1-\eta) \cdot U_{p-1} . \qquad \#$$

Now we shall define a series of Λ-lattices:

$V_l := \{(x,y,z) \in \Lambda \mid y \in (1-\delta)^{p-l-1} R[\delta]\}$, $l = 0,1,\ldots,p-2$,

$W_l := AU_l$, $l = 1,\ldots,p-2$,

$X_1 := \{(x,y) \in R \oplus R[\delta] \mid x-y \in (1-\delta)R[\delta]\}$,

$X_2 := \{(x,z) \in R \oplus R[\eta] \mid x-z \in (1-\eta)R[\eta]\}$,

$\Gamma_l := \{(x,y,z) \in R \oplus R[\delta] \oplus R[\eta] \mid (x,y) \in X_1, (y,z) \in U_l\}$ for $l=1,\ldots,p-1$.

The process described in the introduction for computing almost split sequences yields

<u>Lemma 2:</u> (i) $0 \to W_l \to V_l \oplus \Gamma_l \to U_l \to 0$, $l = 1,\dots,p-2$ are almost split sequences.

(ii) $AV_l \simeq \Gamma_{l+1}$, $l = 0,1,\dots,p-2$.
Since the Γ_l are local over-orders of Λ , this implies that both V_l and Γ_{l+1} are indecomposable Λ-lattices for $l = 0,\dots,p-2$.

(iii) $AR \simeq U_{p-1}$; $AU_{p-1} \simeq R$; $AR[\delta] \simeq X_2$; $AR[\eta] \simeq X_1$.

(iv) $0 \to X_2 \to V_o \to R[\delta] \to 0$,
$0 \to X_1 \to V_o \to R[\eta] \to 0$ are almost split sequences. #

<u>Theorem</u>: The AR-graph of Λ has the following form:

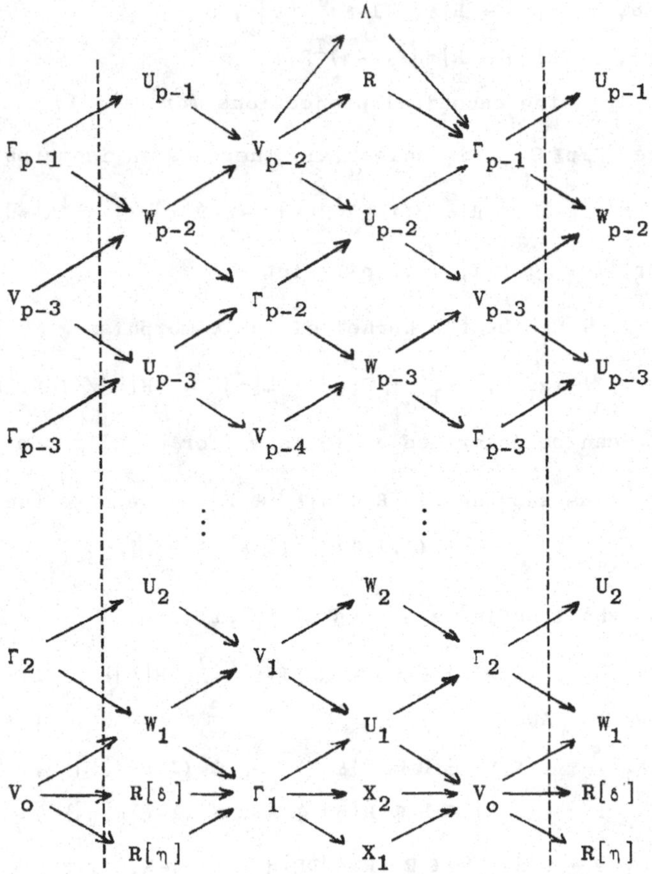

(Identify along the dotted lines.)

In fact, from usual arguments about irreducible morphisms one finds the following part of the AR-graph of Λ:

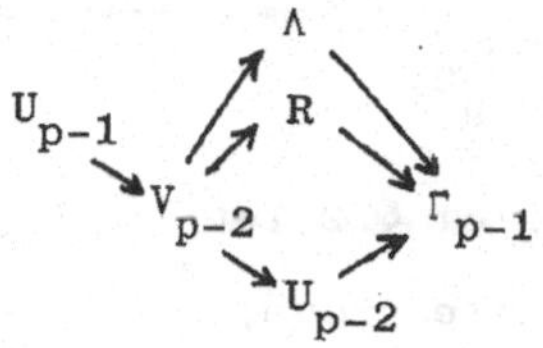

Starting with this, the AR-graph of Λ is then constructed by using successively Lemma 2. #

This result implies immediately with Theorem § 1 the well-known result of Heller and Reiner [HR]:

Corollary: $\Lambda = RC_{p^2}$ has up to isomorphism exactly $4p+1$ indecomposable lattices:

$\Lambda, R, R[\delta], R[\eta], X_1, X_2$	:	6
$U_l, \quad l = 1,\dots,p-1$	:	$p-1$
$V_l, \quad l = 0,\dots,p-2$	:	$p-1$
$W_l, \quad l = 1,\dots,p-2$	:	$p-2$
$\Gamma_l, \quad l = 1,\dots,p-1$	:	$p-1$
		$4p+1$

#

Lemma 3: Let S be an unramified extension of $R = \hat{\mathbb{Z}}_p$ then the AR-graph of $S\Gamma$ arises from that one of Λ by "pointwise" tensoring with S over R.

In fact, all the simple quotients of the endomorphism rings of all indecomposable Λ-lattices are $\mathbb{Z}/p\mathbb{Z}$. #

Remark: Obviously the stable AR-graph of RC_{p^2} is a two-fold covering of the Dynkin-diagram $\mathbb{D}_{2p}$ in the sense of Riedtmann [Rm]. In connection with the integral representations of the cyclic group of order p^2, this diagram occured for the first time in Butler's paper [Bu], where he proved that RC_{p^2} has $4p+1$ indecomposable lattices using representations of $\mathbb{D}_{2p}$ over the residue class field R/pR .

§ 3 Normal defect group C_{p^2}

In addition to the notation of § 2 let

N	be a finite group,
P	$\simeq C_{p^2}$ a normal subgroup of N ,
C	$= C_N(P)$ the centralizer of P in N ,
B	a block of RN with defect group P , and
b	a block of RC such that $B = b^N$,
	i.e. b is a root of B .

Lemma 1: There exists an unramified extension S of R and an $m \in \mathbb{N}$ such that

$$b \simeq (SP)_m = m \times m - \text{matrix ring over } SP.$$

Proof: Since b has only one class of simple modules [M1], we conclude that

$$b \simeq (E)_m ,$$

where E is the endomorphism ring of the unique indecomposable projective b-module.

Then RP can be identified with a subring in the centre of E. Now E is of finite lattice type, and therefore KE contains at most three primitive orthogonal idempotents [RR]. Hence the primitive idempotents of E lie already in RP and are all central.

Rank considerations show that

$$E \simeq RP^{(r)}$$

as RP-module with $r = |S:R|$. This altogether implies by ramification arguments that $SP \subset E$ and

$$KE = KSP .$$

Comparing the number of indecomposable lattices we can conclude that

$$E = SP \quad . \qquad \#$$

Corollary: b has - up to isomorphism - 4p+1 indecomposable lattices. Identifying b with $(SP)_m$ we can explictely give the following complete list:

$$\mathfrak{Q} = \begin{pmatrix} SP \\ \vdots \\ SP \end{pmatrix}_m \quad , \quad \mathfrak{S}_o = \begin{pmatrix} S \\ \vdots \\ S \end{pmatrix}_m , \quad \mathfrak{S}_1 = \begin{pmatrix} S[\delta] \\ \vdots \\ S[\delta] \end{pmatrix}_m ,$$

$$\mathfrak{S}_2 = \begin{pmatrix} S[\eta] \\ \vdots \\ S[\eta] \end{pmatrix}_m ,$$

$$\mathfrak{X}_i = \begin{pmatrix} S \otimes_R X_i \\ \vdots \\ S \otimes_R X_i \end{pmatrix}_m , \qquad i = 1,2 ,$$

$$\mathfrak{U}_l = \begin{pmatrix} S \otimes_R U_l \\ \vdots \\ S \otimes_R U_l \end{pmatrix}_m , \qquad l = 1,\dots,p-1 ,$$

$$\mathfrak{V}_l = \begin{pmatrix} S \otimes_R V_l \\ \vdots \\ S \otimes_R V_l \end{pmatrix}_m , \qquad l = 0,\dots,p-2 ,$$

$$\mathfrak{W}_l = \begin{pmatrix} S \otimes_R W_l \\ \vdots \\ S \otimes_R W_l \end{pmatrix}_m , \qquad l = 1,\dots,p-2 ,$$

$$\mathfrak{G}_l = \begin{pmatrix} S \otimes_R \Gamma_l \\ \vdots \\ S \otimes_R \Gamma_l \end{pmatrix}_m , \qquad l = 1,\dots,p-1 \quad . \#$$

Now let $\underline{\underline{E}}^1,\dots,\underline{\underline{E}}^t$ be the non-isomorphic simple B-modules. Then by [F,VI,1.3] $\mathfrak{Q}\!\uparrow^N \simeq (\bigoplus_{i=1}^{t} \mathfrak{Q}^{\underline{\underline{i}}})^{(s)}$, where $\mathfrak{Q}^{\underline{\underline{i}}}$ is an indecomposable projective B-lattice with $\mathfrak{Q}^{\underline{\underline{i}}}/\mathrm{rad}_B \mathfrak{Q}^{\underline{\underline{i}}} \simeq \underline{\underline{E}}^i$, $i = 1,\dots,t$.

The central idempotent e in RN corresponding to B decomposes rationally into three central idempotents $e_o, e_1, e_2 \in KN$ which are induced to N from the three rational central idempotents of Kb from the inertia group of b .

Let $B_k := B e_k$, $k = 0,1,2$.

Since C is normal in N , the kernel of the projection

$$\mathfrak{O}\!\uparrow^N \to \mathfrak{S}_k\!\uparrow^N \to 0$$

is contained in $\mathrm{rad}_B \mathfrak{O}\!\uparrow^N$. By [Ro2] B_k is a hereditary order, which implies that $\mathfrak{S}_k\!\uparrow^N$ is a progenerator of B_k and decomposes as follows:

$$\mathfrak{S}_k\!\uparrow^N \simeq \Big(\bigoplus_{i=1}^{t} \mathfrak{S}_k^{\underline{\underline{i}}}\Big)^{(s)} ,$$

where $\mathfrak{S}_k^{\underline{\underline{i}}}$, $i=1,\dots,t$ are the non-isomorphic indecomposable B_k-lattices with $\mathfrak{S}_k^{\underline{\underline{i}}}/\mathrm{rad}_B \mathfrak{S}_k^{\underline{\underline{i}}} \simeq E^{\underline{\underline{i}}}$. Since B/pB is generalized uniserial we may assume that $\mathrm{rad}_B \mathfrak{S}_k^{\underline{\underline{i}}}/\mathrm{rad}^2_B \mathfrak{S}_k^{\underline{\underline{i}}} \simeq E^{\underline{\underline{i+1}}}$. (In the following the index $\underline{\underline{i}}$ has to be taken modulo t.) Moreover, this allows to define B-lattices similarly as in § 2 for $i=1,\dots,t$:

Let $\mathfrak{U}_l^{\underline{\underline{i}}}$ be the kernel of the "projection"

$$\mathfrak{S}_1^{\underline{\underline{i}}} \oplus \mathfrak{S}_2^{\underline{\underline{i}}} \to \mathfrak{S}_2^{\underline{\underline{i}}}/(1-\eta)^l \mathfrak{S}_2^{\underline{\underline{i}}} , \quad l = 1,\dots,p-1,$$

and let $\mathfrak{X}_j^{\underline{\underline{i}}}$ be the kernel of the "projection"

$$\mathfrak{S}_o^{\underline{\underline{i}}} \oplus \mathfrak{S}_j^{\underline{\underline{i}}} \to \mathfrak{S}_j^{\underline{\underline{i}}}/(1-\delta_j)\mathfrak{S}_j^{\underline{\underline{i}}} , \quad j = 1,2 ,$$

where $\delta_1 = \delta$ and $\delta_2 = \eta$.

(Note that
$$\mathfrak{S}_1^{\underline{\underline{i}}}/(1-\delta)^l \mathfrak{S}_1^{\underline{\underline{i}}} \simeq \mathfrak{S}_2^{\underline{\underline{i}}}/(1-\eta)^l \mathfrak{S}_2^{\underline{\underline{i}}} \quad \text{and}$$
$$\mathfrak{S}_o^{\underline{\underline{i}}}/p\mathfrak{S}_o^{\underline{\underline{i}}} \simeq \mathfrak{S}_j^{\underline{\underline{i}}}/(1-\delta_j)\mathfrak{S}_j^{\underline{\underline{i}}} , \quad l = 1,\dots,p-1 \text{ and } j = 1,2.)$$

Moreover, we put

$$\mathfrak{W}_l^{\underline{\underline{i}}} := A\,\mathfrak{U}_l^{\underline{\underline{i}}} , \quad l = 1,\dots,p-2 ,$$

$$\mathfrak{B}_l^{\underline{\underline{i}}} := \{(x,y,z) \in \mathfrak{O}^{\underline{\underline{i+1}}} \subset \mathfrak{S}_o^{\underline{\underline{i+1}}} \oplus \mathfrak{S}_1^{\underline{\underline{i+1}}} \oplus \mathfrak{S}_2^{\underline{\underline{i+1}}} \mid y \in (1-\delta)^{p-l-1}\mathfrak{S}_1^{\underline{\underline{i+1}}}\}, l=0,\dots,p-2,$$

$$\mathfrak{G}_l^{\underline{\underline{i}}} := \{(x,y,z) \in \mathfrak{S}_o^{\underline{\underline{i}}} \oplus \mathfrak{S}_1^{\underline{\underline{i}}} \oplus \mathfrak{S}_2^{\underline{\underline{i}}} \mid (x,y) \in \mathfrak{X}_1^{\underline{\underline{i}}} \text{ and } (y,z) \in \mathfrak{U}_l^{\underline{\underline{i}}}\} , \quad l = 1,\dots,p-2.$$

Altogether we have defined $(4p+1)t$ B-lattices.

Computations completely analoguous to those described in § 2 show that all these B-lattices are indecomposable and finally yield the AR-graph of B :

<u>Theorem:</u> The Auslander-Reiten-graph of B has the following form:

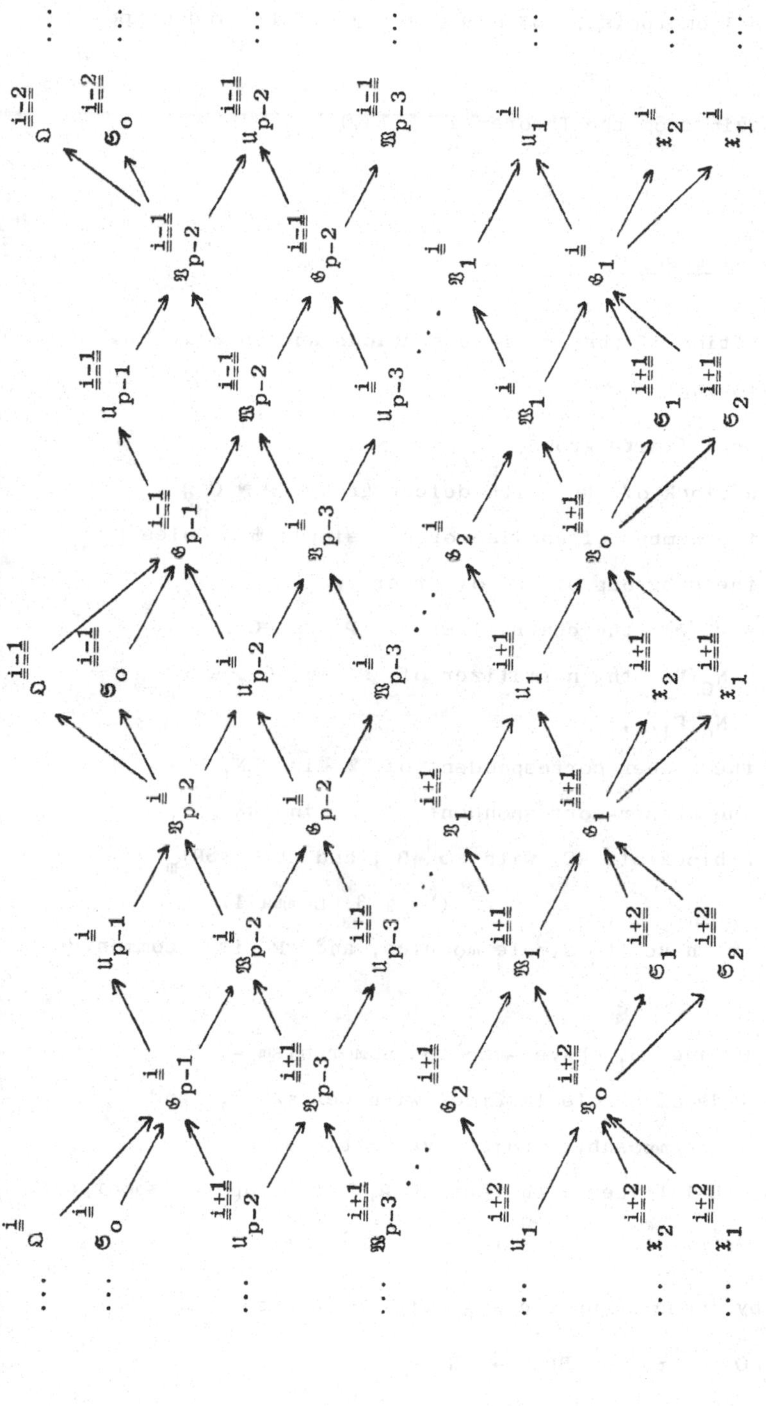

. #

<u>Corollary:</u> Up to isomorphism, B has exactly (4p+1)t indecomposable lattices. #

(This follows again from the Theorem in § 1 .)

§ 4 The general case.

We keep the notation of the previous sections and in addition introduce the following:

Let G be a finite group,

$\mathfrak{B}$ a block of RG with defect group $P \simeq C_{p^2}$,

t the number of non-isomorphic simple $\mathfrak{B}$-modules,

P_1 the subgroup of P of order p ,

$C = C_G(P)$ the centralizer of P in G ,

$N = N_G(P)$ the normalizer of P in G ,

$N_1 = N_G(P_1)$,

$\mathfrak{B}_1$ the Brauer-correspondent of $\mathfrak{B}$ in RN_1 ,

B the Brauer-correspondent of $\mathfrak{B}$ in RN ,

b a block of RC with $b^N = B$, and $b \simeq (SP)_m$ (by § 3, Lemma 1) .

Then B, $\mathfrak{B}_1$ and $\mathfrak{B}$ have t simple modules, and P is a common defect group [M2].

<u>Lemma 1:</u> Each B and $\mathfrak{B}_1$ have - up to isomorphism -

2t indecomposable lattices with vertex P_1 and

t indecomposable projective lattices.

The indecomposable B-lattices with vertex P_1 are - up to isomorphism

$$\mathfrak{S}_2{}^{\underline{i}} , \quad \mathfrak{T}_1{}^{\underline{i}} , \qquad i = 1,\dots,t .$$

This follows by inducing up the augmentation sequence

$$0 \to \mathfrak{p}_1 \to RP_1 \to R \to 0$$

from P_1 to N and then multiplying with B. Note that $\mathfrak{p}_1 B$ lies in $\mathrm{Rad}\, B$, and hence $B/\mathfrak{p}_1 B$ has also t simple modules.

The same works for $\mathfrak{B}_1$.

The last statement follows from

$$\mathfrak{p}_1 b\!\uparrow^N = (0,0,1-\delta)(SP)_m\!\uparrow^N \simeq \mathfrak{S}_2^{(m)}\!\uparrow^N \simeq (\bigoplus_{i=1}^{t} \mathfrak{S}_2^{\underline{i}})^{(ms)},$$

and
$$\text{vertex}\,(\mathfrak{X}_1^{\underline{i}}) = \text{vertex}\,(A\mathfrak{S}_2^{\underline{i}}) = P_1 \quad . \qquad \#$$

The Green-correspondence between B and $\mathfrak{B}_1$ gives a bijection between the isomorphism classes of indecomposable B-lattices with vertex P and indecomposable $\mathfrak{B}_1$-lattices with vertex P ; moreover, it preserves irreducible morphisms between those. (cf.[M2,Ro2])

We summarize as follows:

Lemma 2: The full subgraphs of the AR-graphs of B resp. $\mathfrak{B}_1$ containing the lattices with vertex P correspond bijectively to each other. They are isomorphic to

$$\mathbb{Z}\mathbb{A}_{2p-1}/\langle A^{2t}\rangle \quad .$$

(Here $\mathbb{Z}\mathbb{A}_{2p-1}$ is the universal covering of the Dynkin diagram $\mathbb{A}_{2p-1}$, and $\langle A^{2t}\rangle$ is its admissible automorphism group generated by the 2t-th power of the Auslander-translation. (cf.[Rm])) #

By [Rm,HPR] the connected components of the stable Auslander-Reiten-Graph $\mathfrak{A}_s(\mathfrak{B}_1)$ of $\mathfrak{B}_1$ are of the form

$$\mathbb{Z}\Delta / G$$

for Δ an oriented Dynkin diagram, and G an admissible automorphism group of $\mathbb{Z}\Delta$.

Lemma 3: (i) The stable AR-graph of $\mathfrak{B}_1$ is connected.
(ii) Δ is ramified, i.e. has type $\mathbb{D}$ or $\mathbb{E}$.

The proof of (i) uses that $\mathfrak{p}_1\mathfrak{B}_1$ and $\mathfrak{B}_1/\mathfrak{p}_1\mathfrak{B}_1$ correspond to different central idempotents of $K\mathfrak{B}_1$ which annihilate each other.

One then shows that there is an irreducible morphism between an indecomposable $\mathfrak{B}_1$-lattice with vertex P and one with vertex P_1. (ii) follows from Lemma 2 and a similar argument as used in (i) which again involves the rational idempotents of $K\mathfrak{B}_1$. (For details see [W2].) #

<u>Corollary:</u> The stable AR-graph $\mathfrak{A}_s(\mathfrak{B}_1)$ of $\mathfrak{B}_1$ is isomorphic to

$\mathbb{Z}\,\mathbb{D}_{2p}/\langle A^{2t}\rangle$ or possibly but unlikely to

$\mathbb{Z}\,\mathbb{E}_6/\langle A^{2t}\rangle$ in case $p = 3$.

This follows from Lemma 3 and a case by case investigation of the remaining possibilities described in [Rm, Anhang 2]. #

<u>Theorem:</u> (i) $\mathfrak{B}$ has - up to isomorphism - $(4p+1)t$ indecomposable lattices.

(ii) The stable Auslander-Reiten-graph of $\mathfrak{B}$ is a $2t$-fold covering of the Dynkin diagram

$\mathbb{D}_{2p}$ or possibly

$\mathbb{E}_6$ in case $p = 3$.

In fact, the Green-correspondence between $\mathfrak{B}$ and $\mathfrak{B}_1$ is a stable equivalence; on the other hand $\mathfrak{B}$ and $\mathfrak{B}_1$ have the same number of non-isomorphic indecomposable projectives.

Since a stable equivalence preserves the stable AR-graph, (ii) follows from the above Corollary. #

References:

[A1] Auslander, M.: "Existence theorems for almost split sequences"; Ring Theory II, Proc. Second Oklahoma Conference; Marcel Dekker (1977), 1-44.

[A2] Auslander, M.: "Applications of morphisms determined by modules"; Proc.Conf. on Representation Theory, Philadelphia 1976, Marcel Dekker (1978), 245-327.

[Be] Bessenrodt, Chr.: Thesis, Essen 1980 resp. "On blocks of finite lattice type"; these Proceedings.

[Bu] Butler, M.C.R.: "On the classification of local integral representations of finite abelian p-groups"; Proc.ICRA (1974), Springer Lecture Notes 488, 54-71.

[F] Feit, W.: "Representations of finite groups"; Lecture Notes, Yale University New Haven, Connecticut (1969).

[HPR] Happel, D. - U. Preiser - C.M. Ringel: "A characterization of Dynkin-diagrams using subadditive functions with application to DTr-periodic modules"; Proc.Second Intern.Conf. on Representation Theory of Algebras, Ottawa 1979.

[HR] Heller, A. - I. Reiner: "Representations of cyclic groups in rings of integers I"; Ann. of Math. 76 (1962), 73-92.

[M1] Michler, G.: "Blocks and centers of group algebras"; Lectures on rings and modules, Springer Lecture Notes 246 (1972), 430-552.

[M2] Michler, G.: " Green correspondence between blocks with cyclic defect groups II"; Proc.ICRA 1974, Springer Lecture Notes 448 (1975), 210-235.

[Rm] Riedtmann, Chr.: "Algebren, Darstellungsköcher, Überlagerungen und zurück"; Comment.Math. Helvetici 55 (1980), 199-224.

[Ri] Ringel, C.M.: " Report on the Brauer-Thrall conjectures: Rojter's theorem and the theorem of Nazarova and Rojter"; Proc.Second Intern. Conf. on Representation Theory of Algebras, Ottawa 1979.

[RR] Ringel, C.M. - K.W.Roggenkamp: "Diagrammatic methods in the representation theory of orders"; J.Algebra 60 (1979), 11-42.

[Ro1] Roggenkamp, K.W.: "The construction of almost split sequences for integral group rings and orders"; Comm.Algebra 5 (1977), 1363-1373.

[Ro2] Roggenkamp, K.W.: "Integral representations and structure of finite group rings"; Les Presses de l'Université de Montréal, 1980.

[W1] Wiedemann, A.: "Orders with loops in their Auslander-Reiten graph"; to appear in Comm.Algebra 1981.

[W2] Wiedemann, A.: "Auslander-Reiten-Graphen von Ordnungen und Blöcke mit zyklischem Defekt zwei"; Thesis, Stuttgart 1980.

ON THE REPRESENTATION TYPE OF TWISTED GROUP RINGS

Th. Theohari-Apostolidi
Department of Mathematics
University of Thessaloniki
Thessaloniki-Greece

Let R be a complete discrete valuation ring with quotient field K, L a Galois extension of K with Galois group G and S the valuation ring of L.

In the first paragraph of this note we shall examine the representation type of the twisted group ring Λ=SoG. In the second paragraph we shall deal with the integral representations of Λ, where $|G|=2$ and the extension L/K is wildly ramified, and the computation of the ζ-function of the Λ-lattice Λ,([1], [7]).

Throughout this note we shall follow the terminology of [6] and we shall sketch the proofs of some propositions.

§1. We reduce the problem of the representation type of Λ to that of Λ_1=SoG$_1$, where G_1 is the first ramification group of the extension L/K.

Proposition 1. Λ is of finite representation type if and only if Λ_1 is of finite representation type.

Sketch of proof: If Λ is of finite representation type and M is a Λ_1- lattice then M is a direct summand of the restriction of $\Lambda\otimes_{\Lambda_1} M$ as Λ_1-lattice, hence Λ_1 is of finite representation type. Similar is the machinery of the proof of the other direction.

Proposition 2. Λ_1 is a primary order.

Proposition 3. If Γ is the intersection of all maximal orders in the twisted group algebra LoG$_1$ containing Λ_1, then Γ is a hereditary order.

Sketch of the proof. We consider the canonical map:

$$\rho : \Gamma_o \to \bar{\Gamma}_o,$$

where $\Gamma_o = \mathrm{End}_{R_1} S$, R_1 is the valuation ring of the first ramification field K_1 of the extension L/K and $\bar{\Gamma}_o = \Gamma_o/\mathrm{rad}\Gamma_o$.

Let $E: S/_{\pi_1 S} > \pi S/_{\pi_1 S} > \ldots > \pi^{p^m} S/_{\pi_1 S} = 0$ be a chain of the $S/_{\pi S}$-vector space $S/_{\pi_1 S}$, where π (resp. π_1) is a prime element of L(resp. K_1) and $p^m = |G_1|$. We form the chain ring O(E) of E, which is a heredi-

tary ring ([6], 39.7).
It is proved that $\Gamma/\mathrm{rad}\Gamma_o = O(E)$ and this leads to the fact that Γ is hereditary.
The following proposition gives two useful properties of Γ and Λ_1.

Proposotion 4. a) $\mathrm{rad}\Gamma = \Gamma\pi$, b) $\mathrm{rad}\Lambda_1 \subset \mathrm{rad}\Gamma$.

Proposition 5. The minimal number of generators of the Λ_1- module Γ/Λ_1 is p^m-1.

Proposition 6. The minimal number of generators of the Λ_1- module $\mathrm{rad}(\Gamma/\Lambda_1)$ is $p^m - \mu_k \dfrac{J}{J \cap J^2\Gamma}$ where:

$k = S/\pi S$, $J = \mathrm{rad}\Lambda_1$ and $\mu_k X$ is the minimal number of generators of the k-module X.

Sketch of the proof. The following exact sequences of k-vector spaces:

$$0 \longrightarrow \frac{J\Gamma}{J+J^2\Gamma} \longrightarrow \frac{\Lambda_1+J\Gamma}{J+J^2\Gamma} \longrightarrow \frac{\Lambda_1+J\Gamma}{J\Gamma} \longrightarrow 0,$$

$$0 \longrightarrow \frac{J+J^2\Gamma}{J^2\Gamma} \longrightarrow \frac{J\Gamma}{J^2\Gamma} \longrightarrow \frac{J\Gamma}{J+J^2\Gamma} \longrightarrow 0$$

and the properties of $\mu_{\Lambda_1}(\mathrm{rad}\,\frac{\Gamma}{\Lambda_1})$, ([4], p. 58), Λ_1 being a primary order, lead to the desired result.

The criterion of Deozd-Kirichenko ([8]) and the propositions 2-6 lead us to the following:

Theorem 1. The order Λ is of finite representation type if and only if $|G_1| \leq 2$.

Therefore Λ is of finite representation type if and only if $|G_1| \leq 2$.
It was known that if $|G_1| = 1$ then Λ is hereditary and hence of finite representation type ([1]). By the above result we give the answer about the representation type of twisted group rings in the local case without restriction of the extension L/K.

§2. Let us consider the case of $\Lambda = S \circ G$ in which $|G| = 2$ and the extension L/K is wildly ramified. By the result of the §1, Λ is of finite representation type.

The following proposition is the key of the computation of the non-isomorphic indecomposable lattices:

Proposition 1.

$$\mathrm{Ext}^1_\Lambda(P^i, P^j) \simeq \begin{cases} 0 & i \neq j \\ Z/_{2Z} & i = j \end{cases} \qquad i,j\ \{0,1\}$$

where P is the prime ideal of S and Z is the ring of rational integers.

Theorem 1. The Λ-lattices Λ,S and P are the non isomorphic indecomposable Λ-lattices.

The following properties of Λ are very useful to the computation of the ζ-function of the Λ-lattice Λ.

Proposition 2. i) $\mathrm{Hom}_\Lambda(\Gamma_0,\Lambda) = \mathrm{rad}\Gamma_0$

ii) $\mathrm{rad}\Gamma = \mathrm{rad}\Lambda$

iii) $\mathrm{Hom}_\Lambda(\Gamma,\Lambda) = \mathrm{rad}\Lambda$

iv) $\Gamma^x = \Lambda^x$,

where Λ^x (resp. Γ^x) is the group of units of Λ (resp.Γ)

Remark: The relation ii) holds only in the case $|G| = 2$, as we saw in the first paragraph, in general $\mathrm{rad}\Lambda \subset \mathrm{rad}\Gamma$, prop. 4.

By ([1], §1) $\zeta_\Lambda(\Lambda;s) = \sum\limits_M Z_\Lambda(\Lambda,M;s)$,

the sum extending over a full set of representatives of the isomorphic classes of the full Λ-lattices in LoG, and

$$Z_\Lambda(\Lambda,M;s) = \sum_{\substack{N\subset\Lambda \\ N\simeq M}} (\Lambda:N)^{-s}$$

the sum extending over all full sublattices N of Λ, such that $N \simeq M$, $(\Lambda:N)$ denotes the index of N in Λ and s is a complex variable.

The non-isomorphic classes of full Λ-lattices in LoG are the classes of the Λ-lattices:

$$S \oplus S \simeq \Gamma_0 = \mathrm{End}_R S,\quad P \oplus P \simeq \Gamma_1 = \mathrm{End}_R P,\quad S \oplus P \simeq \Gamma, \Lambda,$$

because the Λ-lattices Λ,S,P are the non-isomorphic indecomposable Λ-lattices, hence

$$\zeta_\Lambda(\Lambda;s) = Z_\Lambda(\Lambda,\Gamma_0;s) + Z_\Lambda(\Lambda,\Gamma_1;s) + Z_\Lambda(\Lambda,\Gamma;s) + Z_\Lambda(\Lambda,\Lambda;s).$$

From the computation of the ζ-function of a maximal order and the prop. 2 of this paragraph we obtain

$$Z_\Lambda(\Lambda,\Gamma_0;s) = Z_\Lambda(\Lambda,\Gamma_1;s) = \frac{2^{2s}(3-2^{1-2s})}{(1-2^{1-2s})(2^{2s}-1)} .$$

On the other hand, if we work with the order

$$\Delta = \begin{bmatrix} 1 & 1 \\ 0 & 1 \end{bmatrix} \Gamma \begin{bmatrix} 1 & 1 \\ 0 & 1 \end{bmatrix}^{-1}$$

instead of the order Γ, after a few computations, we obtain:

$$Z_\Gamma(\Gamma,\Gamma;s) = Z_\Delta(\Delta,\Delta;s) = \frac{2+2^{2s}}{(1-2^{1-2s})(2^{2s}-1)},$$

$$Z_\Lambda(\Lambda,\Gamma;s) = 2^s(-1+Z_\Delta(\Delta,\Delta;s)) \text{ and}$$

$$Z_\Lambda(\Lambda,\Lambda;s) = Z_\Gamma(\Gamma,\Gamma;s).$$

These relations lead us to the computation of $\zeta_\Lambda(\Lambda;s)$,

$$\zeta_\Lambda(\Lambda;s) = \frac{3.2^{1-2s}+2^{2s}+5.2^s-2^{1-s}-2}{(1-2^{1-2s})(2^{2s}-1)} .$$

REFERENCES

[1] M. Auslander and D.S. Rim, Ramification index and mureltiplicity, Illinois J.Math. 7 (1963), 566-581.

[2] C.J. Bushnell and I. Reiner, Zeta functions of arithmetic orders and Solomon's conjectures, to appear.

[3] C.W. Curtis and I. Reiner, Representation Theory of Finite Groups and Associative Algebras, Interscience, New York, Second Edition, 1966.

[4] E.L. Green and I. Reiner, Integral representations and diagrams, Michigan Math.J.25 (1978), 53-84.

[5] A. Jones, Groups with a finite number of indecomposable representations, Michigan Math. J. 10 (1963), 257-261.

[6] I. Reiner, "Maximal Orders", Academic Press, London, 1975.

[7] L. Solomon, Zeta-functions and integral representation theory, Advances in Mathematics 26, 1977, 306-326.

[8] Yu. A. Drozd and V.V. Kirichenko, Primary orders with a finite number of indecomposable representations, Izv. Akad. Nauk. SSSR ser. Mat. 36 (1972) 328-370.

The Complexity and Varieties of Modules

by Jon F. Carlson[1]

1 Introduction.

Let K be a field of characteristic $p > 0$ and let G be a finite group. The complexity $c_G(M)$ of a KG-module M is the polynomial rate growth of the cohomology of M. More specifically, if

$$\cdots \to P_1 \to P_0 \to M \to 0$$

is a minimal projective KG-resolution of M, then $c_G(M)$ is the least integer $s \geq 0$ such that

$$\lim_{n\to\infty} \frac{\operatorname{Dim} P_n}{n^s} = 0 .$$

For any KG-module N let $c_G(M,N)$ be the least integer $s \geq 0$ such that

$$\lim_{n\to\infty} \frac{\operatorname{Dim} \operatorname{Ext}^n_{KG}(M,N)}{n^s} = 0$$

It is not difficult to show that $c_G(M)$ is the largest of the integers $c_G(M,N)$. It is also the maximum of the integers $c_G(M,N)$ as N runs through the set of irreducible KG-modules.

J. L. Alperin first defined the complexity of a module in [1]. The ideal was not entirely new, for it is related to the degrees of certain polynomial functions arrising from the cohomology of M. In particular it is the maximum of the orders of the poles at $t = 1$ of the various Poincare series

$$P_{M,N}(t) = \sum_{n=0}^{\infty} t^n \operatorname{Dim} \operatorname{Ext}^n_{KG}(M,N)$$

where N runs through the set of irreducible modules (see for example [14]). However Alperin proposed complexity as an object of study because he saw it as the logical generalization of the idea of periodicity in modules. Indeed, it has now been shown that a KG-module is periodic

1. This work was particially supported by NSF Grants Nos. MCS 7801685 and MCS 8002509

(i.e. has a periodic projective resolution) if and only if its complexity is one [8].

In this paper we will review recent work in this area. Particular emphasis will be placed on the case in which G is an elementary abelian p-group. In this situation the complexity is the same as the dimensions of certain varieties associated to the module. As a consequence we obtain a new proof of O. Kroll's Theorem which specifies the complexity in terms of the order of a "shifted" subgroup that acts freely on the module. Since these results are not complete we will only sketch the proofs. A full exposition will appear later.

To establish rational for this work, we summarize, in section 2, results on modules over more general groups. Most of these concern the relationship between the cohomology of a module and that of its restrictions to elementary abelian p-subgroups of G.

2 Restrictions to Elementary Abelian p-Subgroups

The complexity of a KG-module is never larger than the complexity of the trivial module K_G. For a projective resolution of M can be obtained by tensoring M over K with a projective resolution of K_G. Quillen's Dimension Theorem [11] may be interpreted to say that the complexity of the trivial module is the p-rank of G, i.e. the exponent on p of the order of the maximal elementary abelian p-subgroup of G. Now a KG-module is projective if and only if its complexity is zero. In [6], Chouinard proved that a KG-module is projective if and only if its restriction to every elementary abelian p-subgroup is projective. The first major general result on complexity was J. Alperin and L. Evens' generalization of Chouinard's Theorem.

<u>Theorem 2.1</u> [2]. If M is a KG-module, then

$$c_G(M) = \max_{E \in EA(G)} \{c_E(M_E)\}$$

where the maximum is taken over the set EA(G) of all elementary abelian p-subgroups of G.

Here M_E is the restriction of M to a KE-module. This theorem has been reworked and generalized to the point where, at this time, there exist four distinct proofs of it. All of them share some common elements. Using standard restriction and induction techniques it is easy to reduce to the case in which G is a p-group. Then one applies an induction argument using the Lyndon-Hochschild-Serre spectral sequence and a theorem of Serre concerning the vanishing of a certain cup product of Bocksteins. These techniques were first employed by D. Quillen and

B. Venkov [13] to give an algebraic proof of Quillen's Theorem [12] on the Krull dimension of the cohomology ring of G. Chouinard, for his proof, applied them in the case where the spectral sequence collapses. I have been informed that O. Kroll has a fifth proof of the Alperin-Evens result and that his proof does not use the spectral sequences.

The Alperin-Evens Theorem can be derived from a more basic result on the structure of modules. If H is a subgroup of G, and if M is a KG-module then its restriction can be expressed as

$$M_H \cong f_H(M) \oplus \rho_H(M)$$

where $f_H(M)$ is a projective KH-module and $\rho_H(M)$ has no projective submodules. These two modules are unique up to isomorphism.

Theorem 2.2 [5]. For any group G there exists a number B_G such if M is a KG-module which has no projective submodules then

$$\text{Dim } M \leq B_G \cdot \max_{E \in EA(G)} \{\text{Dim } \rho_E(M_E)\}.$$

The Alperin-Evens Theorem follows from this and the observation that, for some elementary abelian p-subgroup E, the dimensions of the terms in the minimal projective resolution of M can be no bigger than B_G times the dimensions of the corresponding terms in a KE-projective resolution of $\rho_E(M_E)$. It is interesting to note that P. Webb has recently extended both Theorems 2.1 and 2.2 to the case where the coefficient ring is the ordinary integers.

Some more recent work has focused on the annihilator in $H^{ev}(G,K) = \sum_{n=0}^{\infty} H^{2n}(G,K)$ of the cohomology of M. Let $X(G,M) = \bigoplus_U \text{Ext}^*_{KG}(M,U)$ where the sum is taken over all isomorphism classes of indecomposable KG-modules U. Let Y(G,M) be the same except with the sum taken only over the irreducible modules. It is not difficult to show that the radicals of the annihilators in $H^{ev}(G,K)$ of X(G,M) and Y(G,M) are the same. For any subgroup H of G and any KG-module V, $\text{Ext}^*_{KH}(M_H,V)$ is a finitely generated $H^{ev}(G,K)$-module under restriction. Independently G. Avrunin [3] and Alperin and Evens have shown that the radical of the annihilator of Y(G,M) is the intersection of those of $Y(E,M_E)$ where the intersection is taken over EA(G). Not only does this imply Theorem 2.1, but Alperin and Evens have used it to get information about the prime ideal spectrum of the factor ring $H^{ev}(G,K)$ modulo the annihilator of Y(G,M).

Finally we should observe that it is possible to copy Quillen's approach to the subject [12]. For $H^*(G,\text{Hom}_K(M,M)) \cong \text{Ext}^*_{KG}(M,M)$ is a ring. It can be shown that $c_G(M)$ is the same as $c_G(\text{Hom}_K(M,M))$ and also the same as the Krull dimension of the center of $\text{Ext}^*_{KG}(M,M)$. The Quillen-Venkov proof may be generalized to show that an element of $\mathrm{Ext}^*_{KG}(M,M)$ is nilpotent if and only if its restriction to every elementary abelian p-subgroup is nilpotent. Part of the generalization is contained in a lemma in [2]. Theorem 2.1 follows from the fact that the nil radical of the center of $\mathrm{Ext}^n_{KG}(M,M)$ is an intersection of prime ideals each of which is the pull back of a prime ideal in the center of $\mathrm{Ext}^*_{KE}(M,M)$ for some $E \in EA(G)$.

3 Modules over Elementary Abelian p-Groups.

For this section let $G = \langle x_1,\ldots,x_n\rangle$ be an elementary abelian group of order p^n and let K be an algebraically closed field of characteristic p. Let $A = (\alpha_{ij})$ be a non-singular $n \times n$ matrix with entries in K. Then A induces an automorphism or "shifting" of KG, by appropriate extension of the map which sends x_i to $1 + \sum_{j=1}^{n} \alpha_{ij}(x_j - 1)$. That is, suppose that for $i = 1,\ldots,n$, the elements $A_i = (\alpha_{i1},\ldots,\alpha_{in}) \in K^n$ form a basis for K^n. Let $u_i = 1 + \sum_{j=1}^{n} \alpha_{ij}(x_j - 1)$. Each u_i is a unit of order p in KG and the group $G' = \langle u_1,\ldots,u_n\rangle$ is elementary abelian of order p^n. The embedding of G' into KG induces an isomorphism of KG' onto KG, by linear extension. If $A_1,\ldots,A_t$ are linearly independent elements of K^n, $t \leq n$, then the group $H = \langle u_1,\ldots,u_t\rangle$ is called a shifted subgroup of KG. Every shifted subgroup H has the property that KG is free as a KH-module. When $\alpha = (\alpha_1,\ldots,\alpha_n)$ is a nonzero element of K^n, let $u_\alpha = 1 + \Sigma\, \alpha_i(x_i - 1)$.

If M is a KG-module, then let V(M) denote the set consisting of the zero element and of all nonzero $\alpha \in K^n$ such that $M_{\langle u_\alpha\rangle}$ is not free as a $K\langle u_\alpha\rangle$-module. As motivation for this definition we present the following results.

<u>Dade's Lemma 3.1</u> [7]. $V(M) = \{0\}$ if and only if M is projective.

<u>Theorem 3.2</u> [4]. If M is an indecomposable periodic module then V(M) is a line.

It is not difficult to prove the following.

<u>Lemma 3.3</u>. V(M) is a homogeneous affine variety (though not necessarily irreducible).

An example illustrates both the nature of the variety and the proof of the lemma. Let $G = \langle x_i, x_2, x_3\rangle$ be elementary abelian of order 8. Here K is algebraically closed of characteristic 2. Let M be a KG-module of dimension 4 and with a basis, relative to which the group element act by the following matrices:

$$x_1 \leftrightarrow \begin{bmatrix} 1 & & & \\ & 1 & & \\ 1 & & 1 & \\ & 1 & & 1 \end{bmatrix}, \quad x_2 \leftrightarrow \begin{bmatrix} 1 & & & \\ & 1 & & \\ A & & 1 & \\ & B & & 1 \end{bmatrix}, \quad x_3 \leftrightarrow \begin{bmatrix} 1 & & & \\ & 1 & & \\ & C & 1 & \\ 1 & & & 1 \end{bmatrix}$$

$A, B, C \in K$. If $u_\alpha = 1 + \Sigma\, \alpha_j(x_j - 1)$ then the matrix of u_α is

$$\begin{bmatrix} 1 & & & \\ & 1 & & \\ \alpha_1 + A\alpha_2 & C\alpha_3 & 1 & \\ \alpha_3 & \alpha_1 + B\alpha_2 & & 1 \end{bmatrix}.$$

Now $M_{<u_\alpha>}$ is free if and only if $M_{<u_\alpha>} = K<u_\alpha> \oplus K<u_\alpha>$, in which case the rank of the operator $u_\alpha - 1$ must be 2. Hence $M_{<u_\alpha>}$ is free if and only if the determinant of the 2×2 block in the lower left hand corner is nonzero. Thus $\alpha \in V(M)$ if and only if

$$(\alpha_1 + A\alpha_2)(\alpha_1 + B\alpha_2) + C\alpha_3^2 = 0$$

If, for example, $C = 0$ then $V(M)$ is not an irreducible variety. For many choices of A, B and C the variety $V(M)$ is not linear.

In general $V(M)$ is a variety simply because an element $\alpha \in K^n$ is in $V(M)$ if and only if the rank of the matrix of $u_\alpha - 1$ is strictly less than $((p-1)/p)$ Dim M.

Suppose that $p > 2$. By the Kunneth formula the cohomology ring of G is given by

$$H^*(G,K) \cong \mathrm{Ext}^*_{KG}(K,K) \cong K[\zeta_1,\ldots,\zeta_n] \otimes \Lambda[\mu_1,\ldots,\mu_n),$$

where $P(G,K) = K[\zeta_1,\ldots,\zeta_n]$ is a polynomial ring in variables $\zeta_1,\ldots,\zeta_n \in H^2(G,K)$ and $\Lambda(\mu_1,\ldots,\mu_n)$ is an exterior algebra. We wish to choose the element $\zeta_1,\ldots,\zeta_n$ very carefully. In particular we insist that the restriction of ζ_i to the subgroup $<x_j \mid j \neq i>$ be zero. Now for any $\alpha = (\alpha_1,\ldots,\alpha_n) \in K^n$, $H^2(<u_\alpha>,K)$ has dimension 1 and we can choose a cannonical generator γ_α. With careful choice of the elements ζ_i, we get that the change of rings map

$$\mathrm{res}_{G,<u_\alpha>}: H^*(G,K) \to H^*(<u_\alpha>,K)$$

has the property that $\mathrm{res}_{G,<u_\alpha>}(\zeta_i) = \alpha_i^p\gamma_\alpha$. Of course $\mathrm{res}_{G,<u_\alpha>}$ is not really a restriction since $<u_\alpha>$ is not a subgroup of G. But for cohomology with coefficients in K the map does commute with cup product (see [11], p. 232). Consequently if $\zeta = f(\zeta_1,\ldots,\zeta_n) \in H^{2t}(G,K)$ then

$$\mathrm{res}_{G,<u_\alpha>}(\zeta) = f(\alpha_1^p,\ldots,\alpha_n^p)\cdot\gamma_\alpha^t.$$

When $p = 2$, $H^*(G,K) = P(G,K) = K[\zeta_1,\ldots,\zeta_n]$ for $\zeta_1,\ldots,\zeta_n \in H^1(G,K)$. In this situation we may show that for $\zeta = f(\zeta_1,\ldots,\zeta_n) \in H^t(G,K)$,

$$\mathrm{res}_{G,<u_\alpha>}(\zeta) = f(\alpha_1,\ldots,\alpha_n)\gamma_\alpha^t.$$

If M is a KG-module, let J(M) be the annihilator in P(G,K) of the identity element $I \in \mathrm{Ext}^0_{KG}(M,M)$. Then J(M) is a homogeneous ideal. Let W(M) be the set of all $\alpha \in K^n$ such that $\mathrm{res}_{G,<u_\alpha>}\zeta = 0$ for all $\zeta \in J(M)$. Hence if $p = 2$, then W(M) is the zero set of the polynomials in J(M). For $p > 2$, W(M) is the translation of the zero set by the inverse of the Frobenius automorphism on K^n. In either case W(M) is a homogeneous variety in K^n.

The main result of this section is the following.

Theorem 3.4. $c_G(M) = \dim V(M) = \dim W(M)$.

By the dimension of the variety we mean the largest dimension of any of its components. The proof of the theorem consists of verifying each of the following three statements.

1) $c_G(M) \geq \dim W(M)$.

2) $W(M) \supseteq V(M)$.

3) $\dim V(M) \geq c_G(M)$.

It is known that $c_G(M) \geq c_G(\mathrm{Hom}_K(M,M))$ (see [1]). Cup product with $I \in H^*(G,\mathrm{Hom}_K(M,M))$ induces an injection of $P(G,K)/J(M)$ into $H^*(G,\mathrm{Hom}_K(M,M))$. The first statement follows from the fact that the ring $P(G,K)/J(M)$ contains $c_G(M)$ algebraically independent elements.

We give only a heuristic argument for statement (2). Suppose that $\zeta \in J(M)$ and that $\alpha \in V(M)$. Then $\zeta\cdot I = 0$ and

$$\mathrm{res}_{G,<u_\alpha>}(\zeta)\cdot\mathrm{res}_{G,<u_\alpha>}(I) = 0.$$

Since $<u_\alpha>$ is cyclic and since $M_{<u_\alpha>}$ is not free, this is impossible unless $\mathrm{res}_{G<u_\alpha>}(\zeta) = 0$. Hence $\alpha \in W(M)$. The above argument is not reall valid since it assumes that $\mathrm{res}_{G,<u_\alpha>}$ commutes with cup product. A correct proof involves showing by lengthy but elementary techniques that no respresentative of $\zeta \cdot I$ can be a coboundary unless $res_{G,<u_\alpha>}(\zeta) = 0$ for all $\alpha \in V(M)$.

Let $s = \dim V(M)$. By basic results from algebraic geometry there is a linear subspace $U \subseteq K^n$ such that $U \cap V(M) = \{0\}$ and $\operatorname{Dim} U = n - s$. Corresponding to a basis $\alpha(1),\ldots,\alpha(n - s)$ of U there is a shifted subgroup $H = <u_{\alpha(1)},\ldots,u_{\alpha(n-s)}>$ which, by Dade's Lemma (3.1), must act freely on M. If we take a minimal projective resolution of M and reduce each term module Rad KH, then we get a minimal projective KG/(KG·Rad KH)-resolution of M/Rad KH·M. Moreover the two resolutions have the same rate of growth. Now KG/(KG·Rad KH) is isomorphic to a group algebra K(G'/H) where G' is the image of G under a shifting in KG, and $H \subseteq G'$. So $c_G(M) = c_{G'/H}(M/\text{Rad } KH\cdot M)$. Hence $c_G(M) \leq s$ since $|G'/H| = p^s$.

Two facts may be derived directly from the proof of the theorem. The first arises in the proof of statement (1), but as noted in the previous section it may be demonstrated independently.

<u>Corollary 3.5</u>. $c_G(M) = c_G(\text{Hom}_K(M,M))$.

The second corollary is a consequence of the argument on statement (3). It was first proved by O. Kroll using induction on the order of G. Kroll considered the annihilator in H*(G,K) of H*(G,M) and its images under restrictions to maximal shifted subgroups.

<u>Corollary 3.6</u> [9]. Let s be the largest integer such that there exists a shifted subgroup which has order p^s and which acts freely on M. Then $c_G(M) = n - s$.

Many questions remain. For example, is V(M) = W(M)? The answer is yes when M is a periodic module but it is unknown otherwise. It is known that for KG-modules M and N, $V(M \otimes N) = V(M) \cap V(N)$. If the same statement were true for W then it would necessitate that V(M) = W(M) for all M. It would be helpful to know what sort of varieties can occur as V(M) or W(M) when M is indecomposable. Clearly there are some limitations (see for example Theorem 3.2), but the nature of these limitations is not understood.

References

1. J. L. Alperin, Periodicity in groups, Illinois J. Math. 23(1977), 776-783

2. J. Alperin and L. Evens, Representations, resolutions and Quillen's dimension theorem, (to appear).

4. G. S. Avrunin, Annihilators of cohomology modules, (to appear).

4. J. F. Carlson, The dimensions of periodic modules over modular group algebras, Illinois J. Math. 23(1979), 295-306.

5. J. F. Carlson, The dimensions of modules and their restrictions over modular group algebras, J. of Algebra (to appear).

6. L. G. Chouinard, Projectivity and relative projectivity over group rings, J. Pure Appl. Algebra 7(1976), 278-302.

7. E. C. Dade, Endo-permutation modules over p-groups II, Ann. of Math. 108(1978), 317-346.

8. D. Eisenbud, Homological algebra on a complete intersection, with an application to group representations, Trans. Amer. Math. Soc. 260(1980), 35-64.

9. O. Kroll, Complexity and elementary abelian p-groups, Thesis, U. of Chicago (1980).

10. S. MacLane, Homology, Springer-Verlag, New York, 1963.

11. D. Quillen, The spectrum of an equivariant cohomology ring, I, II, Ann. of Math. 94(1971), 549-602.

12. D. Quillen and B. B. Venkov, Cohomology of finite groups and elementary abelian subgroups, Topology 11(1972), 317-318.

13. R. G. Swan, Groups with no odd dimensional cohomology, J. of Algebra 17(1971), 401-403.

Department of Mathematics
University of Georgia
Athens, Georgia 30602

RESTRICTING $\mathbb{Z}G$-LATTICES TO ELEMENTARY ABELIAN SUBGROUPS

P. J. Webb

1.

In recent years a number of results in modular representation theory have been inspired by the Quillen-Venkov proof in [QV] that the Krull dimension of the mod p cohomology ring of a finite group G equals the maximum rank of an elementary abelian p-subgroup of G. These results are due to Chouinard [Ch], Alperin and Evens [AE] and Carlson [Ca]. The strongest of these results is Carlson's, and since it is also one of the easiest to state, we will do so here. We use the notation that k is a field of characteristic $p > 0$, with $p \mid |G|$, and if M is a kG-module we write $M = \text{core}(M) \oplus \text{proj}(M)$ where proj(M) is projective and core(M) has no non-zero projective summands.

Theorem 0 (J. Carlson [Ca]) Let G be a finite group. There exists a constant C with the property that if M is any finitely generated kG-module with no non-zero projective summands, there exists an elementary abelian p-subgroup E of G such that

$$\dim_k M \leq C.\dim_k \text{core}(M\downarrow_E)$$

It is not difficult to formulate versions of the above results for the integral group ring $\mathbb{Z}G$, and our aim in this article is to show that all of these integral versions are true. There are two points to note in translating the modular statements to the integers. Firstly, one must restrict attention to $\mathbb{Z}G$-lattices and use the $\mathbb{Z}$-rank of a lattice instead of its dimension. The second and more important point is that in removing the dependence on some particular prime one must state one's theorem in terms of the collection of all elementary abelian p-subgroups for all prime divisors p of $|G|$. Thus our main theorem is the following:

Theorem 1 Let G be a finite group. There exists a constant B with the property that if M is any $\mathbb{Z}G$-lattice with no non-zero projective summands, there exists an elementary abelian p-subgroup E of G for some prime p such that

$$\mathrm{rank}_{\mathbb{Z}}(M) \leq B.\mathrm{rank}_{\mathbb{Z}}(\mathrm{core}(M\downarrow_E)) .$$

We are now using the notation $M = \mathrm{core}(M) \oplus \mathrm{proj}(M)$ when M is a lattice, and although the choice of the sublattice core(M) may not be unique in this case, its rank is uniquely determined by M.

There is an integral version of the theorem of Alperin and Evens [AE] which is implied by Theorem 1. To state this we must first define the complexity of a $\mathbb{Z}G$-lattice M. Following the definition given by Alperin [A] in the case of a kG-module, let

$$\cdots \to P_2 \to P_1 \to P_0 \to M \to 0 \qquad (1)$$

(with $P_2 \twoheadrightarrow K_1 \hookrightarrow P_1$, $P_1 \twoheadrightarrow K_0 \hookrightarrow P_0$)

be a minimal $\mathbb{Z}G$-projective resolution of M. This means that each projective module P_d has minimal rank subject to having K_{d-1} as an image, or equivalently that the K_d have no non-zero projective summands (see [Sw]). Then we say that the complexity of M is c provided that c is the least non-negative integer such that there exists $\lambda > 0$ with $\mathrm{rank}_{\mathbb{Z}}(P_d) \leq \lambda.d^{c-1}$ for all sufficiently large d. We will write $c_{\mathbb{Z}G}(M)$ for c to distinguish it from the complexity $c_{kG}(k \otimes_{\mathbb{Z}} M)$ of the kG-module $k \otimes_{\mathbb{Z}} M$ as defined by Alperin [A]. The method Swan used to prove Proposition 6.1 in [Sw] allows one to deduce the ranks of the P_d from modular information, and it is possible to show by this method that the complexity $c_{\mathbb{Z}G}(M)$ exists, given that $c_{kG}(k \otimes M)$ always exists.

Corollary 2 Let M be a $\mathbb{Z}G$-lattice. Then there exists a prime $p \mid |G|$ and an elementary abelian p-subgroup E of G such that $c_{\mathbb{Z}G}(M) = c_{\mathbb{Z}E}(M) = c_{kG}(k \otimes M)$ where $k = \mathbb{Z}/p\mathbb{Z}$.

In section 2 we will show how this Corollary may be deduced from Theorem 1, but it can also be proved in a different manner by building on Swan's ideas in [Sw]. It is interesting that both this alternative proof and the proof we shall give of Theorem 1 seem to need the use of the non-singularity of the Cartan matrix at one stage.

Just as in the modular case described in [AE], lattices of complexity 0 and 1 are respectively projective and periodic. In the case of complexity 0, Corollary 2 becomes the following integral version of Chouinard's theorem [Ch]:

Corollary 3 The $\mathbb{Z}G$-lattice M is projective if and only if it is projective on restriction to all elementary abelian p-subgroups of G, for all $p||G|$.

For complexity 1 we recover the following result which is essentially due to Olympia Talelli [T]:

Corollary 4 The following are equivalent for the $\mathbb{Z}G$-lattice M:
(i) M is periodic
(ii) M is periodic on restriction to every elementary abelian p-subgroup of G, for each prime $p||G|$
(iii) For every prime $p||G|$, M/pM is periodic as a $(\mathbb{Z}/p\mathbb{Z})E$-module for every elementary abelian p-subgroup E of G.

Before passing to the proofs of these results we should remark that in particular cases they can be deduced without difficulty from the corresponding modular results just by applying standard lifting techniques. This is so if we restrict G always to be a p-group, or alternatively if we wish to prove analogous results for $\mathbb{Z}_{(p)}G$ where $\mathbb{Z}_{(p)}$ is the ring of p-adic integers and G may be any finite group. Thus we can prove a p-adic version of our main theorem by means of the observation that a $\mathbb{Z}_{(p)}G$-lattice M has a projective summand if and only if M/pM has a projective summand. So if M has no projective summands then

$$\begin{aligned} \text{rank}(M) = \dim(M/pM) &\leq C.\dim \text{core}((M/pM)\downarrow_E) \\ &= C.\text{rank}(\text{core}(M\downarrow_E)) \end{aligned} \tag{2}$$

for some elementary abelian p-subgroup E of G, where C is the constant in Carlson's theorem. The real problem in proving Theorem 1 is that of linking these p-adic statements together to give a global one. Note, however, that in all cases Corollary 3 may be deduced directly from Chouinard's original theorem [Ch], since a $\mathbb{Z}G$-lattice M is projective if and only if M/pM is a projective $(\mathbb{Z}/p\mathbb{Z})G$-module for all prime divisors p of $|G|$ (see 3.13 and 8.6 of [G]). Also, one may prove results such as the following:

all lattices in a block of $\mathbb{Z}_{(p)}G$ with cyclic defect group are periodic.

This is just the combination of the corresponding statement over $\mathbb{Z}/p\mathbb{Z}$ with Proposition 2.2 of [T].

2. Proofs of Theorem 1 and Corollary 2

If M is a $\mathbb{Z}G$-lattice we let $M_{(p)} = \mathbb{Z}_{(p)} \otimes_{\mathbb{Z}} M$ and put $\sigma_p(M) = \text{rank}_{\mathbb{Z}_{(p)}}(\text{core } M_{(p)})$, $\sigma(M) = \max\{\sigma_p(M) : p||G|\}$. The following result will allow us to deduce Theorem 1 from the local information:

Proposition 5 There exists a constant B_1 such that whenever M is a $\mathbb{Z}G$-lattice with no projective summands then $\text{rank}_{\mathbb{Z}}(M) \leq B_1.\sigma(M)$.

Proof We produce some inequalities on the values of the character χ of M. Firstly, if x is any non-identity element of G and p is a prime dividing the order of x then $\chi = \chi_1 + \chi_2$ where χ_1 and χ_2 are the characters of core$(M_{(p)})$ and proj$(M_{(p)})$ respectively. Because $\chi_2(x) = 0$ (Theorem 4 of [Sw2] is a convenient reference) $|\chi(x)| = |\chi_1(x)| \leq |\chi_1(1)| = \sigma_p(M)$, and so $|\chi(x)| \leq \sigma(M)$ holds for all non-identity x in G. Here we are regarding the values of the characters as lying in some subfield K of the complex numbers which is a splitting field for G, and the absolute value is the usual one for complex numbers.

If $\eta_1,\ldots,\eta_s$ are the characters of the indecomposable projective $\mathbb{Z}_{(p)}G$-modules we may write $\chi_2 = \lambda_1\eta_1 + \ldots + \lambda_s\eta_s$ for some $\lambda_i \in \mathbb{Z}$. For any non-identity element $x \in G$,

$$|\chi_2(x)| \leq |\chi(x)| + |\chi_1(x)| \leq \sigma(M) + \sigma_p(M) \leq 2.\sigma(M).$$

Hence

$$\sigma(M)^{-1}.|\sum_{i=1}^{s}\lambda_i\eta_i(x)| \leq 2$$

for non-identity x, since $\sigma(M) \neq 0$ if, as we may suppose, $M \neq 0$. If we let $1 = x_1, \ldots ,x_r$ be representatives of the p-regular conjugacy classes of G we may interpret this inequality as saying that the vector

$\sigma(M)^{-1}(\lambda_1, \dots, \lambda_s).H$ lies a bounded distance from the line $(1,0, \dots, 0)K$ in an r-dimensional K-vector space, where $H = (\eta_i(x_j))$. The $s \times r$ matrix H has rank s (c.f. p.599 of [CR]), and this means that $\sigma(M)^{-1}(\lambda_1, \dots, \lambda_s)$ lies a bounded distance from the inverse image of $(1,0, \dots, 0)K$ under H, i.e. from the line $(\mu_1, \dots, \mu_s)K$ where $\mu_1\eta_1 + \dots + \mu_s\eta_s$ is the character of the regular representation.

If $\mathrm{proj}(M_{(p)})$ were to contain a copy of the regular representation for every prime divisor p of |G| then M would have a non-trivial projective summand (c.f. 6.3 and 6.4 of [G]). Since M has no non-trivial projective summand there is a prime p||G| with $\lambda_j < \mu_j$ for some j. This extra condition means that $\sigma(M)^{-1}(\lambda_1, \dots, \lambda_s)$ actually lies in a bounded region of K^s for this p, and moreover the bound is independent of the module M. It is now easy to deduce that

$$\begin{aligned} \mathrm{rank}(M) = \chi_1(1) + \chi_2(1) &= \sigma_p(M) + \sum_{i=1}^{s} \lambda_i\eta_i(1) \\ &\leq \sigma(M) + \max\{\lambda_1, \dots, \lambda_s\}.\sum_{i=1}^{s} \eta_i(1) \\ &\leq B_1.\sigma(M) \end{aligned}$$

for some constant B_1 which depends only on G.

Proof of Theorem 1 Choose a prime p so that $\sigma(M) = \sigma_p(M)$. By Proposition 5 and inequality (2) there exists an elementary abelian p-subgroup E of G for which

$$\begin{aligned} \mathrm{rank}_{\mathbb{Z}}(M) \leq B_1.\sigma(M) &\leq B_1.C.\mathrm{rank}_{\mathbb{Z}_{(p)}} \mathrm{core}(M_{(p)}\downarrow_E) \\ &= B_1.C.\mathrm{rank}_{\mathbb{Z}}\mathrm{core}(M\downarrow_E), \end{aligned}$$

the latter equality holding since E is a p-group (c.f. 6.3 and 6.4 of [G]). Now take $B = B_1.C$.

Proof of Corollary 2 Minimality of the resolution (1) is equivalent to the requirement that each kernel K_d has no non-zero projective summand. Furthermore, whenever H is a subgroup of G, $\mathrm{core}(K_d\downarrow_H)$ is the d-th kernel in a minimal $\mathbb{Z}H$-projective resolution of $M\downarrow_H$ and thus $c_{\mathbb{Z}G}(M) \geq c_{\mathbb{Z}H}(M)$ always.

On the other hand, if $c = \max\{c_{\mathbb{Z}E}(M\downarrow_E)\}$ where the maximum is taken over all elementary abelian p-subgroups E, for all p||G|, then for all sufficiently large d,

$$\mathrm{rank}(K_d) \leq B.\mathrm{rank}\ \mathrm{core}(K_d\downarrow_{E_d}) \leq B.\lambda_{E_d} d^{c-1}$$

for some elementary abelian subgroup E_d and constant λ_{E_d}. Since G has only finitely many subgroups we may choose a $\lambda \geq \lambda_{E_d}$ for all d, and then $\mathrm{rank}(K_d) \leq B.\lambda d^{c-1}$. Therefore

$$\mathrm{rank}(P_d) = \mathrm{rank}(K_d) + \mathrm{rank}(K_{d-1}) \leq 2.B.\lambda.d^{c-1}$$

and $c_{\mathbb{Z}G}(M) \leq c$, as required for the first equality. For the subgroup E we may take any elementary abelian p-subgroup which realizes the maximum value $c = c_{\mathbb{Z}E}(M{\downarrow_E})$.

Since E is a p-group, $k \otimes_{\mathbb{Z}} (\mathrm{core}(K_d{\downarrow_E}))$ has no non-zero projective summands (6.5 of [G]) and so $c_{kE}(k \otimes M) = c_{\mathbb{Z}E}(M)$. Since $c_{kE}(k \otimes M) \leq c_{kG}(k \otimes M) \leq c_{\mathbb{Z}G}(M)$, we have $c_{\mathbb{Z}E}(M) = c_{kG}(k \otimes M) = c_{\mathbb{Z}G}(M)$.

3. An Example

It is not hard to produce indecomposable non-projective lattices M for particular groups G for which the quantity $\mathrm{rank}(M) - \sigma(M)$ can become arbitrarily large. We will illustrate this phenomenon with some lattices which arise in a natural way. Note that by Proposition 5 as $\mathrm{rank}(M) - \sigma(M)$ becomes large so $\mathrm{rank}(M)$ must also become large in proportion, and to find such lattices we must deal with a group of infinite lattice type.

To fix our example let $G = A_5 \times A_5$ be the product of two copies of the alternating group on five letters and let

$$\cdots \to P_2 \to P_1 \to P_0 \to \mathbb{Z} \to 0 \qquad (P_2 \searrow K_1 \nearrow P_1, \quad P_1 \searrow K_0 \nearrow P_0)$$

be a minimal projective resolution of $\mathbb{Z}$ over $\mathbb{Z}G$. Minimality of the resolution means that the K_d have no projective summands, and by Chapter 8 of [G] these modules are indecomposable since the prime graph of G is connected. Now for each prime p, $\mathrm{core}((K_d)_{(p)})$ is the d-th kernel in a minimal $\mathbb{Z}_{(p)}G$-resolution of $\mathbb{Z}_{(p)}$. Since $c_{kG}(k) = 2$ when $k = \mathbb{Z}/p\mathbb{Z}$ for $p = 3$ and 5 and $c_{\mathbb{Z}G}(\mathbb{Z}) = 4$, it follows that $\mathrm{rank}_{\mathbb{Z}}(K_d) - \mathrm{rank}\ \mathrm{core}((K_d)_{(p)})$ becomes arbitrarily large as $d \to \infty$ for $p = 3$ or 5. Modulo 2, A_5 has a non-principal block and hence G has a non-principal block also. It is clear that the only projectives which appear in a minimal $\mathbb{Z}_{(2)}G$-resolution of $\mathbb{Z}_{(2)}$ belong to the principal block, and so for each d the non-principal summands of $(P_d)_{(2)}$ must also appear as summands of either $(K_d)_{(2)}$ or $(K_{d-1})_{(2)}$. By Swan's structure theorem for projectives (4.8 of [G]), $(P_d)_{(2)} \cong \mathbb{Z}_{(2)}G^n$ for some n, so that since $c_{\mathbb{Z}G}(\mathbb{Z}) = 4$ the number of non-principal summands of $(P_d)_{(2)}$

increases without bound as $d \to \infty$. This means we can find a subsequence of the modules K_d so that $\mathrm{rank}(K_d) - \mathrm{rank\ core}(K_d)_{(2)}$ becomes arbitrarily large, and putting that together with the information about the primes 3 and 5 we have that $\mathrm{rank}(K_d) - \sigma(K_d)$ becomes arbitrarily large.

REFERENCES

[A] J. Alperin, Periodicity in groups, Ill. J. Math. 21 (1977), 776-783

[AE] J. Alperin and L. Evens, Representations, resolutions and Quillen's dimension theorem, to appear.

[Ca] J. F. Carlson, The dimensions of modules and their restrictions over modular group algebras, to appear.

[Ch] L. G. Chouinard, Projectivity and relative projectivity over group rings, J. Pure Appl. Algebra 7 (1976), 278-302

[CR] C. W. Curtis and I. Reiner, Representation theory of finite groups and associative algebras (Wiley Interscience, New York, 1962)

[G] K. W. Gruenberg, Relation modules of finite groups, Amer. Math. Soc. Regional Conf. Ser. 25 (1975)

[QV] D. Quillen and B. B. Venkov, Cohomology of finite groups and elementary abelian subgroups, Topology 11 (1972), 317-318

[Sw] R. G. Swan, Minimal resolutions for finite groups, Topology 4 (1965), 193-208

[Sw2] R. G. Swan, The Grothendieck ring of a finite group, Topology 2 (1963), 85-110

[T] O. Talelli, On cohomological periodicity of ℤG-lattices, Math. Z. 169 (1979), 119-126

D.P.M.M.S
16, Mill Lane
Cambridge CB2 1SB
England

THE LATTICE TYPE OF ORDERS II : AUSLANDER-REITEN QUIVERS

K.W.Roggenkamp (Stuttgart)

This is a continuation of [Ro,80,5], and we shall use the same notation and refer frequently to it.

R is a complete valuation ring with maximal ideal πR, field of quotients K and residue field $\mathfrak{k}$. Λ is an R-order in the separable K-algebra A. By ${}_{\Lambda}\mathfrak{M}^{o}$ we denote the category of left Λ-lattices.

§ 3 Irreducible maps and Auslander-Reiten quivers.

In the first part let

$$\mathfrak{M}^{o} = {}_{\Lambda}\mathfrak{M}^{o} \quad \text{or}$$

$$\mathfrak{M}^{o} = \mathfrak{T}(\mathfrak{A}) \quad ,$$

where $\mathfrak{T}(\mathfrak{A})$ are the torsionfree $\mathfrak{A}$-modules - $\mathfrak{A}$ a finite dimensional $\mathfrak{k}$-algebra with respect to a hereditary torsion theory. Moreover,

$\operatorname{ind}\mathfrak{M}^{o}$ are the indecomposable objects of $\mathfrak{M}^{o}$.

3.1 Definition: (i) Let $X, Y \in \operatorname{ind}\mathfrak{M}^{o}$, then a homomorphism $\varphi: X \to Y$ is said to be an irreducible map if φ is not an isomorphism and, whenever there is a factorization

$$\begin{array}{ccc} X & \xrightarrow{\varphi} & Y \\ & {\scriptstyle\alpha}\searrow \quad \nearrow{\scriptstyle\beta} & \\ & Z & \end{array} \quad , \quad Z \in \mathfrak{M}^{o} \quad ,$$

then either α is a split monomorphism or β is a split epimorphism.

(ii) We say that an exact sequence in $\mathfrak{M}^o$ is an <u>Auslander-Reiten sequence</u>

$$\mathfrak{E}: \quad 0 \to X' \xrightarrow{\varphi} X \xrightarrow{\psi} X'' \to 0 \qquad \text{if}$$

a) $\mathfrak{E}$ is not split exact,

b) $X', X'' \in \operatorname{ind} \mathfrak{M}^o$,

c) whenever there is a homomorphism $\beta: Z \to X''$ $(\alpha: X' \to Z)$ in $\mathfrak{M}^o$ which is not a split epimorphism (split monomorphism) then there is a factorization

$$\begin{array}{ccc} X & \xrightarrow{\psi} & X'' \\ & \nwarrow \;\; \nearrow_{\beta} & \\ & Z & \end{array} \qquad\qquad \begin{array}{ccc} X' & \xrightarrow{\varphi} & X \\ & {}_{\beta}\searrow \;\; \swarrow & \\ & Z & \end{array} .$$

We say that <u>$\mathfrak{M}^o$ has Auslander-Reiten sequences</u> if, whenever X'' is not projective in $\mathfrak{M}^o$ (X' is not injective in $\mathfrak{M}^o$), then there exists an Auslander-Reiten sequence $\mathfrak{E}$. Since then $\mathfrak{E}$ is uniquely determined, we use the following <u>notation</u>

$$X'' = \tau_{\mathfrak{M}^o}(X') = \tau(X')$$

$$X' = \tau^-_{\mathfrak{M}_o}(X'') = \tau^-(X'') .$$

<u>3.2 Theorem:</u> $\mathfrak{M}^o$ has Auslander-Reiten sequences.

<u>3.3 Remark:</u> If $\mathfrak{M}^o = {}_\Lambda\mathfrak{M}^o$ is the category of Λ-lattices, the existence was established by Auslander-Reiten [Au,77, Au-Re,77] and in [Ro-Sch, 76]. For $\mathfrak{M}^o = \mathfrak{J}(\mathfrak{A})$ in special cases the existence was shown by Bautista-Martinez [Ba-Ma, 79] and in [Ro,80,1]. The general case was established by Auslander Smalø [Au-Sm, 80].

There is a close connection between irreducible maps and Auslander-Reiten sequences: Let

$$0 \to N \xrightarrow{\psi_i} \bigoplus_{i=1}^{t} E_i \xrightarrow{\varphi_i} M \to 0 \quad , \quad E_i \in \operatorname{ind}_\Lambda \mathfrak{M}^o$$

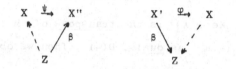

be an Auslander-Reiten sequence, then the φ_i and ψ_i are irreducible maps. Moreover, for every irreducible map $\sigma: X \to M$, $\tau: N \to Y$, $X \simeq E_i$, $Y \simeq E_j$ for some i,j and $\sigma(\tau)$ is "essentially" $\varphi_i(\psi_j)$.

3.4 Construction of Auslander-Reiten sequences in ${}_\Lambda\mathfrak{M}^o$ [Ro,75].

Let M be an indecomposable non-projective Λ-lattice and let

$$P \xrightarrow{\varphi} M \to 0$$

be its projective cover sequence. Then we obtain the exact sequence of right Λ-lattices

$$0 \to \mathrm{Hom}_\Lambda(M,\Lambda) \to \mathrm{Hom}_\Lambda(P,\Lambda) \to (M) \to 0 ,$$

where $(M) \subset \mathrm{Hom}_\Lambda(\mathrm{Ker}\,\varphi,\Lambda)$ is the transpose of M. We denote the functor $-^* = \mathrm{Hom}_R(-,R)$ as dual, D(-). Then we obtain the exact sequence

$$0 \to tr(M)^* \to \mathrm{Hom}_\Lambda(P,\Lambda)^* \to \mathrm{Hom}_\Lambda(M,\Lambda)^* \to 0 ,$$

and

$$D\,tr(M) = tr(M)^* = \tau^-_{{}_\Lambda\mathfrak{M}^o}(M) =: \tau^-_\Lambda(M)$$

is the kernel of the Auslander-Reiten sequence.

In order to compute the Auslander-Reiten sequence itself, we apply $\mathrm{Hom}_\Lambda(M,-)$, and get the exact sequence

$$0 \to \mathrm{Hom}_\Lambda(M,\tau^-_\Lambda(M)) \to \mathrm{Hom}_\Lambda(M, \mathrm{Hom}_\Lambda(P,\Lambda)^*)$$
$$\xrightarrow{\rho} \mathrm{Hom}_\Lambda(M,\mathrm{Hom}_\Lambda(M,\Lambda)^*) \xrightarrow{\sigma} \mathrm{Ext}^1_\Lambda(M,\tau^-_\Lambda(M)) \to 0 .$$

The image of ρ is - up to natural isomorphism - given by $\mathrm{Hom}_\Lambda(M,M)^*$, which turns out to be an injective lattice over $\mathrm{End}_\Lambda(M)$, and as such has a unique minimal overmodule X. Then σX is the simple socle of $\mathrm{Ext}^1_\Lambda(M,\tau^-_\Lambda(M))$ as $\mathrm{End}_\Lambda(M)$-module. The Auslander-Reiten sequence of M is then represented by an extension in this socle.

3.5 Remarks: (i) One should note the difference of the construction for lattices over Λ and for modules over an artinian $\mathfrak{k}$-algebra $\mathfrak{A}$: Let X be an indecomposable non-projective $\mathfrak{A}$-module.

Then we take the projective cover sequence of X :

$$P_2 \to P_1 \to X \to 0 ,$$

and construct the exact sequence

$$0 \to \mathrm{Hom}_{\mathfrak{A}}(X,\mathfrak{A}) \to \mathrm{Hom}_{\mathfrak{A}}(P_1,\mathfrak{A}) \to \mathrm{Hom}_{\mathfrak{A}}(P_2,\mathfrak{A}) \to \mathrm{tr}_{\mathfrak{A}}(X) \to 0 .$$

Then

$$\tau_{\mathfrak{A}}^{-}(X) = \mathrm{Hom}_{\mathfrak{k}}(\mathrm{tr}_{\mathfrak{A}}(X), \mathfrak{k}) .$$

(ii) In case of orders, $\tau_{\Lambda}^{-}(M)$ is easily computed: We view $\Lambda \subset \Gamma$, where Γ is the "natural" maximal order $\Gamma = \prod_{i=1}^{t} (\Omega_i)_{n_i}$ and hence every Λ-lattice M is contained in

$$\Gamma M = \bigoplus_i (\Omega_i)_{n_i \times m_i} .$$

E.g., if

$$\Lambda = \begin{bmatrix} R & R \\ \pi^t & R \end{bmatrix} ,$$

and $M = \begin{bmatrix} R \\ \pi \end{bmatrix}$, then the projective resolution of M is

$$0 \to \begin{bmatrix} \pi \\ \pi^t \end{bmatrix} \to \begin{bmatrix} \pi \\ \pi \end{bmatrix} \oplus \begin{bmatrix} R \\ \pi^t \end{bmatrix} \to \begin{bmatrix} R \\ \pi \end{bmatrix} \to 0 .$$

Applying $\mathrm{Hom}_{\Lambda}(-,\Lambda)$, we get the exact sequence

$$0 \to (\pi^{t-1} R) \to \begin{pmatrix} \pi^{t-1} & \pi^{-1} \\ R & R \end{pmatrix} \to (R \; \pi^{-1}) \to 0$$

and so

$$\tau_{\Lambda}^{-}(M) \simeq \binom{R}{\pi} ,$$

and the Auslander-Reiten sequence of $\binom{R}{\pi}$ has the form

$$0 \to \begin{bmatrix} R \\ \pi \end{bmatrix} \to \begin{bmatrix} R \\ \pi^2 \end{bmatrix} \oplus \begin{bmatrix} R \\ R \end{bmatrix} \to \begin{bmatrix} R \\ \pi \end{bmatrix} \to 0 .$$

Similar computations show, that for the indecomposable non-projective Λ-lattices $\binom{R}{\pi^i}$, $1 \le i \le t-1$, one has the Auslander-Reiten sequences

$$0 \to \begin{pmatrix} R \\ \pi^i \end{pmatrix} \begin{array}{c} \nearrow \begin{pmatrix} R \\ \pi^{i+1} \end{pmatrix} \searrow \\ \searrow \begin{pmatrix} R \\ \pi^{i-1} \end{pmatrix} \nearrow \end{array} \begin{pmatrix} R \\ \pi^i \end{pmatrix} \to 0 \quad .$$

(iii) By studying Auslander-Reiten sequences one obtains not only information on the indecomposable Λ-lattices, but also on the interrelation between them via irreducible maps. (Note one of the main features of Auslander-Reiten sequences is, that if X is not projective (not injective) in $\mathrm{ind}_\Lambda \mathfrak{M}^o$, then $\tau_\Lambda(X)$ and $\tau_\Lambda^-(X)$ is indecomposable.

The study of Auslander-Reiten sequences has lead to the following results (just to name some):

<u>3.6 Theorem:</u> (i) Schmidt [Sch, 76]: Λ is of finite lattice type if and only if for the finitely presented functors

$$\mathbb{F} : {}_\Lambda\mathfrak{M}^o \to {}_R\mathfrak{M}^o$$

the Jordan-Zassenhaus theorem holds.

(ii) Auslander [Au, 77], Schmidt [Sch, 76]:

Λ is of finite lattice type if and only if for every $M \in {}_\Lambda\mathfrak{M}^o$, the functor $\underline{\mathrm{Hom}}_\Lambda(-,M)$ has finite length, where $\underline{\mathrm{Hom}}_\Lambda(-,M)$ is the quotient of $\mathrm{Hom}_\Lambda(-,M)$ modulo homomorphisms which factor via projective Λ-lattices.

<u>3.7 Theorem</u> (Bautista-Brenner [Ba-Br, 81]): Let Λ be an order of finite lattice type and $M \in \mathrm{ind}_\Lambda\mathfrak{M}^o$ non-projective. If

$$0 \to \tau^-(M) \to \bigoplus_{i=1}^{t} E_i \to M \to 0$$

is its Auslander-Reiten sequence, with $E_i \in \mathrm{ind}_\Lambda\mathfrak{M}^o$, then

$$t \leq 4 \text{ , and if } t = 4$$

then at least one of the E_i must be a projective and injective Λ-lattice.

Proof: The proof of the above result by S.Brenner is formulated for artinian algebras; however, if one replaces the length function in the artinian case by the rank over R in case of lattices, it carries over. #

The Auslander-Reiten quiver of Λ consists of the "glueing together" of the various Auslander-Reiten sequences. More precisely:

3.8 Definition: (i) Let $M,N \in \mathrm{ind}_\Lambda \mathfrak{M}^o$. Then $\mathrm{Irr}(M,N) = \mathrm{rad}\,\mathrm{Hom}_\Lambda(M,N)/\mathrm{rad}^2\,\mathrm{Hom}_\Lambda(M,N)$ is called the bimodule of irreducible maps; it is a left $\mathrm{End}_\Lambda(M)$, right $\mathrm{End}_\Lambda(N)$ bimodule. Since $\mathrm{Irr}(M,N)$ is an artinian module, we have the natural numbers

$$a_{M,N} = \text{length of } \mathrm{Irr}(M,N) \text{ as } \mathrm{End}_\Lambda(M)\text{-module} ,$$
$$a'_{M,N} = \text{length of } \mathrm{Irr}(M,N) \text{ as } \mathrm{End}_\Lambda(N)\text{-module.}$$

The Auslander-Reiten quiver $\mathfrak{A}(\Lambda)$ of Λ has as vertices the isomorphism classes in $\mathrm{ind}_\Lambda \mathfrak{M}^o$, and there exists a valued arrow

$$[M] \xrightarrow{(a_{M,N},\ a'_{M,N})} [N] ,$$

if $\mathrm{Irr}(M,N) \neq 0$ and $a_{M,N}$ and $a'_{M,N}$ are defined as above. Here $\mathrm{rad}\,\mathrm{Hom}_\Lambda(M,N)$ is to be understood as follows: We have the functors $\mathrm{Hom}_\Lambda(-,N)$ and $\mathrm{Hom}_\Lambda(-,M)$ with radicals (the radical of a functor is the intersection of its maximal subfunctors) $\mathrm{rad}(\mathrm{Hom}_\Lambda(-,N))$ and $\mathrm{rad}(\mathrm{Hom}_\Lambda(M,-))$, and the functorial description of Auslander-Reiten sequences shows that

$$[\mathrm{rad}(\mathrm{Hom}_\Lambda(-,N))/\mathrm{rad}^2\mathrm{Hom}_\Lambda(-,N)](M)$$
$$\simeq [\mathrm{rad}\,\mathrm{Hom}_\Lambda(M,-)/\mathrm{rad}^2\mathrm{Hom}_\Lambda(M,-)](N) .$$

For brevity we write

$$\mathrm{rad}(M,N)/\mathrm{rad}^2(M,N)$$

for the above isomorphic modules [Au 78].

(ii) $M \in \mathrm{ind}_\Lambda \mathfrak{M}^o$ is said to be <u>stable</u> if for every $n \in \mathbb{N}$,

$$\tau_\Lambda^n(M) = \tau_\Lambda(\tau_\Lambda^{n-1}(M)) \quad \text{and}$$

$$\tau_\Lambda^{-n}(M) = \tau_\Lambda^-(\tau_\Lambda^{-(n-1)}(M))$$

are defined.

$\mathfrak{A}_s(\Lambda)$ consists of the full subquiver of $\mathfrak{A}(\Lambda)$ the vertices of which are the isomorphism classes of the stable modules; $\mathfrak{A}_s(\Lambda)$ is called the <u>stable Auslander-Reiten quiver</u>.

(iii) $M \in \mathrm{ind}_\Lambda \mathfrak{M}^o$ is said to be <u>periodic</u>, if it is not an injective Λ-lattice and $\tau^n(M) \simeq M$ for some $n \in \mathbb{N}$.

It follows directly from the functorial description of Auslander-Reiten sequences that $\mathrm{Irr}(M,N)$ are indeed the irreducible maps; in particular $a_{M,N}$ is - provided N is not projective - the number of times M occurs in the Auslander-Reiten sequence of N, and $a'_{M,N}$ is - provided M is not an injective Λ-lattice - the number of times N occurs in the Auslander-Reiten sequence of $\tau(M)$. If one has an Auslander-Reiten sequence

$$0 \to N \xrightarrow{\varphi_{ij}} \oplus E_i^{(n_i)} \xrightarrow{\psi_{ij}} M \to 0$$

then the $\{\varphi_{ij}\}_{1 \le j \le n(i)}$ constitute a basis for $\mathrm{Irr}(N,E_i)$ and dually.

We point out one feature which distinguishes Auslander-Reiten quivers of orders of finite type from those of artinian algebras. In the finite representation type case, any homomorphism between indecomposable objects is a sum of compositions of irreducible maps (3.6 ii). If one applies this to $\varphi_\pi : M \to M$, $M \in \mathrm{ind}_\Lambda \mathfrak{M}^o$ and φ_π multiplication by π, then there is a path of irreducible maps from M to M; hence the Auslander-Reiten quiver is never flat.

<u>3.9 Examples:</u> (i) If $P \in \mathrm{ind}_{\Lambda}\mathfrak{M}^{o}$ is projective (injective), then the irreducible maps $X \xrightarrow{\varphi} P$ are precisely the embedding of the indecomposable direct summands of $\mathrm{rad}\,P$ into P (for injective lattices $I \to X$ exactly the canonical maps from I to the indecomposable direct summands of its unique minimal overmodule). Hence if $\Gamma = \prod_{i=1}^{t} (\Omega_i)_{n_i}$ then

$$\mathfrak{A}(\Gamma) = \underset{1}{\circlearrowleft\!\bullet} \quad \underset{2}{\circlearrowleft\!\bullet} \quad \cdots \quad \underset{t}{\circlearrowleft\!\bullet}\ , \qquad \mathfrak{A}_s(\Gamma) = \emptyset \ .$$

If Γ is hereditary in $(\Omega)_n$ with t non-isomorphic indecomposable projectives $P_1,\ldots,P_t$ numbered in such a way that $P_{i-1} = \mathrm{rad}\,P_i$ $(i \bmod (t))$, then

$$\mathfrak{A}(\Gamma) \quad \underset{1}{\bullet} \rightarrow \underset{2}{\bullet} \ \cdots \rightarrow \underset{t}{\bullet} \ (t \rightarrow 1)\ , \qquad \mathfrak{A}_s(\Gamma) = \emptyset \ .$$

The converse is not true as shows

$$\Lambda = \begin{bmatrix} R & R & R \\ \pi & R & R \\ \pi^2 & \pi & R \end{bmatrix} \ , \qquad \text{when} \quad \mathfrak{A}_s(\Lambda) = \emptyset \ .$$

(ii) If $\Lambda = \begin{pmatrix} R & R \\ \pi^t & R \end{pmatrix}$ (cf. 3.5 ii), then

$\mathfrak{A}(\Lambda)$:

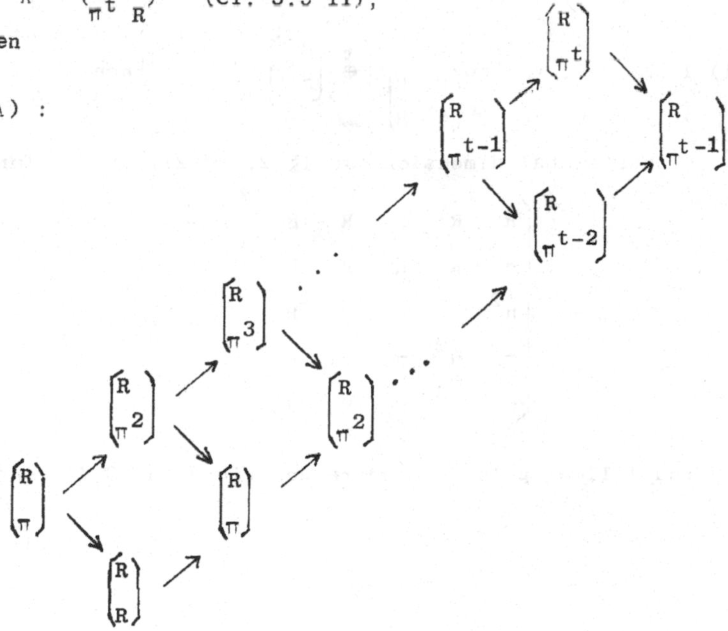

where the same modules have to be identified; i.e. $\mathfrak{A}(\Lambda)$ is wrapped on a cylinder

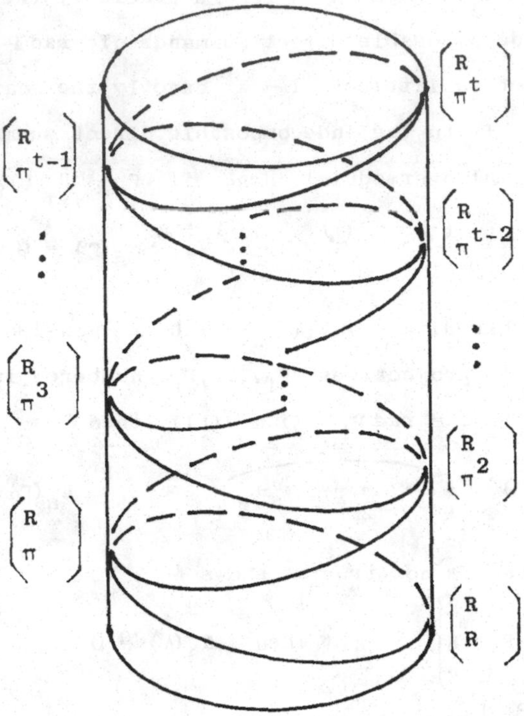

(iii) If $\Lambda = \operatorname{End}_{\begin{pmatrix} R & R \\ \pi^t & R \end{pmatrix}} \left(\bigoplus_{i=o}^{t} \begin{pmatrix} R_i \\ \pi \end{pmatrix} \right)$, then

Λ has global dimension two (§ 2, (2.2)) , for $t = 4$,

$$\Lambda = \begin{pmatrix} R & R & R & R & R \\ \pi & R & R & R & R \\ \pi^2 & \pi & R & R & R \\ \pi^3 & \pi^2 & \pi & R & R \\ \pi^4 & \pi^3 & \pi^2 & \pi & R \end{pmatrix} ,$$

and, following the procedure described in (3.4), one finds

$\mathfrak{A}(\Lambda)$:

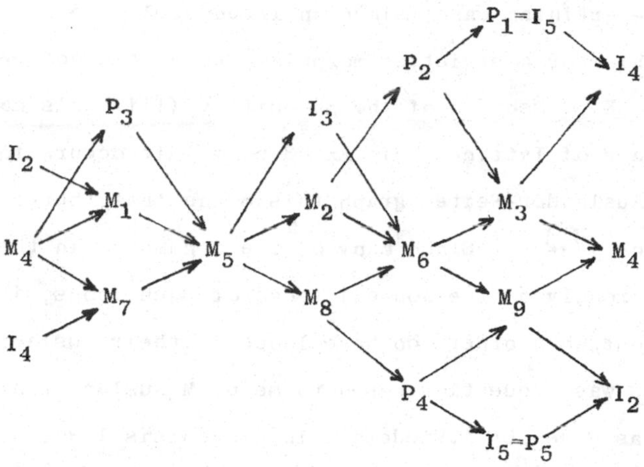

Here $P_1, \ldots, P_5$ are the indecomposable projective Λ-lattices, numbered according to the columns, $I_1, \ldots, I_5$ are the indecomposable injective Λ-lattices numbered according to the rows. The rows in $\mathfrak{A}(\Lambda)$ constitute the orbits under τ, and the same modules must be identified. In particular, $\mathfrak{A}(\Lambda)$ lives on a Möbius-strip.

$\mathfrak{A}_s(\Lambda)$:

.

For irreducible maps between lattices we have the following restriction:

<u>3.10 Lemma:</u> Let $M,N \in \mathrm{ind}_\Lambda \mathfrak{M}^o$ and $\varphi : M \to N$ an irreducible map, then

(i) either φ is surjective

(ii) or φ is injective and coker φ is torsionfree, i.e. $M \simeq KM\varphi \cap N$,

(iii) or φ is a monomorphism onto a maximal submodule.

This <u>follows</u> immediately from the definition of irreducible maps and holds equaly well for $\mathfrak{Z}(\mathfrak{A})$ - torsionfree $\mathfrak{A}$-modules.

In the artinian case, since an irreducible map is either a strict monomorphism or a strict epimorphism, there can not be an irreducible map $X \xrightarrow{\varphi} X$. Because of the possibility (iii) this can not be ruled out in case of lattices (cf. Ex. 3.9 i) - it occurs for maximal orders. For the Auslander-Reiten graph this means that there is a <u>loop</u> of length one  . Since many of the arguments in the artinian case depend strongly on the non-existence of such loops, it is necessary to find out, when orders do have loops in their Auslander-Reiten quiver. This was a question posed to me by M.Auslander around 1976. The answer was given by A.Wiedemann in his thesis 1980.

<u>3.11 Theorem</u> [Wi80]: Let Λ be a two-sidedly indecomposable R-order. Then the following statements are equivalent:

(i) There is $M \in \text{ind}_\Lambda \mathfrak{M}^o$ and an irreducible map $\varphi: M \to M$.

(ii) The Auslander-Reiten quiver of Λ has the following form

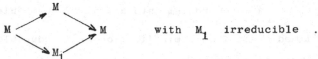

If $R/\pi R$ is a finite field, then

(iii) Λ is Morita equivalent to a Bass-order Γ of the following form [R II, IX, 6.17]: Γ is a Bass-order in the separable skewfield D with maximal order Ω and $\Omega/\text{rad}_T \Omega \simeq \Gamma/\text{rad}_T \Gamma[\varepsilon]$, $\varepsilon^2 = 0$, or Γ is maximal.

The <u>proof</u> uses the lemma of Harada - Sai , which can be applied since by a result of Maranda, then exists an integer $s = s(\Lambda)$ such that for $M \in {}_\Lambda\mathfrak{M}^o$, $M \in \text{ind}_\Lambda \mathfrak{M}^o$ if and only if $M/\pi^s M$ is indecomposable. The next step is to show that if $\varphi: M \to M$ is irreducible, M is an irreducible Λ-lattice and its Auslander-Reiten sequence is of the form

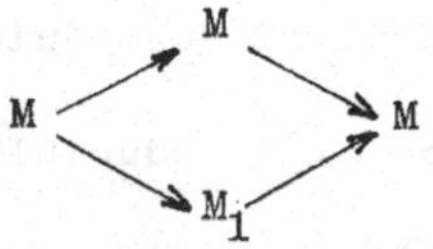

with M_1 irreducible .

The final ingredient then is the following

3.12 Theorem [Wi 80]: Assume Λ is two-sidedly indecomposable. Let Δ be a connected component of $\mathfrak{A}(\Lambda)$ and assume that the R-ranks of the Λ-lattices in Δ are bounded, then Λ is of finite lattice type and $\Delta = \mathfrak{A}(\Lambda)$.

3.13 Remarks: (i) Let me recall the first Brauer-Thrall conjecture for orders: If there exists an integer s such that for $M \in \mathrm{ind}_{\Lambda}\mathfrak{M}^o$, $\mathrm{rk}\, M \leq s$, then Λ is of finite lattice type. If R has finite residue field, this is an immediate consequence of the Jordan-Zassenhaus theorem. However, if the residue field of R is infinte (e.g. $R = L[[T]]$, L an infinite field), then (3.12) settles this conjecture positively. (For artinian algebras, this was proved by Roiter [Na-Roi,70]

(ii) An important application of (3.12) is the following: Assume one has a two-sided indecomposable order Λ and by some technique one has constructed a finite number of indecomposable Λ-lattices which constitute a connected component of the Auslander-Reiten quiver, then (3.12) implies that one has found all indecomposable Λ-lattices.

(iii) The second Brauer-Thrall conjecture asserts that, provided $R/\pi R$ is infinite and Λ is of infinite type, then there are infinitely many integers n_i such that for each n_i there are infinitely many indecomposable Λ-lattices of rank n_i . This was verified in [Ri-Ro, 79] provided the second Brauer-Thrall conjecture holds in the artinian category $\mathfrak{C}$ (§ 1, (1.3)) - which is very likely.

(iv) It is surprising to me that loops in the Auslander-Reiten quiver can only occur in case of finite lattice type, and moreover, that one can give an explicit description of the orders with loops in their Auslander-Reiten quiver.

(v) The result of (3.11) allows to use the results of Riedtmann[Rie,80, 1,2] in the version of Happel-Preiser-Ringel, to get the structure of connected components of the stable Auslander-Reiten quiver containing a periodic vertex.

In order to do so we have to introduce a considerable amount of notation:

3.14 Definitions [HPR,80]: (i) A quiver $\Gamma = (\Gamma_o, \Gamma_1)$ consists of the set of vertices Γ_o and the set of arrows Γ_1. We shall always assume that Γ does not have loops or double arrows. If $x \in \Gamma_o$ is a vertex, then we denote by x^+ the set of endpoints of arrows with starting point x and x^- the set of starting points of arrows with endpoint x. If for all x, the sets x^+ and x^- are finite, the quiver Γ is said to be locally finite.

(ii) A Riedtmann quiver $\Delta = (\Gamma_o, \Gamma_1, \tau)$ is a quiver (Γ_o, Γ_1) together with an injective function

$$\tau : \Gamma_o' \to \Gamma_o \quad ,$$

defined on a subset Γ_o' of Γ_o, satisfying $(\tau x)^- = x^+$. So given an arrow $\alpha: x \to y$ there exists a unique arrow $y \to \tau x$, this arrow will be denoted by $\sigma\alpha$. A Riedtmann quiver Δ is called stable provided τ is defined on all of Γ_o and is also surjective. Any Riedtmann quiver has a unique maximal stable subquiver. A vertex $x \in \Gamma_o$ will be called periodic provided $\tau^n(x) = x$ for some $n \in \mathbb{N}$.

(iii) For a quiver (Γ_o, Γ_1), a function $a: \Gamma_1 \to \mathbb{N} \times \mathbb{N}$ will be called a valuation and $\Gamma = (\Gamma_o, \Gamma_1, a)$ a valued quiver. The image of $\alpha: x \to y$ will be denoted by (a_α, a_α') or (a_{xy}, a_{xy}'). A valued Riedtmann quiver is a Riedtmann quiver $(\Gamma_o, \Gamma_1, \tau)$ together with a valuation a for (Γ_o, Γ_1) such that $a_{\sigma\alpha} = a'_{\alpha}$, $a'_{\sigma\alpha} = a_{\alpha}$ for all $\alpha: y \to x$ with $x \in \Gamma'_o$. This valued Riedtmann quiver will be denoted by $\Delta = (\Gamma_o, \Gamma_1, \tau, a)$.

<u>3.15 Examples:</u> (i) Let Λ be an R-order, which has no irreducible map $M \to M$ (3.11). Then $\mathfrak{A}(\Lambda)$ is a valued Riedtmann quiver. In this case $\tau(x) = y$, provided x corresponds to $[M]$, $M \in \mathrm{ind}_{\Lambda}\mathfrak{M}^o$ and $[\tau(M)]$ corresponds to y under this Auslander-Reiten sequence (3.1, ii), and x^+ corresponds to the non-isomorphic indecomposable modules in the middle term of the Auslander-Reiten sequence of M. Moreover, τ is defined on the non-injective modules in $\mathrm{ind}_{\Lambda}\mathfrak{M}^o$. The valuation is given in (3.8) and the conditions $(\tau x)^+ = x^-$ and $a_{\sigma\alpha} = a'_{\alpha}$, $a'_{\sigma\alpha} = a_{\alpha}$ follow from (3.3).

(ii) Let Γ be an oriented tree (i.e. a quiver whose underlying graph is a tree) and define $\mathbb{Z}\Gamma$ as follows: The vertices are the elements of $\mathbb{Z} \times \Gamma_o$, and given an arrow $\alpha: x \to y$ in Γ , there are arrows

$$(n,\alpha) : (n,x) \to (n,y) \quad \text{and}$$

$$\sigma(n,\alpha) : (n,y) \to (n+1,x) \quad \text{for all } n \in \mathbb{Z}.$$

Finally we put $\tau(n,x) = (n+1,x)$.

This way one obtains stable Riedtmann quivers.

E.g. if Γ  , then

$\mathbb{Z}\Gamma$

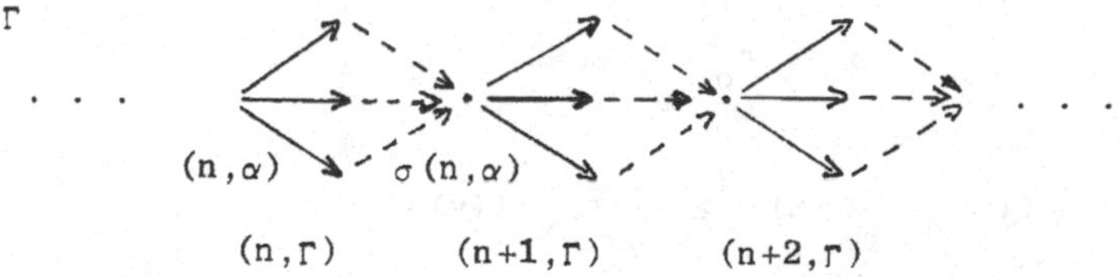

If now $\Gamma = (\Gamma_o, \Gamma_1, a)$ is a valued tree, then on $\mathbb{Z}\Gamma$ a valuation is defined by

$$a_{(n,\alpha)} = a_\alpha = a'_{\sigma(n,\alpha)} \qquad \text{and}$$

$$a'_{(n,\alpha)} = a'_\alpha = a'_{\sigma(n,\alpha)} \qquad .$$

This way $\mathbb{Z}\Gamma$ becomes a valued Riedtmann quiver $\mathbb{Z}\Gamma = \mathbb{Z}(\Gamma_o, \Gamma_1, a)$.

<u>3.16 Theorem</u> (Riedtmann [Rie 80, 1,2], Happel-Preiser-Ringel [HPR 80]): Given any stable valued Riedtmann quiver Δ , there is a valued oriented tree Γ and group G of automorphisms of $\mathbb{Z}\Gamma$ such that Δ is isomorphic to $\mathbb{Z}\Gamma/G$. Moreover, Γ is uniquely determined by Δ . It is called the <u>tree class of Γ</u> .

Note that this result depends heavily on the fact that Δ has no loops.

<u>3.17 Example:</u> Let $\Lambda = \begin{pmatrix} R & R \\ \pi^t & R \end{pmatrix}$, then $\mathfrak{A}_s(\Lambda)$ is $\mathbb{Z}\Gamma/G$, where $\Gamma = \underset{1}{\cdot} \longrightarrow \underset{2}{\cdot} \longrightarrow \cdots \longrightarrow \underset{t-1}{\cdot}$ is an A_{t-1} and

$\mathbb{Z}\Gamma$:

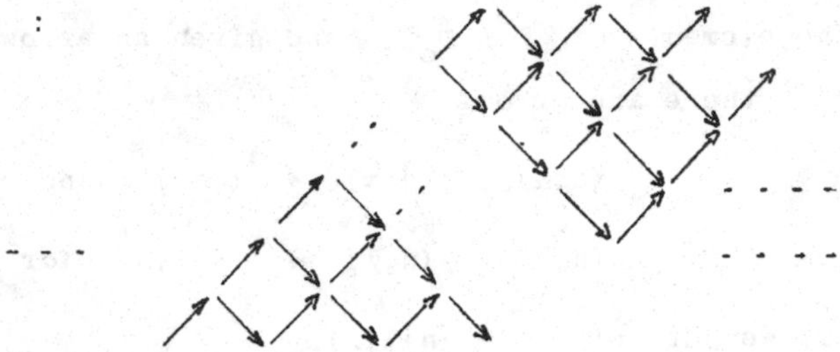

If $G = \langle\tau\rangle$, then $\mathfrak{A}_s(\Lambda) \simeq \mathbb{Z}\Gamma/\langle\tau\rangle$.

<u>3.18 Definition:</u> Let Δ be a stable valued Riedtmann quiver, $\Delta = (\Gamma_o, \Gamma_1, a, \tau)$, a <u>subadditive function</u> on Δ is a function

$$f : \Gamma_o \to \mathbb{N}$$

satisfying

$$f(x) + f(\tau x) \geq \sum_{y \in x^+} f(y) a'_{xy} \qquad .$$

f is said to be <u>additive</u>, provided we always have equality. We say that f is <u>periodic with respect to τ</u> , if there exists $n \in \mathbb{N}$

such that $f(x) = f(\tau^n x)$ for all x .

The following theorem was proved by Happel-Preiser-Ringel for connected stable Riedtmann quivers having a periodic vertex. Webb has noted - for his applications to group rings - that the result also holds for periodic functions.

3.19 Theorem (Happel-Preiser-Ringel [HPR 80]; Webb [We, 80, 2]): Let Δ be a stable valued Riedtmann quiver, which is locally finite and connected. Assume there is a subadditive function f on Δ which is periodic with respect to τ .

(i) The tree class of Δ is either a Dynkin diagram, or a Euclidian diagram or one of A_∞, A_∞^∞, B_∞, C_∞, D_∞ . (cf. the following list.)

(ii) If f is not additive, then the tree class of Δ is a Dynkin diagram or A_∞ .

(iii) If f is unbounded, then the tree class of Δ is A_∞ .

3.20 The diagrams:

(i) Dynkin diagrams

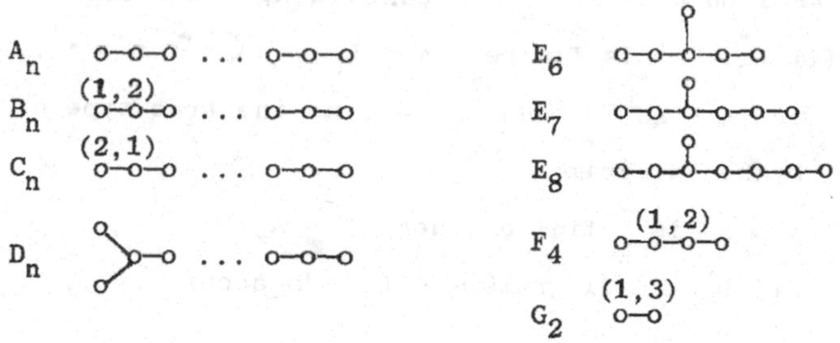

(ii) Euclidian diagrams

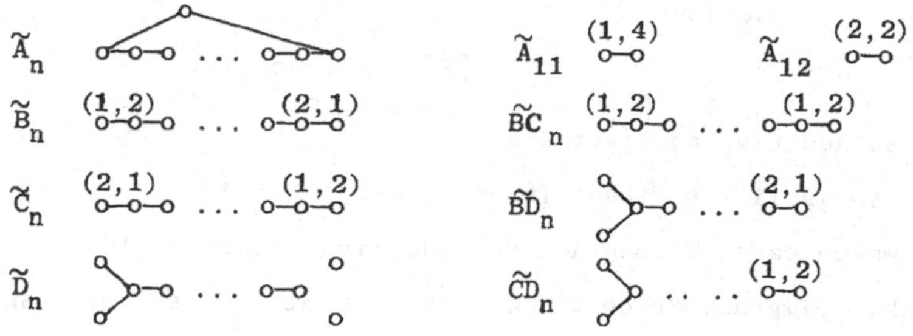

$\widetilde{E}_6$ o—o—o—o—o

$\widetilde{E}_7$ o—o—o—o—o—o—o

$\widetilde{E}_8$ o—o—o—o—o—o—o—o

$\widetilde{F}_{41}$ $\overset{(1,2)}{\text{o—o—o—o—o}}$

$\widetilde{F}_{42}$ $\overset{(2,1)}{\text{o—o—o—o—o}}$

$\widetilde{G}_{21}$ $\overset{(1,3)}{\text{o—o—o}}$ $\widetilde{G}_{22}$ $\overset{(3,1)}{\text{o—o—o}}$

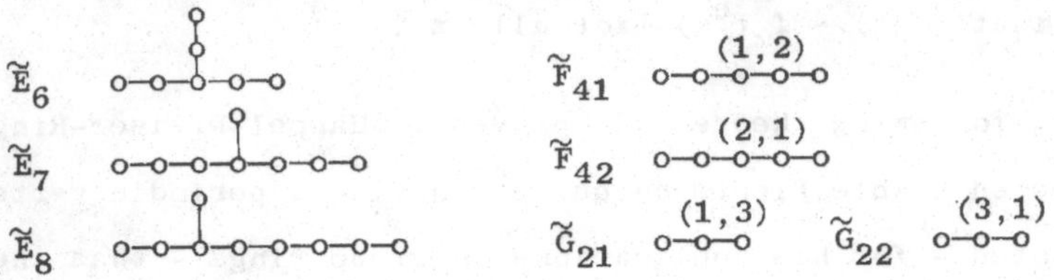

(iii) Infinite diagrams

A_∞ o—o—o ... o—o ...

B_∞ $\overset{(1,2)}{\text{o—o—o}}$ ... o—o ...

C_∞ $\overset{(2,1)}{\text{o—o—o}}$ ... o—o ...

D_∞ >o—o ... o—o ...

A_∞^∞ .... o—o—o ....

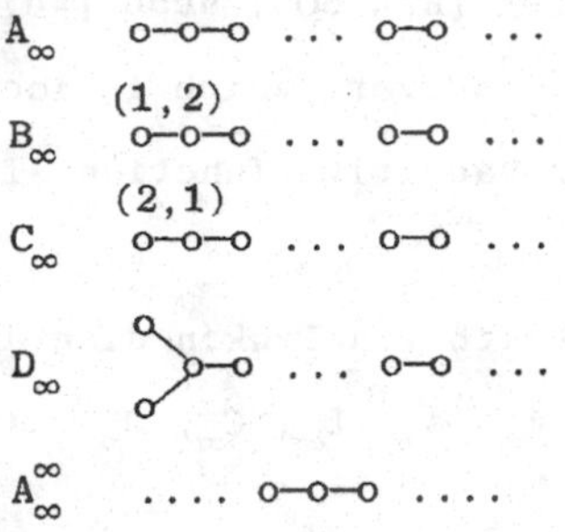

Using this and (3.11) one obtains

3.21   Theorem (Happel-Preiser-Ringel [HPR, 80], Riedtmann [Rie,80,1,2],
        Wiedemann [Wi, 80, 2]): Let Λ be an R-order, which is two-
sided indecomposable, and Δ a connected component of the stable
Auslander-Reiten quiver $\mathfrak{A}_s(\Lambda)$ containing a periodic vertex. Then
either   (i) if Δ is finite, $\Delta = L_n = $ $C \cdot \overset{\leftarrow}{\leftrightarrow} \cdot \overset{\leftarrow}{\leftrightarrow} \cdot \ \ldots \ \cdot \overset{\leftarrow}{\leftrightarrow} \cdot_n$
        or $\Delta = \mathbb{Z}\Gamma/G$ and Γ - i.e. the tree type of Δ - is a
        Dynkin diagram;
       (ii) if Δ is infinite, then $\Gamma = A_\infty$.
Moreover, all Dynkin diagrams and L_n do occur.

We prove this using (2.19): Since Δ contains a periodic vertex,
the rank-function

$$f([M]) = \text{rank}_R(M)$$

is subadditive and periodic.

If Δ is finite, then either $\Delta = L_n$ (3.11) or $\Delta = \mathfrak{A}_s(\Lambda)$ (3.12),
in which case f can not be additive, so by (3.19) Δ must be a
Dynkin diagram. Hence we can assume that Δ is Euclidian or one

of the infinite diagrams. Moreover, by (3.12) f must be unbounded and so $\Gamma = A_\infty$ by (3.19).

<u>3.22 Remarks:</u> (i) If Λ is of finite lattice type this implies in particular, that in the middle term of an Auslander-Reiten sequence there occur at most 4 non-projective direct summands. Looking back at the result of Bautista and Brenner (3.7), one sees that their result is much better in this respect, since it not only concerns $\mathfrak{A}_s(\Lambda)$ but $\mathfrak{A}(\Lambda)$.

(ii) Naturally the restriction that Δ contains a periodic vertex is quite severe. In case of group rings, P.Webb has refined in an ingenious way the above methods (cf.§ 4) to get informations in the general situation.

At the end of this section we turn to <u>preprojective partitions</u>, a concept which is motivated from the study of artinian algebras. Let $S = \{\Gamma, f_i, {}_iX_j\}$ be a $\mathfrak{k}$-species (§ 1, Dl-Ri,76) and let $\mathfrak{A}$ be the tensor algebra of S over $\mathfrak{k}$. Then there is a natural notion of preprojective and preinjective $\mathfrak{A}$-modules: The $\mathfrak{A}$-modules lying in the same connected component - in this case all projectives lie in the same connected component - as the projective modules (the injective $\mathfrak{A}$-modules). The simple minded generalization to arbitrary $\mathfrak{k}$-algebras does not make sense, since in general the projective $\mathfrak{A}$-modules (injective $\mathfrak{A}$-modules) lie in different components. The natural generalization from tensor algebras of species to arbitrary $\mathfrak{k}$-algebras was given by Auslander-Smalø [Au-Sm 80,81] and their results have a natural generalization to orders. (The idea of all this goes back to the proof of the first Brauer-Thrall conjecture for artinian algebras [Na-Roi, 70].)

So we start again with definitions:

<u>3.23 Definitions:</u> (i) Let $\mathfrak{D}$ be a subcategory of ${}_\Lambda\mathfrak{M}^o$; a <u>cover</u>

$\mathfrak{C}$ of $\mathfrak{D}$ is a subcategory $\mathfrak{C} \subset \text{ind}_{\Lambda}\mathfrak{M}^o$, such that for every $D \in \mathfrak{D}$ there exists a surjective map

$$\bigoplus_{i=1}^{t} C_i^{(\alpha_i)} \to D \quad , \qquad C_i \in \mathfrak{C} .$$

A minimal cover is a cover, in which no lattice is superfluous. (If such a minimal cover exists, it is unique.)

(ii) A preprojective partition of $_{\Lambda}\mathfrak{M}^o$ is a sequence of subcategories $\mathbb{P}_o, \mathbb{P}_1, \ldots, \mathbb{P}_i, \quad , 0 \le i \le \infty$ such that

α.) $\text{ind}_{\Lambda}\mathfrak{M}^o = \bigcup_{i \geq 0} \mathbb{P}_i$,

β.) $\mathbb{P}_i$ is a finite minimal cover for $\bigcup_{j \geq i} \mathbb{P}_j$.

(iii) $\text{add}(\bigcup_{i<\infty} \mathbb{P}_i)$ are called preprojective, where add(-) denotes the additive category generated by (-) .

The dual notion is that of a preinjective partition and of preinjective lattices.

The crucial point in this is the word "finite" in β.).

3.24 Theorem (Auslander-Smalø [Au-Sm 80]): $_{\Lambda}\mathfrak{M}^o$ has a preprojective partition: and it also has a preinjective partition, which is dual to the preprojective partition.

The importance of these preprojective (preinjective) partitions shows the next result.

3.25 Theorem (Auslander-Smalø [Au-Sm, 80]):

(i) $\mathbb{P}_i = \emptyset$ for some $i < \infty$ if and only if Λ is of finite lattice type.

(ii) $\mathbb{P}_\infty = \emptyset$ if and only if Λ is of finite lattice type.

(iii) If $\mathbb{P}_\infty \neq \emptyset$, then it does not have a finite minimal cover.

For the details we refer to the contribution of Auslander and Smalø in this collection.

§ 4 Auslander-Reiten quivers for group rings and special classes of orders.

4. A.) Group rings

In this section let B be a block of RG, G a finite group such that (char K, $|G|$) = 1 . Since the projective B-lattices are exactly the injective B-lattices, the stable Auslander-Reiten quiver of B is obtained from the Auslander-Reiten quiver of B by just omitting the points corresponding to the indecomposable projective B-lattices.

Let R be an unramified extension of $\hat{\mathbb{Z}}_p$, the p-adic integers.

4.1 Theorem ([Ja 79,Ro,80,4]): Let B be a block of defect 1 and $M_o \in \operatorname{ind}_B \mathfrak{M}^o$, and choose a projective resolution

$$\mathbb{P} : 0 \to M_o \to P_t \searrow M_{t-1} \nearrow \cdots P_2 \searrow M_1 \nearrow P_1 \to M_o \to 0,$$

then the P_i are indecomposable, their distribution is obtained by "walking around" the Brauer tree [GJA, 74] and so $\mathbb{P}$ is the Auslander-Reiten quiver of B , after identifying the two copies of M_o.

4.2 Examples: (i) Assume that the defect group of B is a normal cyclic defect group of order p ; and assume that B has t non isomorphic indecomposable projective lattices.

Then the Auslander-Reiten quiver of B has the form

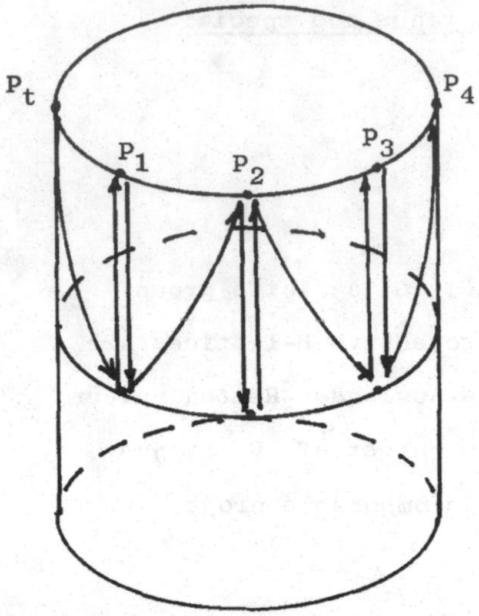

The Auslander-Reiten quiver of $B/\pi B$, i.e. the modular Auslander-Reiten quiver has the form

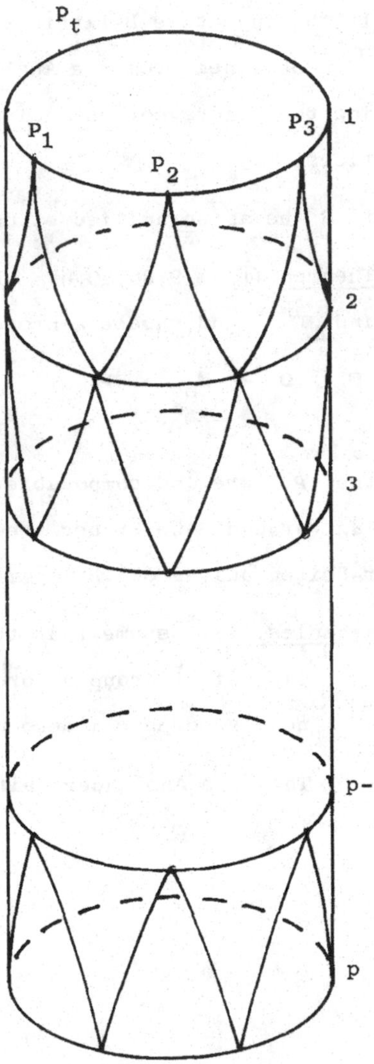

The Brauer tree is

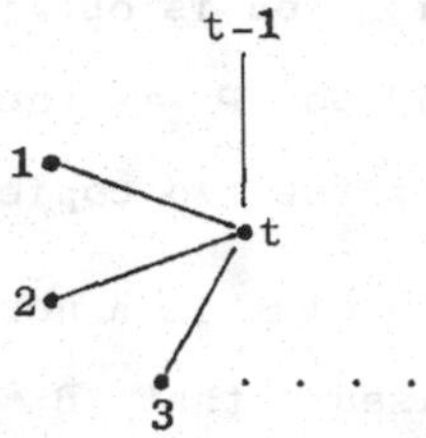

(ii) Let M_{11} be the Mathieu group and B the principal block at $p = 11$.

Then

$B =$

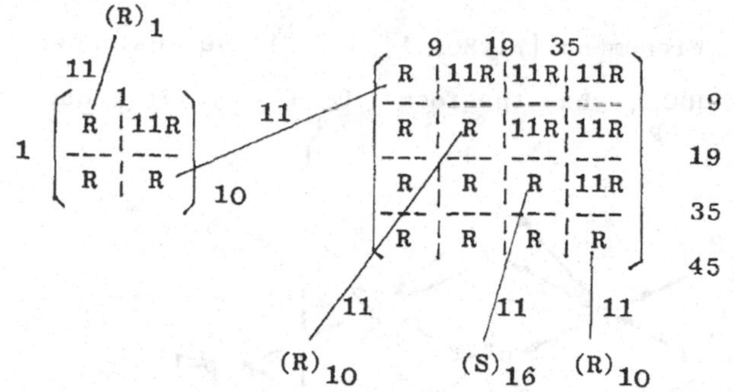

,

where $S = \mathrm{Fix}_{C_5} R(\sqrt[11]{T})$, C_5 the cyclic group of order 5, and the congruences are given as pullbacks

$$\begin{array}{ccc} R \xrightarrow{11} R & \to & R \\ \downarrow & & \downarrow \\ R & \to & \mathbb{F}_{11} \end{array} \quad \text{and} \quad \begin{array}{ccc} R \xrightarrow{11} S & \to & S \\ \downarrow & & \downarrow \\ R & \to & \mathbb{F}_{11} \end{array} .$$

The Brauer tree is 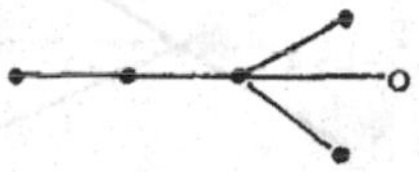 ,

and the Auslander-Reiten graph has the form

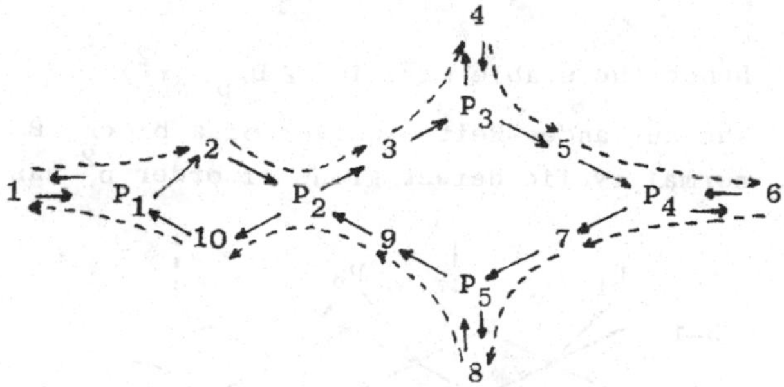

Where the dotted lines indicate the translation τ.

We recall that blocks of defect 1 are Bäckström orders (1.6) and so there should be a close connection between the Auslander-Reiten quiver of B and the Auslander-Reiten quiver of the underlying species (§ 1). We elaborate on this under 4.B.)

We next turn to blocks with cyclic defect p^2. R is unramified over $\hat{\mathbb{Z}}_p$.

<u>4.3 Theorem</u> (Wiedemann [Wi 80, 1]): (i) The Auslander-Reiten quiver of RC_{p^2} has the form (C_{p^2} = cyclic group of order p^2)

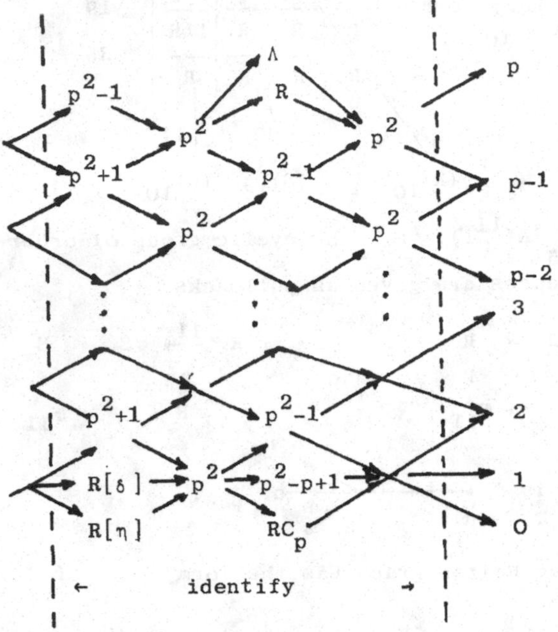

R[η], RC_p of vertex p. η a primitive p^2-th root of unity, θ a primitive p-th root of units.

Hence the stable part is $\mathbb{Z}D_{2p}/\langle\tau^2\rangle$.

(ii) The Auslander-Reiten quiver of a block B of RG with normal cyclic defect group of order p^2 has the form

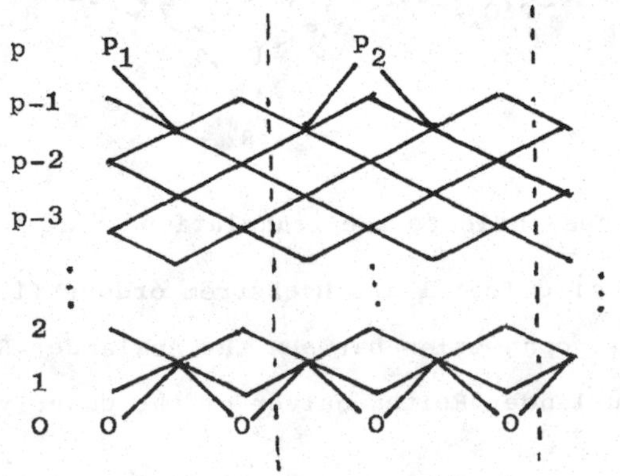

O = modules with vertex p.

4.4 Remark: If B is an arbitrary block with cyclic defect group of order p^2, then one passes with Brauer and Green correspondence from the block b of $RN_G(D)$ – D the defect group of B – to the block B. This preserves irreducible maps between modules of vertex D, and in our case it also preserves modules of vertex p. However, nothing can be said about irreducible map between modules of vertex D and modules of vertex p; equally, nothing can be said about the position of the projective modules. But one still can prove

4.5 Theorem ([Wi,80]): Let B be a block of cyclic defect of order p^2. Then the tree class of the stable Auslander-Reiten quiver $\mathfrak{A}_s(B)$ is D_{2p} (or possibly E_6 in case $p = 3$), i.e. $\mathfrak{A}_s(B) \simeq \mathbb{Z}D_{2p}/G$.

We now return to a general R but assume $\operatorname{char} K = 0$, $\operatorname{char} \mathfrak{k} = p \,|\, |G|$. For a block B with cyclic defect group, every $M \in \operatorname{ind}_B \mathfrak{M}^o$ has a periodic resolution, and since for Gorenstein order $\tau(M) = \Omega^{-1}(M)$, where $\Omega^i(M)$ is the i-th syzygy of M (cf. 3.4), every connected Δ component of $\mathfrak{A}_s(B)$ has a periodic module, and so we can apply – noting that Δ can not have loops – (3.21) to conclude:

4.6 Lemma: Let B be a block of RG with cyclic defect group but of infinite lattice type, and Δ a connected component of the stable Auslander-Reiten quiver of B. Then the tree class of B is A_∞.

4.7 Remark: If Δ is a connected component of $\mathfrak{A}_s(RG)$, G a finite group, then Δ belongs to $\mathfrak{A}_s(B)$, where B is a block with defect group D. If D is not cyclic, then by (3.12) Δ is infinite and the ranks of the B-lattices in Δ are unbounded and so the rank function is not periodic with respect to τ (3.18), and so it can not be used to apply (3.19) in order to find the tree class of Δ.

P. Webb has nevertheless constructed a subadditive function on Δ and so (3.19) is applicable. We shall now describe Webb's construction.

The rank function is periodic with respect to τ^- on M, $M \in \operatorname{ind}_{RG} \mathfrak{M}^o$

if and only if M has a periodic resolution (note that RG is a Gorenstein order). Now Alperin [AL, 77], also Carlson [Ca, 81] have defined the <u>complexity of $M \in {}_{RG}\mathfrak{M}^o$</u> as a measure of how far off M is from having a periodic resolution: Let

$$\dots \to P_2 \to P_1 \to P_o \to M \to 0$$

be a minimal projective resolution of M then the complexity of M, $C_{RG}(M)$ is the least integer $s \geq 0$ such that

$$\lim_{n\to\infty} \frac{\mathrm{rank}_R(P_n)}{n^s} = 0$$

this integer always exists [AL, 77] and it measures the polynomial growth of $\mathrm{rank}(P_n)$. If $M,N \in {}_{RG}\mathfrak{M}^o$, then let $C_{RG}(M,N)$ be the least integer $s \geq 0$ such that

$$\lim_{n\to\infty} \frac{d_R(\mathrm{Ext}^n_{RG}(M,N))}{n^s} = 0 \quad ,$$

where $d_R(-)$ denotes the minimal number of generators as R-module. Then $C_{RG}(M)$ is the largest of the integers $C_{RG}(M,N)$ for all N, and $C_{RG}(M)$ is the maximum of the orders of the poles at $t = 1$ of the various Poincaré series

$$P_{M,N}(t) = \sum_{n=0}^{\infty} t^n \cdot d_R(\mathrm{Ext}^n_{RG}(M,N))$$

as N runs over all indecomposable RG-lattices [Ca, 81]. The complexity is the logical generalization of the idea of periodicy in modules; indeed an R-lattice is periodic if and only if its complexity is one [Ei, 80].

Webb's idea was to use the complexity to define a subadditive function on Δ. A function $\theta : \mathbb{Z} \to \mathbb{Z}$ is -following G.Higman - called an <u>almost PORC-function</u> (polynomial on residue classes), if there exist polynomials $g_1, \dots, g_m \in \mathbb{Z}[t]$ such that for almost all $n \in \mathbb{N}$

$$\theta(n) = g_r(n) \quad , \quad \text{where } r \equiv n \bmod (m) .$$

m is said to be the <u>period of θ</u> and the <u>degree</u> is

$$\delta(\theta) = \max_{1\leq i\leq m} \delta(g_i) ,$$

where $\delta(g_i)$ is the ordinary degree. If $M \in \mathrm{ind}_{RG}\mathfrak{M}^o$ is not projective and has a periodic projective resolution, then

$$\theta(n) = \mathrm{rank}_R(\tau^{-n} M)$$

is a PORC-function with period s, where s is the smallest integer with $M \simeq \tau^{-s} M$. Moreover, its degree is zero.

<u>4.8 Proposition</u> (Webb [We, 81, 2]):

$$\theta_M(n) = \mathrm{rank}_R(\tau^{-n} M)$$

for $M \in \mathrm{ind}_{RG}\mathfrak{M}^o$ non-projective, is an almost PORC-function. Moreover, the period of θ_M is independent of M, and $\delta(\theta_M) \leq C_{RG}(R) - 1$, where R is the trivial R-lattice.

The <u>proof</u> uses at a crucial point the finite generation of the cohomology rings of lattices over group rings [EVG]. This arises naturally because of the close connection between the rank function of the τ-orbits and the complexity and the relation of the complexity with Poincaré series.

<u>4.9 Theorem</u> (Webb [We, 81, 2]): Let Δ be a connected component of $\mathfrak{A}_s(RG)$. Then there exists a non-zero subadditive function

$$f : \Delta \to \mathbb{N}$$

which is periodic with respect to τ.

Further if there exists a lattice M in Δ, which is not a periodic vertex, then f can be taken to be additive.

<u>f is defined as follows:</u> Since $C_{RG}(M) \leq C_{RG}(R)$ - note that a projective resolution of M is obtained from that of R by tensoring with M - we can choose a lattice M in Δ of <u>maximal complexity</u> $c = C_{RG}(M)$. Let m be the common period of θ_N (4.8), $N \in \Delta$.

If $g_1, \dots, g_m$ are the polynomials for θ_N, we have for sufficiently large $n \in \mathbb{N}$,

$$\operatorname{rank}(\tau^{-n} N) = g_r(n) , \qquad r \equiv n \bmod (m) .$$

Let α_r be the coefficient of $t^{(c-1)}$ in $g_r(t)$, and put

$$f(\tau^{-n} N) = \alpha_r \qquad \text{if } r \equiv n \bmod (m) .$$

Then f is obviously periodic with respect to τ. #

Now one is in the position to apply (3.19).

<u>4.10 Theorem</u> (Webb [We, 81, 2]): Let Δ be a connected component of a block B of RG, B of infinite lattice type. Then the tree class of Δ is either an Euclidian diagram or one of the infinite types. If Δ does not contain a projective lattice, then the tree class is one of the infinite types.

Finally we report on some other results, Webb has obtained in this connection

<u>4.11 Theorem</u> (Webb[We, 81, 2]): (i) All RG-lattices in Δ have the same complexity.

(ii) If $R \in \Delta$, and the p-Sylow-subgroup P of G is not cyclic, then all lattices in Δ have the same vertex.

<u>Remark:</u> (ii) is definitely false for $G = RC_{p^2}$ (4.3).

4. B.) Auslander-Reiten quivers of Bäckström orders.

We recall from § 1 that an R-order Λ in A^{*} is said to be a Bäckström order provided there exists a hereditary order Γ in A with

$$\mathrm{rad}\,\Lambda = \mathrm{rad}\,\Gamma \subset \Lambda \subset \Gamma .$$

We put $\mathfrak{A} = \Lambda/\mathrm{rad}\,\Lambda$ and $\mathfrak{B} = \Gamma/\mathrm{rad}\,\Gamma$, and then $\mathfrak{B}$ is a two-sided $\mathfrak{A}$-module via the natural injection $\mathfrak{A} \rightarrow \mathfrak{B}$, and

$$\mathfrak{D} = \begin{pmatrix} \mathfrak{B} & {}_{\mathfrak{B}}\mathfrak{B}_{\mathfrak{A}} \\ 0 & \mathfrak{A} \end{pmatrix}$$

is a k-algebra with $\mathrm{rad}\,\mathfrak{D} = \begin{pmatrix} 0 & \mathfrak{B} \\ 0 & 0 \end{pmatrix}$ and $\mathrm{rad}^2\mathfrak{D} = 0$. The species S of $\mathfrak{D}$ is the same as the species of the Bäckström order Λ as defined in § 2 [Dl-Ri, 76, Ri-Ro,79, Gr, 77], and the Auslander-Reiten quiver of $\mathfrak{D}$ is well understood:

4.12 Notation: (i) The simple $\mathfrak{D}$-modules come in two classes: $S_1, \dots, S_s$, the simple $\mathfrak{B}$-modules; these are simple projective $\mathfrak{D}$-modules and the simple $\mathfrak{A}$-modules $S_{s+1}, \dots, S_t$ which are simple injective $\mathfrak{D}$-modules.

(ii) The non-simple indecomposable projective $\mathfrak{D}$-modules have Loewy-length 2; they are $\overline{P}_{s+1}, \dots, \overline{P}_t$ where $\overline{P}_j/\mathrm{rad}\,\overline{P}_J \simeq S_j$, $s+1 \le j \le t$.

(iii) The non-simple indecomposable injective $\mathfrak{D}$-modules have Loewy-length 2; they are $E_1, \dots, E_s$, where $\mathrm{Soc}\,E_i \simeq S_i$, $1 \le i \le t$.

4.13 The Auslander-Reiten quiver of $\mathfrak{D}$: It is easily seen that the only possible irreducible maps starting from S_i, $1 \le i \le s$, (ending in S_j, $s+1 \le j \le t$) go to $\overline{P}_j$, $s+1 \le j \le t$ (come from I_i, $s \le i \le t$), moreover,

* we assume Λ to be two-sidedly indecomposable

$Irr(S_i, \overline{P}_j) = Hom_{\mathfrak{D}}(S_i, \overline{P}_j)$, $1 \le i \le s$, $s+1 \le j \le t$ and

$Irr(E_i, S_j) = Hom_{\mathfrak{D}}(E_i, S_j)$, $1 \le i \le s$, $s+1 \le j \le t$.

The preprojective component of $\mathfrak{A}(\mathfrak{D})$ i.e. the component containing the projective $\mathfrak{D}$-modules is generated by $Irr(S_i, \overline{P}_j)$, $1 \le i \le s$, $s+1 \le j \le t$ in the following sense:

We denote for $X \in ind_{\mathfrak{D}}\mathfrak{M}^f$, X not projective (not injective), by $c(X)$ $(c^-(X))$ the left (right) hand side of the Auslander-Reiten sequence of X in the category of $\mathfrak{D}$-modules. Since $\mathfrak{D}$ is a $\mathfrak{k}$-algebra, the function $dim_{\mathfrak{k}}$ is additive on exact sequences. Starting now with S_i , we have the irreducible maps starting from S_i , and $c^-(S_i)$ is the cokernel of all these irreducible maps. Note that this generates in the first step a valued Riedtmann quiver (3.14). Once this is done with all simple projective modules, we will have valued arrows - corresponding to irreducible maps - leaving $\overline{P}_j$. The cokernel of all irreducible maps leaving a fixed $\overline{P}_j$ will be $c^-(\overline{P}_j)$. This way one constructs inductively a valued Riedtmann quiver. The process has to be stopped, if the dimension function is not positive any more.

The preprojective component is constructed dually.

The remaining components are called regular components - they can be described in the tame case [Dl-Ri, 76].

4.14 Examples: (i) Let $\mathfrak{D}$ be the $rad^2 = 0$ algebra corresponding to the quiver - all valuations are 1:

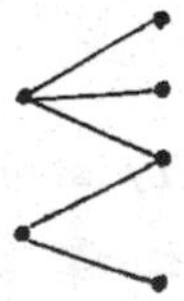

Then with the above process the Auslander-Reiten quiver of $\mathfrak{D}$ is constructed as follows - the numbers indicate the $\mathfrak{k}$-dimensions of the corresponding $\mathfrak{D}$-modules.

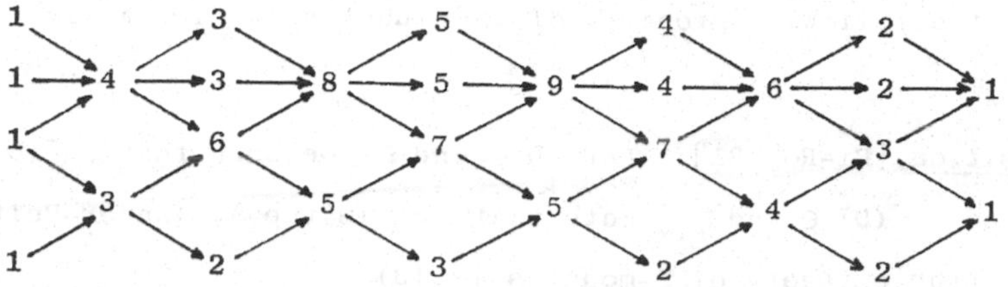

Hence $\mathfrak{D}$ has 30 non-isomorphic indecomposable modules.

It should be noted that if $\mathfrak{D}$ has finite representation type, the preprojective and preinjective components coincide, and the component constitutes the entire Auslander-Reiten quiver. Hence the outer sections correspond to the simple modules, and the second section to the projective and the injective modules of Loewy-length 2 resp.

(ii) Let $\mathfrak{D}$ be the $rad^2=0$ algebra corresponding to the quiver.

Then the Auslander-Reiten quiver is symbolically described as follows

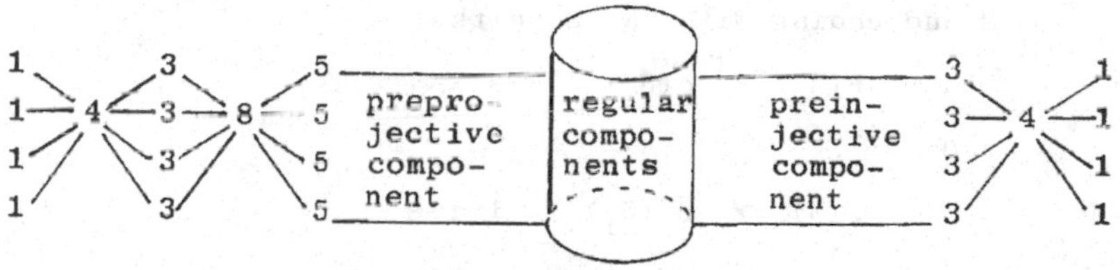

We shall now construct the Auslander-Reiten quiver of Λ from that of $\mathfrak{D}$ [Ri-Ro, 81] (§2, 1.8). Recall that the functor

$$\mathbb{F} : {}_{\Lambda}\mathfrak{M}^{o} \to {}_{\infty}\mathfrak{M}^{f} \qquad \text{induced by}$$

$$M \to \begin{pmatrix} \Gamma M/(\operatorname{rad}\Gamma)M \\ M/(\operatorname{rad}\Gamma)M \end{pmatrix}$$

gives a representation equivalence between ${}_{\Lambda}\mathfrak{M}^{o}$ and the $\mathfrak{D}$-modules which do not have simple direct summand [Ri-Ro, 79] . We denote the image category by $\mathfrak{C}$.

By observing that Γ-lattices are projective and $\mathfrak{A}$ and $\mathfrak{B}$ are semi-simple, the following property of the functor $\mathbb{F}$ is easily proved:

4.15 Proposition [Ri-Ro, 81]: Let $D \in \operatorname{ind} \mathfrak{T}$ be such that $c^-(D)$ is defined and $c^-(D) \in \operatorname{ind} \mathfrak{T}$ (note that c is the Auslander-Reiten translate in the category of $\mathfrak{D}$-modules (4.13)).

If

$$0 \to D \to D_1 \to c^-(D) \to 0$$

is the Auslander-Reiten sequence in ${}_{\mathfrak{D}}\mathfrak{M}^f$, then

$$0 \to \mathbb{F}^{-1}(D) \to \mathbb{F}^{-1}(D_1) \to \mathbb{F}^{-1}(c^-(D)) \to 0$$

is the Auslander-Reiten sequence in the category of Λ-lattices; in particular, if $M \in \operatorname{ind}{}_{\Lambda}\mathfrak{M}^o$ is such that $c^-(\mathbb{F}(M))$ is defined and lies in $\mathfrak{C}$, then $\tau^-(M) \simeq \mathbb{F}^{-1}(c^-(\mathbb{F}(M)))$. A similar statement holds if for $D \in \operatorname{ind} \mathfrak{T}$, $c(D)$ is defined and lies in $\mathfrak{C}$.

The above result describes completely:

(i) the regular components of the Auslander-Reiten quiver of Λ,

(ii) all Auslander-Reiten sequences starting on the left with M and ending with N such that

$$\mathbb{F}(M) \not\simeq c(S_j), \quad s+1 \le j \le t$$

and

$$\mathbb{F}(N) \not\simeq c^-(S_i), \quad 1 \le i \le s .$$

Note that

$$\mathbb{F}^{-1}(\overline{P}_j), \quad s+1 \le j \le t$$

are the indecomposable projective Λ-lattices. Moreover, $\tau^-(\mathbb{F}^{-1}E_i)$ is not yet defined.

4.16 Proposition ([Ri-Ro, 81]):

(i) $\mathbb{F}^{-1}(c(S_j)) = I_j$, $s+1 \le j \le s$ are the indecomposable injective Λ-lattices.

(ii) $\mathbb{F}^{-1}(\overline{P}_j) = P_j$, $s+1 \le j \le s$ are the indecomposable projective Λ-lattices.

(iii) $\mathbb{F}^{-1}(E_i) = Q_i$, $1 \leq i \leq s$ are the indecomposable Γ-lattices.

Note that for technical reasons we have a different numbering than in § 2.

In order to define the remaining Auslander-Reiten sequences, we use the following notation: We have the non-isomorphic indecomposable Γ-lattices $Q_1, \ldots, Q_s$. Then $\operatorname{rad} Q_i$ is projective, say $\operatorname{rad} Q_i = Q_{\varphi^-(i)}$. Then φ^- is a permutation of $\{1, \ldots, s\}$. Let φ be the inverse permutation.

4.17 Proposition ([Ri-Ro, 81]): Let

$$0 \to S_{\varphi(i)} \to \oplus \overline{P}_{k(i)} \to c^-(S_{\varphi(i)}) \to 0$$

be the almost split sequence in ${}_{\mathfrak{D}}\mathfrak{M}^f$. Then in ${}_{\Lambda}\mathfrak{M}^o$ we have the almost split sequence

$$0 \to \mathbb{F}^-(E_i) \to \oplus P_{k(i)} \to \mathbb{F}^-(c(S_{\varphi(i)})) \to 0 , \quad 1 \leq i \leq s .$$

Together we have

4.18 Theorem: Let Λ be a Bäckström order. Then the Auslander-Reiten quiver of ${}_{\Lambda}\mathfrak{M}^o$ is obtained from that of ${}_{\mathfrak{D}}\mathfrak{M}^f$ by identifying

$$E_i \text{ with } S_{\varphi(i)} , \quad 1 \leq i \leq s .$$

Remark: (i) In particular, the preprojective and the preinjective components of $\mathfrak{A}(\mathfrak{D})$ are glued together to one component in $\mathfrak{A}(\Lambda)$. Moreover, at the points of glueing, there are holes in the Auslander-Reiten quiver; in particular, it is not simply connected any more.

(ii) If Γ is a maximal order, then $\mathfrak{D}$ does not decompose two-sidedly and the permutation is the identity, and so one has to identify E_i with its socle.

4.19 Examples: (i) Let for brevity $\pi = \pi R$ and put

$$\Lambda = \{ (\alpha)(\alpha')\begin{pmatrix}\alpha'' & \pi \\ \pi & \beta\end{pmatrix}(\beta') \ , \ \alpha \equiv \alpha' \equiv \alpha'' \bmod(\pi), \ \beta \equiv \beta' \bmod(\pi)\} .$$

Then

$$\Gamma = R \ \Pi \ R \ \Pi \ \begin{pmatrix}R & R \\ R & R\end{pmatrix} \ \Pi \ R$$

is maximal and so the quiver of the Bäckström order Λ

(§ 2) is

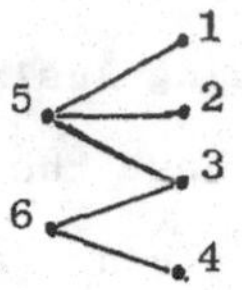

The Auslander-Reiten quiver of $\mathfrak{D}$ is given in (4.14(i)).
Hence we obtain the Auslander-Reiten quiver of Λ as

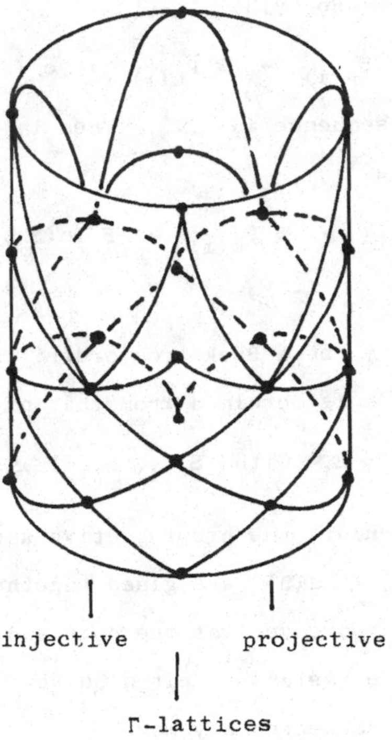

In this picture it should be noted that the meshes between the injective Λ-lattices and the projective ones do not constitute Auslander-Reiten sequences.

The $\mathfrak{k}$-dimension in (4.13(i)) now represents the number of simple A-modules into which KM , $M \in \mathrm{ind}_\Lambda \mathfrak{M}^o$, decomposes.

(ii) Put

$$\Lambda = \left\{ \begin{bmatrix} \alpha & R & R & R & R \\ \pi & \alpha' & R & R & R \\ \pi & \pi & \alpha'' & \pi & R \\ \pi & \pi & \pi & \beta & R \\ \pi & \pi & \pi & \pi & \beta' \end{bmatrix} , \quad \alpha \equiv \alpha' \equiv \alpha'' \bmod(\pi),\ \beta \equiv \beta' \bmod(\pi) \right\}$$

Then $\Gamma = \begin{bmatrix} R & R & R & R & R \\ \pi & R & R & R & R \\ \pi & \pi & R & R & R \\ \pi & \pi & R & R & R \\ \pi & \pi & \pi & \pi & R \end{bmatrix}$,

and so the quiver of the Bäckström order Λ (cf.§ 2) is again

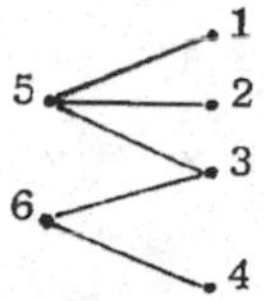

But now the permutation φ is given by (1 2 3 4), and the Auslander-Reiten quiver of $\mathfrak{D}$ is given in (4.14(i)). Hence we have to make the following identifications:

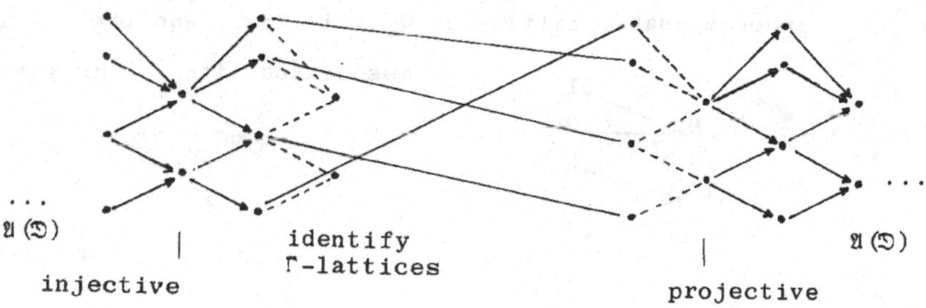

It is seen that this Auslander-Reiten quiver can not be realized on a cylinder.

If we now insert at each module the number of simple A-modules which it spans, then this is an additive function and

we get (cf. 4.14(i))

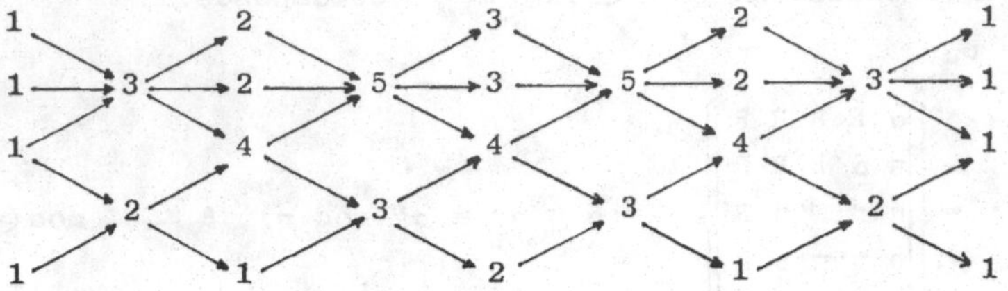

Γ-latt. proj.Λ-latt. inj.Λ-latt.

(iii) Let

$$\Lambda = \left\{ \begin{pmatrix} \alpha & R & R & R & R & R & R & R & R \\ \pi & \alpha' & R & R & R & R & R & R & R \\ \pi & \pi & \alpha'' & \pi & R & R & R & R & R \\ \pi & \pi & \pi & \beta & R & R & R & R & R \\ \pi & \pi & \pi & \pi & \beta' & R & R & R & R \\ \pi & \pi & \pi & \pi & \pi & \gamma & R & R & R \\ \pi & \pi & \pi & \pi & \pi & \pi & \gamma' & R & R \\ \pi & \pi & \pi & \pi & \pi & \pi & \pi & \gamma'' & R \\ \pi & \pi & \pi & \pi & \pi & \pi & \pi & \pi & \gamma''' \end{pmatrix} \quad \begin{matrix} \alpha \equiv \alpha' \equiv \alpha'' & \bmod(\pi) \\ \beta \equiv \beta' & \bmod(\pi) \\ \gamma \equiv \gamma' \equiv \gamma'' \equiv \gamma''' & \bmod(\pi) \end{matrix} \right\}$$

Then Λ has 3 non-isomorphic indecomposable projective Λ-lattices P_9, P_{10}, P_{11} and Γ has 8 non-isomorphic indecomposable lattices, Q_i, $1 \le i \le 8$, and the species associated with Λ is given by

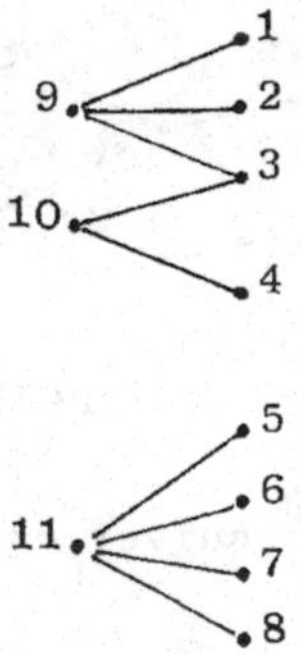

So in particular, $\mathfrak{D}$ decomposes two-sidedly. The permutation φ is given as $\varphi = (1\,2\,3\,4\,5\,6\,7\,8)$. Hence Λ has the following

Auslander-Reiten quiver

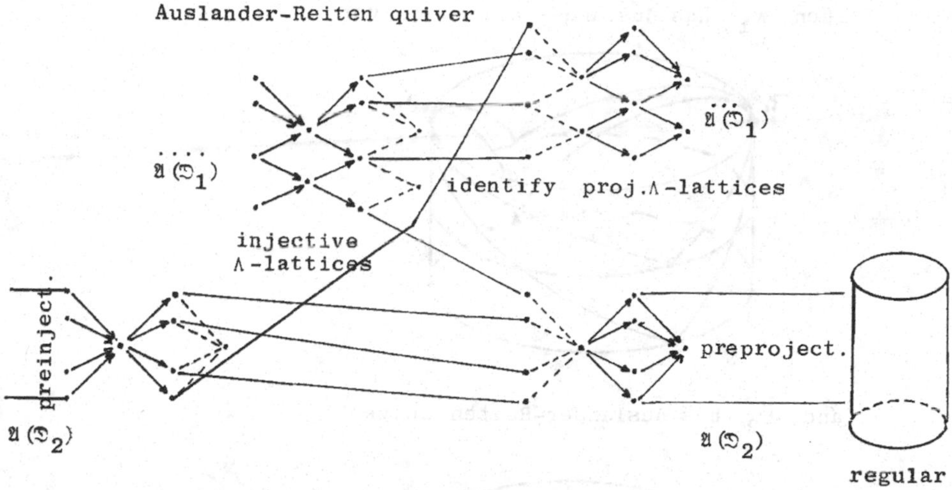

So all preprojective and preinjective components of $\mathfrak{A}(\mathfrak{D}_1 \amalg \mathfrak{D}_2)$ become one component in $\mathfrak{A}(\Lambda)$.

(iii) This example shall once more demonstrate the importance of the permutation φ .

Let

$$\Lambda_1 = \left\{ \begin{pmatrix} \alpha & R \\ \pi & \beta \end{pmatrix} \quad \begin{pmatrix} \beta' & R \\ \pi & \alpha' \end{pmatrix} \quad , \quad \begin{matrix} \alpha \equiv \alpha' \bmod(\pi) \\ \beta \equiv \beta' \bmod(\pi) \end{matrix} \right\}$$

$$\Lambda_2 = \left\{ \begin{pmatrix} \alpha & R & R & R \\ \pi & \alpha' & R & R \\ \pi & \pi & \beta & R \\ \pi & \pi & \pi & \beta' \end{pmatrix} \quad , \quad \begin{matrix} \alpha \equiv \alpha' \bmod(\pi) \\ \beta \equiv \beta' \bmod(\pi) \end{matrix} \right\} .$$

Then the species of both orders is

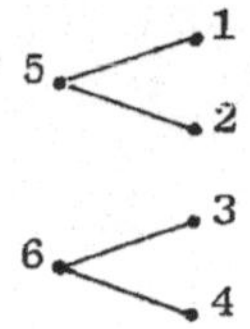

i.e. a disjoint union of two quivers of type A_3 .

For Λ_1 we have $\varphi_1 = (14)\,(23)$ and for Λ_2 we have $\varphi_2 = (1\,2\,3\,4)$.

Then Λ_1 has Auslander-Reiten quiver

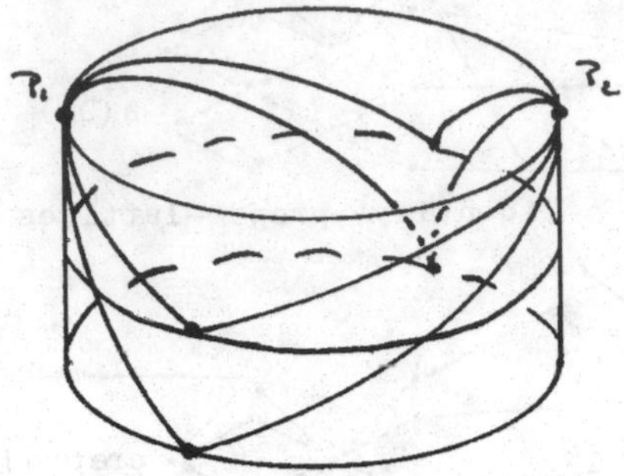

and Λ_2 has Auslander-Reiten quiver

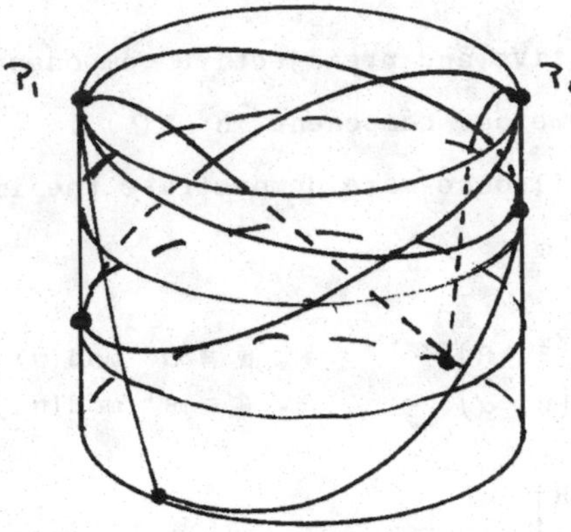

4. C.) Auslander-Reiten quivers of orders of global dimension 2.

A characterization of orders Λ of global dimension at most two, which contain the radical of a maximal order has been given in (§2,2.4). In the meantime there has been made some further progress in classifying orders of global dimension two. So let me first report on this:

4.20 Definition: We say that an R-order Λ in A is Schurian provided

$$\mathrm{Hom}_\Lambda(P_i, P_j) = \mathfrak{a}_{ij} ,$$

where $\{P_i\}_{1 \leq i \leq s}$ is a complete set of non-isomorphic indecomposable projective Λ-lattices, and $\mathfrak{a}_{ij}$ is an ideal in R.

Remark: . (i) This definition is in accordance with the definition of Schurian algebras.

(ii) It should be noted that a Schurian order is two-sided indecomposable if and only if $\mathfrak{a}_{ij} \neq 0$ for all (i,j) ; in this case $A = (K)_n$; and if we assume that Λ is basic, then $n = s$.

From now on we shall therefore assume that a Schurian order is two-sided indecomposable and basic.

4.21 The singly valued quiver of a Schurian order

(i) A singly valued quiver is a quiver $Q = (\Gamma_o, \Gamma_1)$ together with a map $\nu: \Gamma_1 \to \mathbb{N}$, where Γ_1 is the set of arrows in Γ. In addition we assume that each pair of vertices of Q is connected by some oriented path, but there should be no path of length one from $x \in \Gamma_o$ to x ; i.e. there should be no loops in Γ_1.

(ii) The path order $\Lambda(Q)$ of Q is defined as follows: Let $\{1, \ldots, s\} = \Gamma_o$, then $\Lambda(Q)$ is an order in $(K)_s$ with diagonal entries R and (i,j)-entry $\pi^{\alpha(i,j)}$, where $\alpha(i,j)$ is the minimal integer m such that there exists an oriented path from i to j of valued path length m. Obviously $\Lambda(Q)$ is a Schurian order, it is two-sided indecomposable, but not necessarily basic.

(iii) Let now Λ be a Schurian order basic and indecomposable, $\{e_i\}_{1\le i\le s}$ a complete set of orthogonal primitive idempotents in Λ . Then $\Gamma_o = \{\Lambda e_i\}$, and there is an arrow from Λe_i to Λe_j provided Λe_i occurs as a direct summand in the projective cover of $\operatorname{rad}\Lambda e_j$. We then put

$$\nu(\Lambda e_i \to \Lambda e_j) = \alpha_{ij}$$

provided $\operatorname{Hom}_\Lambda(\Lambda e_i, \Lambda e_j) = \pi^{\alpha_{ij}}R$.

<u>4.22 Proposition:</u> If Λ has finite global dimension and is Schurian, then (i) $Q = (\Gamma_o, \Gamma_1, \nu)$ as defined in (iii) is a singly valued quiver,

(ii) $\Lambda(Q) \simeq \Lambda$.

<u>4.23 Example:</u> (i) The order considered in (3.9 (ii)) $\Lambda = \begin{pmatrix} R & R \\ \pi^t & R \end{pmatrix}$ has quiver $1 \overset{t}{\underset{0}{\rightleftarrows}} 2$.

(ii) The order considered in (3.9(iii)) has quiver

$$1 \overset{1}{\underset{0}{\rightleftarrows}} 2 \overset{1}{\underset{0}{\rightleftarrows}} 3 \overset{1}{\underset{0}{\rightleftarrows}} 4 \overset{1}{\underset{0}{\rightleftarrows}} 5 \quad .$$

(iii) Let Q be given as

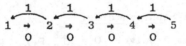

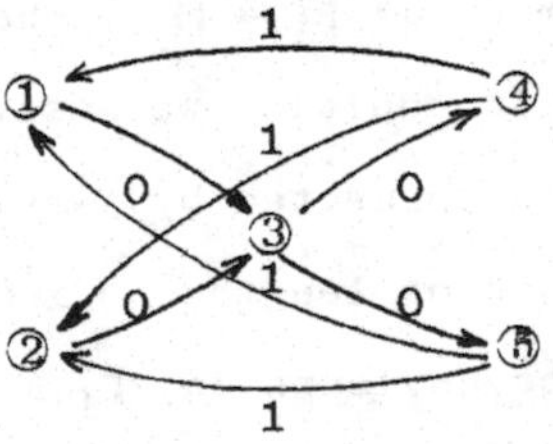

then

$$\Lambda(Q) = \begin{pmatrix} R & \pi & R & R & R \\ \pi & R & R & R & R \\ \pi & \pi & R & R & R \\ \pi & \pi & \pi & R & \pi \\ \pi & \pi & \pi & \pi & R \end{pmatrix}$$

In [Wi-Ro, 81] a necessary and sufficient condition is given for $\Lambda(Q)$ to have global dimension two. Thus characterizing Schurian orders of global dimension two. Since these conditions are rather technical, we only give sufficient conditions on Q to give rise to an order $Q(\Lambda)$ of global dimension two.

Let $Q = (\Gamma_o, \Gamma_1, \nu)$ be a single valued quiver, and assume that the subquiver $\overline{Q}$ consisting of the vertices Γ_o and only those arrows which have valuation zero, that this subquiver is a tree. In addition, assume

a) $\nu(\Gamma_o) \subset \{0,1\}$;

b) Q has no zero cycles;

c) through each point x there is a 1-cycle (i.e. of valued cycle length one);

d) each arrow is part of a one cycle;

e) there are no "superfluous" arrows (e.g. 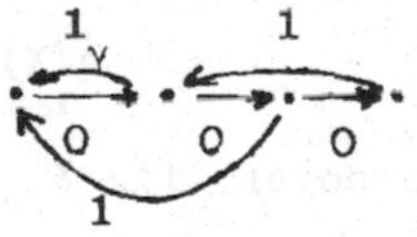

then γ is superfluous);

f) Q does not contain a full subquiver, which can be "shrunk" to (4.23 (iii)).

<u>4.24 Theorem</u> ([Wi-Ro, 81]): Let Q be a single valued quiver satisfying the above conditions, then $\Lambda(Q)$ has global dimension two, and $\Lambda(Q)$ is basic Schurian and two-sided indecomposable.

<u>4.25 Examples</u>: (i) (4.23 (ii)) has global dimension two.

(ii) If Q = 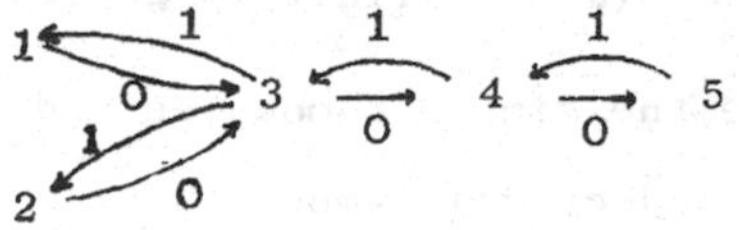

,

then

$$\Lambda(Q) = \begin{pmatrix} R & \pi & R & R & R \\ \pi & R & R & R & R \\ \pi & \pi & R & R & R \\ \pi^2 & \pi^2 & \pi & R & R \\ \pi^3 & \pi^3 & \pi^2 & \pi & R \end{pmatrix} \quad \text{has global dimension two.}$$

We shall next turn to the Auslander-Reiten quiver of an order Λ of global dimension two. Recall gl.dim $\Lambda = 2$ is equivalent to gl.dim ${}_\Lambda\mathfrak{M}^o = 1$, and so in some sense, orders of global dimension two should behave similarly to hereditary artinian algebras. For hereditary artinian algebras Auslander-Platzek [Au-Pl, 78] have defined a natural "Coxeter-transformation" as follows:

4.26 Let $\mathfrak{A}$ be a hereditary $\mathfrak{k}$-algebra and let $X \in \operatorname{ind}_{\mathfrak{A}}\mathfrak{M}^f$ be non-projective, if

$$0 \to P_1 \to P_o \to X \to 0$$

is a minimal projective resolution, then we have the exact sequence

$$0 \to \operatorname{Hom}_{\mathfrak{A}}(X, \mathfrak{A}) \to \operatorname{Hom}_{\mathfrak{A}}(P_o, \mathfrak{A}) \to \operatorname{Hom}_{\mathfrak{A}}(P_1, \mathfrak{A}) \to \operatorname{Ext}^1_{\mathfrak{A}}(X,\mathfrak{A}) \to 0,$$

and so

$$c(X) \simeq \operatorname{Hom}_{\mathfrak{k}}(\operatorname{Ext}^1_{\mathfrak{A}}(X,\mathfrak{A}), \mathfrak{k}) .$$

Moreover, if X' is indecomposable non-projective, and if

$$0 \to X' \to X \to X'' \to 0$$

is an exact sequence, then

$$0 \to c(X') \to c(X) \to c(X'') \to 0$$

is exact, since $\operatorname{Hom}_{\mathfrak{A}}(X',\mathfrak{A}) = 0$, and $\operatorname{Ext}^2_{\mathfrak{A}}(X'',\mathfrak{A}) = 0$.
Hence c is compatible with exact sequences and "acts" therefore on the Grothendieck group $G(\mathfrak{A})$ of finitely generated $\mathfrak{A}$-modules.

Let now S_i, $1 \le i \le t$ be the simple $\mathfrak{A}$-modules and P_i, (E_i) the indecomposable projective (injective) $\mathfrak{A}$-modules with

$P_i/\mathrm{rad}\,P_i \simeq S_i$ $(\mathrm{soc}\,E_i \simeq S_i)$, $1 \leq i \leq s$. Then the above observations lead to

4.27 Theorem ([Au-Pl, 78]): The map

$$c : [P_i] \to -[E_i]$$

induces a linear transformation on $G(\mathfrak{A})$ such that

(i) If $X \in \mathrm{ind}_{\mathfrak{A}}\mathfrak{M}^f$ is non-projective, then

$$c[X] = [c(X)] \qquad \text{in} \quad G(\mathfrak{A}) \quad ,$$

(ii) $\mathfrak{A}$ is of finite representation type if and only if $c^m = \mathrm{id}$ for some $m \in \mathbb{N}$. In that case for every $X \in \mathrm{ind}_{\mathfrak{A}}\mathfrak{M}^f$, there exists an i,t , $1 \leq i \leq s$, $0 \leq t \leq m$ such that $X \simeq c^t(E_i)$.

(iii) If $\mathfrak{A}$ is of finite type then for $X, Y \in \mathrm{ind}_{\mathfrak{A}}\mathfrak{M}^f$, $X \simeq Y$ if and only if $[X] = [Y]$.

One would hope for a similar nice result for orders of global dimension two. So let us look at the analogue of (4.26). Let Λ be an order of global dimension two, and $M \in \mathrm{ind}_{\Lambda}\mathfrak{M}^o$ non-projective. If we choose a minimal projective resolution

$$0 \to P_1 \to P_o \to M \to 0 \quad ,$$

then we obtain the exact sequence (3.4)

$$\begin{array}{ccccccccc} 0 \to \mathrm{Hom}_{\Lambda}(M,\Lambda) & \to & \mathrm{Hom}_{\Lambda}(P_1,\Lambda) & & \to & & \mathrm{Hom}_{\Lambda}(P_o,\Lambda) & \to & \mathrm{Ext}^1_{\Lambda}(M,\Lambda) \to 0 \\ & & & \searrow & & \nearrow & & & \\ & & & & \mathrm{tr}(M) & & & & \\ & & & \nearrow & & \searrow & & & \\ & & 0 & & & & 0 & & \end{array}$$

and so we get the exact sequences

$$0 \to \tau(M) \to \mathrm{Hom}_R(\mathrm{Hom}_{\Lambda}(P,\Lambda),R) \to \mathrm{Hom}_R(\mathrm{Hom}_{\Lambda}(M,\Lambda),R) \to 0$$

and

$$0 \to \mathrm{Hom}_R(\mathrm{Hom}_{\Lambda}(P_o,\Lambda),R) \to \tau(M) \to \mathrm{Hom}_R(\mathrm{Ext}^1_{\Lambda}(M,\Lambda),R) \to 0 \quad .$$

Moreover, $\mathrm{Hom}_{\Lambda}(M,\Lambda) \neq 0$, and this shows that there is no reason why τ should act linearly on the Grothendieck group $G(\Lambda)$ of the Λ-lattices.

We substanciate this with

4.28 Examples:

(i) In the example (3.9) we have the Auslander-Reiten quiver of the order Λ of global dimension 2 corresponding to the single valued quiver $\bullet \underset{0}{\overset{1}{\rightleftarrows}} \bullet \underset{0}{\overset{1}{\rightleftarrows}} \bullet \underset{0}{\overset{1}{\rightleftarrows}} \bullet \underset{0}{\overset{1}{\rightleftarrows}} \bullet$ (cf. 4.23 (ii)). In this quiver one sees that τ(injective)=projective and τ^-(projective)=injective. Hence there is no hope for orders of global dimension two to get all indecomposable Λ-lattices in the orbits under τ starting with the indecomposable projective Λ-lattices. In addition, this example shows that τ does not operate linearly on the Grothendieck group of Λ-lattices.

(ii) Let

$$\Lambda = \left\{ \begin{bmatrix} R & R & R & R & R & R & R \\ \pi & R & R & R & R & R & R \\ \pi & \pi & R & R & R & R & R \\ \pi & \pi & \pi & R & \pi & \pi & \pi \\ \pi & \pi & \pi & \pi & R & \pi & \pi \\ \pi & \pi & \pi & \pi & \pi & R & R \\ \pi & \pi & \pi & \pi & \pi & \pi & \alpha \end{bmatrix} \begin{bmatrix} R & R \\ \pi & \alpha' \end{bmatrix} , \ \alpha \equiv \alpha' \operatorname{mod}(\pi) \right\}$$

Then gl.dim Λ = 2 by (§ 2, 2.4), and computations show that the Auslander-Reiten quiver of Λ has the following form

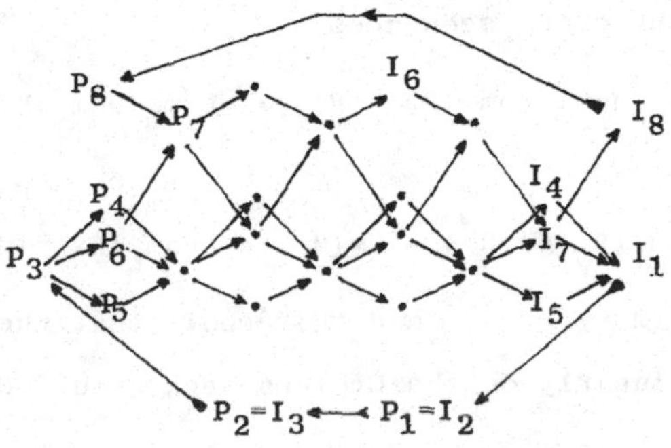

Now, here it should be noted that every indecomposable Λ-lattice is of the form $\tau^k(I_j)$ for some injective Λ-lattices I_j. Moreover, τ operates on the Grothendieck group $G({}_\Lambda\mathfrak{M}^o)$ of Λ-lattices as a linear transformation. The matrix of linear transformation τ^- with respect to the basis of the indecomposable projective lattices is given by

$$\tau^- = \begin{pmatrix} -1 & 0 & 0 & 0 & 0 & 0 & 0 & 0 \\ 0 & -1 & 0 & 0 & 0 & 0 & 0 & 0 \\ 0 & 0 & -1 & 1 & 1 & 1 & 0 & 0 \\ 0 & 0 & -1 & 0 & 1 & 1 & 0 & 0 \\ 0 & 0 & -1 & 1 & 0 & 1 & 0 & 0 \\ 0 & 0 & -1 & 1 & 1 & 0 & 1 & 0 \\ 0 & 0 & -1 & 1 & 1 & 1 & 0 & -1 \end{pmatrix}$$

and $\tau^{-10} = \text{id}$.

Despite (4.28 (i)) we have (cf. § 2, 2.4 ff.)

4.29 Theorem

Let Λ be an R-order satisfying

(i) $\text{gl.dim}\,\Lambda \leq 2$,

(ii) there exists a maximal R-order Γ with $\text{rad}\,\Gamma \subset \Lambda \subset \Gamma$,

(iii) $\Lambda/\text{rad}\,\Gamma$ is a hereditary k-algebra.

Then the following are equivalent:

α.) Λ is of finite lattice type,

β.) every $M \in \text{ind}\,{}_\Lambda\mathfrak{M}^o$ is of the form $\tau^{-t}(P)$ for some indecomposable projective Λ-lattice P.

However, τ does not act linearly on $G({}_\Lambda\mathfrak{M}^o)$, in general.

References:

[AL, 77] Alperin, J.L.: "Periodicity in groups", Illinois Math.J. 23 (1977), 776-783.

[Au, 77] Auslander, M.: "Existence theorems for almost split sequences", Proc. Conference Ring Theory II, Oklahoma, Marcel Dekker (1977).

[Au, 78] Auslander, M.: "Functors and morphisms determined by objects and applications", Proc.Conf. Representation Theory, Philadelphia 1976, Marcel Dekker (1978), 1-244, 245-327.

[Au-Pl, 78] Auslander, M. - M.I.Platzeck: "Representation theory of hereditary artin algebras", Proc.Conf. Representation Theory, Philadelphia 1976, Marcel Dekker (1978), 389-424.

[Au-Re, 77] Auslander, M. - I.Reiten: "Representation theory of artin algebras, IV, V, VI", Com. in Algebra 5 (1977), 443-518, 5 (1977), 519-554, 6 (1978), 257-300.

[Au-Sm, 80] Auslander, M. - S.O. Smalø: "Almost split sequences in subcategories", Math.Institut, Univ.Trondheim, Preprint.

[Au-Sm, 81] Auslander, M. - S.O. Smalø: "Preprojective lattices over orders", these Springer Lecture Notes.

[Ba-Ma, 79] Bautista, R. - R.Martinez: "Representations of partially ordered sets and 1-Gorenstein Artin algebras", Proc.Conf.Ring Theory (Antwerpen 1978), Marcel Dekker (1979).

[Ba-Br, 81] Bautista, R. - S.Brenner: "On the number of terms in the middle of an almost split sequence", Preprint (1981).

[Bu, 74] Butler, M.C.R.: "On the classification of local integral representations of finite abelian p-groups", Proc.ICRA 1974, Springer Lecture Notes 488, 54-71.

[Bu, 79] Butler, M.C.R.: "The construction of almost split sequences II, lattices over orders", Bulletin London Math.Soc. 11 (1979), 155-160.

[Ca, 81] Carlson, J.F.: "The complexity and varities of modules", these Springer Lecture Notes.

[Dl-Ri, 76] Dlab,V. - C.M.Ringel: "Indecomposable representations of graphs and algebras", Mem. Amer.Math.Soc. 173, Providence (1976).

[EGV] Evens, L.: "The cohomology ring of a finite group", Trans. AMS 101 (1961), 224-239.

Golod, E.: "The cohomology ring of a finite p-group", Dokl.Akad. Nauk 125 (1959), 703-706.

Venkov, B.B.: "Cohomology algebras for some classifying spaces", Dokl.Akad.Nauk 127 (1959), 943-944.

[Ei, 80] Eisenbud, D.: "Homological algebra on a complete intersection with an application to group representations", Trans. AMS 260 (1980), 35-64.

[GJA, 74] Green, J.A.: "Walking around the Brauer tree", J.Austral.Math.Soc. 17 (1974), 197-213.

[Gr, 77] Green, E.L.: "Diagrammatic techniques in the study of indecomposable modules", Proc.Conf.Ring Theory II, Oklahoma, Marcel Dekker (1977), 149-169.

[HPR, 80] Happel, D.- U.Preiser - C.M.Ringel: "Vinberg's characterization of Dynkin diagrams using subadditive functions - with applications to DTr-periodic modules", Proc. ICRA II, Ottawa 1979, Springer Lecture Notes.

[Ja, 79] Jacobinski, H.: "Hereditary covers and blocks", Lecture given at ICRA II, Ottawa 1979.

[Na-Roi, 70] Nazarova, L.A. - A.V.Roiter: "Matrix questions and the Brauer-Thrall conjectures on algebras with an infinite number of indecomposable representations. Representation theory of finite groups and related topics", Proc.Symp. Pure Math. XXI, AMS 1971.

[Rie, 80, 1] Riedtmann, C.: "Algebren, Darstellungsköcher, Überlagerungen und zurück", Comment.Math.Helvetici 55 (1980), 199-224.

[Rie, 80, 2] Riedtmann, C.: "Representations of finite selfinjective algebras of class A_n", Proc. ICRA II, Springer Lecture Notes in Math. 831 (1980).

[Ri-Ro, 79] Ringel, C.M. - K.W.Roggenkamp: "Diagrammatic methods in the representation theory of orders", J.Algebra 60, 1 (1979), 11-42.

[Ri-Ro, 80] Ringel, C.M. - K.W.Roggenkamp: "Socle determined categories of representations of artinian hereditary tensor algebras", J.Algebra 64, 1 (1980), 249-269.

[Ri-Ro, 81] Ringel, C.M. - K.W.Roggenkamp: "The Auslander-Reiten quiver of Bäckström orders", to be published.

[Ro, 70] Roggenkamp, K.W.: "Lattices over orders II", Springer Lecture Notes in Math. 142 (1970).

[Ro, 76] Roggenkamp, K.W.: "Almost split sequences for group rings", Mitt.Math.Sem.Giessen 21 (1976), 1-25.

[Ro, 77] Roggenkamp, K.W.: "The construction of almost split sequences for integral group rings and orders", Com.in Algebra 5 (13) (1977), 1363-1373.

[Ro, 78] Roggenkamp, K.W.: "Orders of global dimension two", Math.Zeit. 160 (1978), 63-67.

[Ro, 80, 1] Roggenkamp, K.W.: "Auslander-Reiten sequences for 'nice' torsion theories of artinian algebras", Canad.Math.Bull. 23 (1) (1980), 61-65.

[Ro, 80, 2] Roggenkamp, K.W.: "Indecomposable representations of orders of global dimension two", J.of Algebra 64 (1) (1980), 230-248.

[Ro, 80, 3] Roggenkamp, K.W.: "Representation theory of finite groups" Presses de l'Université de Montréal, 18e Session Séminaire de Mathématiques Supérieures (1980), 1-165.

[Ro, 80, 4] Roggenkamp, K.W.:"Representation theory of blocks of defect 1", Proc. ICRA II, Springer Lecture Notes in Math. 832 (1980), 521-544.

[Ro, 80, 5] Roggenkamp, K.W.: "The lattice type of orders: A diagrammatic approach. I", Proc.of May Conf.Ring Theory, Antwerpen, Springer Lecture Notes in Math. 825 (1980), 104-129.

[Ro-Sch, 76] Roggenkamp, K.W. - J.Schmidt: "Almost split sequences for integral group rings and orders", Com. in Algebra 4 (10) (1976), 893-917.

[Sch, 76] Schmidt, J.: "Beinahe zerfallende Sequenzen für Gitter über Ordnungen", Dr.-Arbeit, Universität Stuttgart (1976).

[We, 81, 1] Webb, P.J.: "Restricting $\mathbb{Z}G$-lattices to elementary abelian subgroups", these Springer Lecture Notes in Math.

[We, 81, 2] Webb, P.J.: "The Auslander-Reiten quiver of a finite group", Preprint 1981.

[Wi, 80, 1] Wiedemann, A.: "Auslander-Reiten Graphen von Ordnungen und Blöcke mit zyklischem Defekt 2", Dr.-Arbeit, Universität Stuttgart (1980).

[Wi, 80, 2] Wiedemann, A.: "Orders with loops in their Auslander-Reiten graph", Com. in Algebra 9 (1981), 641-656.

[Wi-Ro, 81] Wiedemann, A. - K.W.Roggenkamp: "Path orders of global dimension two", to be published.

AUTHOR INDEX